Animal and Translational Models for CNS Drug Discovery

Animal and Translational Models for CNS Drug Discovery
Volume I Psychiatric Disorders (ISBN: 978-0-12-373856-1)
Volume II Neurological Disorders (ISBN: 978-0-12-373855-4)
Volume III Reward Deficit Disorders (ISBN: 978-0-12-373860-8)

(ISBN set: 978-0-12-373861-5)

Animal and Translational Models for CNS Drug Discovery

VOLUME III
Reward Deficit Disorders

Edited by

Robert A. McArthur, PhD
Associate Professor of Research
Consultant Behavioral Pharmacologist
McArthur and Associates GmbH, Basel, Switzerland

Franco Borsini, PhD
Head, Central & Peripheral Nervous System and
General Pharmacology Area – R&D Department
sigma-tau S.p.A., Pomezia (Rome), Italy

AMSTERDAM • BOSTON • HEIDELBERG • LONDON • NEW YORK • OXFORD
PARIS • SAN DIEGO • SAN FRANCISCO • SINGAPORE • SYDNEY • TOKYO
Academic Press is an imprint of Elsevier

Academic Press is an imprint of Elsevier
30 Corporate Drive, Suite 400, Burlington, MA 01803, USA
360 Park Avenue South, Newyork, NY 10010-1710, USA
525 B Street, Suite 1900, San Diego, CA 92101-4495, USA
32 Jamestown Road, London NW1 7BY, UK

⊗ This book is printed on acid-free paper.

Library of Congress Cataloging-in-Publication Data
A catalog record for this book is available from the Library of Congress.

British Library Cataloguing-in-Publication Data
A catalogue record for this book is available from the British Library.

ISBN: 978-0-12-373861-5 (set)
ISBN: 978-0-12-373860-8 (vol 3)

For information on all Academic Press publications
visit our web site at www.elsevierdirect.com

Typeset by Charon Tec Ltd., A Macmillan Company.
(www.macmillansolutions.com)
Printed and bound in Great Britain by
CPI Antony Rowe, Chippenham and Eastbourne
Transferred to Digital Printing, 2010

Working together to grow
libraries in developing countries

www.elsevier.com | www.bookaid.org | www.sabre.org

ELSEVIER   BOOK AID
            International   Sabre Foundation

This book is dedicated to that happy band of behavioral pharmacologists who over the generations have occasionally seen their compound progress into clinical development, and more rarely still seen it used to treat patients. New skills are being learned and new species creeping into the lab, including the ones "without tails." These offer new opportunities and challenges, but equally so greater satisfaction working at the interface. May all your compounds be winners!

Contents

Volume 3 Animal and Translational Models for CNS Drug Discovery: Reward Deficit Disorders

CHAPTER 5   **Pharmacotherapy of Alcohol Dependence: Improving Translation from the Bench to the Clinic 91**

Hilary J. Little, David L. McKinzie, Beatrice Setnik,
Megan J. Shram and Edward M. Sellers

CHAPTER 7   Development of Medications for Heroin and Cocaine Addiction and Regulatory Aspects of Abuse Liability Testing 221

Beatriz A. Rocha, Jack Bergman, Sandra D. Comer, Margaret Haney and Roger D. Spealman

CHAPTER 8   Anti-obesity Drugs: From Animal Models to Clinical Efficacy 271

Colin T. Dourish, John P.H. Wilding and Jason C.G. Halford

What Do *You* Mean by "Translational Research"? An Enquiry Through Animal and Translational Models for CNS Drug Discovery: Reward Deficit Disorders

Robert A. McArthur[1] and Franco Borsini[2]

[1]McArthur and Associates GmbH, Basel, Switzerland
[2]sigma-tau S.p.A., Pomezia (Rome), Italy

In the 50-odd years since the introduction of clinically effective medications for the treatment of behavioral disorders such as depression,[1] anxiety[2] or schizophrenia[3] there has recently been growing unease with a seeming lack of substantive progress in the development of truly innovative and effective drugs for behavioral disorders; an unease indicated by escalating research and development expenditure associated with diminishing returns (e.g.,[4] and discussed by Hunter[5] in this book series). There are a number of reasons that may account for this lack of new drugs for CNS disorders (cf.,[6]), but according to the US Food and Drug Administration's (FDA) white paper on prospects for 21st century drug discovery and development,[7] one of the main causes for failure in the clinic is the discrepancy between positive outcomes of candidate drugs in animal models and apparent lack of efficacy in humans, that is, the predictive validity of animal models. Consequently, there have been a number of initiatives from the US National Institutes of Health (NIH) (http://nihroadmap.nih.gov/) and The European Medicines Agency (EMEA)[8] to bring interested parties from Academia and Industry together to discuss, examine and suggest ways of improving animal models of behavioral disorders.[9-14] The value of NIH-supported initiatives, even to the point of participating directly in drug discovery from screening to registration is not to be underestimated, as evidenced by the successful registration of buprenorphine (Subutex®) and buprenorphine/naloxone (Suboxone®) by Reckitt-Benckiser in collaboration with the National Institute on Drug Abuse (NIDA)[15], see also[16].[i]

Translational research and experimental medicine are closely related activities that have evolved in answer to the need of improving the attrition rate of novel drugs between the preclinical and clinical stage of development.[5,19-22] In general, translational research defines the *process* through which information and insights flow from

---

[i] For a comprehensive discussion of NIH-sponsored initiatives and collaborations and opportunities, please refer to Winsky and colleagues[17] and Jones and colleagues[18] for specifics on NIH-Academic-Industrial collaborations in schizophrenia.

Animal and Translational Models for CNS Drug Discovery,
Vol. 3 of 3: Reward Deficit Disorders
Robert McArthur and Franco Borsini (eds), Academic Press, 2008

clinical observations to refine the development of animal models *as well as* the complementary flow of information and insights gained from animal models to the clinical setting, be it through improved diagnosis, disease management or treatment; including pharmacological treatment.[23] Experimental medicine, in terms of drug discovery, refers to studies in human volunteers to (1) obtain mechanistic and pharmacological information of compounds entering into development, (2) explore and define biological markers with which the state and progress of a disorder can be monitored, as well as the effects of pharmacological interventions on its progress and (3) establish models and procedures with which to obtain initial signals of efficacy test.[5,15,22] Though claimed as an innovative paradigm shift, translational research nevertheless, is not a new concept, as pointed out by Millan in this book series.[24] The origins of psychopharmacology abound with numerous examples of how pharmaceutical or medicinal chemists interacted directly with their clinical colleagues to "test their white powder", or clinicians who would knock at the chemists' door for anything new. Kuhn and Domenjoz, for example, describes the initial "Phase II" trials of the novel "sleeping pill" forerunner of imipramine.[25,26] Paul Janssen tells how the observation of the paranoid schizophrenia-like hallucinations experienced by cyclists who were consuming amphetamine to stay alert, led him to search for better amphetamine antagonists, one of which was haloperidol. This compound was subsequently given to a young lad in the midst of a psychotic episode by a local psychiatrist with good results.[27] Though largely overtaken in sales and prescription rates by 2nd generation atypical antipsychotics, Haloperidol (Haldol®) remains one of the standard drugs used in the treatment of schizophrenia.[18,28,29]

Translational research is a two-way process which, nonetheless can lead to differences in emphasis and agenda. We have gathered a number of definitions from different sources listed in Table 1 below to help us determine what one of our authors asked us to do when he was contacted to contribute to this book project, "What do *you* mean by 'translational research'?"

These definitions may emphasize the clinical, or top-down approach to translational research,[20,30] or the bottom-up approach of "bench-to-bedside".[21,31] It is clear though, that translational research has a purpose of integrating basic and clinical research for the benefit of the patient in need. While we welcome this as a general definition of translational research, we acknowledge, as do others (e.g.,[31,32]), that a more pragmatic, working definition is required. Consequently, we define translational research, in the context of drug discovery and research, as the partnership between preclinical and clinical research to align not only "... basic science discoveries into medications",[31] but also the information derived from the clinic during the development of those medications. The purpose of this reciprocal definition is to refine the model systems used to understand the disorder by identifying the right targets, interacting with those targets pharmacologically in both animals and humans and monitoring the responses in each throughout a compound's development (cf.,[5 and 15]). Central to this definition is the acknowledgement that the etiology of behavioral disorders and their description are too diffuse to attempt to model or simulate in their entirety. Consequently, emphasis must be placed on identifying specific symptoms or core features of the disorder to model, and to define biological as well as behavioral responses as indices of state, changes in state and response to pharmacological treatment. This process is made easier if, at the same time, greater effort is made to identify procedures used to

Table 1 Selected definitions of translational research

Definition	Reference
Translational medicine may also refer to the wider spectrum of patient-oriented research that embraces innovations in technology and biomedical devices as well as the study of new therapies in clinical trials. It also includes epidemiological and health outcomes research and behavioral studies that can be brought to the bedside or ambulatory setting.	30
…connotes an attempt to bring information that has been confined to the laboratory into the realm of clinical medicine.	
To the extent that clinical studies could be designed to answer such questions (generated by information from the laboratory), they would represent types of translational clinical research.	20
…a two-way street where the drive to cure should be complemented by the pursuit to understand human diseases and their complexities.	21, 162
1. Basic science studies which define the biological effects of therapeutics in humans 2. Investigations in humans which define the biology of disease and provide the scientific foundation for development of new or improved therapies for human disease 3. Non-human or non-clinical studies conducted with the intent to advance therapies to the clinic or to develop principles for application of therapeutics to human disease 4. Any clinical trial of a therapy that was initiated based on #1–3 with any endpoint including toxicity and/or efficacy.	M. Sznol cited by [21]
…research efforts intended to apply advances in basic science to the clinical research setting. For drug discovery and development, the term refers to research intended to progress basic science discoveries into medications.	31
By bringing together top-down and bottom-up approaches, there is potential for a convergence of unifying explanatory constructs relating aetiology to brain dysfunction and treatment.	37, 163
…information gathered in animal studies can be translated into clinical relevance and vice versa, thus providing a conceptual basis for developing better drugs.	
…the application of scientific tools and method to drug discovery and development … taking a pragmatic or operational rather than a definitional approach, a key to a successful translation of non-human research to human clinical trials lies in the choice of biomarkers.	32
…two-way communication between clinical and discovery scientists during the drug development process are likely to help in the development of more relevant, predictive preclinical models and biomarkers, and ultimately a better concordance between preclinical and clinical efficacy.	82

measure these biological and behavioral responses that are consistent within and between species.[23,24] Brain imaging is one technique that has cross-species consistency (e.g.,[33-36]), as do various operant conditioning procedures.[37,38]

There are at least two aspects of translational research to be considered as a result of the definition proposed above. First is the concept of specific symptoms, or core features of the disorder to model. Attempts to simulate core disturbances in behavior formed the basis of early models of behavioral disorders. McKinney and Bunney, for example, describe how they sought to "translate" the clinically observed changes in human depressed behavior (secondary symptoms) with analogous changes in animals induced by environmental or pharmacological manipulations.[39]

Whereas modelers have traditionally referred to diagnostic criteria such as DSM-IV[40] or ICD-10[41] the consensus to be found in this book series and other sources is that these diagnostic criteria do not lend themselves easily to basic or applied research. The etiology of behavioral disorders is unclear, and there is considerable heterogeneity between patients with different disorders but similar symptoms. Nevertheless, attempts to model particular behavioral patterns have been and are being done. Thus, for example, the construct of anhedonia (the loss of ability to derive pleasure), or the construct of social withdrawal, may be diagnostic criteria for a number of behavioral disorders including depression, schizophrenia, as well as a number of other disorders (cf.,[42]). There is considerable momentum to establish a dimension – rather than diagnostic-based classification or to "deconstruct" syndromes into "symptom-related clusters" that would help guide neurobiological research.[ii][18,43,44] In order to define these "symptom-based clusters", however, the symptoms have to be defined. Previously, these were identified as behavioral patterns, though lately they have been referred to variously as behavioral endophenotypes or exophenotypes (e.g.,[45-49]). It is appropriate here to review the definitions of both. Exophenotype and endophenotypes have been defined by Gottesman and Shields[50] as:

> *John and Lewis (1966) introduced the useful distinction between* exophenotype (external phenotype) *and* endophenotype (internal), *with the latter only knowable after aid to the naked eye, e.g. a biochemical test or a microscopic examination of chromosome morphology (p. 19).*[iii]

Subsequently, endophenotypes have been more rigorously defined[51] as:

> 1. *The endophenotype is associated with illness in the population.*
> 2. *The endophenotype is heritable.*
> 3. *The endophenotype is primarily state-independent (manifests in an individual whether or not illness is active).*
> 4. *Within families, endophenotype and illness co-segregate.*
> 5. *The endophenotype found in affected family members is found in nonaffected family members at a higher rate than in the general population. (p. 639)*

---

ii For reviews of the initiatives deconstructing a complex disorder like schizophrenia, the reader is invited to consult the following 2 issues of *Schizophrenia Bulletin*, where these initiatives are thoroughly discussed: *Schizophr Bull*, 2007, 33:1 and *Schizophr Bull*, 2007, 33:4.

iii See also Tannock *et al.*,[61] for definitions of endophenotypes and biomarkers.

And that "…The number of genes involved in a phenotype is theorized to be directly related to both the complexity of the phenotype and the difficulty of genetic analysis" (*op cit.*, p. 637). On the other hand, exophenotypes have been defined by Holzman[52] (and others) as:

> … *the external symptoms of a disorder that clinicians detect during an examination. An endophenotype, on the other hand, is a characteristic that requires special tools, tests, or instruments for detection. (p. 300)*

It behooves the unwary researcher to be careful with terminology and thus not fall into the trap of pretending greater accuracy by changing the name of the phenomenon being studied. Finally, to quote Hyman's *caveat*[43],

> *The term "endophenotype" has become popular for describing putatively simpler or at least objectively measurable phenotypes, such as neuropsychological measures that might enhance diagnostic homogeneity. I find this term less than ideal, because it implies that the current diagnostic classification is basically correct, and that all that is lacking is objective markers for these disorders. If, however, the lumping and splitting of symptoms that gave rise to the current classification was in error, then the search for biological correlates of these disorders will not prove fruitful. (p. 729).*

The second aspect to be considered in translational research is the concept of biomarkers. Biomarkers are crucial to translational research and serve as the interface between preclinical research, experimental medicine and clinical development. As with endophenotypes above, however, biomarkers also require some discussion. The FDA, NIH and EMEA have been at the forefront in helping define and establish biomarkers, surrogate markers and clinical endpoints[53-57] (http://ospp.od.nih.gov/biomarkers/); an initiative now being carried out in partnership with private enterprise[58] (http://ppp.od.nih.gov/pppinfo/examples.asp). Lesko and Atkinson have provided summary definitions of various markers that are worth considering:[55]

> *A synthesis of some proposed working definitions is as follows:* (a) biological marker *(biomarker) - a physical sign or laboratory measurement that occurs in association with a pathological process and that has putative diagnostic and/or prognostic utility;* (b) surrogate endpoint *- a biomarker that is intended to serve as a substitute for a clinically meaningful endpoint and is expected to predict the effect of a therapeutic intervention; and* (c) clinical endpoint *- a clinically meaningful measure of how a patient feels, functions, or survives. The hierarchical distinction between biomarkers and surrogate endpoints is intended to indicate that relatively few biomarkers will meet the stringent criteria that are needed for them to serve as reliable substitutes for clinical endpoints (p. 348).*

An important characteristic of biomarkers is that they should also be capable of monitoring disease progression.[54] It is interesting more over that the establishment of biomarkers should also be subject to the same concepts of validity as defined by Willner initially for models of behavioral disorders, that is, face, construct and predictive validity.[59] Lesko and Atkinson further indicate that biomarkers must be evaluated and validated for (1) clinical relevance (face validity in being able to reflect physiologic/pathologic processes), (2) sensitivity and specificity (construct validity that it is capable to measure changes though a given mechanism in a target population) and (3) must ultimately be validated in terms of clinical change, that is, predictive validity. Biomarkers also have other criteria that they need to fulfill such as: their accuracy, precision and reproducibility; an estimated rate of false positive and false negative probability; and practicality and simplicity of use. In addition, pharmacological isomorphism is used to establish a biomarker's predictive validity where response to a known clinically effective standard is ultimately required, especially if drugs of different mechanisms of action produce the same response in the biomarker. These criteria are very familiar to the animal modeler and highlight the shared interests and expertise that the preclinical researcher brings to the clinical arena. Biomarkers for behavioral disorders thus share many of the problems inherent to their animal models.[60] Nevertheless, it is among the most active pursuits in Pharma today (cf.,[61-70]).

It is clear from the previous discussion that translational research demands the combined efforts of a number of participants, each of which contributes a particular expertise to achieve a common goal. Translational research cannot be done effectively using the "tried and true" process of compartmentalization prevalent up to the end of the last century, that is, the splitting of R from D, or maintaining the preclinical from clinical, academic from industrial divides. For the past decade Pharma has fostered cross-disciplinary collaboration with the creation of Project teams in which participants from preclinical, clinical and marketing sections of the Industry are brought together in relation to the maturity of the Project. The concept of "pitching the compound over the fence" is no longer tolerated, and preclinical participation even in mature Projects is expected. This creates a much more stimulating environment for all the participants, who not only learn from the experiences of others, but also maintain a sense of ownership even when their particular expertise is no longer required for a Project's core activities. Nevertheless, creation of and participation in Project teams is not always an easy task as group dynamics evolve. Team members are assigned to a Project by line managers, and can be removed depending on priorities. Some team members contribute more than their share, while others coast. The skills of the Project Leader must go beyond scientific expertise in order to forge an effective team and deliver a successful drug.

The use of animal models is an essential step in the drug discovery and indeed the translational research process. Use of appropriate models can minimize the number of drug candidates that later fail in human trials by accurately predicting the pharmacokinetic and dynamic (PK/PD) characteristics, efficacy and the toxicity of each compound. Selection of the appropriate models is critical to the process. Primary diseases such as those caused by infections, genetic disorders or cancers are less problematic to model using both *in vitro* and *in vivo* techniques. Similarly some aspects of degenerative diseases have also been successfully modeled. However, modeling of disorders with a strong behavioral component has been less successful. This is not to say that there are no models for various aspects of these disorders. Many models have been proposed, validated pharmacologically with standard, clinically effective drugs and extensively reviewed. Indeed, these models have become so standardized that their use to characterize mechanisms of action and lead novel compounds in CNS drug discovery projects is mandatory, and positive outcomes are required before these compounds are considered for further development. However, it has become clear that positive outcome in these models is no guarantee that these new compounds will be efficacious medicines in humans. Refinements of existing models and development of new models relevant to drug discovery and clinical outcome are being pursued and documented (e.g.,[71-74]). Advancements in genetic aspects of disease are also being aided through the development and use of genetically modified animals as model systems. However, even though these techniques are more precise in modeling aspects of a disease such as amyloid overexpression in Alzheimer's disease, the ability of procedures used to assess the changes in behavior, and relating them to altered human behavior remains uncertain.

Books on animal models of psychiatric and neurological diseases have tended to be compendia of so-called "standard" procedures developed over the years. Some of these books have formed part of classic reference texts for behavioral pharmacologists (e.g.,[75,76]). Others – more pragmatic in their approach – describe the application of these models and are useful as "cookbook" manuals (e.g.,[77,78]), while yet others have been very specific in their focus; for example, books entirely with models for a particular disorder, for example, depression or schizophrenia. It could be argued, however, that these books address a very circumscribed audience, and need not be necessarily so. Clinicians might and do claim that animal models are intellectually interesting, but of no relevance to their daily work of (1) demonstrating proof of concept, (2) showing efficacy or (3) treating their patients. Nevertheless, clinicians are constantly on the watch for potentially new pharmacological treatments with which to treat their patients, for example, new chemical entities that have reached their notice following extensive profiling in animal models. Academics develop a number of procedures or models to help them study neural substrates and disorders of behavior, and may use pharmacological compounds as tools to dissect behavior. The industrial scientist is charged with the application of these methods and models, establishing them in the lab at the request of the Project team and Leader. There is thus a shared interest in the development, use and ability of animal models to reflect the state of a disorder and predict changes in state following pharmacological manipulation. This shared interest has generated much collaboration between academics, clinicians and the industry (cf.,[79]).

Paradoxically in view of shared interest, close ties and general agreement on the need for bidirectional communication, the integration of the perspective and experience of the participants in the drug discovery and development process is not always apparent and is a source of concern (e.g.,[6,21,31,32,80-82]). Although we do not necessarily agree entirely with Horrobin's description of biomedical research scientists as latter day Castalians,[80] we suggest that there is a certain truth to the allusion that considerable segregation between the academic, clinician and industrial researcher exists (see also[21,81]). There have been numerous attempts to break down these barriers, such as having parallel sessions at conferences, or disorder-specific workshops organized by leading academics, clinicians and industrial scientists (e.g., *op. cit.*,[83]). With few exceptions, however, academics will talk to academics, clinicians to clinicians and industrial scientists will talk to either academics or clinicians; depending on at which stage their Project is. Willner's influential book[84], "*Behavioral Models in Psychopharmacology: Theoretical, Industrial and Clinical Perspectives*" represents one of the first published attempts that brings together academics, pharmaceutical researchers and clinicians to discuss the various aspects of the animal models of behavioral disorders. Yet even in Willner's book with alternating chapters expounding the academic, industrial and clinical perspective on a subject, the temptation is always to go to the "more interesting", that is, directly relevant chapter and leave the others for later.

This three volume book series aims to bring objective and reasoned discussion of the relative utility of animal models to all participants in the process of discovery and development of new pharmaceuticals for the treatment of disorders with a strong behavioral component, that is, clearly psychiatric and reward deficits, but also neurodegenerative disorders in which changes in cognitive ability and mood are important characteristics. Participants include the applied research scientists in the Pharma industry as well as academics who carry out animal research, academic and industry clinicians involved in various aspects of clinical development, government officials and scientists setting funding priorities, and industrial, academic and clinical opinion leaders, who very clearly influence and help shape the decisions determining what therapeutic areas and molecular targets are to be pursued (or dropped) by Pharma. Rather than a catalog of existing animal models of behavioral disorders, the chapters of the book series seek to explore the role of these models within CNS drug discovery and development from the shared perspective of these participants in order to move beyond the concept of animal behavioral assays or "gut baths",[85,86] to stimulate the development of animal models to support present research of the genetic basis of behavioral disorders and to improve the ability to translate findings and concepts between animal research and clinical therapeutics.

As indicated, the aim and scope of this book series has been to examine the contribution of the animal models of behavioral disorders to the process of CNS drug discovery and development rather than a simple compendium of techniques and methods. This book goes beyond the traditional models book published in that it is more a considered review of *how* animal models of behavioral disorders are used rather than *what* they are. In order to achieve this goal, leading preclinical and clinical investigators from both Industry and Academia involved in translational research were identified and asked to participate in the Project. First, a single author was asked to write an introductory chapter explaining the role of animal and translational models for CNS drug discovery from their particular perspective. Each volume thus starts with an industry perspective from a senior Pharma research executive, which sets a framework. Considering the prominent role assumed by Governmental agencies such as the FDA or NIH in fostering translational research, NIH authors were asked to discuss animal and translational models from the perspective of the Government. A leading academic author was contacted to provide a general theoretical framework of how animal and translational models are evolving to provide the tools for the study of the neural substrates of behavior and how more efficient CNS drug discovery may be fostered. Finally, leading clinicians involved in the changing environment of clinical trial conductance and design were asked to discuss how issues in clinical trial design and conductance have affected the development and registration of CNS drugs in their area of specialty, and how changes are likely to affect future clinical trials.

Following these 4 introductory chapters, there are therapeutic area chapters in which a working party of at least 3 (industrial preclinical, academic and clinical) authors were identified and asked to write a consensual chapter that reflects the view of the role of animal models in CNS translational research and drug development in their area of expertise. We deliberately created our chapter teams with participants who had not necessarily worked together before. This was done for three reasons. First, we were anxious to avoid establishing teams with participants who had already evolved a conceptual framework *a priori*. Second, we felt that by forcing people to "brainstorm" and develop new ideas and concepts would be more stimulating both for the participants and the readers. Thirdly, we wanted to simulate the conditions of the creation of an industrial Project team, where participants need not know each other initially or indeed may not even like each other, but who are all committed to achieving the goals set out by consensus. We sought to draw upon the experiences of industrial and academic preclinical and clinical investigators who are actively involved in CNS drug discovery and development, as well as translational research. These therapeutic chapter teams have contributed very exciting chapters reflecting the state of animal models used in drug discovery in their therapeutic areas, and their changing roles in translational research. For many, this has been a challenging and exhilarating experience, forcing a paradigm shift from how they have normally worked. For some teams, the experience has been a challenge for the same reason. One is used to write for one's audience and usually on topics with which one is comfortable. For some authors, it was not easy to be asked to write with other equally strong personalities with different perspectives, and then to allow someone else to integrate this work into a consensual chapter. Indeed, some therapeutic area chapter teams were not able to establish an effective team. As a consequence, not all the therapeutic areas envisioned to be covered initially in this Project were possible. Nevertheless, as translational research becomes more established, what appears to be a novel and unusual way of working will become the norm for the benefit not only of science, but for the patient in need.

VOLUME OVERVIEW AND CHAPTER SYNOPSES

This volume comprises contributions by different authors on some reward deficit disorders, such as alcohol dependence, nicotine dependence, heroin and cocaine addiction, obesity, and pathological gambling and other impulse disorders. Furthermore, Rocha and colleagues present a very valuable discussion on the assessment of and regulatory aspects of abuse liability potential of candidate drugs.[16] The study of addictions has evolved considerably since the early theories of mesolimbic dopamine reward and motivated behavior (cf.,[87-89]). Central to this evolving conceptual framework is the cyclical nature of addiction and associated neural adaptations in which impulsivity and loss of control play a central role.[90-94] Furthermore, the role of neurochemical substrates other than the monoamines, norepinephrine, dopamine and serotonin, or nicotine is being investigated. The role of hypothalamic peptides in motivated behavior, for example, is of particular interest,[93,95,96] as is the role of GABA and glutamate.[93,94,97-99]

Common themes that emerge from these chapters is the role played by stress, changes in the hypothalamic–pituitary–adrenal axis (HPA), and how the brain adapts to chronic stress through the process of allostasis (cf.,[90,100-104], and discussed by Doran and colleagues in Volume 1, Psychiatric Disorders[104]). Response to chronic stress and vulnerability to stress mediated by neurodevelopmental factors are key risk factors inherent in not only in psychiatric disorders (e.g.,[18,61,105]), but reward deficit disorders as well.[92-94,106,107] Early exposure to substances of abuse (cf.,[93,94]), or gambling[107] increases the risk of subsequent abuse and dependence, and pathological gambling later in life.

Acute and chronic stressors induce well-known neuroendocrinological homeostatic adaptations.[108] The cumulative effects of various stressors, however, can lead to an allostatic state, or "…a state of chronic deviation of the regulatory system from its normal (homeostatic) operating level; and allostatic load is the cost to the brain and body of the deviation, accumulating over time, and reflecting in many cases pathological states and accumulation of damage."[90] Maladaptive neural changes can be traced chronologically from the acute reinforcing properties of drugs associated with an increase in mesolimbic activity; reinstatement of drug-seeking behavior associated with changes in the medial prefrontal cortex, nucleus accumbens, ventral pallidum and basolateral amygdala; and drug-seeking and compulsive behaviors associated with the ventral striatal, ventral pallidus, thalamic and orbitofrontal loop,[92] see also Joel and colleagues[109] and Williams *et al.*[107]

The theme of treatment resistance has been discussed in other chapters of the Psychiatry and Neurology volumes of this book series (e.g.,[61,110-113]). Treatment resistance in reward deficit disorders takes on a particular meaning in terms of effects of withdrawal from the substance of abuse and the reinstatement of drug-seeking behavior. Cycles of withdrawal and reinstatement can set up a kindling cycle contributing to the allostatic nature of these disorders.[93,110,112,114] The effects of withdrawal from alcohol, cocaine, heroin and other abused substances are well documented. Opiate withdrawal, for example, is associated with somatic symptoms such as vomiting, pain, diarrhea, insomnia; and psychological symptoms such as irritability, dysphoria and anxiety. Cocaine and alcohol withdrawal, on the other hand, are associated more with psychological symptoms such as depression, fatigue, anxiety and irritability, although sleep disturbances such as insomnia are prominent.[16,93,104] These are powerful inducers to relapse,[90] and the prevention of relapse is a major therapeutic goal.[15,93,94]

As discussed in the first volume of this book series, allostatic changes in response to chronic stress is thought to underlie many psychiatric disorders and consequently there is a great degree of comorbidity between psychiatric and reward deficit disorders.[92-94,106,107,109,111,115] Psychiatric disorders can be a risk factor for reward and impulse deficit disorders, and impair successful treatment.[94] Equally, moreover, the high degree of comorbidity between reward deficit, psychiatric and neurological disorders has meant that (a) insights into the neurobiology of impulse control disorders have been gleaned, for example, by the disinhibited behavior observed in Parkinson's disease[107] and (b) new indications can and have been found for drugs originally developed for psychiatric (e.g.,[94,116]) and neurological indications (e.g.,[93]), (See also Williams *et al.*[107] for further discussion of psychopharmacological treatments for impulse control disorders including gambling). Koob, considering the confluence of psychiatric and reward deficit disorders, has contributed an analysis of validity of the chronic mild stress model for depressed like behaviors, the drug withdrawal model of dependence, as well as the behavioral procedures used to assess the effects of these manipulations.[90]

Comorbidity can lead to very heterogeneous patient populations, which must be addressed when selecting subjects for clinical trials of pharmacological treatments of various reward deficit and impulse control disorders.[106] Some of the problems have already been discussed by McEvoy and Freudenreich,[19] and by Schneider[117] in this book series but some problems are germane to reward deficit disorders. For example, Gardner and colleagues[106] refer to the requirement that subjects must be previous drug abusers for inclusion in drug administration studies, and drug abusers rarely will restrict themselves to one drug. Varying degrees of motivation of the participants may also confound subject selection and trial outcome (see also[117]). An important development in clinical trial design is the incorporation of adaptive design through which an ongoing trial may be monitored, and if needs be, changed, without compromising the statistical integrity of the study. The use of Bayesian statistics helps the investigator use ongoing information.[118] This statistical approach is discussed in other chapters of the book series;[5,19,94,106,112] including the use of Bayesian statistics to plan large scale preclinical studies with transgenic mice.[119]

The maintenance of abstinence, or prevention of relapse, is a clinical endpoint in clinical trials for substance abuse treatment. Modification or prevention of further abuse is another important endpoint.[93,94,106] Similar clinical goals are sought for the treatment of impulse control disorders.[92,107,109] Clinical trials for obesity, on the other hand, assess the rate of body weight loss and maintenance of body weight loss in addition to changes in obesity-related metabolic alterations such as blood lipids and glucose.[116] Some clinically effective drugs for alcohol abuse such as naltrexone (ReVia®) have shown to reduce drinking (although not total abstinence[93,120] acamprosate (Campral®) prevented relapse in abstinent drinkers).[121] These effects are variable and have modest efficacy. Nevertheless, naltrexone and acamprosate are FDA approved.[93] On the other hand, varenicline (Chantix®/Champix®) and bupropion (Zyban®) maintain abstinence from smoking[122] and are approved for smoking cessation along with various forms of nicotine replacement strategies such as gums or patches.[94] Buprenorphine (Subutex®), buprenorphine/naloxone (Suboxone®) and methadone are substitution or replacement therapies for opiate (heroin) abuse, and there are currently no effective medications to treat cocaine addiction, although $GABA_B$ receptor agonists such as baclofen (Kemstro®/Lioresal®), the $\alpha 1$ adrenergic agonist (*inter alia*) modafinil (Provigil®), dopamine D_2 partial agonists (aripiprazole, Abilify®), or D_1 antagonists (ecopipam) have been evaluated.[16,19] In general though, the treatment of reward deficit and impulse control disorders combine pharmacological interventions with various forms of psychosocial counseling and behavioral therapy such as cognitive behavioral therapy.[16,24,106,107]

In addition to the NIH initiatives supporting academic, industrial and government collaborations described above, NIDA has interacted directly with academia and industry for 20 years in the search for effective pharmacological treatments of drug abuse. Initially this collaboration consisted of soliciting compounds from various pharmaceutical companies with the commitment of screening them for anti-drug abuse potential; promising compounds would then be passed on to academic and commercial groups contracted by NIDA for secondary evaluation, and toxicological and pharmacokinetic testing. The guarantee of anonymity and uncompromised intellectual property, coupled with the potential of developing compounds for a hitherto unthought-of alternate therapeutic indication at no direct cost made this collaboration attractive. Since then over 3000 new compounds have been evaluated, 3 new molecular entities have reached IND

(Investigational New Drug http://www.fda.gov/cder/Regulatory/applications/ind_page_1.htm), 1 new molecular entity has reached NDA (New Drug Application http://www.fda.gov/cder/regulatory/applications/nda.htm) and 4 new products [levomethadyl acetate (Orlaam®), buprenorphine (Subutex®), buprenorphine/naloxone (Suboxone®) and depot naltrexone (Vivitrol®)] have been registered through NIDA aegis.

Modeling aspects of reward deficit disorders involving substance abuse has a considerable advantage over modeling either endophenotypes or behavioral traits in psychiatric indications,[123] or causative neurodegenerative factors.[124] Substances being abused are self-administered directly by both animals and humans.[15,16,93,94,106,116] However, modeling of the neurobiological and psychological substrates and the neurological consequences of reward deficit disorders relies greatly on procedures developed for psychiatry and neurology. In Volume 1 of this book series Miczek[86] proposes eight conceptual principles for translational research. Principle 8 is particularly germane to this discussion: "*It is more productive to focus on behaviorally defined symptoms when translating clinical to preclinical measures, and vice versa. Psychological processes pertinent to affect and cognition are hypothetical constructs that need to be defined in behavioral and neural terms.*" This principle reminds us we are ultimately referring to hypothetical constructs when we talk about motivation, mood, affect, cognition, etc. (cf.,[115]). Consequently discussion of aspects of validity of models is an important theme in reward deficit disorders.[15,16,90,92,93,106,107]

Miczek's 1st conceptual principle for translational psychopharmacological research, "*The translation of preclinical data to clinical concerns is more successful when the development of experimental models is restricted in their scope to a cardinal or core symptom of a psychiatric disorder.*" This principle is well illustrated by Little and colleagues'[93] by considering the DSM-IV diagnostic criteria for alcohol dependence, the behaviors associated with such criteria, and analyzing how closely they can be modeled. The criteria of development of tolerance, acute (and protracted) withdrawal symptoms and degree of work or motivation to get alcohol, respectively, can be modelled successfully. Other criteria are not so easy to model. Criterion 4, for example, which is the desire to reduce alcohol consumption and unsuccessful attempts to do so is considerably more difficult to model because it involves hypothetical constructs such as intention or volition (see also[107]). Technical difficulty to demonstrate behaviors such as the increased levels of alcohol consumption after prolonged excess drinking (criterion 3), or the general lack of interest and attention from researchers to model such criteria as the reduction in general activity (criterion 6) or continuation of drug intake despite knowledge of the adverse consequences (criterion 7) may account for the lack of models of these aspects of DSM-IV diagnostic criteria. Furthermore, not all alcohol (and other substance) abuse behaviors are being described in DSM-IV, such as craving, priming, drug-associated cues, cognitive deficits, comorbidity, genetic, neurodevelopmental and environmental factors, but which nevertheless are being modeled.

Various direct, self-administration paradigms, which are meant to model the voluntary intake of abused substances, are used. These include preference paradigms in which the subject selects between two drugs, or between a drug and a neutral substance. Other paradigms involve the self-administration of the drug that is either continuously available or limited to a few hours within a 24-hour period.[93] Voluntary oral intake of a substance is most commonly used in alcohol abuse research, however, or as a measure of the reward sensitivity for a substance like sucrose and how this changes in response to models for depressed-like behaviors like anhedonia.[90,111] Nicotine, cocaine, heroin and other abused substances – including alcohol – are more usually self-administered systemically using operant conditioning schedules.[16,90,92-94] Operant conditioning is also used to assess the relative motivation of a subject to obtain access to a drug of abuse. The progressive-ratio schedule of reinforcement is used in which the subject is required to work increasingly more for a reward (i.e., make successively more lever presses for a single reward), until the subject quits responding.[16,90,92-94] Animals will readily self-administer small electrical impulses through intra-cranial electrodes.[125] This technique is used to determine the rewarding properties of a substance and behavioral changes during acquisition of or withdrawal from an abused substance,[90,94] and indeed the abuse potential of a novel drug.[16] The subjective similarity of a novel drug with another, that is, the operant technique of drug discrimination is another procedure used in the assessment of the abuse potential of a novel drug.[16] While operant techniques are used extensively to assess the effects of abused drugs, the role of genetics[93] and/or environment[90] on drug abuse, the conditioned place preference or aversion open-field procedure is also used in which the interoceptive effects of drug are paired with the cues or stimuli of a previously neutral environment (cf.,[92-94,126]).

While the procedures discussed above are used to assess behavioral properties of drugs of abuse, assessing the *intention* or propensity to carry out an action, or the assessment of *relative worth* in choice situations requires a perceptual shift in model development.[16,90,92,106,107,127] These are much more difficult as we have indicated in Miczek's 8th principle quoted above. Models of impulsive or compulsive behaviors have been discussed by Joel and colleagues[109] and by Tannock *et al.*[61] in the first volume, Psychiatric Disorders, of this book series. The application of behavioral economics to impulse control and reward deficit disorders is discussed mostly in terms of delayed discounting in which subjects tend to chose smaller rewards rather than a large, but delayed reward.[16,92,94,107,128,129]

As discussed above, there are many reasons why a compound that has shown a promising preclinical behavioral profile may fail either to get into clinical development, or fail when in one of the clinical phases. Determination of the ADME (absorption, distribution, metabolism and excretion) characteristics of a candidate drug is one of the fundamental steps in drug discovery and development.[5,24,92,116,130,131] Correct PK/PD assessment in humans is crucial to determine whether the compound is present at the right receptors at a concentration sufficient to have the desired effect, and that there is adequate drug exposure (see also[5,94,111,112,132] for further discussion on this theme). Heidbreder[92] points out a number of key questions in PK/PD modeling.

1. What is the expected clinical response for a treatment strategy?
2. What is the level of certainty surrounding the predicted response?
3. How do different treatment strategies impact response?
4. What dose is required to achieve a target response?
5. What is the probability of achieving a specific efficacy target while keeping probability for adverse events below a certain level?
6. What is optimal strategy to balance safety and efficacy?

The incorporation of micro-dosing at Phase 0 for early human pharmacokinetic studies represents an innovation in clinical development,[19] as is the use of radio-labeled compounds for brain imaging.[5,15,92]

Theoretically all drugs designed to interact with the CNS are associated with some type of behavioral adverse side effect that can limit or restrict their use. Common examples include the amphetamines and fenfluramine for use in the treatment of obesity,[116] and the use of benzodiazepines for the treatment of anxiety,[115] or sleep disorders.[104] The FDA requires that every candidate drug being submitted for an NDA be evaluated for potential risks, which can include abuse potential.[16] As indicated previously, abuse liability potential is assessed using the self-administration, or drug discrimination procedures described. If there is any suggestion of abuse potential generated from these animal studies, then the candidate drug is likely to be assessed in humans.[16] Obviously, the association of a candidate drug with abuse potential and subsequent scheduling as a controlled substance has major financial implications for the sponsoring company. We have seen limitations imposed upon one such drug recently approved for the treatment of obesity in the EU, the endocannabinoid CB_1 receptor antagonist rimonabant (Acomplia®)/Zimulti®). However, this drug has also been associated with anxiety and depressive disorders, seizures and suicide, which has prompted the FDA to withhold approval until more safety and efficacy data are available[116] (see also http://www.fda.gov/ohrms/dockets/ac/07/briefing/2007-4306b1-fda-backgrounder.pdf). The endocannabinoid system is an important drug target in psychiatric,[24,133] neurological[134,135] and reward deficit[136] disorders, and rimonabant is currently being used as a pharmacological tool and investigated as a potential therapy for alcohol[137], but see[93] and nicotine[94,136] abuse. As both these indications are associated with mood disorders, and alcohol abuse withdrawal with seizures,[138,139] therapeutic use of rimonabant in either of these disorders should be carefully evaluated.

Experimental medicine plays a key role in translational research being at the interface of preclinical and clinical research,[5,22,24,92,140,141] and human laboratory studies are being used frequently to bridge animal and human reward disorders research.[15,16,106,116] Many of the procedures such as self-administration, drug discrimination, delay-discounting discussed previously have their human analogs. For example, drug discrimination has its direct analog simply by asking the subject to discriminate between one drug and another.[16] Operant conditioning procedures such as the progressive-ratio schedules are used with human volunteers.[16,90] Assessment of risk-reward decision making in humans typically use the Iowa Gambling Task,[142] which has a rodent analog.[143] Obesity is both a metabolic and reward deficit disorder,[116] and although objective methods for analyzing the behavioral processes of hunger and satiety in both rodents[144,145] and humans[146,147] have been in use for over 30 years,[148-151] very few candidate drugs for the treatment of obesity have been evaluated behaviorally;[116] this perhaps being a function of the metabolic emphasis given to these drugs. Nevertheless, the combined use of these procedures in a translational research environment not only would provide an extensive behavioral profile of the effects of candidate drugs on eating behavior, but also provide information as to the side effect profile of the drugs.

Brain imaging is far and beyond one the most effective translational techniques being used extensively in early development[5,24,68] not only in neurology[17,110,132,152,153] but psychiatry[18,61,104,105,111,112,154] and reward deficit disorders.[90,92,106,107,116] These are techniques that lend themselves very well to translational research (e.g.,[82,155-159]).

Perhaps obesity research is most advanced in the use of biomarkers. Some of these are simple measures that do not require laboratory space or elaborate equipment such as body weight, or indices of body mass like the BMI and measures of waist circumference and waist hip ratio. Various laboratory-based techniques like underwater weighing, CT or MRI scanning of fat distribution; plus diverse energy expenditure exist as well as measurement of cardiovascular and metabolic factors such as blood pressure, blood lipids and glucose.[116] Metabolite detection methods are used to monitor treatment effects and compliance in reward deficit clinical trials. These include urine assays for the cocaine metabolite benzoylecognine,[106] or the nicotine metabolite cotinine, or blood chemistry assays for carbohydrate-deficient transferrin and liver amino transferases[160,161] for alcohol abuse and plasma nicotine for nicotine abuse.[94]

As in psychiatric and neurological disorders, concepts of validity figure prominently in the chapters of this Volume, the authors have indicated the face and construct validity of their models (cf.,[90,94]). These confirm the Miczek's 8th conceptual principle described previously.[86] Predictive validity of the models presented in this volume is being established by pharmacological isomorphism,[15,16] and by the ability to apply procedures and paradigms originally designed for rodents to non-human primates and human subjects. This ability to "translate" from one species to the next is a notable strength for CNS drug discovery for reward deficits disorders, and illustrates Lindner and colleagues addendum to Miczek's[86] conceptual principles of translational psychopharmacology research, *"Therapeutic effects should also be produced on measures of cognitive and behavioral/functional abilities most similar to the primary endpoints required for establishing efficacy in clinical trials."*[124,152] Pharmacological isomorphism is an important if not essential step for establishing predictive validity of models for CNS disorders, but, as discussed extensively in Volume 3, Psychiatric Disorders, reliance on drugs with varying efficacy, or restricted mechanisms of action may create a tautology that would limit the ability to discover more effective drugs. For example, there is no clinically approved drug for the treatment of cocaine addiction, pathological gambling and other impulse deficit disorders. Opiate addiction is currently being treated by replacement therapy with levomethadyl acetate, methadone and buprenorphine. Alcoholism is being treated with Campral® and ReVia®, while smoking is treated with Chantix®/ Champix®, Zyban® and nicotine replacement. Obesity is being treated with Reductil®/ Meridia, Xenical®/Alli® or Acomplia®. These latter clinical treatments are of varying efficacy, and not without their problems or limitations. Xenical®/Alli®, for example, is a mechanical barrier to block the digestion of fats. Acomplia®, though effective enough to be registered, is limited by worrying side effects that may limit its use not only in the treatment of obesity, but in other reward deficit disorders where it is being evaluated. It would be instructive at this point therefore to recall Miczek's 6th principle, *"Preclinical experimental preparations can be useful screens that are often validated by predicting reliably treatment success with a prototypic agent. These screens detect "me-too" treatments that are based on the same principle as an existing treatment that has been used to validate the screen"*[86]; but with the hope that these models will help in the discovery of new and more effective treatments for these very complex disorders.

REFERENCES

1. Kline, N.S. (1958). Clinical experience with iproniazid (marsilid). *J Clin Exp Psychopathol*, 19(2, Suppl 1):72–78. discussion 8–9.
2. Selling, L.S. (1955). Clinical study of a new tranquilizing drug; use of miltown (2-methyl-2-n-propyl-1,3-propanediol dicarbamate). *JAMA*, 157(18):1594–1596.
3. Delay, J. and Deniker, P. (1955). Neuroleptic effects of chlorpromazine in therapeutics of neuropsychiatry. *J Clin Exp Psychopathol*, 16(2):104–112.
4. Kola, I. and Landis, J. (2004). Can the pharmaceutical industry reduce attrition rates? *Nat Rev Drug Discov*, 3(8):711–715.
5. Hunter, A.J. (2008). Animal and translational models of neurological disorders: An industrial perspective. In McArthur, R.A. and Borsini, F. (eds.), *Animal and Translational Models for CNS Drug Discovery: Neurologic Disorders*. Academic Press: Elsevier, New York.
6. McArthur, R. and Borsini, F. (2006). Animal models of depression in drug discovery: A historical perspective. *Pharmacol Biochem Behav*, 84(3):436–452.
7. FDA (2004). Innovation or Stagnation: Challenge and Opportunity on the Critical Path to New Medical Products. US Department of Health and Human Services, Food and Drug Administration, Washington, DC.
8. EMEA. (2005). *The European Medicines Agency Road Map to 2010: Preparing the Ground for the Future*. The European Medicines Agency, London.
9. Nestler, E.J., Gould, E., Manji, H., Buncan, M., Duman, R.S., Greshenfeld, H.K. *et al.* (2002). Preclinical models: Status of basic research in depression. *Biol Psychiatry*, 52(6):503–528.
10. Shekhar, A., McCann, U.D., Meaney, M.J., Blanchard, D.C., Davis, M., Frey, K.A. *et al.* (2001). Summary of a National Institute of Mental Health workshop: Developing animal models of anxiety disorders. *Psychopharmacology (Berl)*, 157(4):327–339.
11. Bromley, E. (2005). *A Collaborative Approach to Targeted Treatment Development for Schizophrenia: A Qualitative Evaluation of the NIMH-MATRICS Project. Schizophr Bull*, 31(4):954–961.
12. Winsky, L. and Brady, L. (2005). Perspective on the status of preclinical models for psychiatric disorders. *Drug Discov Today: Disease Models*, 2(4):279–283.
13. Stables, J.P., Bertram, E., Dudek, F.E., Holmes, G., Mathern, G., Pitkanen, A. *et al.* (2003). Therapy discovery for pharmacoresistant epilepsy and for disease-modifying therapeutics: Summary of the NIH/NINDS/AES models II workshop. *Epilepsia*, 44(12):1472–1478.
14. Stables, J.P., Bertram, E.H., White, H.S., Coulter, D.A., Dichter, M.A., Jacobs, M.P. *et al.* (2002). Models for epilepsy and epileptogenesis: Report from the NIH workshop, Bethesda, Maryland. *Epilepsia*, 43(11):1410–1420.
15. McCann, D.J., Acri, J.B., and Vocci, F.J. (2008). Drug discovery and development for reward disorders. In McArthur, R.A. and Borsini, F. (eds.), *Animal and Translational Models for CNS Drug Discovery: Reward Deficit Disorders*. Academic Press: Elsevier, New York.
16. Rocha, B., Bergman, J., Comer, S.D., Haney, M., and Spealman, R.D. (2008). Development of medications for heroin and cocaine addiction and Regulatory aspects of abuse liability testing. In McArthur, R.A. and Borsini, F. (eds.), *Animal and Translational Models for CNS Drug Discovery: Reward Deficit Disorders*. Academic Press: Elsevier, New York.
17. Winsky, L., Driscoll, J., and Brady, L. (2008). Drug discovery and development initiatives at the National Institute of Mental Health: From cell-based systems to Proof-of-Concept. In McArthur, R.A. and Borsini, F. (eds.), *Animal and Translational Models for CNS Drug Discovery: Psychiatric Disorders*. Academic Press: Elsevier, New York.
18. Jones, D.N.C., Gartlon, J.E., Minassian, A., Perry, W., and Geyer, M.A. (2008). Developing new drugs for schizophrenia: From animals to the clinic. In McArthur, R.A. and Borsini, F. (eds.), *Animal and Translational Models for CNS Drug Discovery: Psychiatric Disorders*. Academic Press: Elsevier, New York.

19. McEvoy, J.P. and Freudenreich, O. (2008). Issues in the design and conductance of clinical trials. In McArthur, R.A. and Borsini, F. (eds.), *Animal and Translational Models for CNS Drug Discovery: Psychiatric Disorders*. Academic Press: Elsevier, New York.

20. Schuster, D.P. and Powers, W.J. (2005). *Translational and Experimental Clinical Research*. Lippincott Williams & Wilkins, Philadelphia, PA.

21. Mankoff, S.P., Brander, C., Ferrone, S., and Marincola, F.M. (2004). Lost in translation: Obstacles to translational medicine. *J Transl Med*, 2(1):14.

22. Littman, B.H. and Williams, S.A. (2005). The ultimate model organism: Progress in experimental medicine. *Nat Rev Drug Discov*, 4(8):631-638.

23. Robbins, T.W. (1998). Homology in behavioural pharmacology: An approach to animal models of human cognition. *Behav Pharmacol*, 9(7):509-519.

24. Millan, M.J. (2008). The discovery and development of pharmacotherapy for psychiatric disorders: A critical survey of animal and translational models, and perspectives for their improvement. In McArthur, R.A. and Borsini, F. (eds.), *Animal and Translational Models for CNS Drug Discovery: Psychiatric Disorders*. Academic Press: Elsevier, New York.

25. Domenjoz, R. (2000). From DDT to imipramine. In Healey, D. (ed.), *The Psychopharmacologists III: Interviews with David Healy*. Arnold, London, pp. 93-118.

26. Kuhn, R. (1999). From imipramine to levoprotiline: The discovery of antidepressants. In Healey, D. (ed.), *The Psychopharmacologists II: Interviews with David Healy*. Arnold, London, pp. 93-118.

27. Janssen, P. (1999). From haloperidol to risperidone. In Healy, D. (ed.), *The Psychopharmacologists II: Interviews by David Healy*. Arnold, London, pp. 39-70.

28. Almond, S. and O'Donnell, O. (2000). Cost analysis of the treatment of schizophrenia in the UK. A simulation model comparing olanzapine, risperidone and haloperidol. *Pharmacoeconomics*, 17(4):383-389.

29. Gasquet, I., Gury, C., Tcherny-Lessenot, S., Quesnot, A., and Gaudebout, P. (2005). Patterns of prescription of four major antipsychotics: A retrospective study based on medical records of psychiatric inpatients. *Pharmacoepidemiol Drug Safety*, 14(11):805-811.

30. Pizzo P. (2002). Letter from the Dean, *Stanford Medicine Magazine*. Stanford University School of Medicine.

31. Lerman, C., LeSage, M.G., Perkins, K.A., O'Malley, S.S., Siegel, S.J., Benowitz, N.L. *et al.* (2007). Translational research in medication development for nicotine dependence. *Nat Rev Drug Discov*, 6(9):746-762.

32. Horig, H. and Pullman, W. (2004). From bench to clinic and back: Perspective on the 1st IQPC Translational Research conference. *J Transl Med*, 2(1):44.

33. Beckmann, N., Kneuer, R., Gremlich, H.U., Karmouty-Quintana, H., Ble, F.X., and Muller, M. (2007). In vivo mouse imaging and spectroscopy in drug discovery. *NMR Biomed*, 20(3):154-185.

34. Risterucci, C., Jeanneau, K., Schoppenthau, S., Bielser, T., Kunnecke, B., von Kienlin, M. *et al.* (2005). Functional magnetic resonance imaging reveals similar brain activity changes in two different animal models of schizophrenia. *Psychopharmacology (Berl)*, 180(4):724-734.

35. Tamminga, C.A., Lahti, A.C., Medoff, D.R., Gao, X.-M., and Holcomb, H.H. (2003/11/1). Evaluating Glutamatergic Transmission in Schizophrenia. *Ann NY Acad Sci*, 1003(1):113-118.

36. Shah, Y.B. and Marsden, C.A. (2004). The application of functional magnetic resonance imaging to neuropharmacology. *Curr Opin Pharmacol*, 4(5):517-521.

37. Robbins, T.W. (2005). Synthesizing schizophrenia: A bottom-up, symptomatic approach. *Schizophr Bull*, 31(4):854-864.

38. Porrino, L.J., Daunais, J.B., Rogers, G.A., Hampson, R.E., and Deadwyler, S.A. (2005). Facilitation of task performance and removal of the effects of sleep deprivation by an ampakine (CX717) in nonhuman primates. *PLoS Biol*, 3(9):e299.

39. McKinney, W.T.J. and Bunney, W.E.J. (1969). Animal model of depression I: Review of evidence: Implications for research. *Arch Gen Psychiatry*, 21(2):240-248.

40. American Psychiatric Association. (1994). Diagnostic and Statistical Manual of Mental Disorders, 4th edition. American Psychiatric Association, Washington, DC.

41. World Health Organization. (2007). International Statistical Classification of Diseases, 10th Revision, 2nd Edition. World Health Organization, Geneva.

42. Silverstone, P.H. (1991). Is anhedonia a good measure of depression? *Acta Psychiatr Scand*, 83(4):249–250.

43. Hyman, S.E. (2007). Can neuroscience be integrated into the DSM-V? *Nat Rev Neurosci*, 8(9):725–732.

44. Kupfer, D.J., First, M.B., and Regier, D.A. (eds.) (2002). *A Research Agenda for DSM-V*. American Psychiatric Association, Washington, DC.

45. Eisenberg, D.T., Mackillop, J., Modi, M., Beauchemin, J., Dang, D., Lisman, S.A. *et al.* (2007). Examining impulsivity as an endophenotype using a behavioral approach: A DRD2 TaqI A and DRD4 48-bp VNTR association study. *Behav Brain Funct*, 3:2.

46. Hasler, G., Drevets, W.C., Manji, H.K., and Charney, D.S. (2004). Discovering endophenotypes for major depression. *Neuropsychopharmacology*, 29(10):1765–1781.

47. Cannon, T.D. and Keller, M.C. (2006). Endophenotypes in the genetic analyses of mental disorders. *Ann Rev Clin Psychol*, 2(1):267–290.

48. Meehl, P.E. (1972). A critical afterword. In Gottesman, I.I. and Schields, J. (eds.), *Schizophrenia and Genetics: A Twin Study Vantage Point*. Academic Press, New York, pp. 367–415.

49. Breiter, H., Gasic, G., and Makris, N. (2006). Imaging the neural systems for motivated behavior and their dysfunction in neuropsychiatric illness. In Deisboeck, T.S. and Kresh, J.Y. (eds.), *Complex Systems Science in Biomedicine*. Springer, New York, pp. 763–810.

50. Gottesman, I.I. and Shields, J. (1973). Genetic theorizing and schizophrenia. *Br J Psychiatry*, 122(566):15–30.

51. Gottesman, I.I. and Gould, T.D. (2003). The endophenotype concept in psychiatry: Etymology and strategic intentions. *Am J Psychiatry*, 160(4):636–645.

52. Holzman, P.S. (2001). Seymour, S. Kety and the genetics of schizophrenia. *Neuropsychopharmacology*, 25(3):299–304.

53. Biomarkers Definitions Working Group. (2001). Biomarkers and surrogate endpoints: Preferred definitions and conceptual framework. *Clin Pharmacol Ther*, 69(3):89–95.

54. Katz, R. (2004). Biomarkers and surrogate markers: An FDA perspective. *NeuroRx*, 1(2):189–195.

55. Lesko, L.J. and Atkinson, A.J.J. (2001). Use of biomarkers and surrogate endpoints in drug development and regulatory decision making: Criteria, validation, strategies. *Ann Rev Pharmacol Toxicol*, 41:347–366.

56. De Gruttola, V.G., Clax, P., DeMets, D.L., Downing, G.J., Ellenberg, S.S., Friedman, L. *et al.* (2001). Considerations in the evaluation of surrogate endpoints in clinical trials. Summary of a National Institutes of Health workshop. *Control Clin Trials*, 22(5):485–502.

57. EMEA. (2007). *Innovative Drug Development Approaches: Final Report from the EMEA/CHMP-Think-Tank Group on Innovative Drug Development*. The European Medicines Agency, London.

58. Zerhouni, E.A., Sanders, C.A., and von Eschenbach, A.C. (2007). The biomarkers consortium: Public and private sectors working in partnership to improve the public health. *Oncologist*, 12(3):250–252.

59. Willner, P. (1991). Methods for assessing the validity of animal models of human psychopathology. In Boulton, A., Baker, G., and Martin-Iverson, M. (eds.), *Neuromethods Vol 18: Animal Models in Psychiatry I*. Humana Press, Inc., pp. 1–23.

60. Kraemer, H.C., Schultz, S.K., and Arndt, S. (2002/12/1). Biomarkers in psychiatry: Methodological issues. *Am J Geriatr Psychiatry*, 10(6):653–659.

61. Tannock, R., Campbell, B., Seymour, P., Ouellet, D., Soares, H., Wang, P. *et al.* (2008). Towards a biological understanding of ADHD and the discovery of novel therapeutic approaches. In McArthur, R.A. and Borsini, F. (eds.), *Animal and Translational Models for CNS Drug Discovery: Psychiatric Disorders.* Academic Press: Elsevier, New York.

62. Gordon, E., Liddell, B.J., Brown, K.J., Bryant, R., Clark, C.R., Das, P. *et al.* (2007). Integrating objective gene-brain-behavior markers of psychiatric disorders. *J Integr Neurosci,* 6(1):1-34.

63. Turck, C.W., Maccarrone, G., Sayan-Ayata, E., Jacob, A.M., Ditzen, C., Kronsbein, H. *et al.* (2005). The quest for brain disorder biomarkers. *J Med Invest,* 52(Suppl):231-235.

64. Gomez-Mancilla, B., Marrer, E., Kehren, J., Kinnunen, A., Imbert, G., Hillebrand, R. *et al.* (2005). Central nervous system drug development: An integrative biomarker approach toward individualized medicine. *NeuroRx,* 2(4):683-695.

65. Javitt, D.C., Spencer, K.M., Thaker, G.K., Winterer, G., and Hajos, M. (2008). Neurophysiological biomarkers for drug development in schizophrenia. *Nat Rev Drug Discov,* 7(1):68-83.

66. Cho, R.Y., Ford, J.M., Krystal, J.H., Laruelle, M., Cuthbert, B., and Carter, C.S. (2005). Functional neuroimaging and electrophysiology biomarkers for clinical trials for cognition in schizophrenia. *Schizophr Bull,* 31(4):865-869.

67. Choi, D.W. (2002). Exploratory clinical testing of neuroscience drugs. *Nat Neurosci,* 5(Suppl):1023-1025.

68. Pien, H.H., Fischman, A.J., Thrall, J.H., and Sorensen, A.G. (2005). Using imaging biomarkers to accelerate drug development and clinical trials. *Drug Discov Today,* 10(4):259-266.

69. Thal, L.J., Kantarci, K., Reiman, E.M., Klunk, W.E., Weiner, M.W., Zetterberg, H. *et al.* (2006). The role of biomarkers in clinical trials for Alzheimer disease. *Alzheimer Dis Assoc Disord,* 20(1):6-15.

70. Phillips, M.L. and Vieta, E. (2007). Identifying functional neuroimaging biomarkers of bipolar disorder: Toward DSM-V. *Schizophr Bull,* 33(4):893-904.

71. Schaller, B. (ed.) (2004). *Cerebral Ischemic Tolerance: From Animal Models to Clinical Relevance.* Nova Science Publishers, Hauppauge, NY.

72. Carroll, P.M. and Fitzgerald, K. (eds.) (2003). *Model Organisms in Drug Discovery.* John Wiley and Sons, Chichester, UK.

73. Offermanns, S. and Hein, L. (eds.) (2003). *Transgenic Models in Pharmacology (Handbook of Experimental Pharmacology).* Springer, Heidelberg, Germany.

74. Levin, E.D. and Buccafusco, J.J. (eds.) (2006). *Animal Models of Cognitive Impairment.* Taylor & Francis CRC Press.

75. Boulton, A.A., Baker, G.B., and Martin-Iverson, M.T. (eds.) (1991). *Neuromethods: Animal Models in Psychiatry I.* The Humana Press, Clifton, New Jersey.

76. Olivier, B., Mos, J., and Slangen, J.L. (eds.) (1991). *Animal Models in Psychophramacology.* Birkhaüser-Verlag, Basel.

77. Myers, R.D. (ed.) (1971). *Methods in Psychobiology: Laboratory Techniques in Neuropsychology and Neurobiology.* Academic Press, New York.

78. Svartengren, J., Modiri, A.-R., and McArthur, R.A. (2005). Measurement and characterization of energy intake in the mouse. *Curr Protocols Pharmacol,* 5(Supplement 28). 5.40.1-5..19

79. Chin-Dusting, J., Mizrahi, J., Jennings, G., Fitzgerald, D. (2005). Finding improved medicines: The role of academic-industrial collaboration, *Nat Rev Drug Discov,* 4(11):891-7.

80. Horrobin, D.F. (2003). Modern biomedical research: An internally self-consistent universe with little contact with medical reality? *Nat Rev Drug Discov,* 2(2):151-154.

81. FitzGerald, G.A. (2005). Anticipating change in drug development: The emerging era of translational medicine and therapeutics, *Nat Rev Drug Discov* 4(10):815-818.

82. Pangalos, M.N., Schechter, L.E., and Hurko, O. (2007). Drug development for CNS disorders: Strategies for balancing risk and reducing attrition. *Nat Rev Drug Discov*, 6(7):521-532.

83. Agid, Y., Buzsaki, G., Diamond, D.M., Frackowiak, R., Giedd, J., Girault, J.-A. *et al.* (2007). How can drug discovery for psychiatric disorders be improved? *Nat Rev Drug Discov*, 6(3):189-201.

84. Willner, P. (ed.) (1991). *Behavioural Models in Psychopharmacology: Theoretical, Industrial and Clinical Perspectives*. Cambridge University Press, Cambridge.

85. Willner, P. (1991). Behavioural models in psychopharmacology. In Willner, P. (ed.), *Behavioural models in psychopharmacology: Theoretical, industrial and clinical perspectives*. Cambridge University Press, Cambridge, pp. 3-18.

86. Miczek, K.A. (2008). Challenges for translational psychopharmacology research – the need for conceptual principles. In McArthur, R.A. and Borsini, F. (eds.), *Animal and Translational Models for CNS Drug Discovery: Psychiatric Disorders*. Academic Press: Elsevier, New York.

87. Stein, L., Belluzzi, J.D., Ritter, S., and wise, C.D. (1974). Self-stimulation reward pathways: Norepinephrine vs dopamine. *J Psychiatry Res*, 11:115-124.

88. Wise, R.A. (2004). Dopamine, learning and motivation. *Nat Rev Neurosci*, 5(6):483-494.

89. Everitt, B.J., Dickinson, A., and Robbins, T.W. (2001). The neuropsychological basis of addictive behaviour. *Brain Res Brain Res Rev*, 36(2-3):129-138.

90. Koob, G.F. (2008). The role of animal models in reward deficit disorders: Views from Academia. In McArthur, R.A. and Borsini, F. (eds.), *Animal and Translational Models for CNS Drug Discovery: Reward Deficit Disorders*. Academic Press: Elsevier, New York.

91. Koob, G.F. and Le Moal, M. (1997). Drug abuse: Hedonic homeostatic dysregulation. *Science*, 278(5335):52-58.

92. Heidbreder, C. (2008). Impulse and reward deficit disorders: Drug discovery and development. In McArthur, R.A. and Borsini, F. (eds.), *Animal and Translational Models for CNS Drug Discovery: Reward Deficit Disorders*. Academic Press: Elsevier, New York.

93. Little, H.J., McKinzie, D.L., Setnik, B., Shram, M.J., and Sellers, E.M. (2008). Pharmacotherapy of alcohol dependence: Improving translation from the bench to the clinic. In McArthur, R.A. and Borsini, F. (eds.), *Animal and Translational Models for CNS Drug Discovery: Reward Deficit Disorders*. Academic Press: Elsevier, New York.

94. Markou, A., Chiamulera, C., and West, R.J. (2008). Contribution of animal models and preclinical human studies to medication development for nicotine dependence. In McArthur, R.A. and Borsini, F. (eds.), *Animal and Translational Models for CNS Drug Discovery: Reward Deficit Disorders*. Academic Press: Elsevier, New York.

95. Horvath, T.L., Diano, S., and Tschop, M. (2004). Brain circuits regulating energy homeostasis. *Neuroscientist*, 10(3):235-246.

96. Wurst, F.M., Rasmussen, D.D., Hillemacher, T., Kraus, T., Ramskogler, K., Lesch, O. *et al.* (2007). Alcoholism, Craving, and Hormones: The Role of Leptin, Ghrelin, Prolactin, and the Pro-Opiomelanocortin System in Modulating Ethanol Intake. *Alcohol Clin Exp Res*, 31(12):1963-1967.

97. Lea, P.M.t. and Faden, A.I. (2006). Metabotropic glutamate receptor subtype 5 antagonists MPEP and MTEP. *CNS Drug Rev*, 12(2):149-166.

98. De Witte, P. (2004). Imbalance between neuroexcitatory and neuroinhibitory amino acids causes craving for ethanol. *Addict Behav*, 29(7):1325-1339.

99. Pulvirenti, L. and Diana, M. (2001). Drug dependence as a disorder of neural plasticity: Focus on dopamine and glutamate. *Rev Neurosci*, 12(2):141-158.

100. Korte, S.M., Koolhaas, J.M., Wingfield, J.C., and McEwen, B.S. (2005). The Darwinian concept of stress: Benefits of allostasis and costs of allostatic load and the trade-offs in health and disease. *Neurosci Biobehav Rev*, 29(1):3-38.

101. McEwen, B.S. and Seeman, T. (1999). Protective and Damaging Effects of Mediators of Stress: Elaborating and Testing the Concepts of Allostasis and Allostatic Load. *Ann NY Acad Sci*, 896(1):30–47.

102. Sterling, P. and Eyer, J. (1988). Allostasis: A new paradigm to explain arousal pathology. In Fisher, S. and Reason, J. (eds.), *Handbook of Life Stress, Cognition and Health*. John Wiley & Sons, New York, pp. 629–649.

103. Koob, G.F. and Le Moal, M. (2001). Drug addiction, dysregulation of reward, and allostasis. *Neuropsychopharmacology*, 24(2):97–129.

104. Doran, S.M., Wessel, T., Kilduff, T.S., Turek, F.W., and Renger, J.J. (2008). Translational models of sleep and sleep disorders. In McArthur, R.A. and Borsini, F. (eds.), *Animal and Translational Models for CNS Drug Discovery: Psychiatric Disorders*. Academic Press: Elsevier, New York.

105. Bartz, J., Young, L.J., Hollander, E., Buxbaum, J.D., and Ring, R.H. (2008). Preclinical animal models of Autistic Spectrum Disorders (ASD). In McArthur, R.A. and Borsini, F. (eds.), *Animal and Translational Models for CNS Drug Discovery: Psychiatric Disorders*. Academic Press: Elsevier, New York.

106. Gardner, T.J., Kosten, T.A., and Kosten, T.R. (2008). Issues in designing and conducting clinical trials for reward disorders. In McArthur, R.A. and Borsini, F. (eds.), *Animal and Translational Models for CNS Drug Discovery: Reward Deficit Disorders*. Academic Press: Elsevier, New York.

107. Williams, W.A., Grant, J.E., Winstanley, C.A., and Potenza, M.N. (2008). Currect concepts in the classification, treatment and modelling of pathological gambling and other impulse control disorders. In McArthur, R.A. and Borsini, F. (eds.), *Animal and Translational Models for CNS Drug Discovery: Reward Deficit Disorders*. Academic Press: Elsevier, New York.

108. Selye, H. (1936). A syndrome produced by diverse nocuous agents. *Nature*, 138:32.

109. Joel, D., Stein, D.J., and Schreiber, R. (2008). Animal models of obsessive-compulsive disorder: From bench to bedside via endophenotypes and biomarkers. In McArthur, R.A. and Borsini, F. (eds.), *Animal and Translational Models for CNS Drug Discovery: Psychiatric Disorders*. Academic Press: Elsevier, New York.

110. Klitgaard, H., Matagne, A., Schachter, S.C., and White, H.S. (2008). Animal and translational models of the epilepsies. In McArthur, R.A. and Borsini, F. (eds.), *Animal and Translational Models for CNS Drug Discovery: Neurologic Disorders*. Academic Press: Elsevier, New York.

111. Cryan, J.F., Sánchez, C., Dinan, T.G., and Borsini, F. (2008). Developing more efficacious antidepressant medications: Improving and aligning preclinical and clinical assessment tools. In McArthur, R.A. and Borsini, F. (eds.), *Animal and Translational Models for CNS Drug Discovery: Psychiatric Disorders*. Academic Press: Elsevier, New York.

112. Large, C.H., Einat, H., and Mahableshshwarkar, A.R. (2008). Developing new drugs for bipolar disorder (BPD): From animal models to the clinic. In McArthur, R.A. and Borsini, F. (eds.), *Animal and Translational Models for CNS Drug Discovery: Psychiatric Disorders*. Academic Press: Elsevier, New York.

113. Montes, J., Bendotti, C., Tortarolo, M., Cheroni, C., Hallak, H., Speiser, Z. *et al.* (2008). Translational research in ALS. In McArthur, R.A. and Borsini, F. (eds.), *Animal and Translational Models for CNS Drug Discovery: Neurologic Disorders*. Academic Press: Elsevier, New York.

114. Post, R.M. (1992). Transduction of psychosocial stress into the neurobiology of recurrent affective disorder. *Am J Psychiatry*, 149(8):999–1010.

115. Steckler, T., Stein, M.B., and Holmes, A. (2008). Developing novel anxiolytics: Improving preclinical detection and clinical assessment. In McArthur, R.A. and Borsini, F. (eds.), *Animal and Translational Models for CNS Drug Discovery: Psychiatric Disorders*. Academic Press: Elsevier, New York.

116. Dourish, C.T., Wilding, J.P.H., and Halford, J.C.G. (2008). Anti-obesity drugs: From animal models to clinical efficacy. In McArthur, R.A. and Borsini, F. (eds.), *Animal and Translational Models for CNS Drug Discovery: Reward Deficit Disorders.* Academic Press: Elsevier, New York.

117. Schneider, L.S. (2008). Issues in design and conduct of clinical trials for cognitive-enhancing drugs. In McArthur, R.A. and Borsini, F. (eds.), *Animal and Translational Models for CNS Drug Discovery: Neurologic Disorders.* Academic Press: Elsevier, New York.

118. Berry, D.A. (2006). Bayesian clinical trials. *Nat Rev Drug Discov*, 5(1):27–36.

119. Wagner, L.A., Menalled, L., Goumeniouk, A.D., Brunner, D.P., and Leavitt, B.R. (2008). Huntington Disease. In McArthur, R.A. and Borsini, F. (eds.), *Animal and Translational Models for CNS Drug Discovery: Neurologic Disorders.* Academic Press: Elsevier, New York.

120. Volpicelli, J.R., Alterman, A.I., Hayashida, M., and O'Brien, C.P. (1992). Naltrexone in the treatment of alcohol dependence. *Arch Gen Psychiatry*, 49(11):876–880.

121. Lhuintre, J.P., Moore, N., Tran, G., Steru, L., Langrenon, S., Daoust, M. *et al.* (1990). Acamprosate appears to decrease alcohol intake in weaned alcoholics. *Alcohol Alcohol*, 25(6):613–622.

122. Nides, M., Glover, E.D., Reus, V.I., Christen, A.G., Make, B.J., Billing, C.B. *et al.* (2008). Varenicline Versus Bupropion SR or Placebo for Smoking Cessation: A Pooled Analysis. *Am J Health Behav*, 32(6):664–675.

123. McArthur, R.A. and Borsini, F. (2008). What Do You Mean By "Translational Research"? An Enquiry through Animal and Translational Models for CNS Drug Discovery: Psychiatric Disorders. In McArthur, R.A. and Borsini, F. (eds.), *Animal and Translational Models for CNS Drug Discovery: Psychiatric Disorders.* Academic Press: Elsevier, New York.

124. McArthur, R.A. and Borsini, F. (2008). What Do You Mean By "Translational Research"? An Enquiry through Animal and Translational Models for CNS Drug Discovery: Neurological Disorders. In McArthur, R.A. and Borsini, F. (eds.), *Animal and Translational Models for CNS Drug Discovery: Neurological Disorders.* Academic Press: Elsevier, New York.

125. Markou, A. and Koob, G.F. (1992). Construct validity of a self-stimulation threshold paradigm: Effects of reward and performance manipulations. *Physiol Behav*, 51(1): 111–119.

126. Tzschentke, T.M. (2007). Measuring reward with the conditioned place preference (CPP) paradigm: Update of the last decade. *Addict Biol*, 12(3–4):227–462.

127. Bickel, W.K., Marsch, L.A., and Carroll, M.E. (2000). Deconstructing relative reinforcing efficacy and situating the measures of pharmacological reinforcement with behavioral economics: A theoretical proposal. *Psychopharmacology (Berl)*, 153(1):44–56.

128. Winstanley, C.A., Eagle, D.M., and Robbins, T.W. (2006). Behavioral models of impulsivity in relation to ADHD: Translation between clinical and preclinical studies. *Clin Psychol Rev*, 26(4):379–395.

129. Ainslie, G. (1975). Specious reward: A behavioral theory of impulsiveness and impulse control. *Psychol Bull*, 82(4):463–496.

130. Liu, X., Chen, C., and Smith, B.J. (2008). Progress in Brain Penetration Evaluation in Drug Discovery and Development. *J Pharmacol Exp Ther*, 325(2):349–356.

131. Alavijeh, M.S., Chishty, M., Qaiser, M.Z., and Palmer, A.M. (2005). Drug metabolism and pharmacokinetics, the blood–brain barrier, and central nervous system drug discovery. *NeuroRx*, 2(4):554–571.

132. Merchant, K.M., Chesselet, M.-F., Hu, S.-C., and Fahn, S. (2008). Animal models of Parkinson's Disease to aid drug discovery and development. In McArthur, R.A. and Borsini, F. (eds.), *Animal and Translational Models for CNS Drug Discovery: Neurologic Disorders.* Academic Press: Elsevier, New York.

133. Vinod, K.Y. and Hungund, B.L. (2006). Cannabinoid-1 receptor: A novel target for the treatment of neuropsychiatric disorders. *Expert Opin Ther Targets*, 10(2):203–210.

134. Guindon, J. and Hohmann, A.G. (2008). Cannabinoid CB2 receptors: A therapeutic target for the treatment of inflammatory and neuropathic pain. *Br J Pharmacol*, 153(2):319–334.

135. Fernandez-Ruiz, J., Romero, J., Velasco, G., Tolon, R.M., Ramos, J.A., and Guzman, M. (2007). Cannabinoid CB2 receptor: A new target for controlling neural cell survival? *Trends Pharmacol Sci*, 28(1):39–45.

136. Lancaster, T., Stead, L., and Cahill, K. (2008). An update on therapeutics for tobacco dependence. *Expert Opin Pharmacother*, 9(1):15–22.

137. Colombo, G., Orru, A., Lai, P., Cabras, C., Maccioni, P., Rubio, M. *et al.* (2007). The cannabinoid CB1 receptor antagonist, rimonabant, as a promising pharmacotherapy for alcohol dependence: Preclinical evidence. *Mol Neurobiol*, 36(1):102–112.

138. Rathlev, N.K., Ulrich, A.S., Delanty, N., and D'Onofrio, G. (2006). Alcohol-related seizures. *J Emerg Med*, 31(2):157–163.

139. Ahmed, S., Chadwick, D., and Walker, R.J. (2000). The management of alcohol-related seizures: An overview. *Hosp Med*, 61(11):793–796.

140. Dawson, G.R. and Goodwin, G. (2005). Experimental medicine in psychiatry. *J Psychopharmacol*, 19(6):565–566. 2005/11/1

141. Duyk, G. (2003). Attrition and translation. *Science*, 302(5645):603–605.

142. Bechara, A., Damasio, A.R., Damasio, H., and Anderson, S.W. (1994). Insensitivity to future consequences following damage to human prefrontal cortex. *Cognition*, 50(1–3):7–15.

143. van den Bos, R., Lasthuis, W., den Heijer, E., van der Harst, J., and Spruijt, B. (2006). Toward a rodent model of the Iowa gambling task. *Behav Res Methods*, 38(3):470–478.

144. Halford, J.C., Wanninayake, S.C., and Blundell, J.E. (1998). Behavioral satiety sequence (BSS) for the diagnosis of drug action on food intake. *Pharmacol Biochem Behav*, 61(2):159–168.

145. Halford, J.C.G. (1997). J.E. Blundell. Direct, continuous behavioral analysis of drug action. Unit 8-6C. In Wellman, P. (ed.), *Current Protocols in Neuroscience*. John Wiley and Sons, Inc., New York.

146. Kissileff, H.R. and Guss, J.L. (2001). Microstructure of eating behavior in humans. *Appetite*, 36(1):70–78.

147. Kissileff, H.R., Walsh, B.T., Kral, J.G., and Cassidy, S.M. (1986). Laboratory studies of eating behavior in women with bulimia. *Physiol Behav*, 38(4):563–570.

148. Blundell, J.E. and McArthur, R.A. (1981). Behavioural flux and feeding: Continuous monitoring of food intake and food selection, and the video-recording of appetitive and satiety sequences for the analysis of drug action. In Garattini, S. and Samanin, R. (eds.), *Anorectic Agents: Mechanisms of Action and Tolerance*. Raven Press, New York, p. 19.

149. Blundell, J.E. and Leshem, M.B. (1973). Dissociation of the anorexic effects of fenfluramine and amphetamine following intrahypothalamic injection. *Br J Pharmacol*, 47(1):183–185.

150. Rogers, P.J. and Blundell, J.E. (1979). Effect of anorexic drugs on food intake and the microstructure of eating in human subjects. *Psychopharmacology (Berl)*, 66(2):159–165.

151. Kissileff, H.R., Klingsberg, G., and Van Itallie, T.B. (1980). Universal eating monitor for continuous recording of solid or liquid consumption in man. *Am J Physiol*, 238(1):R14–R22.

152. Lindner, M.D., McArthur, R.A., Deadwyler, S.A., Hampson, R.E., and Tariot, P.N. (2008). Development, optimization and use of preclinical behavioral models to maximise the productivity of drug discovery for Alzheimer's Disease. In McArthur, R.A. and Borsini, F. (eds.), *Animal and Translational Models for CNS Drug Discovery: Neurologic Disorders*. Academic Press: Elsevier, New York.

153. Brooks, D.J., Frey, K.A., Marek, K.L., Oakes, D., Paty, D., Prentice, R. *et al.* (2003). Assessment of neuroimaging techniques as biomarkers of the progression of Parkinson's disease. *Exp Neurol*, 184(Suppl 1):S68–S79.

154. Airan, R.D., Meltzer, L.A., Roy, M., Gong, Y., Chen, H., and Deisseroth, K. (2007). High-speed imaging reveals neurophysiological links to behavior in an animal model of depression. *Science*, 317(5839):819–823.

155. Rauch, S.L. and Savage, C.R. (1997). Neuroimaging and neuropsychology of the striatum. Bridging basic science and clinical practice. *Psychiatr Clin North Am*, 20(4):741-768.

156. Andreasen, N.C. (1997). Linking mind and brain in the study of mental illnesses: A project for a scientific psychopathology. *Science*, 275(5306):1586-1593.

157. Karitzky, J. and Ludolph, A.C. (2001). Imaging and neurochemical markers for diagnosis and disease progression in ALS. *J Neurol Sci*, 191(1-2):35-41.

158. Wise, R.G. and Tracey, I. (2006). The role of fMRI in drug discovery. *J Magn Reson Imaging*, 23(6):862-876.

159. Bacskai, B.J., Kajdasz, S.T., Christie, R.H., Carter, C., Games, D., Seubert, P. *et al.* (2001). Imaging of amyloid-beta deposits in brains of living mice permits direct observation of clearance of plaques with immunotherapy. *Nat Med*, 7(3):369-372.

160. Anton, R.F., Lieber, C., and Tabakoff, B. (2002). Carbohydrate-deficient transferrin and gamma-glutamyltransferase for the detection and monitoring of alcohol use: Results from a multisite study. *Alcohol Clin Exp Res*, 26(8):1215-1222.

161. Conigrave, K.M., Degenhardt, L.J., Whitfield, J.B., Saunders, J.B., Helander, A., and Tabakoff, B. (2002). CDT, GGT, and AST as markers of alcohol use: The WHO/ISBRA collaborative project. *Alcohol Clin Exp Res*, 26(3):332-339.

162. Marincola, F.M. (2003). Translational Medicine: A two-way road. *J Transl Med*, 1(1):1.

163. Fray, P.J., Robbins, T.W., and Sahakian, B.J. (1996). Neuropsychiatric applications of CANTAB. *Int J Geriatr Psychiatry*, 11(4):329-336.

Acknowledgements

We would like to thank our many colleagues who have pooled their knowledge and who have contributed to the creation of this book. We have enjoyed this experience of working with them on a "Global Project Team." Hopefully this sharing of our experiences will "translate" into a more efficient and fruitful use of animals to model the devastating disorders we are trying to understand, and to help us discover and develop the drugs needed to alleviate them.

We would especially like to thank Stephanie Diment, Keri Witman, Kirsten Funk and Renske van Dijk of Elsevier for their cheerful encouragement throughout this project, and without their help this book would have never been completed. We thank as well the members of our "advisory board": Professors Trevor Robbins, Tamas Bartfai, Bill Deakin; and Doctors Danny Hoyer, David Sanger, Julian Gray and Markus Heilig for their productive interventions and thoughtful discussions at various times over the past 18 months.

Finally we would like to thank all those who provided the space and opportunity so that this book could become a reality. You know who you are....

List of Contributors

Jane B. Acri, PhD National Institute on Drug Abuse, National Institutes of Health, Department of Health and Human Services, Bethesda, MD, USA

Franco Borsini, PhD sigma-tau Industrie Farmaceutiche Riunite S.p.A., Via Pontina km 30,400, 00040 Pomezia (Rome), Italy

Jack Bergman, PhD Harvard Medical School, ADARC-McLean Hospital, 115 Mill Street, Belmont, MA 02178-9106, USA

Christian Chiamulera, PhD Department of Medicine and Public Health, University of Verona, Piazzale L. Scuro 10, 37100 Verona, Italy

Sandra Comer, PhD College of Physicians and Surgeons of Columbia University, Department of Psychiatry, 1051 Riverside Drive, Unit 120, New York, NY 10032, USA

Colin T. Dourish, PhD, DSc P1vital, Department of Psychiatry, University of Oxford, Warneford Hospital, Headington, Oxford, OX3 7JX, UK

Tracie J. Gardner, PhD Division of Alcohol and Addictive Disorders, Menninger Department of Psychiatry and Behavioral Sciences, Baylor College of Medicine, Michael E. DeBakey VA Medical Center (151), 2002 Holcombe Blvd, Houston, TX 77030, USA

Jon E. Grant, JD, MD University of Minnesota Medical Center, 2450 Riverside Avenue, Minneapolis, MN 55454, USA

Jason Halford, PhD University of Liverpool, Kissileff Laboratory for the Study of Human Ingestive Behaviour, School of Psychology, Eleanor Rathbone Building, Bedford Street South, Liverpool, L69 7ZA, UK

Margaret Haney, PhD College of Physicians and Surgeons of Columbia University, Department of Psychiatry, 1051 Riverside Drive, Unit 120, New York, NY 10032, USA

Christian A. Heidbreder, PhD GlaxoSmithKline S.p.A., Centre of Excellence for Drug Discovery in Psychiatry, Via A. Fleming 4, 37135 Verona (VE), Italy

George F. Koob, PhD Committee on the Neurobiology of Addictive Disorders, The Scripps Research Institute, 10550 North Torrey Pines Road, SP30–2400, La Jolla, CA 92037, USA

Therese Kosten, PhD Menninger Department of Psychiatry, Baylor College of Medicine, Michael E. DeBakey VA Medical Center, 2002 Holcombe Blvd, Houston, TX 77030, USA

Thomas Kosten, MD VA National Substance Use Disorders, Quality Enhancement Research Initiative (QUERI), Michael E. DeBakey VA Medical Center, Research 151 – BLDG 110, Room 229, 2002 Holcombe Boulevard, Houston, TX 77030, USA

Hilary J. Little, PhD Departments of Basic Medical Sciences and Mental Health, St George's, University of London, Cranmer Terrace, London, UK

Athina Markou, PhD Department of Psychiatry, M/C 603, Room 2052, University of California, San Diego, School of Medicine, 9500 Gilman Drive, La Jolla, CA 92093-0603, USA

Robert A. McArthur, PhD McArthur and Associates, GmbH, Ramsteinerstrasse 28, CH-4052, Basel, Switzerland

David J. McCann, PhD National Institute on Drug Abuse, National Institutes of Health, Department of Health and Human Services, Bethesda, MD, USA

David L. McKinzie, PhD Lilly Research Laboratories, Eli Lilly and Company, Indianapolis, Indiana 46285, USA

Marc N. Potenza, MD, PhD Yale University School of Medicine, Connecticut Mental Health Center, SAC, Room S-104, 34 Park Street, New Haven, CT 06519, USA

Beatriz A. Rocha, MD, PhD Merck Research Laboratories, Worldwide Regulatory Affairs, RY33-200; PO Box 2000, Rahway, NJ 07065, USA

Edward M. Sellers, MD, PhD, FRCP(C) Decision Line Clinical Research Corporation, 720 King St. W., 7th Floor, Toronoto, ON M5V 2T3, Canada

Beatrice Setnik, PhD DecisionLine Clinical Research Corporation, 720 King St. W., 7th Floor, Toronto, ON M5V 2T3, Canada

Megan J. Shram, PhD DecisionLine Clinical Research Corporation, 720 King St. W., 7th Floor, Toronto, ON M5V 2T3, Canada

Roger D. Spealman, PhD Harvard Medical School, New England Primate Research Center, One Pine Hill Drive, P.O. Box 9102, Southborough, MA 01772-9102, USA

Frank Vocci, PhD National Institute on Drug Abuse, National Institutes of Health, Department of Health and Human Services, Bethesda, MD, USA

Robert J. West, PhD Cancer Research UK Health Behaviour Unit, Department of Epidemiology and Public Health, University College London, 2-16 Torrington Place, London, WC1E 6BT, UK

John P.H. Wilding, MD School of Clinical Sciences, Clinical Sciences Centre, University Hospital Aintree, Longmoor Lane, Liverpool, L9 7AL, UK

Wendol A. Williams, Sr. MD Departments of Psychiatry and Diagnostic Radiology, Yale PET Center, Yale University School of Medicine, Yale New Haven Psychiatric Hospital, New Haven, CT, USA

Catharine A. Winstanley, PhD Department of Psychiatry, University of British Columbia, 1239 West Mall, Vancouver BC V6T 1Z4, Canada

Paul Willner, PhD, DSc Department of Psychology, University of Swansea, Singleton Park, Swansea, Wales SA2 8D, UK

CHAPTER 1

Impulse and Reward Deficit Disorders: Drug Discovery and Development

Christian Heidbreder

GlaxoSmithKline S.p.A. Centre of Excellence for Drug Discovery
in Psychiatry, Verona (VR), Italy

Animal and Translational Models for CNS Drug Discovery,
Vol. 3 of 3: Reward Deficit Disorders
Robert McArthur and Franco Borsini (eds), Academic Press, 2008

INTRODUCTION

Reinforcements shape the interaction of higher animals, including humans, with their environment. All reinforcers maintain or increase the probability that a response which precedes the reinforcer will subsequently occur. The contingency by which an appetitive, positive reinforcer increases the likelihood of subsequent responding is commonly referred to as a "reward." Where there is a reward, there is also a risk of vulnerability to seeking hedonic stimuli in a compulsive manner. In Western societies, alcoholism and other forms of drug addiction come to mind as the most obvious examples of compulsive addictive disorders, but related diseases may also include "eating disorders," "sexual and gender identity disorders," "anxiety disorders," and "impulse-control disorders not elsewhere classified" such as compulsive gambling, kleptomania, fire-setting, hair-pulling, and "intermittent explosive disorder (IED)." Thus, there is convergence of thinking on the role of dysfunctional reinforcement mechanisms in the development and maintenance of addictive and compulsive disorders. On the one hand, some individuals are highly sensitive to the reinforcing value of either drugs or naturally rewarding stimuli, such as eating, socializing, and engaging in novel activities (reward sensitivity/drive). On the other hand, disinhibited behavior (rashness, impulsiveness) is associated with the inability to discontinue compulsive behavior despite negative consequences. One could therefore argue for a dual component phenomenon, that is, a reward sensitivity/drive component that plays an important role in the initiation of cravings and desire to binge, and a component that incorporates rashness impulsiveness contributing to disinhibited behavior and loss of control during a binge episode, and/or the inability to resist binge cravings. For example, although current diagnostic systems (e.g., DSM-IV) classify eating disorders *primarily* in terms of severity of weight loss (i.e., anorexia versus bulimia), there is growing debate as to whether the conceptualization of eating disorders would be better served by distinguishing between these disorders in terms of *control* (or loss of control) over food intake.[i]

Recent advances in the neurobiology of addiction have provided insight into whether brain activity and biochemistry are affected in the same way in "behavioral" addictions as they are in substance abuse. Thus, addiction research is currently going beyond the earlier conceptual framework of drug abuse *per se*, and is defining a better theory of how the brain processes rewarding events, which includes discovering the algorithms people follow that lead them into compulsive behaviors.

THE KEY MESSAGES OF EPIDEMIOLOGY: WHAT SHOULD BE MODELED?

Drug and behavioral addictions rank among the most common psychiatric disorders in the seven major pharmaceutical markets (United States, France, Germany, Italy, Spain, United Kingdom, and Japan) and are a source of considerable disease burden.[1,2]

---

[i] Please refer to the *Diagnostic and Statistical Manual of Mental Disorders*, 4th edition – Text Revision (DSM-IV-TR), or The International Statistical Classification of Diseases and Related Health Problems 10th Revision, published by the American Psychiatric Association and the World Health Organization, respectively, for current diagnostic criteria manuals in use.

Drug Addiction

According to the World Health Organization (WHO),[3] global trends reflect a general increase in the use of illegal addictive drugs as well as in alcohol abuse; particularly concerning is the increase in use among the youngest sectors of the population. There are about 200 million users of illegal drugs worldwide, which represent 3.4% of the world population. Among other neuropsychiatric diseases, alcohol and drug abuse *per se* costs the American economy an estimated $544.11 billion per year including costs related to crime, loss in productivity, health care, incarceration, and drug enforcement operation.[4] Recent figures indicate societal costs of about €57.274 billion per year in the European Community.[1,2]

The total prevalent cases of nicotine addiction in 2005 were 26.3 million in the United States, 20.25 million in Europe, and 7.2 million in Japan for a total of 53.7 million cases of nicotine addiction in the seven major pharmaceutical markets in 2005, a figure that is expected to increase to 55.94 million by 2015.[5-8] The *worldwide* number of smokers will continue to increase to 1.6 billion by 2025.[3] Alcohol dependence impacts 49.52 million adults in the same market (18.44 million in the United States, 24.51 million in Europe, and 6.57 million in Japan), and predictions indicate that by 2015 the estimated total number of alcohol addiction will increase up to 52.15 million.[3,5,8,9] The total prevalent cases of opioid addiction in 2005 were 1.88 million in this market, a figure that may increase to 1.99 million by 2015.[3,5,8] Finally, the prevalent cases of psychostimulant addiction were estimated at 2.16 million in 2005, and may increase to 2.29 million by 2015.[3,5,8,9]

Behavioral Addictions

Impulse control disorders (ICDs) may be conceptualized as part of an obsessive-compulsive spectrum that may be related to obsessive–compulsive disorders (OCD).[ii] However, rates of OCD in patients with ICDs are not always consistent.[10]

Pathological Gambling

The prevalence of pathological gambling (recurrent and compulsive drive to gamble) is between 1% and 3% of the adult population.[11] Pathological gambling is also characterized by a high level of psychiatric comorbidities, such as depression, bipolar disorders, anxiety, and substance use disorder.[12,13]

Trichotillomania

The prevalence of trichotillomania,[iii] or the urge to pluck out specific hairs, is not known, but estimates from university surveys suggest that up to 1.5% of males and

---

[ii] For further discussion regarding ICDs and OCD, please refer to Williams *et al.*, Current concepts in the classification, treatment and modeling of pathological gambling and other impulse control disorders, in this volume and Joel *et al.*, Animal models of obsessive–compulsive disorder: From bench to bedside via endophenotypes and biomarkers, in Volume 1, Psychiatric Disorders.

[iii] Please refer to Williams *et al.*, Current concepts in the classification, treatment and modeling of pathological gambling and other impulse control disorders, in this volume and Joel *et al.*, Animal models of obsessive–compulsive disorder: From bench to bedside via endophenotypes and biomarkers, in Volume 1, Psychiatric Disorders for further discussion on trichotillomania.

3.4% of females may suffer from hair-pulling behavior, 0.6% of which are eventually diagnosed with trichotillomania.[14,15] Comorbidity data suggest a significant relationship between OCD and trichotillomania.[16]

Pyromania

Very few epidemiological data are currently available for pyromania, or the impulsive, repetitive, and deliberate fire setting without external reward. Studies focus primarily on pyromania in childhood and adolescence, and have reported a 2.4-3.5% prevalence.[17-19]

Intermittent Explosive Disorder

These are recurrent episodes of aggressive behavior that are out of proportion to stressors and/or provocation, and that cannot be accounted for by another underlying psychiatric disorder. There are few studies on the lifetime prevalence of IED. Current rates of IED in psychiatric settings indicate a prevalence of 1-2%.[20,21] Community sample studies, however, report a much higher lifetime prevalence rate of 11.1%. Rates of OCD in patients with IED could be as high as 22%.[22]

Kleptomania

Again, few epidemiological studies on the impulsive stealing behavior without any objective need to do so are available, but estimates indicate that prevalence in the United States is around 0.6%.[23] The rate of OCD in patients suffering from kleptomania is variable from one study to another ranging from 6.5% to 60%.[24,25]

Compulsive–Impulsive Internet Addiction or Problematic Internet Use

The prevalence of compulsive-impulsive Internet use disorder is unknown. However, comorbidity rates between OCD and compulsive-impulsive Internet use may range from 10% to 20% for lifetime OCD up to 15% for current OCD.[26-28]

Compulsive–Impulsive Sexual Behaviors

Prevalence estimates of repetitive sexual acts and compulsive sexual thoughts (compulsive-impulsive sexual behaviors) range from 5% to 6% of the US population.[29,30] The comorbidity rates with OCD could be 12-14%.[31]

Compulsive–Impulsive Shopping

This behavioral addiction is also referred to as compulsive buying is characterized by compulsive thoughts or impulses to buy or shop without the need to do so and/or for items that cost more than can be afforded. Prevalence rates range from 2% to 8% in the US adult population, 80-95% of which are females.[32] The comorbidity rates with OCD are between 2.2% and 30%.[33,34]

Compulsive–Impulsive Skin Picking

Prevalence of compulsive-impulsive skin picking (excoriations on either single or multiple body sites) is unknown, but comorbid OCD is estimated to range from 6% to 19%.[35,36]

From a psychiatric perspective, eating disorders may also include a component of compulsive eating.

Bulimia Nervosa

Bulimia nervosa is characterized by recurrent binge eating followed by recurrent inappropriate compensatory behavior in order to prevent weight gain. These behaviors include self-induced vomiting (purging), misuse of laxatives, diuretics, enemas, or other medications, fasting, and excessive exercise.

Eating Disorders Not Otherwise Specified

Eating disorders not otherwise specified include binge eating without purging, infrequent binge–purge episodes (occurring less than twice a week or such behavior lasting less than 3 months), repeated chewing and spitting without swallowing large amounts of food, or normal weight in people who exhibit anorexic behavior.

Anorexia Nervosa

Anorexia nervosa is an eating disorder universally associated with emaciation and commonly accompanied by marked increases in physical activity. Individuals with anorexia nervosa are unable to maintain a normal healthy body weight, often dropping well below 85% of their ideal weight. Despite increasing cachexia, individuals with anorexia nervosa continue to be obsessive about weight gain, remain dissatisfied with the perceived largeness of their bodies, and engage in an array of behaviors designed to perpetuate weight loss.

Many of the epidemiological studies on eating disorders in Western Europe and the United States yield consistent prevalence rates. In young females, the average prevalence rate for anorexia nervosa is 0.3% and bulimia nervosa 1%. The prevalence of binge eating disorder is at least 1%. Binge eating disorders occur in a significant number of obese patients and increases in frequency with the body mass index. Binge eating behaviors are more prevalent than binge eating disorders as defined in DSM-IV.[iv] It must be assumed that even the studies with the most complete case-finding methods yield an underestimate of the true incidence. One may conclude that the overall incidence of anorexia nervosa is at least 8 per 100,000 population per year and the incidence of bulimia nervosa is at least 12 per 100,000 population per year.[37]

COMORBID ASSOCIATIONS

In 2003, there were 19.6 million adults aged 18 years or older in the United States suffering from severe mental illness (SMI).[v] This represents 9.2% of all adults in the United States. There is a significant comorbidity between drug use and abuse and SMI: adults who used illicit drugs in the past year were more than twice as likely to have SMI as adults who did not use an illicit drug (18.1% and 7.8%, respectively). SMI is also significantly correlated with substance dependence or abuse. Among adults with

---

[iv] For further discussion of binge eating and other disorders, please refer to Dourish *et al.*, Anti-obesity drugs: From animal models to clinical efficacy, in this volume.

[v] SMI is defined as having at some time during the past year a diagnosable mental, behavioral, or emotional disorder that met the criteria specified in the 4th edition of the *Diagnostic and Statistical Manual of Mental Disorders* (DSM-IV) (American Psychiatric Association (APA), 1994) and that resulted in functional impairment substantially interfering with or limiting one or more major life activities.

SMI in 2003, 21.3% were dependent on or abused alcohol or illicit drugs compared with 7.9% adults without SMI.[38] Significant comorbidity between alcohol and nicotine dependence has also been observed. In fact, smoking cessation seems to aid abstinence from alcohol, and combined treatment for both dependencies may achieve the best outcome. Surveys of both inpatient and outpatient treatment for alcohol dependence typically show an 86–97% smoking rate among males and an 82–92% rate among females.[39-44] A 91.5% prevalence rate of nicotine dependence among outpatients who met diagnostic criteria for both nicotine and alcohol dependencies has also been reported.[45] Furthermore, a large epidemiological survey revealed that, in contrast with alcohol abstainers, current heavy drinkers made the fewest cessation attempts and had the least success in quitting smoking.[46]

Both epidemiological and preclinical studies suggest that developmental factors are likely to play an important role in establishing comorbid associations. The preliminary evidence suggests that early exposure to substances of abuse, during periods in which the brain is still undergoing significant changes, could lead to neurobiological changes associated with depression, enhanced sensitivity to stress, and decreased sensitivity to natural reinforcers.[47] Genetic components linking substance abuse and vulnerability to depression and to anxiety disorders were suggested by human (e.g., adoption, twin) and animal (e.g., using genetically altered strains of mice) studies. Several environmental factors, such as family disruption, poor parental monitoring, acute and chronic stress, and low social class of rearing, have also all been found to contribute to the manifestation of substance abuse and certain comorbid mental illnesses.[48]

Drug abuse and HIV/AIDS are intertwined epidemics.[49,50] Injection drug use (IDU) accounts for approximately one-third of all the AIDS cases in the United States, and over 90% of HIV infected injection drug users are also infected with hepatitis C.[51] However, non-IDU drug abusers also show much higher rates of HIV infection than the general population.[50] There are many factors that contribute to such high rates, but chief among them is the risky sexual behaviors that many drug and alcohol abusers engage into as a result of changes in their mental state due to drug intoxication. Basic research shows that most drugs of abuse can also compromise the immune system,[52] putting users at greater risk of contracting an illness or suffering a more severe course of the illness. It is also reasonable to expect a synergistic, deleterious interaction between centrally acting psychoactive drugs and neurotoxic substances released during the course of an HIV infection.[53]

Researchers have identified associations between eating disorders such as bulimia nervosa, anorexia nervosa, eating disorders not otherwise specified, and binge eating disorders, and self-report measures of specific personality traits, such as perfectionism,[54] obsessive-compulsiveness,[55] impulsivity,[56] sensation seeking and narcissism,[57] sociotropy and autonomy.[58] These studies relied on self-report measures of hypothetical constructs and were largely correlational in nature; they should be interpreted with caution, particularly with respect to the direction of causality.

In both anorexia nervosa and bulimia nervosa, major depression is the most commonly diagnosed comorbid disorder.[59] Some evidence suggests that caloric deprivation may contribute to depressive symptomatology in women with anorexia nervosa and bulimia nervosa. Whether starvation-related cognitive and biological changes exacerbate a preexisting predisposition to depression inherent in some individuals with eating disorders, or whether it is the sole cause of depression, remains unclear.

It has also been suggested that anxiety is central to both the etiology and mainte-nance of anorexia nervosa and bulimia nervosa,[60] and that OCD is quite common in those with eating disorders.[61] Many studies suggest that caloric deprivation has a role in either causing obsessional symptoms or creating a physiological environment that exacerbates a preexisting tendency toward obsessionality.[62]

One of the most intensely studied aspects of eating disorders is the comorbidity between substance abuse and anorexia nervosa/bulimia nervosa. Most data come from studies of women with bulimia nervosa. Despite considerable heterogeneity in these data, it appears that there is a relationship between the bulimic behaviors of binge-ing and purging and substance abuse in individuals with bulimia nervosa or anorexia nervosa (bingeing subtype), but not in individuals with anorexia nervosa (restricting type). There is also some evidence that women presenting for treatment for substance abuse disorders have a higher than expected rate of eating disorders.[63] While the evi-dence suggests that there is a "multi-impulsive" syndrome in some individuals with bulimia nervosa, future research must disentangle and clarify the roles of personal-ity pathology and familial factors as mediator/moderator variables in the relationship between eating disorders and substance abuse.

FROM EPIDEMIOLOGY TO UNMET MEDICAL NEEDS TO DRUG DISCOVERY AND DEVELOPMENT

There are a few key messages that can be drawn from the epidemiology of impulse and reward deficit disorders.

First, most studies concur to predict a robust growth in the drug addiction market over the next 10 years. This growth will be driven primarily by increases in the diag-nosis and drug-treatment rates within the nicotine addiction and alcohol addiction populations, taking the total drug-treated population from approximately 3 million people in 2005 to almost 7 million in 2015. The second driver lies in two recent drug launches, varenicline (Pfizer's Chantix®/Champix®) for smoking cessation and once-monthly injectable naltrexone (Alkermes/Cephalon's Vivitrol®) for alcohol depen-dence. These two drugs could jointly account for sales of nearly $1.2 billion in 2015. The expected release of Nabi Biopharmaceuticals' nicotine vaccine (NicVAX®) on the market for nicotine addiction in the United States and Europe in 2010 and 2011, respectively, could increase overall prescription volume thanks to higher use of com-bination therapy (e.g., use with varenicline or bupropion (Wellbutrin®)/Zyban®) to prevent relapse and craving in nicotine-addicted patients.

Second, the small number of patients seeking drug treatment relative to the size of the prevalent populations for drug and behavioral addictions highlights the largely untapped opportunity within the addiction market. There is a desperate need for phar-macotherapies with regulatory approval that would allow physicians to have more options when patients fail in their efforts to overcome their addictions. Therapies with compelling efficacy are still lacking due primarily to the fact that clinical trials are quite different from conditions faced by a drug-addicted patient on a daily basis.

Third, the development of long-acting formulations could be particularly useful in reducing poor patient compliance and adherence to pharmacotherapy. In addition, greater awareness, and education describing addiction as a chronic disease much like depression, hypertension, or diabetes, is important for the future of drug addiction treatment. Such education could contribute to improve diagnosis and increase prescription drug-treatment rates. Therapies that target mechanisms involved in craving are likely to be useful *across* addiction disorders. There is also unmet need for more comprehensive reimbursement of both behavioral therapies and pharmacotherapies for addictive disorders. Finally, in some cases (e.g., binge eating) identifying patients may be particularly important for appropriate diagnosis and treatment, given the higher degree of comorbid psychopathology.

Fourth, it is becoming increasingly clear that studying the relationship between drug and behavioral addictions and other major psychiatric diseases in terms of phenomenological issues and comorbidity patterns will provide the rationale and framework for the discovery, development, and use of pharmacotherapies with or without behavioral interventions.

Fifth, epidemiological findings such as those already reported help construct an empirical basis for building target product profiles and ultimately testing hypothesized mechanisms in early clinical studies. The main goal is to determine how social factors, exogenous agents, and individual factors are linked across time to produce illness. At present, however, laboratory and clinical research is often conducted in isolation from epidemiological research, and epidemiological evidence is often not incorporated into laboratory and clinical research. Since much of what we know about addictive disorders is based on clinical samples, a significant potential exists for selection bias, which may result in a lack of generalizability of findings. Epidemiological studies have the potential to offer new opportunities to address this concern and to inform the selection of participants for laboratory-based research. Study designs that would efficiently combine the advantages of epidemiological samples with more intensive laboratory-based and biological measures would not only be more predictive, but would also be more cost-effective.

CHALLENGES FACING DRUG DISCOVERY AND DEVELOPMENT IN IMPULSE AND REWARD DEFICIT DISORDERS

Many areas of medicine adopt new therapies slowly, but barriers in the area of impulse and reward deficit disorders may be particularly high. Among the general categories that affect acceptance of new therapies are clinician and patient characteristics, the context, including organizational, social, economic, and political elements, and the characteristics of the new therapy or technology itself.

Low overall diagnosis and drug-treatment rates (despite increases in the United States), the complexity of addictive behaviors in general resulting in current therapies with limited efficacy, and a weak late-stage pipeline will remain the major constraints on the addiction market over the next 10 years. Less than 15% of those addicted to drugs or natural reinforcers such as food receive treatment. This is likely a reflection not only of the patient's denial of having a problem, but also of the fear of stigmatization or being labeled as an addict, as well as the lack of access to an acceptable treatment program. In this respect, the lack of reimbursement by private insurance companies hinders a wider involvement of the medical community in the early evaluation and treatment of impulse and reward disorders. For example, stringent reimbursement policies from third-party payers and national health systems limit prescription of drugs to treat nicotine addiction. In the United States and Europe, in particular, limited reimbursement restricts prescription drug use for this large population of patients.

One of the major obstacles pertains to the limited range of medications currently available for the treatment of impulse and reward deficit disorders. This reflects, in part, the limited participation of the pharmaceutical industry in the development of such medications. Understandably, the stigma associated with "addiction" in general, the lack of insurance reimbursement for addiction treatment, and the *perceived* relatively small size of the users' market create an environment that is not conducive to active collaborations with industry. Resistance to the use of pharmacotherapies to treat addiction in favor of cognitive-behavioral therapy (psychosocial support) alone, particularly among addiction counselors and in the primary care setting, is also a significant barrier to growth in the addiction drug market. The popularity of over-the-counter (OTC) nicotine replacement therapies (NRTs) is expected to exert downward pressure on the prescription nicotine addiction market. The switch of the prescription nicotine patch in Japan to OTC and the practice of physician-directed OTC use in Europe are predicted to temper growth in the prescription nicotine addiction market. Limited availability of approved prescription therapies in Japan also constrains the size of the addiction market. The small number of approved therapies for nicotine and alcohol addiction and a lack of approved therapies for opioid and psychostimulant addiction in Japan will limit growth in the addiction market.

The pharmaceutical industry relies too heavily on known pharmacological mechanisms. There is an increasing need for multi-receptorial approaches, especially in the psychiatry domain. A lag in the development of therapies to treat psychostimulant addiction limits market growth. For example, although the target sites for psychostimulants in the brain are relatively well characterized, drug development for this indication is sorely lacking due to the relatively small population affected and the stigma associated with psychostimulant addiction. High-throughput screening in recombinant receptor systems bears little resemblance to the pathological situation *in vivo*. Furthermore, screening of selected compounds in standard pharmacological models is validated by previous pharmacological mechanisms rather than disease processes. This is to say that preclinical models often cannot be translated into Phase I and Phase II clinical testing, and Phase I/II clinical testing is often not transposable to Phase III trials and the general population. Therefore, there is an increasing need for translational approaches (see below). The placebo effect, which can reach up to 60% in some trials, is a major barrier to registration – this is frequently due to inappropriate patient inclusion criteria. Finally, the comorbidity of impulse and reward disorders with other psychiatric diseases restricts the number of patients available for formal clinical trials.[vi]

[vi] For further discussion regarding present developments in current clinical trials for novel drug treatment of behavioral disorders, please refer to Gardner *et al.*, Issues in designing and conducting clinical trials for reward disorders: A clinical view, in this volume; and McEvoy and Freudenreich, Issues in the design and conductance of clinical trials, in Volume 1, Psychiatric Disorders.

TOWARD TRANSLATIONAL VALUE OF ANIMAL MODELS IN IMPULSE AND REWARD DEFICIT DISORDERS

Key Neurosystems Relevant to Drug Discovery and Development

As drug development moves from being chemistry driven to biology driven, the key lies in the understanding of the target's biological function. Significant advances documenting long-lasting changes in the brains of individuals addicted to drugs are currently leading to the conceptualization of addiction as a chronological disease of the brain with the recruitment of specific neural networks, molecular and cellular mechanisms, genetic and environmental factors. The concept of disease stages with transitions from acute rewarding effects to early- and end-stage addiction has had an important impact on the way we are thinking about preclinical animal models and the discovery, validation, and optimization of pharmacotherapeutic approaches.

Three main circuits, each corresponding to a disease stage, can be delineated in the light of recent discoveries in functional neuroanatomy, behavioral neurochemistry and pharmacology, and functional neuroimaging.[vii] A first circuit relates to the *acute reinforcing properties of drugs of abuse* that comprise the binge, intoxication stage of addiction. One of the characteristics shared by drugs of abuse is the significant increase in activity of mesolimbic dopamine neurons originating in the ventral tegmental area and projecting toward limbic forebrain regions, including the nucleus accumbens. Although numerous studies have correlated dopamine release in the nucleus accumbens with the hedonic or aversive value of stimuli, more recent views support the idea that the mesolimbic dopamine system plays a rheostatic role in the learning of the motivational significance (better-than-expected versus worse-than-expected) of a stimulus rather than in the mediation of the hedonic/aversive value of the stimulus *per se*.[64,65] In order to apply this hypothesis to drug addiction, one must thus assume that any response to the drug that occurs during the period of raised extracellular dopamine may have the potential of acquiring incentive salience and contribute to enhanced attentional processing of drug-related cues. A corollary hypothesis is that this attentional bias toward drug-related cues elicits drug craving and contributes to compulsive drug use and relapse to drug-seeking behavior. More recently, specific components of the basal forebrain that have been identified with drug reward have focused on the "extended amygdala," which is comprised of the bed nucleus of the stria terminalis (BNST), the central nucleus of the amygdala and the shell sub-region of the nucleus accumbens. In contrast, the symptoms of acute withdrawal involve decreases in function of the extended amygdala reward system, as well as the recruitment of brain stress neurocircuitry (CRF, norepinephrine, NPY). The extended amygdala may thus represent a common neuroanatomical substrate for acute drug reward and a common neuroanatomical substrate for the negative effect produced by stress.

A second circuit relates to *drug- and cue-induced reinstatement of drug-seeking behavior (craving stage)*. Drug-induced reinstatement mainly involves a circuit

---

[vii] Please refer to Koob, The role of animal models in reward deficit disorders: Views of academia, in this volume for further discussion on the functional neuroanatomy, behavioral neurochemistry and pharmacology, and functional neuroimaging of reward deficit disorders.

comprised of the medial prefrontal cortex, nucleus accumbens, and ventral pallidum circuit. In contrast, cue-induced reinstatement involves the basolateral amygdala as a critical substrate with a possible feed-forward mechanism through the prefrontal cortex system involved in drug-induced reinstatement. Stress-induced reinstatement appears to depend on the activation of both CRF and norepinephrine systems in elements of the extended amygdala (central nucleus of the amygdala and BNST).

Finally, a third circuit implicated in *drug-seeking and compulsive behaviors* involves ventral striatal, ventral pallidal, thalamic, and orbitofrontal loops.[66,67] Importantly, these preclinical findings can be translated into human brain imaging findings showing that (1) drug-related cues produce increased activity in the mesolimbic system;[68,69] (2) drug craving and attentional processes seem to involve similar neural circuits;[70-72] (3) chronic drug use hampers frontal cortex function;[73] (4) deficits in frontal cortex function may contribute to impaired impulse control, as well as lack of judgment and risk assessment,[74] and (5) the personality trait of novelty seeking, of which impulsiveness is one component, is linked to increased extracellular levels of dopamine in the nucleus accumbens[75,76] and addictive propensity.[77]

Key Experimental Paradigms with Translational Potential for Drug Discovery and Development

It is not the intention of this chapter to describe all animal models in the area of impulse and reward deficit disorders. It will focus rather on the potential translational value of a few paradigms and their potential suitability for combined pharmacokinetic and pharmacodynamic (PK/PD) assessments.[viii]

Self-administration

The self-administration model has probably one of the highest face validities as virtually all drugs of abuse, but also natural reinforcers, such as food or sucrose are self-administered by laboratory animals. The key issue relates to the reinforcement contingencies used and the resulting interpretation of the data. The fixed-ratio (FR) paradigm of self-administration has been classically viewed as measuring the fact of reinforcement, but not the degree of reinforcing efficacy.[67,78-80] Yokel and Wise[81,82] proposed that, because animals compensate by increasing their rate of self-administration (under low FR reinforcement conditions) following decreases in unit amount of self-administered drug, such increased rates of FR self-administration must reflect decreased reinforcer efficacy. Yet, 6-hydroxydopamine (6-OHDA) lesions of the mesoaccumbens DA system[83] confound that interpretation as partial depletion of nucleus accumbens DA produced partial inhibition of cocaine self-administration. Since the reinforcing efficacy of cocaine is believed to correlate with its enhancement of nucleus accumbens DA,[67,79,80,84-87] the

---

[viii] Please see also Koob, The role of animal models in reward deficit disorders: Views of academia; Gardner *et al.*, Issues in designing and conducting clinical trials for reward disorders: A clinical view; McCann *et al.*, Drug discovery and development for reward disorders: Views from government; Rocha *et al.*, Development of medications for heroin and cocaine addiction and regulatory aspects of abuse liability testing, in this volume for further detailed discussion of procedures commonly used in reward deficit disorder research.

decreased FR drug self-administration seen during recovery from the 6-OHDA-induced DA depletion has been interpreted as reflecting decreased reinforcer efficacy. However, how can the same alteration in reinforcing efficacy of cocaine (i.e., decreased efficacy) manifest itself equally by opposite patterns of FR drug self-administration (i.e., by either increased or decreased FR drug self-administration)?[88] The answer is that drug self-administration under low response-cost FR reinforcement conditions is an ambiguous measure of reinforcing efficacy, and possibly even an inaccurate measure that is insensitive to changes in reinforcing efficacy.[78,83]

The *progressive ratio (PR) break-point shift paradigm* was specifically developed to measure shifts in reinforcing efficacy and motivation to self-administer drugs.[78,89-92] For example, cocaine self-administration under PR conditions in rats has been shown to be dose-dependent,[78,89,90] to be sensitive to manipulations affecting brain reward systems,[78,89,90] to yield dose–response functions that reflect addictive potential,[78,90,93,94] and to measure not only the reinforcing efficacy of cocaine, but also cocaine-induced craving.[90,91,95] This paradigm benefits from translational value into clinical trials as the reinforcing effects of cocaine, D-amphetamine, caffeine, and methylphenidate are influenced by behavioral demands following drug administration in humans.[96-98]

Although the PR break-point shift paradigm measures the reinforcing efficacy and motivation to self-administer drugs, the sequential build-up of the drug following repeated self-administration may lead to unwanted results including impaired instrumental responding that might be misinterpreted as impairment of reinforcement or incentive motivation. To address this potential issue *conditioned reinforcement paradigms* have been developed. These paradigms typically combine instrumental and Pavlovian (classical) conditioning procedures. For example, *self-administration under second-order schedules* for both natural and drug reinforcers have examined how a response sequence is maintained by intermittent reinforcement of instrumental behavior (typically a lever press) by the environmental stimulus that has acquired conditioned reinforcement properties.[99-101] One of the most attractive features of this paradigm is that it may potentially dissociate brain mechanisms underlying drug-seeking versus drug-taking behaviors.[102]

Conditioned Preferences

Conditioned preferences have been used extensively as measures of reward (conditioned place preference – CPP) or aversion (conditioned place aversion – CPA). The CPP paradigm relies on the capacity of rewarding stimuli to elicit approach responses and maintenance of contact. Thus, the CPP paradigm uses the phenomenon of secondary conditioning in which a neutral stimulus that has been paired with a reward acquires the ability to serve as a reward itself. Consequently, a drug treatment and its presumed internal effects are paired with the external neutral stimuli of one environment. If during a subsequent test the animal increases the time that it spends approaching and maintaining contact with the stimuli in that environment, it is inferred that the drug treatment was rewarding. CPP is a sensitive technique, capable of indicating rewarding effects of low doses of morphine, D-amphetamine, cocaine, nicotine, heroin, or food. CPP can be produced using only a single drug pairing, thus allowing an assessment of the affective consequences of the first drug experience. This also avoids dealing with problems of physical dependence that may be produced by repeated drug administration. Furthermore, the CPP paradigm can be used to evaluate both rewarding and aversive effects permitting an assessment of the affective properties of drugs that block a CPP response. Finally, testing can be conducted under drug-free conditions, and the paradigm is suitable to PK/PD relationships, a key aspect in drug discovery and development. Although one may describe the preference for an environment previously associated with a drug injection as drug seeking, this measure lacks construct validity to the human condition of craving. In other words, time spent in the drug-paired compartment provides no indication of how hard animals will work to obtain the drug. In addition, negative results in this paradigm are difficult to interpret as they do not prove an absence of reward.[103]

Reinstatement of Drug-Seeking Behavior

The propensity for relapse during abstinence can be modeled in the so-called *reinstatement paradigms*.[104] Reinstatement of drug-seeking behavior can be induced by several manipulations including priming injections of drugs, presentation of drug associated stimuli or cues, and following brief periods of intermittent footshock stress. Although these paradigms already have some kind of predictive validity (e.g., effects of acamprosate (Campral®) and naltrexone in cue-triggered reinstatement of alcohol-seeking behavior), they also suffer from limitations. For example, a typical reinstatement experiment in rodents is conducted under drug-free conditions; by definition, true relapse can only be observed when compulsive drug consumption follows a period of abstinence. Furthermore, extinction of drug-seeking behavior usually plays only a minor role in drug addicts trying to achieve and maintain abstinence. Thus, the preclinical reinstatement model may not reflect accurately the situation of abstinent addicts experiencing craving and/or relapse.

Compulsive/Impulsive Behaviors

We have previously mentioned that loss of control is one of the key features of both drug and behavioral addictions. However, despite their potential translational value, none of the paradigms described in the previous paragraphs do address dysfunctional decision-making processes or alterations in impulsive behavior. Although employed less frequently, choice responding under concurrent schedules of reinforcement may provide valuable insight into drug seeking because the impact of competing reinforcers, and the work required to obtain each, can be measured simultaneously. For example, a previous history of cocaine self-administration decreases the ability of an aversive cue to suppress cocaine seeking.[105] Thus, compulsive drug use may reflect a decreased sensitivity to adverse outcomes associated with drug intake.

Impulsivity, on the other hand, is a multifactorial construct that encompasses different processes including the inability to delay gratification, the inability to withhold a response, acting before all of the relevant information is provided, and inappropriate decision making. People suffering from impulse and reward deficit disorders show a greater tendency to choose small, immediate rewards (e.g., drugs or money) over larger, delayed rewards,[106-109] and discount their hedonic stimuli of choice faster than they do monetary rewards of equal value.[110] In addition, some drug addicts show deficits in a gambling task that models the uncertainty of real-life decision making.[111]

In this context one must mention the translational value of behavioral economics models. Behavioral economics is "the study of the allocation of behavior within a system of constraint".[112] Behavioral economics brings forward two important concepts. The first concept is *elasticity of demand* that refers to the proportional change in consumption resulting from a proportional change in price.[113] The second concept is that of how delayed reinforcers are discounted. *Discounting of delayed reinforcers* refers to the observation that the value of a delayed reinforcer is either reduced in value or considered to be worth less compared to the value of an immediate reinforcer. Thus, the notion of discounting of delayed rewards provides readout of impulsivity and its related corollary of loss of control. Recent animal studies investigating the link between abnormal information processing in the mesocorticolimbic system and changes in responding for delayed or intermittent reinforcement are thus extremely valuable.[114,115] Similarly, procedures examining delay discounting in human subjects present these subjects with a choice between a standard larger-later reward (e.g., $1000 delivered in 1 year) and an immediate reward whose magnitude is adjusted until the participant subjectively considers the two rewards to be of approximately equal worth.[116] This point of equivalence is the *indifference point* for that particular delay interval. When indifference points are obtained for a variety of delays, then an *indifference curve* may be plotted. The main challenges for future delay discounting research will be to understand its relationship to elasticity of demand and whether it may be meaningfully related to clinical outcomes. For example, future research will have to address the questions of whether discounting can be used as an outcome measure to assess the effects of a treatment, whether changes in discounting can be a target of therapeutic efforts, and whether discounting relates to biological markers of impulsivity (e.g., reduced serotoninergic neurotransmission).[ix]

TARGET VALIDATION AND THROUGHPUT LIMITATIONS

Increased screening capacity in drug discovery is creating a greater demand for novel validated disease targets. This has led to new opportunities for the use of high-throughput technologies at numerous stages of the drug discovery process. More rapid understanding of the biological function of the target will become increasingly important as genetic and genomic approaches identify novel approaches. A recent study estimated that target validation currently represents approximately one-fifth of the cost and time involved in drug discovery and development. Target validation represents a current bottleneck and although assays are the key drivers, they are often limited in throughput.

To date, new technologies developed to expedite the drug discovery process have not been applied successfully to behavioral pharmacology. While some areas of the drug discovery process have become highly automated, most aspects of the process that involve *in vivo* behavioral testing using animal models still depend on relatively slow paradigms. This bottleneck is especially pronounced for preclinical drug

---

[ix] Please refer to Williams *et al.*, Current concepts in the classification, treatment and modeling of pathological gambling and other impulse control disorders, in this volume for further discussion of models of impulse control disorders.

testing in impulse and reward deficit disorders, where behavioral tests (see previous section) must be conducted for target validation and lead optimization. In addition, new tools in molecular biology have led to an explosion of research using genetically modified mice and rats, either by targeted approaches or by applying ethylnitrosourea (ENU) to induce point mutations in sperm. Thus, a high-throughput, automated behavioral research platform integrating behavioral and physiological data sets would substantially alleviate this critical and costly bottleneck for CNS drug discovery and development.

TARGET VALIDATION, VALIDITY, RELIABILITY, AND AUTOMATION

A given preclinical paradigm should respond to clinically effective agents belonging to various pharmacological classes and interfering with different mechanisms. Care should be taken not to exclude the validity of a model based on an as yet untested clinical principle. Standardization of apparatus size and testing protocol would be desirable to optimize the chances of reproducing similar effects between laboratories. Drug discovery history is telling us that even minor changes in apparatus size and testing protocols can have major consequences on the behavioral outcome and sensitivity of pharmacological agents in the test. Thus, all reports on behavioral data should include as much detailed information as possible with regard to laboratory and testing conditions. Automation should be investigated where possible in order to speed up both behavioral testing and data analysis and to remove inter-rater subjective bias. In addition the following issues should be carefully considered in experimental designs: are there relevant strain, species, or gender differences? Is the paradigm amenable to high-throughput screen (HTS) methodology? Can we use the paradigm in ENU-mutagenesis screens and quantitative trait linkage (QTL) analyses?

CONCLUSIONS

The substantial co-occurrence of behavioral addictions and substance use disorders, as well as other psychiatric disorders, suggests that improved understanding of their relationship will have important implications not only for further understanding of their neurobiological underpinnings, but also for the discovery and development of relevant pharmacotherapeutic strategies. This notion is further supported by epidemiological studies that are currently moving into an integrative era by combining elements of new sampling methods, biological measurements, and qualitative analysis of social networks to explain the dynamics of disease transmission better. These approaches will allow us to develop scientific knowledge to clarify what needs to be modeled and how, with clear applications to clinical practice and public policy.

We have seen that future animal models of impulse and reward deficit disorders will have to address dysfunctional decision-making processes or alterations in impulsive behavior (e.g., behavioral economics) to improve translatability. Animal modeling also requires a constant balance among throughput, validity, and reliability. In addition, integration of PK/PD modeling is critical to aid PK at site of action and dose selection for transition to human studies. Thus, new strategies will have to be implemented to investigate pharmacological activity in humans that is similar in type and magnitude to that observed in preclinical efficacy models (PD and/or disease model). The latter point relates again to the validity of preclinical efficacy models and the extent to which they increase our confidence or decrease associated risks.

The relationship between marker endpoint and intervention should have biologically plausible explanations. For example, biological markers should either correlate with the causal pathway of the disease (surrogate marker) or recognize the mechanism of action of the potential new therapy (PD marker); the latter being used for determining central penetration and optimize dosing (in conjunction with PK/PD modeling). Thus, some of the key questions for PK/PD modeling are as follows: What is the expected clinical response for a treatment strategy? What is the level of certainty surrounding the predicted response? How do different treatment strategies impact response? What dose is required to achieve a target response? What is the probability of achieving a specific efficacy target while keeping probability for adverse events below a certain level? What is optimal strategy to balance safety and efficacy?

On the one hand, better clinical study designs should address questions of treatment efficacy, predictors of response, and outcomes in different subtypes of impulse and reward deficit disorders. Further clinical research should also look more closely at specific symptoms and specific symptom clusters within a syndrome, in terms of pharmacological response. Basic laboratory science, on the other hand, should focus on modeling very specific clinical features of impulse and reward deficit disorders. Animal models should focus on construct validity and reliability; face validity is probably an unrealistic goal. Predictive validity is clearly ideal, but we need to have better information on differential responses to treatment in humans. Finally, animal models should pay greater attention to the basal state of targeted neural systems. Administration of pharmacological treatment to non-pathological condition in humans almost certainly does not elicit the same neural changes as when applied to someone suffering from this pathology.

REFERENCES

1. Andlin-Sobocki, P. and Rehm, J. (2005). Cost of addiction in Europe. *Eur J Neurol*, 12(Suppl 1):28–33.
2. Andlin-Sobocki, P., Jonsson, B., Wittchen, H.U., and Olesen, J. (2005). Cost of disorders of the brain in Europe. *Eur J Neurol*, 12(Suppl 1):1–27.
3. World Health Organization (WHO) (2006). The WHO Family of International Classifications. http://www.who.int/classifications/en
4. Uhl, G.R. and Grow, R.W. (2004). The burden of complex genetics in brain disorders. *Arch Gen Psychiatry*, 61:223–229.
5. European Monitoring Centre for Drugs and Drug Addiction (EMCDDA). Annual Report 2005: The state of the Drugs Problem in Europe. http://www.emcdda.europa.eu
6. Ministry of Health Law (MHLW) (2006). The MHLW National Nutrition Survey (with smoking rates from the Japan Tobacco Survey) http://www.dbtk.mhlw.go.jp/toukei/youran/data17k/2-03.xls
7. National Institute on Drug Abuse (2005). US Department of Health and Human Services. Epidemiologic trends in drug abuse. *Proceedings of the Community Epidemiology Work Group*, Volume I, Bethesda, Maryland, USA. http://www.drugabuse.gov/PDF/CEWG/Vol 1_605.pdf

8. Substance Abuse and Mental Health Services Administration (SAMHSA). Results from the 2005 National Survey on Drug Use and Health: National Findings. Rockville, MD: Office of Applied Studies, NSDUH Series H-30, DHHS Publication No. SMA 06-4194; 2006.

9. NIAAA, National Institute on Alcohol Abuse and Alcoholism. National Epidemiologic Study of Alcohol and Related Conditions (NESARC). 15 June 2004. http://www.niaaa.census.gov/data.html

10. Castle, D.J. and Phillips, K.A. (2006). Obsessive-compulsive spectrum of disorders: A defensible construct? *Aust NZ J Psychiatry*, 40:114-120.

11. Welte, J., Barnes, G., Wieczorek, Q., Tidwell, M.C., and Parker, J. (2001). Alcohol and gambling pathology among US adults: Prevalence, demographic patterns and comorbidity. *J Stud Alcohol*, 62:706-712.

12. National Research Council (1999). *Pathological gambling: A critical review*. National Academy Press, Washington, DC.

13. Roy, A., Custer, R., Lorenz, V., and Linnoila, M. (1988). Depressed pathological gamblers. *Acta Psychiatry Scand*, 77:163-165.

14. Christenson, G.A., Pyle, R.L., and Mitchell, J.E. (1991). Estimated lifetime prevalence of trichotillomania in college students. *J Clin Psychiatry*, 52:415-417.

15. Stanley, M.A., Hannay, H.J., and Breckenridge, J.K. (1997). The neuropsychology of trichotillomania. *J Anxiety Disord*, 11:473-488.

16. Christenson, G.A. and Mansueto, C.S. (1999). Descriptive characteristics and phenomenology. In Stein, D.J., Christenson, G.A., and Hollander, E. (eds), *Trichotillomania*. American Psychiatric Publishing Inc, Washington, DC, pp. 1-42.

17. Jacobson, R.R. (1985). Child firesetters: A clinical investigation. *J Child Psychol Psychiatry*, 26:759.

18. Kosky, R.J. and Silburn, S. (1984). Children who light fires: A comparison between firesetters and non-firesetters referred to a child psychiatrist outpatient service. *Aust NZ J Psychiatry*, 18:251-255.

19. Kolko, D.J. and Kadzin, A.E. (1988). Prevalence of firesetting and related behaviors among child psychiatric patients. *J Consult Clin Psychol*, 56:628-630.

20. Monopolis, S. and Lion, J.R. (1983). Problems in the diagnosis of intermittent explosive disorder. *Am J Psychiatry*, 140:1200-1202.

21. Felthous, A.R., Bryant, S.G., Wingerter, C.B., and Barratt, E. (1991). The diagnosis of intermittent explosive disorder in violent men. *Bull Am Acad Psychiatry Law*, 19:71-79.

22. Coccaro, E.F., Schmidt, C.A., Samuels, J.F., and Nestadt, G. (2004). Lifetime and 1-month prevalence rates of intermittent explosive disorder in a community sample. *J Clin Psychiatry*, 65:820-824.

23. Goldman, M.J. (1991). Kleptomania: Making sense of the nonsensical. *Am J Psychiatry*, 148:986-996.

24. Grant, J.E. and Kim, S.W. (2002). Clinical characteristics and associated psychopathology of 22 patients with kleptomania. *Compr Psychiatry*, 43:378-384.

25. Presta, S., Marazziti, D., Dell'Osso, L., Pfanner, C., Pallanti, S., and Cassano, G.B. (2002). Kleptomania: Clinical features and comorbidity in an Italian sample. *Compr Psychiatry*, 43:7-12.

26. Black, D.W., Belsare, G., and Schlosser, S. (1999). Clinical features, psychiatric comorbidity, and health-related quality of life in persons reporting compulsive computer use behavior. *J Clin Psychiatry*, 60:839-844.

27. Shapira, N.A., Goldsmith, T.D., Keck, P.E., Jr., Khosla, U.M., and McElroy, S.L. (2000). Psychiatric features of individuals with problematic Internet use. *J Affect Disord*, 57:267-272.

28. Shapira, N.A., Lessig, M.C., Goldsmith, T.D., Szabo, S.T., Lazoritz, M., Gold, M.S., and Stein, D.J. (2003). Problematic Internet use: Proposed classification and diagnostic criteria. *Depress Anxiety*, 17:207-216.

29. Coleman, E. (1991). Compulsive sexual behavior: New concepts and treatments. *J Psychol Hum Sexual*, 4:37-52.

30. Schaffer, S.D. and Zimmerman, M.L. (1990). The sexual addict: A challenge for the primary care provider. *Nurse Pract*, 15:25-33.

31. Kafka, M.P. and Prentky, R. (1994). Preliminary observations of DSM-III-R Axis I comorbidity in men with paraphilias and paraphilia-related disorders. *J Clin Psychiatry*, 55:481-487.

32. Black, D.W. (2001). Compulsive buying disorder: Definition, assessment, epidemiology and clinical management. *CNS Drugs*, 15:17-27.

33. Fontenelle, L.F., Mendlowicz, M.V., and Versiani, M. (2005). Impulse control disorders in patients with obsessive-compulsive disorder. *Psychiatry Clin Neurosci*, 59:30-37.

34. Lejoyeux, M., Bailly, F., Moula, H., Loi, S., and Ades, J. (2005). Study of compulsive buying in patients presenting obsessive-compulsive disorder. *Compr Psychiatry*, 46:105-110.

35. Simeon, D., Stein, D.J., Gross, S., Islam, N., Schmeidler, J., and Hollander, E. (1997). A double-blind trial of fluoxetine in pathologic skin picking. *J Clin Psychiatry*, 58:341-347.

36. Arnold, L.M., McElroy, S.L., Mutasim, D.F., Dwight, M.M., Lamerson, C.L., and Morris, E.M. (1998). Characteristics of 34 adults with psychogenic excoriation. *J Clin Psychiatry*, 59:509-514.

37. Hoek, H.W. (2006). Incidence, prevalence and mortality of anorexia nervosa and other eating disorders. *Curr Opin Psychiatry*, 19:389-394.

38. Office of Applied Studies, Substance Abuse and Mental Health Services Administration (2004). Results from the 2003 National Survey on Drug Use and Health: National findings (DHHS Publication No. SMA 04-3964, NSDUH Series H-25). Rockville, MD: Substance Abuse and Mental Health Services Administration, Office of Applied Studies. http://www.oas.samhsa.gov/p0000016.htm#Standard

39. Ayers, J., Ruff, C.F., and Templer, D.I. (1976). Alcoholism, cigarette smoking, coffee drinking and extraversion. *J Studies Alcohol*, 37:983-985.

40. Burling, T.A. and Ziff, D.C. (1988). Tobacco smoking: A comparison between alcohol and drug abuse inpatients. *Addict Behav*, 13:185-190.

41. Burling, T.A., Reilly, P.M., Moltzen, J.O., and Ziff, D.C. (1989). Self-efficacy and relapse among inpatient drug and alcohol abusers: A predictor of outcome. *J Studies Alcohol*, 50:354-360.

42. Dreher, K.F. and Fraser, J.G. (1967). Smoking habits of alcoholic outpatients: I. *Int J Addict*, 2:259-270.

43. Kozlowski, L.T., Jelinek, L.C., and Pope, M.A. (1986). Cigarette smoking among alcohol abusers: A continuing and neglected problem. *Can J Pub Health*, 77:205-207.

44. Walton, R.G. (1972). Smoking and alcoholism: A brief report. *Am J Psychiatry*, 128:1455-1456.

45. Batel, P., Pessione, F., Maitre, C., and Rueff, B. (1995). Relationship between alcohol and tobacco dependencies among alcoholics who smoke. *Addiction*, 90:977-980.

46. Zimmerman, R.S., Warheit, G.J., Ulbrich, P.M., and Auth, J.B. (1990). The relationship between alcohol use and attempts and success at smoking cessation. *Addict Behav*, 15:197-207.

47. Volkow, N.D. (2004). The reality of comorbidity: Depression and drug abuse. *Biol Psychiatry*, 56:714-717.

48. Tarter, R.E. (1995). Genetics and primary prevention of drug and alcohol abuse. *Int J Addict*, 30:1479-1484.

49. Schuster, C.R. (1992). Drug abuse research and HIV/AIDS: A national perspective from the US. *Br J Addict*, 87:355-861.

50. McCoy, C.B., Lai, S., Metsch, L.R., Messiah, S.E., and Zhao, W. (2004). Injection drug use and crack cocaine smoking: Independent and dual risk behaviors for HIV infection. *Ann Epidemiol*, 14:535-542.

51. Amin, J., Kaye, M., Skidmore, S., Pillay, D., Cooper, D.A., and Dore, G.J. (2004). HIV and hepatitis C coinfection within the CAESAR study. *HIV Med*, 5:174-179.

52. Safdar, K. and Schiff, E.R. (2004). Alcohol and hepatitis C. *Semin Liver Dis*, 24:305–315.
53. Nath, A., Hauser, K.F., Wojna, V., Booze, R.M., Maragos, W., Prendergast, M., Cass, W., and Turchan, J.T. (2002). Molecular basis for interactions of HIV and drugs of abuse. *J Acquir Immune Defic Syndr*, 31(Suppl 2):S62–S69.
54. Shafran, R., Cooper, Z., and Fairburn, C.G. (2002). Clinical perfectionism: A cognitive-behavioral analysis. *Behav Res Ther*, 40:773–791.
55. Anderluh, M.B., Tchanturia, K., Rabe-Hesketh, S., and Treasure, J. (2003). Childhood obsessive-compulsive personality traits in adult women with eating disorders: Defining a broader eating disorder phenotype. *Am J Psychiatry*, 160:242–247.
56. Claes, L., Vandereycken, W., and Vertommen, H. (2004). Personality traits in eating-disordered patients with and without self-injurious behaviors. *J Personal Disord*, 18:399–404.
57. Steiger, H., Jabalpurwala, S., Champagne, J., and Stotland, S. (1997). A controlled study of trait narcissism in anorexia and bulimia nervosa. *Int J Eat Disord*, 22:173–178.
58. Narduzzi, K.J. and Jackson, T. (2002). Sociotropy-dependency and autonomy as predictors of eating disturbance among Canadian female college students. *J Genet Psychol*, 163:389–401.
59. Herzog, D.B., Keller, M.B., Sacks, N.R., Yeh, C.J., and Lavori, P.W. (1992). Psychiatric comorbidity in treatment-seeking anorexics and bulimics. *J Am Acad Child Adolesc Psychiatry*, 31:810–818.
60. Bulik, C.M., Sullivan, P.F., Weltzin, T.E., and Kaye, W.H. (1995). Temperament in eating disorders. *Int J Eat Disord*, 17:251–261.
61. Halmi, K.A., Eckert, E., Marchi, P., Sampugnaro, V., Apple, R., and Cohen, J. (1991). Comorbidity of psychiatric diagnoses in anorexia nervosa. *Arch Gen Psychiatry*, 48:712–718.
62. Kaye, W.H., Weltzin, T.E., McKee, M., McConaha, C., Hansen, D., and Hsu, L.K. (1992). Laboratory assessment of feeding behavior in bulimia nervosa and healthy women: Methods for developing a human-feeding laboratory. *Am J Clin Nutr*, 55:372–380.
63. Grilo, C.M., Martino, S., Walker, M.L., Becker, D.F., Edell, W.S., and McGlashan, T.H. (1997). Psychiatric comorbidity differences in male and female adult psychiatric inpatients with substance use disorders. *Compr Psychiatry*, 38:155–159.
64. Salamone, J.D., Correa, M., Mingote, S.M., and Weber, S.M. (2005). Beyond the reward hypothesis: Alternative functions of nucleus accumbens dopamine. *Curr Opin Pharmacol*, 5:34–41.
65. Wise, R.A. (2004). Dopamine, learning and motivation. *Nat Rev Neurosci*, 5:483–494.
66. Koob, G.F. (2006). The neurobiology of addiction: A neuroadaptational view relevant for diagnosis. *Addiction*, 101(Suppl 1):23–30.
67. Wise, R.A. and Gardner, E.L. (2002). Functional anatomy of substance-related disorders. In D'haenen, H., Boer den, J.A., and Willner, P. (eds.), *Biological Psychiatry*. Wiley, New York, pp. 509–522.
68. Garavan, H., Pankiewicz, J., Bloom, A., Cho, J.K., Sperry, L., Ross, T.J., Salmeron, B.J., Risinger, R., Kelley, D., and Stein, E.A. (2000). Cue-induced cocaine craving: Neuroanatomical specificity for drug users and drug stimuli. *Am J Psychiatry*, 157:1789–1798.
69. Kosten, T.R., Scanley, B.E., Tucker, K.A., Oliveto, A., Prince, C., Sinha, R., Potenza, M.N., Skudlarski, P., and Wexler, B.E. (2006). Cue-induced brain activity changes and relapse in cocaine-dependent patients. *Neuropsychopharmacology*, 31:644–650.
70. Childress, A.R., Mozley, P.D., McElgin, W., Fitzgerald, J., Reivich, M., and O'Brien, C.P. (1999). Limbic activation during cue-induced cocaine craving. *Am J Psychiatry*, 156:11–18.
71. Grant, S., London, E.D., Newlin, D.B., Villemagne, V.L., Liu, X., Contoreggi, C., Phillips, R.L., Kimes, A.S., and Margolin, A. (1996). Activation of memory circuits during cue-elicited cocaine craving. *Proc Natl Acad Sci USA*, 93:12040–12045.
72. Tamminga, C.A. (1999). The anterior cingulate and response conflict. *Am J Psychiatry*, 156:1849.
73. Volkow, N.D. and Fowler, J.S. (2000). Addiction, a disease of compulsion and drive: Involvement of the orbitofrontal cortex. *Cereb Cortex*, 10:318–325.

74. Bechara, A. (2005). Decision making, impulse control and loss of willpower to resist drugs: A neurocognitive perspective. *Nat Neurosci*, 8:1458-1463.

75. Boileau, I., Assaad, J.M., Pihl, R.O., Benkelfat, C., Leyton, M., Diksic, M., Tremblay, R.E., and Dagher, A. (2003). Alcohol promotes dopamine release in the human nucleus accumbens. *Synapse*, 49:226-231.

76. Boileau, I., Dagher, A., Leyton, M., Gunn, R.N., Baker, G.B., Diksic, M., and Benkelfat, C. (2006). Modeling sensitization to stimulants in humans: An [11C]raclopride/positron emission tomography study in healthy men. *Arch Gen Psychiatry*, 63:1386-1395.

77. Cloninger, C.R., Sigvardsson, S., and Bohman, M. (1988). Childhood personality predicts alcohol abuse in young adults. *Alcohol Clin Exp Res*, 12:494-505.

78. Arnold, J.M. and Roberts, D.C.S. (1997). A critique of fixed ratio and progressive ratio schedules used to examine the neural substrates of drug reinforcement. *Pharmacol Biochem Behav*, 57:441-447.

79. Gardner, E.L. (2000). What we have learned about addiction from animal models of drug self-administration. *Am J Addict*, 9:285-313.

80. Gardner, E.L. (2005). Brain reward mechanisms. In Lowinson, J.H., Ruiz, P., Millman, R.B., and Langrod, J.G. (eds), *Substance Abuse: A Comprehensive Textbook, 4th edition.* Lippincott Williams & Wilkins, Philadelphia, PA, pp. 48-97.

81. Yokel, R.A. and Wise, R.A. (1975). Increased lever pressing for amphetamine after pimozide in rats: Implication for a dopamine theory of reward. *Science*, 187:547-549.

82. Yokel, R.A. and Wise, R.A. (1976). Attenuation of intravenous amphetamine reinforcement by central dopamine blockade in rats. *Psychopharmacology*, 48:311-318.

83. Roberts, D.C.S. (1989). Breaking points on a progressive ratio schedule reinforced by intravenous apomorphine increase daily following 6-hydroxydopamine lesions of the nucleus accumbens. *Pharmacol Biochem Behav*, 32:43-47.

84. de Wit, H. and Wise, R.A. (1977). Blockade of cocaine reinforcement in rats with the dopamine receptor blocker pimozide, but not with the noradrenergic blockers phentolamine or phenoxybenzamine. *Can J Psychol*, 31:195-203.

85. Ettenberg, A., Pettit, H.O., Bloom, F.E., and Koob, G.F. (1982). Heroin and cocaine intravenous self-administration in rats: Mediation by separate neural systems. *Psychopharmacology*, 78:204-209.

86. Spyraki, C., Nomikos, G.G., and Varonos, D.D. (1987). Intravenous cocaine-induced place preference: Attenuation by haloperidol. *Behav Brain Res*, 26:57-62.

87. Wise, R.A. and Rompré, P.-P. (1989). Brain dopamine and reward. *Annu Rev Psychol*, 40:191-225.

88. Roberts, D.C.S. and Zito, K.A. (1987). Interpretation of lesion effects on stimulant self-administration. In Bozarth, M.A. (ed.), *Methods of Assessing the Reinforcing Properties of Abused Drugs.* Springer-Verlag, New York, pp. 87-103.

89. Richardson, N.R. and Roberts, D.C.S. (1996). Progressive ratio schedules in drug self-administration studies in rats: A method to evaluate reinforcing efficacy. *J Neurosci Meth*, 66:1-11.

90. Stafford, D., LeSage, M.G., and Glowa, J.R. (1998). Progressive-ratio schedules of drug delivery in the analysis of drug self-administration: A review. *Psychopharmacology*, 139:169-184.

91. Rowlett, J.K. (2000). A labor-supply analysis of cocaine self-administration under progressive-ratio schedules: Antecedents, methodologies, and perspectives. *Psychopharmacology*, 153:1-16.

92. Morgan, D. and Roberts, D.C.S. (2004). Sensitization to the reinforcing effects of cocaine following binge-abstinent self-administration. *Neurosci Biobehav Rev*, 27:803-812.

93. Roberts, D.C.S. and Bennett, S.A.L. (1993). Heroin self-administration in rats under a progressive ratio schedule of reinforcement. *Psychopharmacology*, 111:215-218.

94. French, E.D., Lopez, M., Peper, S., Kamenka, J.-M., and Roberts, D.C.S. (1995). A comparison of the reinforcing efficacy of PCP, the PCP derivatives TCP and BTCP, and cocaine using a progressive ratio schedule in the rat. *Behav Pharmacol*, 6:223-228.

95. Markou, A., Weiss, F., Gold, L.H., Caine, S.B., Schulteis, G., and Koob, G.F. (1993). Animal models of drug craving. *Psychopharmacology*, 112:163–182.

96. Rush, C.R., Essman, W.D., Simpson, C.A., and Baker, R.W. (2001). Reinforcing and subject-rated effects of methylphenidate and D-amphetamine in non-drug-abusing humans. *J Clin Psychopharmacol*, 21:273–286.

97. Stoops, W.W., Glaser, P.E., Fillmore, M.T., and Rush, C.R. (2004). Reinforcing, subject-rated, performance and physiological effects of methylphenidate and D-amphetamine in stimulant abusing humans. *J Psychopharmacol*, 18:534–543.

98. Stoops, W.W., Lile, J.A., Robbins, C.G., Martin, C.A., Rush, C.R., and Kelly, T.H. (2007). The reinforcing, subject-rated, performance, and cardiovascular effects of D-amphetamine: Influence of sensation-seeking status. *Addict Behav*, 32:1177–1188.

99. Everitt, B.J. and Robbins, T.W. (2001). Second order schedules of drug reinforcement in rats and monkeys: Measurement of reinforcing efficacy and drug-seeking basis behavior. *Psychopharmacology*, 153:17–30.

100. Goldberg, S.R. (1973). Comparable behavior maintained under fixed-ratio and second-order schedules of food presentation, cocaine injection or D-amphetamine injection in the squirrel monkey. *J Pharmacol Exp Ther*, 186:18–30.

101. Goldberg, S.R. and Tang, A.H. (1977). Behavior maintained under second-order schedules of intravenous morphine injection in squirrel and rhesus monkeys. *Psychopharmacology*, 51:235–242.

102. Whitelaw, R.B., Markou, A., Robbins, T.W., and Everitt, B.J. (1996). Excitotoxic lesions of the basolateral amygdala impair the acquisition of cocaine-seeking behavior under a second order schedule of reinforcement. *Psychopharmacology*, 127:213–224.

103. Tzschentke, T.M. (1998). Measuring reward with the conditioned place preference paradigm: A comprehensive review of drug effects, recent progress and new issues. *Prog Neurobiol*, 56:613–672.

104. Shaham, Y., Shalev, U., Lu, L., de Wit, H., and Stewart, J. (2003). The reinstatement model of drug relapse: History, methodology and major findings. *Psychopharmacology*, 168:3–20.

105. Vanderschuren, L.J.M.J. and Everitt, B.J. (2004). Drug seeking becomes compulsive after prolonged cocaine self-administration. *Science*, 305:1017–1019.

106. Kirby, K.N., Petry, N.M., and Bickel, W.K. (1999). Heroin addicts have higher discount rates for delayed rewards than non-drug using controls. *J Exp Anal Behav*, 128:78–87.

107. Moeller, F.G., Dougherty, D.M., Barratt, E.S., Oderinde, V., Mathias, C.W., Harper, R.A., and Swann, A.C. (2002). Increased impulsivity in cocaine dependent subjects independent of antisocial personality disorder and aggression. *Drug Alcohol Depend*, 68:105–111.

108. Petry, N.M. (2001). Substance abuse, pathological gambling, and impulsiveness. *Drug Alcohol Depend*, 63:29–38.

109. Vuchinich, R.E. and Simpson, C.A. (1998). Hyperbolic temporal discounting in social drinkers and problem drinkers. *Exp Clin Psychopharmacol*, 6:292–305.

110. Madden, G.J., Petry, N.M., Badger, G.J., and Bickel, W.K. (1997). Impulsive and self-control choices in opioid-dependent patients and non-drug-using control participants: Drug and monetary rewards. *Exp Clin Psychopharmacol*, 5:256–262.

111. Bechara, A. and Damasio, H. (2002). Decision-making and addiction (part I): Impaired activation of somatic states in substance dependent individuals when pondering decisions with negative future consequences. *Neuropsychologia*, 40:1675–1689.

112. Bickel, W.K., Green, L., and Vuchinich, R.E. (1995). Behavioral economics. *J Exp Anal Behav*, 64:257–262.

113. Bickel, W.K., Madden, G.J., and Petry, N.M. (1998). The price of change: The behavioral economics of drug dependence. *Behav Therapy*, 29:545–565.

114. Cardinal, R.N., Winstanley, C.A., Robbins, T.W., and Everitt, B.J. (2004). Limbic corticostriatal systems and delayed reinforcement. *NY Acad Sci*, 1021:33–50.

115. Wakabayashi, K.T., Fields, H.L., and Nicola, S.M. (2004). Dissociation of the role of nucleus accumbens dopamine in responding to reward-predictive cues and waiting for reward. *Behav Brain Res*, 154:19–30.

116. Green, L., Fry, A.F., and Myerson, J. (1994). Discounting of delayed rewards: A lifespan comparison. *Psychol Sci*, 5:33–36.

CHAPTER 2

Drug Discovery and Development for Reward Disorders: Views from Government

David J. McCann, Jane B. Acri and Frank J. Vocci

National Institute on Drug Abuse, National Institutes of Health, Department of Health and Human Services, Bethesda, MD, USA

INTRODUCTION AND CONTEXTUAL ISSUES

The National Institute on Drug Abuse (NIDA) is the world's largest research institution dedicated to funding research on drugs of abuse, prevention of drug abuse, and treatment of substance abuse disorders. Being an institution of the US Federal Government, NIDA derives its existence and objectives from Congressional and legislative activities. NIDA was created by the Drug Abuse Office and Treatment Act of 1972 (PL 92-255). This act stipulated that NIDA be established within the National Institute on Mental Health and become operational in 1974, at which time NIDA became one of three institutes in the Alcohol, Drug Abuse, and Mental Health Administration (ADAMHA). NIDA started to receive large funding increases in 1986; funding for acquired immune deficiency syndrome (AIDS) research was especially increased. The Anti-Drug Abuse Act of 1988 (PL 100-690) further increased NIDA funding; $8 million was appropriated for medication development projects. With this Congressional impetus and funding, NIDA began its Medications Development Program. Congressional intent was for NIDA to develop pharmacotherapies to treat the symptoms and disease of drug abuse,

Animal and Translational Models for CNS Drug Discovery,
Vol. 3 of 3: Reward Deficit Disorders
Robert McArthur and Franco Borsini (eds), Academic Press, 2008

23

including medications to: block the effects of abused drugs; reduce craving for abused drugs; moderate/eliminate withdrawal symptoms; block or reverse toxic effects of abused drugs; and prevent relapse in detoxified persons. The organizational structure of NIDA was restructured in 1990 to create the Medications Development Division, now called the Division of Pharmacotherapies and Medical Consequences of Drug Abuse (DPMC). In 1992, The ADAMHA Reorganization Act (PL 102-321) transferred NIDA to the National Institutes of Health (NIH). This act also stipulated that the NIDA Medications Development Program should pursue biological and pharmacological approaches to develop medications for treatment of heroin and cocaine addiction; establish a close working relationships with the pharmaceutical industry; conduct studies to gain approval of new medicines for addiction treatment; and develop a working relationship with Food and Drug Administration (FDA) to assure that efficacy of compounds is expeditiously evaluated and approved.

The Anti-Drug Abuse Act of 1988, which initiated funding for the NIDA Medications Development Program, was a tacit acknowledgement that the pharmaceutical industry needed incentives to consider medications development in the area of addictive disorders. Congress, in the ADAMHA Reorganization Act of 1992, further stipulated that the Federal Government should contract with the National Academy of Sciences (NAS) to establish a committee of the Institute of Medicine (IOM) to review issues impacting the development of medications for addiction. An IOM committee was convened in 1993. This committee examined the extent to which the limitations of the (then) current scientific knowledge hindered the development of pharmacological treatments, reviewed the progress of the NIDA Medications Development Program, considered the role of the FDA and other government entities in the addiction medications approval process, surveyed the incentives and disincentives to private sector development of medications to treat addictive disorders, and attempted to determine the (then) current role of the private sector in such development.

To gain a better understanding of how industry viewed medications development for addictive disorders, the IOM committee sent a survey to pharmaceutical companies. Nineteen companies responded. Although a myriad of issues were discussed, the state of the science, the stigma associated with medications for treatment of addiction, the difficulty of performing clinical studies in substance abusing patients, and reimbursement issues were noted to be of concern to companies (Appendix D, The Development of Medications for the Treatment of Opiate and Cocaine Addictions: Issues for the Government and Private Sector, 1995).

DIVISION OF PHARMACOTHERAPIES AND MEDICAL CONSEQUENCES OF DRUG ABUSE

Drug Discovery Program – Initial Operations

The DPMC initiated a discovery and development program in 1990. The initial program intended to perform standardized tests and clinical studies to facilitate the discovery and development of addiction medications and to facilitate the involvement of the pharmaceutical industry and academia. Setting up a medications discovery program within a company is an expensive undertaking. Moreover, many companies lack the scientific expertise in the neurobiology of drugs of abuse and behavioral pharmacology of addiction. The NIDA discovery program was shaped with the consultation of outside advisors who were members of the (then) Pharmaceutical Manufacturers Association. The primary objective of this discovery program was to discover putative medications for the treatment of cocaine dependence. Smaller discovery teams were also set up to discover medications for opiate, and later (after 2000) methamphetamine and nicotine dependence. Only the cocaine discovery program will be discussed here. To set up the discovery program and ensure its success, four separate tasks had to be initiated. The first task was to set up an appropriate testing hierarchy, using both *in vitro* and *in vivo* assays. Based on past successes in developing medications for other drug addiction disorders, the initial testing scheme of the NIDA Cocaine Treatment Discovery Program was focused on the discovery of "cocaine agonist therapies" (analogous to methadone for heroin addiction or nicotine replacement therapy for smoking) and "cocaine antagonist therapies" (analogous to naltrexone for heroin addiction). Behavioral tests, comprising locomotor activity, drug discrimination, and self-administration in rodents, formed the first tier of animal testing. These tests were selected as a trio because advisors from both academia and the pharmaceutical industry did not feel that any one of these tests had predictive validity. Drug discrimination and self-administration in monkeys formed the second tier of behavioral testing. The dopamine transporter was featured as a primary target, but it was acknowledged from both a scientific and programmatic standpoint that cocaine interacted with other monoamine transporters and, indirectly, with several different types of biogenic amine receptors. In evaluating potential "cocaine agonist therapies," including compounds targeting the dopamine transporter, the ideal candidate compound was considered to be one that produced less maximum effect than cocaine (especially with regard to stimulation of locomotor activity), with a slow onset and long duration of action. As the science and the program have progressed over the last 17 years, the molecular targets have become more diversified and testing procedures have evolved (see below).

The second task was to determine how to obtain compounds for testing and to establish appropriate policies for compound and data handling. Compounds were purchased from chemical catalogs and obtained from Government chemists, NIDA grantees, and the pharmaceutical industry. Pharmacological testing was to be performed "free of charge," NIDA testing sites were to be blinded to the identity and source of compounds, and the results were to be sent to compound submitters and kept confidential by NIDA. The "licensing" component of the NIDA discovery program differs from that of industry in that NIDA does not pay companies or chemists to "license" compounds. Instead, compound submitters retain their full intellectual property rights. The reliance on compounds from outside sources is another deviation from the usual way that industry works. Most companies that have in-house discovery programs have full control of the discovery process whereas NIDA must rely on collaborating entities for compound supply and re-supply. This arrangement can hamper the pace of discovery.

The third task was to determine the type of agreements that would be used to obtain compounds for testing. The NIH Office of Technology Transfer has developed several documents that can be used for the transfer of materials to and from NIH laboratories or its contractors. Information about a compound or compound series may be transmitted after a Confidential Disclosure Agreement has been executed. Depending on the intellectual property status of a compound and the concerns of the compound submitter, a compound may be obtained under a Material Transfer Agreement, a Material Cooperative Research and Development Agreement (MCRADA) or a Screening Agreement. The Screening Agreement is a custom document written by NIDA that allows compounds to be tested without concern that NIDA or its contractors will patent a compound that has gone through preclinical testing. In order for NIDA to offer such protection, a Declaration of Exceptional Circumstances (DEC) had to be obtained from the US Commerce Department for each contract that supports compound testing. The default position of the Federal Government is that contractors own data generated under their contracts and can file invention reports and patents. The DEC rescinds the default position of the Government, disallowing specified contractors from owning data and patenting. As mentioned above, NIDA testing sites are blinded to the identity and source of compounds; this helps NIDA to control the use of data and ensures that NIDA can abide by its agreements with compound submitters. In addition, the contractors are legally required to treat as confidential any and all compound-related data or other information, whether provided by NIDA or generated during the course of testing.

The fourth task was the development of a decision algorithm for compounds to progress through preclinical development. A detailed discussion of DPMC clinical trial capabilities is beyond the scope of the present chapter; however, taking the ability of NIDA to conduct clinical trials for granted, a discovery program is of no use if promising compounds cannot advance to clinical testing. To facilitate preclinical development, NIDA established contracts to support standard toxicological and pharmacokinetic testing, as well as specialized safety studies. Each post-discovery/early development compound, termed a Safety Assessment Candidate (SAC), undergoes limited toxicological and pharmacokinetic testing to determine whether advancement to full, Investigational New Drug (IND)-enabling development is warranted. For compounds advancing beyond SAC status, special drug interaction studies are conducted to evaluate the safety of potential medications in the presence of drugs of abuse. The need for these special drug interaction studies follows from the fact that many patients will ultimately use drugs of abuse on top of any medication that is prescribed for the treatment of drug addiction. Because patients often abuse multiple drugs, a battery of drug interaction studies is required; for example, all new molecular entities (NMEs) under development as cocaine addiction treatments are evaluated for safety in the presence of cocaine, an opiate (morphine), and ethanol.

Drug Discovery Program – Current Status

To date, over 3000 compounds have been evaluated in the NIDA drug discovery programs. Three compounds have been advanced to IND status and two more are being developed for administration to human subjects. The IND for the first of these should be filed in early 2008.

The structure, objectives, and testing scheme of the NIDA drug discovery programs have changed as a result of the maturing of the science and efforts to improve efficiency. Medication targets have expanded and relapse prevention has become the number one focus. In 2005, the Addiction Treatment Discovery Program (ATDP) was created through the merger of four separate discovery programs that were focused exclusively on cocaine, opiates, methamphetamine, and nicotine. As mentioned above, a major goal of the earlier programs was the discovery of "agonist therapies" to facilitate quit attempts in cocaine and methamphetamine dependence. In contrast, the ATDP has shifted focus to emphasize the discovery of compounds aimed at the clinical endpoint of relapse prevention. Because of the focus on relapse prevention as a clinical endpoint, the program has a number of drug self-administration reinstatement procedures (relapse models) using different drugs of abuse.[i] The ATDP has increased resources to evaluate compounds in models of relapse to cocaine, heroin, or methamphetamine, using stress, conditioned cues, or drug primes to produce reinstatement in rats whose self-administration behavior has been extinguished.

ATDP staff coordinates testing for compound submitters and provide related study reports and feedback. In addition to the relapse models mentioned above, established tests that could be selected for a particular compound include *in vitro* receptor assays, rodent locomotor activity and intra-cranial self-stimulation testing, rodent and/or primate drug discrimination testing, and rodent and/or primate drug self-administration testing. In addition, a series of predictive toxicology tests – such as the hERG channel assay to predict QT prolongation and the Spot Ames test to predict mutagenicity – are available and, as necessary, additional animal models may be used. Compound testing is shaped by existing data in rodents and the sequence of testing is determined in collaboration with the compound submitter. ATDP staff members welcome opportunities to discuss specific testing proposals with potential compound submitters and to present NIDA's capabilities to pharmaceutical company management.

Development of Medications for Opiate and Cocaine Addiction

The motivation for Federal involvement in the development of medications for the treatment of addictive disorders is from a public health viewpoint. This is one obvious difference between NIDA and the pharmaceutical industry. The Federal Government is heavily invested in the treatment of addictive disorders and views new medications to treat these disorders as a way to expand treatment and improve the public health of the United States.

Another major difference between the NIDA Medications Development Program and the pharmaceutical industry is the aspect of commercialization. From a legal

---

[i] Please refer to Heidbreder, Impulse and reward deficit disorders: Drug discovery and development; Koob, The role of animal models in reward deficit disorders: Views of academia; Markou *et al.*, Contribution of animal models and preclinical human studies to medication development for nicotine dependence; and Rocha *et al.*, Development of medications for heroin and cocaine addiction and regulatory aspects of abuse liability testing, in this volume for further description and discussion of animal models of substance abuse and relapse, and procedures used to evaluate the efficacy of compounds to treat reward deficit disorders.

standpoint, the US Government does not compete with its citizens or private commercial enterprises. OMB Circular A-76 establishes Federal policy regarding the performance of commercial activities and implements the statutory requirements of the Federal Activities Inventory Reform Act of 1998, PL 105-270. Suffice it to say that the NIH has interpreted this to mean that they expect its pharmaceutical partners to market any commercial products that are the realization of joint research between NIH and its partner(s). NIDA does not wish to undertake the full medications development process on its own. Therefore, NIDA seeks to enter into collaborative agreements with private sector partners as early as possible in any development project. As collaborative agreements are established, the relative contributions of each party are negotiated on a case-by-case basis and a formal agreement is drafted. NIDA has a Technology Transfer section that negotiates the necessary legal agreements under which joint projects are conducted.

One of the initial challenges for the NIDA Medications Development Program was to demonstrate to the pharmaceutical industry that its grantees and contractors could perform clinical trials that would pass muster at the FDA. This was a two-part challenge: the initial challenge was to design appropriate clinical trial endpoints and the second component was to execute the trial according to Good Clinical Practices guidelines. When the program was just beginning in the early 1990s, there were three potential development candidates for the treatment of opiate dependence. Each will be discussed in turn.

Levomethadyl acetate, or LAAM, had been studied in the 1970s and had two previous New Drug Applications (NDAs) rejected by the FDA. Following a consultation with the FDA, it was agreed that a large (26 site) multi-center trial would be run. Six hundred and twenty-three subjects were recruited in about 1 year. Subjects were allowed to stay on LAAM for 65 weeks. NIDA worked with a contractor, Biometric Research Institute, to file a successful NDA. LAAM was approved for marketing 18 days after the NDA was filed.

Buprenorphine was the second of the three development candidates. Following the successful demonstration of the efficacy of buprenorphine to reduce opiate use in the outpatient setting by NIDA Intramural Research Program,[1] NIDA signed a CRADA with Reckitt-Colman (now Reckitt-Benckiser) to develop buprenorphine as a medication for the management of opiate dependence. A subsequent multi-center trial, run by one of NIDA's grantees, showed dose-related efficacy.[2] During the development process it became apparent that a solid dosage form needed to be developed. Although this was a challenge, it also presented an opportunity to add naloxone (a narcotic receptor antagonist) to prevent abuse by intravenous injection. The buprenorphine to naloxone ratio was decided after reviewing data from three clinical pharmacology studies that were subsequently published.[3-5] A subsequent multi-center trial of the buprenorphine and buprenorphine/naloxone 4:1 ratio combination was compared to placebo responses in the first month of opiate therapy. Both buprenorphine dosage forms outperformed placebo in reducing opiate use in the outpatients setting.[6] These data and other developed by NIDA grantees and contractors were used successfully to obtain NDA approvals for buprenorphine (Subutex®) and buprenorphine/ naloxone (Suboxone®) in October 2002. These are the first two opiates in 90 years that can be prescribed by qualified physicians in office-based settings for the management of opiate dependence. Over 10 000 physicians have qualified, either by training or experience, to prescribe buprenorphine. Reckitt-Benckiser estimates that over 500 000 patients have been prescribed buprenorphine for the management of opiate dependence.

The last development candidate, a depot naltrexone formulation, was initially developed by a NIDA contractor. Subsequently, the contract operation was purchased by Alkermes, who decided to market the depot naltrexone (Vivitrol®) for the treatment of alcohol dependence. Clinical studies with other NIDA-supported depot naltrexone dosage forms have shown efficacy in the treatment of opiate dependence.[7] Although this last study may be regarded as a proof of concept trial, it demonstrates that NIDA and its grantees can successfully perform clinical trials with different dosage forms in this difficult to treat population.

NIDA has tried to address industry concerns by demonstrating the feasibility of performing high-quality, large-scale clinical trials in substance abusing populations, demonstrating that it has developed outcome variables and statistical analyses that are scientifically valid and accepted by FDA as capable of demonstrating efficacy for a drug product, and shown its ability to partner with industry for successful development of drug products. This last point implicitly suggests that NIDA has knowledge of the developmental pathways for successful medications research and development for addictive disorders. The ongoing success of the buprenorphine products is a further demonstration that markets exist where others were skeptical of any degrees of successful marketing of a medication for addiction.

The NIDA Medications Development Program is acutely aware of the need for medications for the management of stimulant addictions. NIDA grantees and contractors have tested over 60 marketed psychopharmaceuticals in cocaine-dependent patients. Several of these medications have shown salutary effects to reduce cocaine use (for a comprehensive review see[8]) and are undergoing confirmatory clinical trials. None of the medications to date, for example, disulfiram, has shown a large effect size. NIDA grantees and contractors are also testing several marketed medications for the treatment of methamphetamine dependence. Recently, two medications, methylphenidate and bupropion, have shown preliminary efficacy in amphetamine-dependent[9] and methamphetamine-dependent subjects,[10] respectively.

There are three challenges that NIDA must meet in order to develop effective medications for stimulant dependence. The first is to continue to conduct confirmatory clinical trials. Medications that demonstrate efficacy in these studies will be the first generation of medications for the treatment of stimulant dependence. The second is to evaluate and develop medications that have a neuroscience-driven basis. Recent examples of mechanisms of interest include cannabinoid receptor antagonists[11] and D_3 dopamine receptor antagonists.[12,13] Testing of the neuroscience-based medications will also give some indication of the importance of modulating appetitive behaviors in the treatment of addictive disorders. The third challenge is to use feedback and feedforward mechanisms to discover the relevant animal and human laboratory models that have predictive validity.

ANIMAL MODELS IN THE DISCOVERY OF DRUG ADDICTION TREATMENTS

The Validity of Animal Models in the Field of Drug Addiction

In the field of medications discovery for drug addiction treatment, it is fortunate to have animal models that meet many criteria for validity.[ii] The fact that alcoholism can only follow the consumption of alcohol or that cocaine addiction can only follow exposure to cocaine is often taken for granted; however, the importance of such knowledge to the "etiological validity" of the animal models should not be overlooked. One can administer the causative agents of drug addiction (the drugs themselves) to animals to produce the disorder under study. Contrast this situation with that of researches who long for similar etiological validity in animal models of psychosis or dementia. Beyond etiological validity, animal models that involve ethanol drinking or lever-pressing to receive intravenous injections of a drug such as cocaine show exemplary face validity. Likewise, the similar symptomatology of opiate withdrawal in monkeys and man is a noteworthy example of construct validity. Finally, preclinical studies of the two most recently approved drug addiction treatment medications, varenicline[14] and buprenorphine,[15] demonstrate the predictive validity of drug self-administration procedures as models of nicotine and opiate addiction, respectively.

Despite the generally high level of validity seen in animal models of addiction, it must be acknowledged that predictive validity has not been established for animal models relevant to the field's most important clinical indications, those for which no effective medications are currently available. Clearly, if there is no medication with well-established efficacy in treating a specific drug addiction disorder (e.g., in the case of cocaine or methamphetamine addiction), there can be no relevant animal model with established predictive validity. The same can be said for novel approaches to drug addiction treatment. For example, an animal model of stress-induced relapse to opiate abuse cannot be appropriately evaluated for its predictive validity if clinical studies have not revealed a medication that effectively prevents stress-induced relapse to opiates. Where predictive validity cannot be established, we must rely on the etiological, construct and/or face validity of our animal models.

Use of Data from Animal Models for "Go/No Go" Decisions

Because it is not possible to obtain data from animal models with established predictive validity when focusing on clinical indications that lack effective medications (the tools for validation), it is fortunate that the FDA does not have a firm requirement for such data. In fact, advancement to clinical development does not necessarily require efficacy-related preclinical data. Preclinical safety data are always required but the case

---

[ii] Please refer to Koob, The role of animal models in reward deficit disorders: Views of academia; in this volume; Large et al., Developing therapeutics for bipolar disorder (BPD): From animal models to the clinic; Steckler et al., Developing novel anxiolytics: Improving preclinical detection and clinical assessment in Volume 1, Psychiatric Disorders; Wagner et al. Huntington Disease, in Volume 2, Neurologic Disorders for further discussion regarding concepts of validity in animal models of behavioral disorders.

for potential efficacy in the clinic can be based on a strong theoretical rationale. If it is determined that the potential benefits outweigh the potential risks, then clinical concept testing may progress without supportive preclinical efficacy data.

If preclinical efficacy data are not essential for advancement to clinical development, then why bother with animal models? How one answers this question depends on projected costs of advancing to clinical trials, available resources, and the willingness to take risks. Advancement of a novel compound into development is always a gamble. It costs millions of dollars to shepherd a single NME through preclinical and Phase I safety testing and, unfortunately, the large majority of NMEs must be dropped before advancement to clinical efficacy trials; unexpected toxicities and/or undesirable pharmacokinetic properties are the most common reasons for failure. Including the cost of failures, average out-of-pocket costs for each successful NDA approval have been estimated at $403 million (in 2000 US dollars) within the pharmaceutical industry.[16] Research and development are no less expensive and the odds of success are no greater within the Government. Given cost considerations, it is not surprising that management usually requires preclinical data suggestive of efficacy before compounds are advanced to development. Thus, while FDA reviewers may not require preclinical efficacy data for their "go/no go" decisions, the financial burden of drug development makes such data important for the drug developer's "go/no go" decision process. Advancing an NME to development is regarded as less risky in the presence of promising data from animal models, even if the predictive validity of the models is unknown.

A compelling case for initiating a development project in the absence of promising preclinical efficacy data can be made when animal models are lacking (this does not appear to be the case in the field of drug addiction) or when few costs would be incurred beyond those associated with the clinical efficacy trial. The latter situation occurs when a compound that is marketed for one indication (e.g., Parkinson's disease, depression, or attention deficit hyperactivity disorder) is considered for evaluation in another (e.g., cocaine or methamphetamine addiction). Within NIDA, we also see this situation when a pharmaceutical company has made the initial investment to advance a compound through preclinical and clinical safety testing and the compound has failed for its primary indication. In fact, the most common reason for a company to initiate contact with NIDA regarding a potential collaboration is the desire to pursue drug addiction treatment as a "rescue indication." If development for the initial indication was not halted due to unacceptable toxicities or insurmountable pharmacokinetic problems, such a collaborative development project can represent an attractive opportunity for NIDA. While the first step in such a collaboration may be to evaluate the compound in animal models of addiction, promising findings may not be essential if the compound has a unique mechanism of action and a strong theoretical rationale for efficacy.

Avoiding False Positives in Drug Self-administration and Relapse Model Testing

When administered as a pretreatment to animals trained to obtain a reinforcer by operant responding (e.g., lever-pressing by a rat, nose-poking by a mouse, or key-pecking by a pigeon), high doses of virtually any central nervous system (CNS)-active drug will decrease responding for the reinforcer. This is true regardless of the nature of the reinforcer (whether it be the delivery of a food pellet or an intravenous infusion of a rewarding drug) and it stems from the ability of CNS-active drugs to cause sedation or ataxia, or to stimulate interfering behaviors, such as stereotypic grooming. For the purpose of the present discussion, this phenomenon will be referred to as a "general suppression of responding." The challenge to those who use the drug self-administration technique in evaluating potential addiction treatment medications is clear: how can favorable findings (e.g., the ability of a pretreatment medication to modify the rewarding effects of a self-administered drug) be differentiated from a general suppression of responding?

The most common approach to ruling out a general suppression of responding is to demonstrate that a potential medication decreases responding for a drug of abuse but does not suppress responding for an alternative reinforcer, such as food. For example, doses of varenicline that decrease nicotine self-administration in rats have been shown to lack effects on responding for food.[17] Likewise, doses of buprenorphine that decrease opiate self-administration in monkeys have been shown to lack effects on responding for food.[15] Many false positives can be avoided if the demonstration of such selectivity is regarded as a positive result and the absence of such selectivity is regarded as a negative result. It must be acknowledged, however, that false negatives will follow from such a rigid definition of a positive result. For example, using the same procedure that demonstrated a positive result for buprenorphine, Mello *et al.*[15] found that methadone caused a non-selective (general) suppression of responding for intravenous opiates. If one keeps in mind that medication screening is not hypothesis testing, but rather a probability endeavor to maximize the likelihood of success given limited resources, such false negatives are acceptable.

In medications discovery, we sometimes experience a conflict between our desire to achieve a low percentage of false positives (which can waste valuable drug development resources) and our desire "not to miss anything" and avoid false negatives from animal models. If we act on the former desire, we strive to achieve a high positive predictive value (defined as the number of true positives divided by the sum of all true and false positives) and if we act on the latter desire, we strive to achieve a high sensitivity (defined as the number of true positives divided by the sum of all true positives and false negatives). In general, researchers within pharmaceutical companies must acknowledge the high costs of drug development and, therefore, they strive to avoid false positives and achieve a high positive predictive value. False negatives are regarded as a necessary evil.[iii] In contrast, most researchers in academia are detached from the costs of drug development and they are more accustomed to hypothesis testing than medication screening. For both of these reasons, researchers in academia may be less accepting of false negatives. Perhaps this is why we often see no attempt to rule out a general suppression of responding in published findings from drug self-administration studies. Although this non-critical use of the drug self-administration technique in medications discovery may achieve high sensitivity, the value of this strategy is highly

---

[iii] Please refer to Markou *et al.*, Contribution of animal models and preclinical human studies to medication: Development for nicotine dependence, for further discussion of the consequences of false negative results in drug discovery.

questionable given the likelihood that most compounds with CNS activity will appear as positives.

The challenge of differentiating desirable findings from a general suppression of responding is also relevant to studies that use the increasingly popular models of relapse to drug self-administration behavior.[18] In these models, animals are trained to self-administer a drug of abuse, their level of responding is then reduced by exposure to repeated extinction sessions in which saline is substituted for the drug of abuse, and their response rates are finally assessed after exposure to "relapse triggers" such as footshock, conditioned cues, or a priming dose of the previously self-administered drug. The relapse triggers stimulate responding (drug seeking behavior) even though the drug of abuse is no longer available. In test sessions, potential medications are evaluated for efficacy in blocking the ability of the relapse triggers to simulate responding. Unfortunately, rates of responding during relapse model test sessions (like the standard drug self-administration test sessions discussed above) can be decreased by sedative drug effects or by the stimulation of interfering behaviors.

One need to only examine some of the most recently published studies using relapse models to find cases[19,20] in which investigators have touted the ability of test compounds to block relapse trigger-induced responding without attempting to rule out a general suppression of responding. Such studies often include data that show the test compound has no effect on rates of responding measured on an inactive lever (one with no programmed consequences for pressing) present within the operant chamber. While it may be tempting to conclude that adequate selectivity has been demonstrated if responding on the drug-associated lever is significantly decreased while responding on an inactive lever is not, this approach is inappropriate for at least two reasons. First, the rate of responding on the inactive lever is much lower than the rate of responding on the formerly drug-associated lever. Because test compounds can produce rate-dependent effects,[21] both increasing low rates of operant responding and decreasing high rates of operant responding at a given test compound dose, it is critically important that, in the absence of test compound, response rates for the selectivity control be at least as high as response rates on the formerly drug-associated lever. Second, it is conceivable that the few "responses" occurring on the inactive lever may represent unintentional pumping of the lever while the animal is grooming or ambulating in the chamber. Such unintentional responses could be unaltered even if intentional lever-pressing is decreased through a general suppression of responding. In fact, if a test compound were to stimulate locomotor activity in the chamber, accidental bumping of the inactive lever could increase while intentional responding on the formerly drug-associated lever could be suppressed by the interfering behavior.

Perhaps the best approach to ruling out a general suppression of responding in relapse model studies is to compare data from different relapse models. For example, in a NIDA contract study using cocaine relapse models,[22] the kappa-opioid receptor antagonist JDTic was shown to block footshock-induced responding completely at a dose that did not suppress cocaine-primed responding. Alternatively, some investigators have taken the approach of comparing data from relapse model studies with data from drug or food self-administration studies; in such cases, investigators claim that a general suppression of responding has been ruled out if a test compound reduces responding during relapse model test sessions but does not directly affect responding for drug or food. This approach appears questionable because drug or food self-administration studies always involve the delivery of a reinforcer according to a schedule that differs from the "no reinforcement" schedule that is in effect during relapse model test sessions. In operant studies, a given drug can decrease rates of responding when one schedule of reinforcement is used and increase rates of responding when a different schedule of reinforcement is used.[23] Thus, in ruling out a general suppression of responding, it can be argued that data from relapse model studies should only be compared with data from other studies that assess test compound effects on non-reinforced responding. In this vein, studies of test compound effects on responding during extinction sessions (following either food or drug self-administration training) could serve as appropriate controls.

TRANSLATIONAL RESEARCH, HUMAN LABORATORY MODELS, AND THE BRIDGE BETWEEN ANIMAL MODELS AND CLINICAL EFFICACY TRIALS

Translational research, involving the combined talents and knowledge of clinical and preclinical researchers, offers the hope of achieving greater success in clinical trials through the development of preliminary efficacy data or "proof of principle" on a compound prior to testing in a full-scale clinical trial.[iv] In fact, for quite a few years, NIDA has gathered subjective effects data as an important part of Phase IB safety interaction studies required by the FDA. In these Phase IB studies, subjects with experience with drugs of abuse are exposed to both the investigational medication and cocaine to insure that test compounds do not exacerbate the cardiovascular effects of cocaine, or affect its pharmacokinetics, given that in Phase II efficacy trials, subjects are likely to use cocaine. In these trials subjects have routinely been asked to rate the subjective effects (euphoria, anxiety, etc.) produced by the treatment drug alone as well as cocaine in the presence and absence of the potential medication in order to gain some preliminary understanding of a compound's ability to modify cocaine seeking and cocaine effects. It is not clear whether this form of "translation" has been helpful because no compound has been clearly successful in a Phase II trial.

For drug development in general, the success rate for new drug molecules in entering Phase II across indications has been low; estimated to be one in five for last few years. In medication development for cocaine abuse disorders, only a handful of new drug molecules have been developed, and few have progressed beyond Phase I for reasons of safety. One exception, ecopipam, originally evaluated as an antipsychotic, is discussed below. This difference from industry experience is attributable to the relatively small number of organizations that are both capable of, and interested in advancing new molecules from preclinical efficacy studies to safety studies necessary for filing an IND for drug abuse indications. Instead, as a result of the urgent need to evaluate

---

[iv] Please refer to McEvoy and Freudenreich, Issues in the design and conductance of clinical trials in Volume 1, Psychiatric Disorders for further description of clinical trials of candidate drugs for the treatment of behavioral disorders, and changes undergoing in clinical trial design.

compounds in clinical trials for drug abuse, many compounds that are marketed for other indications and for which a case can be made for potential clinical efficacy based on mechanism of action have been studied in small clinical trials. Because of the involvement of dopaminergic mechanisms in drug abuse and the ubiquity of the dopamine system throughout the brain, a convincing rationale can be developed for compounds with mechanisms of action that modulate or impinge on the dopaminergic circuitry, and as a result, compounds with clinical status as anticonvulsants, antidepressants, antipsychotics, and anxiolytics have been evaluated in clinical trials for cocaine dependence. Often, these compounds have been advanced to clinical trials in the absence of any relevant animal data from preclinical models of drug abuse. As a result, many Phase II failures cannot be taken to reflect a problem in the validity of animal models of addiction. These failures often occur when compounds showing initial efficacy in small laboratory studies, single site, or open label trials conducted by academic researchers fail to show efficacy in larger, double-blind and multi-site trials. This is in contrast to the industry model, in which most compounds fail in clinical trials after showing positive results in animal models. So the question of Phase II failure after achieving positive results in animal models can, at least in the field of drug abuse, be broadened to include Phase II failures after achievement of positive results in early human trials.

Human laboratory studies hold great promise as "translational" or "bridging" studies to determine, in the case of drug abuse research (1) whether effects seen in animal models will occur in humans and (2) how to evaluate drugs optimally in humans, both in terms of clinical dosing and study design. The challenge in this area of drug development is whether human laboratory studies can be designed to ask the "right" questions to insure clinical success and reduce late phase clinical failures.[24] In the development of cocaine addiction treatments, it is not clear that there is a consensus on what the "right" questions might be: craving, self-administration, side effects, etc. The predictive value of both animal paradigms and human laboratory procedures is severely hampered by the absence of even one effective medication with which to validate the models. There are a number of human clinical laboratory designs that have been developed, many of which were designed to measure abuse liability, but increasingly, human laboratory models are aimed at evaluation of potential treatment medications. These designs typically use rating and visual analog scales to measure craving, subjective effects, perceived value of cocaine, as well as physiological interactions with cocaine on cardiovascular and pharmacokinetic measures.[v] One design that has been used extensively is human laboratory cocaine self-administration, in which subjects sample various doses of cocaine in the presence of a medication, and subsequently make choices to either self-administer the available dose of cocaine or obtain a monetary voucher.[25] Other models have measured craving or wanting following stress, exposure to cues,[26] or withdrawal effects of cocaine, all of which seem to be important endpoints that theoretically should predict clinical success. Although both animal and human laboratory models predicted the success of buprenorphine for

---

[v] Please refer to Rocha *et al.*, Development of medications for heroin and cocaine addiction and regulatory aspects of abuse liability testing in this volume for further discussion regarding translational initiatives in substance abuse treatment and abuse liability potential.

opiate-dependence treatment (e.g.,[27,28]), and it can be used to validate models for opiate treatment, the lack of even one effective (FDA-approved) medication for cocaine dependence means that there have been many false positive predictions of efficacy using these models, but not a single true positive.

So have our animal models in cocaine treatment research failed us, as some have suggested? There are very few examples of compounds that failed Phase II efficacy studies despite positive animal models. Two that can be found will be described for illustrative purposes, but unfortunately, the results of both Phase II clinical trials are not published at the time of this writing, so the details and possible "reasons" for failure are not available for careful analysis and discussion. The compounds are ecopipam, a dopamine D_1 receptor antagonist, and baclofen, a $GABA_B$ receptor agonist.

Ecopipam is a dopamine D_1/D_5 receptor antagonist that was developed by Schering-Plough as an antipsychotic. There was a wealth of published data from animal models in which D_1 ligands have been implicated in the behavioral effects of cocaine. Early self-administration studies in which D_1 receptor antagonists were administered prior to a single dose of cocaine produced robust increases in responding, confirming the importance of these receptors in the reinforcing effects of cocaine.[29,30] When ecopipam was administered as a pretreatment in primate studies using the entire dose–effect curve for cocaine self-administration and other behavioral effects, it was found to produce rightward shifts in the cocaine dose–effect curve.[31,32] It should be noted that both of these early studies observed that the antagonism was surmountable and not irreversible, which was consistent with the earlier studies in rodents.

Human laboratory studies were conducted that indicated that single acute doses of ecopipam (10, 25, and 100 mg) decreased both the desire to take cocaine, and the stimulating effects of a 30 mg/kg infusion of cocaine. In addition, the euphoric and anxiogenic effects of cocaine were reported to be attenuated.[33] A later laboratory study that used a chronic dosing paradigm (100 mg for 5 days) indicated that smoked cocaine self-administration was increased, as were its subjective effects,[34] and another laboratory study reported that ecopipam (10, 25, and 100 mg) failed to alter the subjective effects of intravenous cocaine and potentiated the cardiovascular response to cocaine.[35]

Unfortunately, the results of the multi-site outpatient clinical trial were never published, but it is widely known that the trial was terminated prematurely because an interim data analysis suggested a lack of efficacy. Could its failure have been more accurately predicted by both animal and human laboratory models? Hindsight is 20/20, so a retrospective analysis can point to the slight rightward shifts in cocaine self-administration in animal models that might have predicted that subjects would simply take additional drug to counteract the medication. Similarly, it has been known for years that chronic doses of dopamine receptor antagonists produce increased receptor sensitivity. This may explain both the potentiated cocaine reinforcement following chronic administration in the human laboratory and the negative Phase II outcome. There are a number of possible issues in the interpretation of both human and animal laboratory studies.[36]

Another example that yields no simple answers as to the predictive validity of animal models is the study of baclofen, a $GABA_B$ receptor agonist and clinically available drug used as an antispasmodic and muscle relaxant. This compound was the subject of so many investigations of its effects against cocaine and other drugs of abuse that entire review articles have been written to describe them.[37,38] In general, using a number of different experimental paradigms, baclofen has been widely reported to reduce the self-administration of cocaine in animals using progressive ratio schedules,[39] second-order[40] and fixed ratio schedules[41] with little effect on food-reinforced behavior.

In a human laboratory study that preceded the clinical trial, patients were maintained on baclofen (0, 30, or 60 mg PO) for a period of 7 days. At days 3–4 and 6–7, each volunteer was challenged with smoked cocaine (0, 12, 25, and 50 mg), with subjects subsequently asked to choose to self-administer the available dose of cocaine or to receive a voucher. Baclofen at a dose of 60 mg was found to decrease self-administration of the low (12 mg) dose of cocaine. At 30 mg baclofen, the perceived value of 50 mg cocaine was also decreased.[42]

When baclofen was studied in clinical pilot study of 70 subjects, it was reported that baclofen (20 mg t.i.d. or placebo) for 16 weeks resulted in statistically significant reductions in cocaine use as compared to placebo as indicated by urine benzoylecgonine levels. The effect was most pronounced for subjects with higher levels of cocaine use at baseline, and was used to justify a full-scale multi-site study.[43] Like with ecopipam, the results of the multi-site study have not yet been published although a manuscript is in preparation and may be published soon. Baclofen did not demonstrate efficacy greater than placebo, but the reasons for its lack of efficacy are not completely clear and may be related to the subject population, side effects, or other factors yet to be determined. Whether these results could have been predicted by either animal models or human laboratory studies will be facilitated by the disclosure of the results through publication, which will permit the "bedside-to-bench" feedback that is so urgently needed. So for both ecopipam and baclofen, animal models suggested decreases in cocaine self-administration, a seemingly positive outcome for a potential treatment medication. In both cases, at least some of the human laboratory studies were consistent with animal models, but did not predict the negative Phase II clinical outcome. For ecopipam, two of the three human laboratory studies were negative, but the relationships of the effects reported in those studies to the reasons for trial failure are also unknown. The question of whether it would have been theoretically possible to predict the Phase II clinical failure if the "right" questions had been asked of the laboratory studies or if the trial had been designed differently, perhaps as a relapse prevention trial, is simply not known at this time, although the results of the baclofen trial, when published, may be illustrative. Thus the Phase II failure rate for drug abuse medications does not appear to be the sole result of failure of animal models to predict clinical outcomes accurately. Studies using translational models to extend the animal model data to ask the kinds of questions that would have predicted Phase II failure may not have occurred. So the question of whether the failure rate of clinical trials could be reduced by the judicious use of translational human laboratory studies and small efficacy trials remains to be addressed.

Our perspective is that Phase II clinical trial failures for stimulant dependence stem mainly from our inability to validate any of our animal or human laboratory models with a positive control, meaning an effective medication. The complexity of the disorder, with interactions among genetics, environment, and behavior, as well as the heterogeneity of the populations available for study, may make a successful Phase II outcome in a multi-site clinical trial a particularly difficult challenge. Human laboratory techniques have been developed to predict clinical efficacy, but like the animal models, they appear to have resulted in false positives in at least a few cases. Whether we can design human laboratory studies that will unveil the issues predictive of a successful Phase II trial remains to be determined.

Overall, there have been remarkable successes in the development of medications for opiate addiction, and there are significant challenges ahead for development of medications for stimulant dependence. NIDA has developed a highly effective infrastructure for the evaluation of the efficacy and safety of potential medications through the DPMC and its associated grantees and contractors. Animal models in the field of drug abuse are some of the best behavioral paradigms available for any CNS disorder, and drug abuse researchers have pioneered the use of translational human laboratory studies to guide Phase II evaluations. Although stimulant abuse disorders are complex and challenging, it is anticipated that these resources will ultimately lead to success.

REFERENCES

1. Johnson, R.E., Jaffe, J.H., and Fudala, P.J. (1992). A controlled trial of buprenorphine treatment for opioid dependence. *JAMA*, 267(20):2750-2755.
2. Ling, W., Charuvastra, C., Collins, J.F., Batki, S., Brown, L.S., Jr., Kintaudi, P. *et al.* (1998). Buprenorphine maintenance treatment of opiate dependence: A multicenter, randomized clinical trial. *Addiction*, 93(4):475-486.
3. Fudala, P.J., Yu, E., Macfadden, W., Boardman, C., and Chiang, C.N. (1998). Effects of buprenorphine and naloxone in morphine-stabilized opioid addicts. *Drug Alcohol Depend*, 50(1):1-8.
4. Mendelson, J., Jones, R.T., Welm, S., Baggott, M., Fernandez, I., Melby, A.K. *et al.* (1999). Buprenorphine and naloxone combinations: The effects of three dose ratios in morphine-stabilized, opiate-dependent volunteers. *Psychopharmacology (Berl)*, 141(1):37-46.
5. Harris, D.S., Jones, R.T., Welm, S., Upton, R.A., Lin, E., and Mendelson, J. (2000). Buprenorphine and naloxone co-administration in opiate-dependent patients stabilized on sublingual buprenorphine. *Drug Alcohol Depend*, 61(1):85-94.
6. Fudala, P.J., Bridge, T.P., Herbert, S., Williford, W.O., Chiang, C.N., Jones, K. *et al.* (2003). Office-based treatment of opiate addiction with a sublingual-tablet formulation of buprenorphine and naloxone. *New Engl J Med*, 349(10):949-958.
7. Comer, S.D., Sullivan, M.A., Yu, E., Rothenberg, J.L., Kleber, H.D., Kampman, K. *et al.* (2006). Injectable, sustained-release naltrexone for the treatment of opioid dependence: A randomized, placebo-controlled trial. *Arch Gen Psychiatry*, 63(2):210-218.
8. Vocci, F. and Ling, W. (2005). Medications development: Successes and challenges. *Pharmacol Ther*, 108(1):94-108.
9. Tiihonen, J., Kuoppasalmi, K., Fohr, J., Tuomola, P., Kuikanmaki, O., Vorma, H. *et al.* (2007). A comparison of aripiprazole, methylphenidate, and placebo for amphetamine dependence. *Am J Psychiatry*, 164(1):160-162.
10. Elkashef, A.M., Rawson, R.A., Smith, E., Anderson, A., Kahn, R., Pierce, V., *et al.* (2006). Bupropion for the treatment of methamphetamine dependence. Presented at the *College on Problems of Drug Dependence Meeting*, Scottsdale, Arizona.
11. Le, F.B. and Goldberg, S.R. (2005). Cannabinoid CB1 receptor antagonists as promising new medications for drug dependence. *J Pharmacol Exp Ther*, 312(3):875-883.
12. Vorel, S.R., Ashby, C.R., Jr., Paul, M., Liu, X., Hayes, R., Hagan, J.J. *et al.* (2002). Dopamine D3 receptor antagonism inhibits cocaine-seeking and cocaine-enhanced brain reward in rats. *J Neurosci*, 22(21):9595-9603.

13. Heidbreder, C. (2005). Novel pharmacotherapeutic targets for the management of drug addiction. *Eur J Pharmacol*, 526(1-3):101-112.

14. Rollema, H., Chambers, L.K., Coe, J.W., Glowa, J., Hurst, R.S., Lebel, L.A. *et al.* (2007). Pharmacological profile of the alpha4beta2 nicotinic acetylcholine receptor partial agonist varenicline, an effective smoking cessation aid. *Neuropharmacology*, 52(3):985-994.

15. Mello, N.K., Bree, M.P., and Mendelson, J.H. (1983). Comparison of buprenorphine and methadone effects on opiate self-administration in primates. *J Pharmacol Exp Ther*, 225(2):378-386.

16. DiMasi, J.A., Hansen, R.W., and Grabowski, H.G. (2003). The price of innovation: New estimates of drug development costs. *J Health Econ*, 22(2):151-185.

17. Rollema, H., Chambers, L.K., Coe, J.W., Glowa, J., Hurst, R.S., Lebel, L.A. *et al.* (2007). Pharmacological profile of the alpha4beta2 nicotinic acetylcholine receptor partial agonist varenicline, an effective smoking cessation aid. *Neuropharmacology*, 52(3):985-994.

18. Shaham, Y., Shalev, U., Lu, L., De, W.H., and Stewart, J. (2003). The reinstatement model of drug relapse: History, methodology and major findings. *Psychopharmacology (Berl)*, 168(1-2):3-20.

19. Boutrel, B., Kenny, P.J., Specio, S.E., Martin-Fardon, R., Markou, A., Koob, G.F. *et al.* (2005). Role for hypocretin in mediating stress-induced reinstatement of cocaine-seeking behavior. *Proc Natl Acad Sci USA*, 102(52):19168-19173.

20. Yao, L., McFarland, K., Fan, P., Jiang, Z., Ueda, T., and Diamond, I. (2006). Adenosine A2a blockade prevents synergy between mu-opiate and cannabinoid CB1 receptors and eliminates heroin-seeking behavior in addicted rats. *Proc Natl Acad Sci USA*, 103(20):7877-7882.

21. Dews, P.B. (1958). Studies on behavior. IV. Stimulant actions of methamphetamine. *J Pharmacol Exp Ther*, 122(1):137-147.

22. Beardsley, P.M., Howard, J.L., Shelton, K.L., and Carroll, F.I. (2005). Differential effects of the novel kappa opioid receptor antagonist, JDTic, on reinstatement of cocaine-seeking induced by footshock stressors vs cocaine primes and its antidepressant-like effects in rats. *Psychopharmacology (Berl)*, 183(1):118-126.

23. Dews, P.B. (1955). Studies on behavior. I. Differential sensitivity to pentobarbital of pecking performance in pigeons depending on the schedule of reward. *J Pharmacol Exp Ther*, 113(4):393-401.

24. Lesko, L.J. (2007). Paving the critical path: How can clinical pharmacology help achieve the vision? *Clin Pharmacol Ther*, 81(2):170-177.

25. Fischman, M.W. and Foltin, R.W. (1992). A laboratory model for evaluating potential treatment medications in humans. *NIDA Res Monogr*, 119:165-169.

26. Sinha, R., Fuse, T., Aubin, L.R., and O'Malley, S.S. (2000). Psychological stress, drug-related cues and cocaine craving. *Psychopharmacology (Berl)*, 152(2):140-148.

27. Mello, N.K., Bree, M.P., and Mendelson, J.H. (1983). Comparison of buprenorphine and methadone effects on opiate self-administration in primates. *J Pharmacol Exp Ther*, 225(2):378-386.

28. Greenwald, M.K., Schuh, K.J., Hopper, J.A., Schuster, C.R., and Johanson, C.E. (2002). Effects of buprenorphine sublingual tablet maintenance on opioid drug-seeking behavior by humans. *Psychopharmacology (Berl)*, 160(4):344-352.

29. Britton, D.R., Curzon, P., Mackenzie, R.G., Kebabian, J.W., Williams, J.E., and Kerkman, D. (1991). Evidence for involvement of both D_1 and D_2 receptors in maintaining cocaine self-administration. *Pharmacol Biochem Behav*, 39(4):911-915.

30. Corrigall, W.A. and Coen, K.M. (1991). Cocaine self-administration is increased by both D_1 and D_2 dopamine antagonists. *Pharmacol Biochem Behav*, 39(3):799-802.

31. Bergman, J., Kamien, J.B., and Spealman, R.D. (1990). Antagonism of cocaine self-administration by selective dopamine D(1) and D(2) antagonists. *Behav Pharmacol*, 1(4):355-363.

32. Spealman, R.D. (1990). Antagonism of behavioral effects of cocaine by selective dopamine receptor blockers. *Psychopharmacology (Berl)*, 101(1):142–145.

33. Romach, M.K., Glue, P., Kampman, K., Kaplan, H.L., Somer, G.R., Poole, S. *et al.* (1999). Attenuation of the euphoric effects of cocaine by the dopamine D1/D5 antagonist ecopipam (SCH 39166). *Arch Gen Psychiatry*, 56(12):1101–1106.

34. Haney, M., Ward, A.S., Foltin, R.W., and Fischman, M.W. (2001). Effects of ecopipam, a selective dopamine D1 antagonist, on smoked cocaine self-administration by humans. *Psychopharmacology (Berl)*, 155(4):330–337.

35. Nann-Vernotica, E., Donny, E.C., Bigelow, G.E., and Walsh, S.L. (2001). Repeated administration of the D1/5 antagonist ecopipam fails to attenuate the subjective effects of cocaine. *Psychopharmacology (Berl)*, 155(4):338–347.

36. Cance-Katz, E.F., Kosten, T.A., and Kosten, T.R. (2001). Going from the bedside back to the bench with ecopipam: A new strategy for cocaine pharmacotherapy development. *Psychopharmacology (Berl)*, 155(4):327–329.

37. Cousins, M.S., Roberts, D.C., and de, W.H. (2002). GABA(B) receptor agonists for the treatment of drug addiction: A review of recent findings. *Drug Alcohol Depend*, 65(3):209–220.

38. Roberts, D.C. (2005). Preclinical evidence for GABAB agonists as a pharmacotherapy for cocaine addiction. *Physiol Behav*, 86(1–2):18–20.

39. Roberts, D.C., Andrews, M.M., and Vickers, G.J. (1996). Baclofen attenuates the reinforcing effects of cocaine in rats. *Neuropsychopharmacology*, 15(4):417–423.

40. Di, C.P. and Everitt, B.J. (2003). The GABA(B) receptor agonist baclofen attenuates cocaine- and heroin-seeking behavior by rats. *Neuropsychopharmacology*, 28(3):510–518.

41. Campbell, U.C., Lac, S.T., and Carroll, M.E. (1999). Effects of baclofen on maintenance and reinstatement of intravenous cocaine self-administration in rats. *Psychopharmacology (Berl)*, 143(2):209–214.

42. Haney, M., Hart, C.L., and Foltin, R.W. (2006). Effects of baclofen on cocaine self-administration: Opioid- and nonopioid-dependent volunteers. *Neuropsychopharmacology*, 31(8):1814–1821.

43. Shoptaw, S., Yang, X., Rotheram-Fuller, E.J., Hsieh, Y.C., Kintaudi, P.C., Charuvastra, V.C. *et al.* (2003). Randomized placebo-controlled trial of baclofen for cocaine dependence: Preliminary effects for individuals with chronic patterns of cocaine use. *J Clin Psychiatry*, 64(12):1440–1448.

CHAPTER 3

Issues in Designing and Conducting Clinical Trials for Reward Disorders: A Clinical View

Tracie J. Gardner[1,2], Therese A. Kosten[1,2] and Thomas R. Kosten[1,2]

[1]Division of Alcohol and Addictive Disorders, Menninger Department of Psychiatry and Behavioral Sciences, Baylor College of Medicine, Michael E. DeBakey VA Medical Center (151), Houston, TX, USA
[2]Veterans Affairs Medical Center, Houston, TX, USA

INTRODUCTION

What are reward deficit disorders? For those individuals who abuse substances, reward deficits appear to be a product of their substance abuse and these resulting states might be considered reward deficit disorders. Whether the initial urge to use these substances is a reward disorder because the normal rewards that life offers are insufficient is an interesting question that presents a host of challenges when designing clinical studies to test such a hypothesis. This chapter will examine designing and conducting clinical trials in patients who might be considered to have reward deficit disorders and attempt to enrich the design of these clinical trials by drawing on the pre-clinical and clinical perspectives of this conceptualization of a reward deficit.

Animal and Translational Models for CNS Drug Discovery.
Vol. 3 of 3: Reward Deficit Disorders
Robert McArthur and Franco Borsini (eds), Academic Press, 2008

UTILITY OF RESEARCH USING ANIMAL MODELS

Research using animal models has provided enormous input into developing clinical treatments for drug abuse. This type of research can be used in at least two ways: (1) to investigate and delineate the neuropharmacological effects of the drug exposure and (2) to test whether a pharmacological agent can alter the behavioral effects of the drug in a manner consistent with reduction of drug intake in addicts. While these approaches to illuminating potential new treatments for drug abuse appear straightforward, there are many difficulties inherent in translating findings from animal research to the treatment of humans. In this section, we will discuss the issues involved in animal research including the advantages and limitations of this approach. We will also briefly describe the behavioral techniques used in this field and present the ways in which evidence gained from this approach can inform clinical trials and examine new treatment strategies for addiction.

Advantages of Animal Research

The benefits of animal research including behavioral approaches are that it allows control over various factors such as environmental and genetic influences and the degree of drug exposure.[1,2] This more homogenous set of subjects permits more statistically powerful experiments in animals than humans. In addition, various manipulations directed at investigating neuropharmacological processes involved in psychoactive drug effects can be used in animals but not in humans. While neuro-imaging or obtaining neurotransmitter levels in cerebral spinal fluid via spinal taps, approximate these animal measures, human procedures are costly and have many limitations.

Limitations common to all neuropharmacological human research are the heterogeneity of the subject pool and the inability to administer most psychoactive drugs either acutely or chronically to drug-naïve humans. Because the humans must be previously drug abusers for inclusion in drug administration studies, we cannot examine causal effects. Any characteristics found to correlate with addiction in humans may reflect factors related to vulnerability to addiction just as readily as being due to the consequences of drug use. Further, it is difficult, if not impossible, to study the development of addiction in humans. Such issues can be addressed in animal research. While species differences constrain direct translation of findings from animal research to the human condition, common effects found in similar tests carried out in rodents and in non-human primates would give strong support for their applicability.

Drug Exposure

Control over the length and amount of drug exposure is a great advantage to animal research, where various parameters of drug exposure, such as dose, route of administration, and timing and length of exposure can be varied systematically across studies. The route of administration in animal research can be the same as the route used and preferred by addicts – inhalation for nicotine, oral for alcohol, and intravenous for cocaine and opiates. To mimic the human situation further, the drug can be self-administered as opposed to administered passively by the investigator,[3-5] although this increases the difficulty of conducting the experiments and it may not be necessary.[6] Another option for enhancing human relevance is to study drug administration during the adolescent developmental stage in which drug exposure begins in humans. This is a time when certain areas of the brain are still differentiating.

Another aspect of the neuropharmacological mechanisms involved in addiction that animal models can systematically address is the many neural, hormonal, and behavioral effects after chronic exposure. At least two processes can occur with chronic exposure – tolerance and sensitization. The former reflects a reduction in a specific effect with chronic drug exposure so that *more* drug is needed to achieve the original degree of effect. The latter reflects an enhancement in a specific effect so that *less* drug is needed to obtain the original effect. Both tolerance and sensitization can occur at the same time but be expressed in different effects. For example, chronic exposure to opiates is often associated with a tolerance or a decrease in analgesic effects of the drug but sensitization or an increase in its locomotor effects. The dose of the drug affects whether tolerance or sensitization occurs. For example, low doses of opiates or alcohol cause locomotor stimulation that increases or sensitizes with chronic exposure. Such effects are not associated with chronic administration of higher doses. Whether tolerance or sensitization is seen can also depend upon the pattern or timing of the drug exposure. For example, chronic intermittent administration of psychostimulants leads to sensitization of locomotor effects whereas chronic continuous administration leads to tolerance. It is believed that the subjective effects of drugs that promote addiction are associated with tolerance so that more drug or higher doses are needed over time to achieve the same level of effect. One prominent theory of the development of addiction posits that while tolerance to the rewarding effects of drugs occurs with chronic exposure, sensitization of effects related to "craving" show sensitization.[7] Thus, it is important to understand the mechanisms that contribute to the processes of tolerance and sensitization, which can only be done fully in animals.

Behavioral Techniques in Animal Research

Various behavioral techniques have been used to examine the effects of psychoactive drugs or to test the potential of new treatments. The degree to which these various measures obtained reflect processes involved in addiction in humans likely varies across behavioral procedures. Animal studies can yield assessments of unconditioned effects of drugs as well as complex conditioned and learned effects. Unconditioned effects include withdrawal signs and locomotor activity, while conditioned effects include operant self-administration, drug discrimination, place preference (or aversion), and taste aversion. These various procedures differ in the amount of time necessary to run the study, the complexity and costs of the equipment needed, the expertise required to conduct the study, and the number of animals needed. There are advantages and disadvantages to each behavioral technique used in animal research on drug addiction, and convergence of multiple approaches will give the best estimation of likely effects in humans.

Drug Self-administration

The animal behavioral procedure with the greatest face validity for human addiction is drug self-administration. In drug self-administration, the animal can be implanted with a chronic indwelling venous catheter and is commonly used for studies of cocaine and opiates. Self-administration orally is most relevant for studies on alcohol. In operant self-administration procedures, the rat makes a voluntary response to obtain the drug just like drug addicts do. The voluntary response can be pressing a lever in an operant chamber. If this behavior increases due to the consequences of the drug, it is inferred that the drug was positively reinforcing.[8,9] Many self-administered drugs serve as reinforcers as evidenced by the fact that they increase the likelihood that behaviors preceding the drug infusion occur again. Indeed, more than 20 drugs self-administered by humans serve as reinforcers in animals supporting its use as a model of abuse liability.[10][i]

There are several variations of drug self-administration procedures and these can model various aspects of addiction.[ii] It is thought that studying the development or acquisition of self-administration is a model of vulnerability to addiction.[11] This variation on the drug self-administration procedure can be used to investigate genetic or environmental factors that enhance (or retard) predisposition to addiction. For example, we showed that genetic (or strain) differences[12] and early life stress[13] alter acquisition of cocaine self-administration.

Once the animal has acquired the operant of self-administration and is responding at stable levels over sessions, the maintenance of this behavior can be examined. This is particularly useful to determine whether the behavior is maintained in response to behavioral or pharmacological challenges. An example of a behavioral challenge is to alter the schedule of drug reinforcement, such as employing a progressive ratio schedule of reinforcement. In this method, the number of lever presses (or other operant) required to obtain the drug delivery is increased after each reinforcement so that the animal needs to work harder and harder to receive the drug.[14-16] The point at which responding ceases is known as the "break point." This procedure may reflect the degree of motivation to obtain the drug reinforcement and can be a useful tool for investigating potential pharmacological treatments; that is, if the breaking point is reduced with a particular treatment agent, this would support its use for addiction. In many studies on the maintenance of drug self-administration, the schedule of reinforcement is kept constant but the dose of the drug is varied. Under the commonly used fixed ratio schedule condition, the varying of the dose results in an inverted U-shaped dose–response function. In this situation, as dose increases from that which supports the greatest response level, a decrease in responding is seen; this is known as the descending limb of the dose–response curve. There are difficulties in interpreting effects of a pharmacological

---

[i] Please refer to Rocha *et al.*, Development of medications for heroin and cocaine addiction and regulatory aspects of abuse liability testing, in this volume for further discussion of the use of operant procedures to assess abuse liability potential of novel drugs.

[ii] Please refer to Koob, The role of animal models in reward deficit disorders: Views of academia, in this volume for detailed descriptions of operant procedures used in animal models of reward deficit behaviors.

treatment agent using this method. An agent that increases responding may suggest that it enhances the effects of the abused drug or it may suggest that it decreases the effects of the abused drug because the animal is responding as if the dose were lowered. Mello and Negus have argued that the best predictor for a potential treatment agent is one that shifts the whole dose–response function downward.[17] Perhaps the best approach is to use both approaches to test a potential treatment agent.

Arguably, the most difficult aspect of treating addiction is to prevent relapse to drug use after cessation of use. In recent years, this aspect of addiction has been addressed in animal studies through the use of reinstatement of drug self-administration behavior.[18-20] In this procedure, an animal well trained to self-administer drug in an operant chamber is allowed to extinguish lever pressing by replacing the drug with saline. As in other learning procedures, the response (lever press) can be recovered after extinction if the animal is provided with non-contingent exposure to the unconditioned stimulus (i.e., drug), particularly if the context has not been altered.[21] After extinction, responding can be reinstated not only by non-contingent drug infusions, but by exposure to cues previously conditioned to drug delivery or to foot shock even though the drug is not available.[22-25] Reinstatement of self-administration behavior in animals necessarily uses active extinction. Altering the contingency between lever press responding and cocaine availability is used in the animal procedures in order to reduce response levels so that reinstatement of the behavior can be seen. Since addicts usually do not actively engage in such extinction processes, this limits the direct applicability of the reinstatement procedure. Nonetheless, it is necessary to use active extinction in animals and important information can be gained. It should be noted however that neither the procedural nor predictive validity of the reinstatement procedure in animals for drug craving in human addicts has been demonstrated.[26]

HUMAN TESTING PARADIGMS FOR NEW MEDICATIONS

Animal models offer considerable insight into the potential utility of new medications for the treatment of drug use disorders, but human studies include a combination of biological, psychological, and social determinants that complicate easy translation of these animal findings. Whereas animal research allows for control over environmental and genetic influences and the degree of drug exposure, human drug users cannot be controlled in such a manner. The diversity of drug users therefore results in major methodological problems in designing clinical trials of treatments for drug dependence and abuse. We will now discuss the various paradigms for testing medications, which include new methods that have the potential to offer surrogate efficacy data as well as medical safety data.

Human Laboratory

The human laboratory setting in which cocaine or amphetamine is administered to volunteer subjects has been a critical paradigm for testing potential pharmacotherapies for stimulant dependence.[27-31] Variations on this paradigm have used visual, tactile, aural, or cognitive cues to induce craving for these abused substances. In both experimental settings the outcome measures have been subjective responses such as euphoria, unpleasant feelings, or craving itself, as well as estimates of how much the drug is worth to the subject (e.g., dollar value). The induction of craving for more cocaine after a small to modest dose of cocaine is called the priming effect, and modulation of this priming effect can be an important role for a treatment medication in reducing relapse.[32] This reduction in relapse would occur by preventing a "slip," that is a single use of cocaine in a patient who wants to remain abstinent, from leading into a full relapse to binge cocaine usage.

Self-administration

To measure outcomes more precisely in this human laboratory model, self-administration has been introduced. In that paradigm the subject can self-administer cocaine or amphetamine repeatedly within a range dictated by medical safety considerations. The subject is offered the alternative of getting cocaine or various other rewards that have a range of monetary values. The subject is thereby not only asked to estimate a worth of the cocaine, but to actually choose to get that amount or get the cocaine. In these paradigms the behavior of drug taking can be more clearly approximated and a medication that might block the effects of cocaine could be detected. This blocking effect would presumably lead the subject to prefer the alternative reward over the cocaine after the first test dose, when the medication is present, since the reinforcing effects of cocaine would be reduced.[27] While this model has theoretical appeal and has shown the expected subject behaviors with various doses of cocaine, and it might be tested with two particular blockers: for opiates with naltrexone and perhaps for cocaine with a cocaine vaccine.

In all of these paradigms the key outcome of cocaine or amphetamine interactions with potential pharmacotherapies yields not only surrogate efficacy data, but also medical safety data. Cardiovascular measures in particular can be carefully monitored after both acute and sub-chronic dosing with potential medications. The baseline effects of these treatment medications can be assessed in escalating dose regimes and then dose–response evaluations using escalating doses of cocaine or amphetamine can be examined. Subjective responses may also be important to assess dysphoric interactions between the medication and the abused drugs. These reactions might help in reducing stimulant abuse, although they might also discourage compliance with the medication. Overall, this is a powerful paradigm for medication development because of its potential to provide the clinical trials process with information about how the outpatients on the new medication might respond if they take a stimulant. Its utility as a rapid screening procedure for eliminating medications from further outpatient testing has yet to be demonstrated, but this may be a future use of these highly controlled paradigms as we obtain gold standard agents with demonstrated efficacy in outpatient trials.

Neuroimaging

A newer technology for human laboratory assessment of potential medications is neuroimaging of either functional activity or receptor and transporter occupancy.[33] Functional activity can involve either cerebral blood flow (CBF) or metabolic activity using fluorodeoxyglucose (FDG). The use of FDG as a medication development strategy has been examined in a study of selegiline combined with cocaine administration. In this study selegiline (Eldepryl®) reduced the euphoria from acute cocaine administration, and positron emission tomography (PET) imaging using FDG showed that the cocaine induced changes in metabolic activity were blocked by the administration of selegiline.[34] This surrogate marker provided an interesting correlate of the attenuation of cocaine's subjective effects, since other outpatient work has suggested that selegiline might reduce cocaine abuse in outpatients. Because similar studies of subjective effects alone have not had corresponding predictive validity for outpatient efficacy, these neuroimaging measures may have promise as a more rapid screening tool for medications.

Another medications development approach using neuroimaging focuses on the CBF defects that have been observed in cocaine abusers and linked to neuropsychological deficits even during sustained abstinence.[35-43] These CBF defects may be responsive to pharmacotherapy. The therapeutic implication is that by resolving these CBF defects cognitive functioning might improve and the response to cognitive behavioral therapies (CBT) also enhanced.

A more speculative use of neuroimaging involves "receptor" occupancy for medication dosing to provide optimal individual brain binding in humans. This approach to establishing what doses of medication might be needed for potential efficacy compared to what doses might produce intolerable side effects is particularly relevant to new compounds that might block dopamine receptors or bind to dopamine transporters for substitution agents and "renormalization" of abnormal up-regulation of transporters.[44][iii]

METHODOLOGICAL ISSUES IN CLINICAL TRIALS

In the previous section we discussed human laboratory models involving self-administration and neuroimaging as human testing paradigms for new medications. We will now turn our attention to outpatient, randomized clinical trials, the gold standard for testing medication efficacy. We will discuss selection of trial subjects, discuss limitations in study design, and address how clinical trial endpoints could be improved.

Target Population and Subject Selection

One of the primary objectives of clinical trials is to provide an accurate and reliable clinical evaluation of a study drug for a target population. A representative sample of the target population is selected from which statistical and clinical inference will be drawn. Diagnoses, patterns of use, effects or consequences of use, and psychiatric

---

[iii] Please refer to McEvoy and Freudenreich, Issues in the design and conductance of clinical trials, in Volume 1, Psychiatric Disorders, for further discussion of changes in traditional clinical trial design and the incorporation of techniques such as neuroimaging and micro-dosing to determine adequate drug exposure in subjects.

comorbidities vary considerably among substance abusers, resulting in a very hetero-geneous population. Ignoring such heterogeneity when selecting a study sample will reduce the potential for showing the efficacy of a new treatment. Conversely, over-selecting a very homogeneous sample may limit the ability to generalize findings to community substance abusers and reduce the potential for replication of treatment results in other populations. Thoughtful consideration of these characteristics is there-fore imperative in selection of appropriate study subjects.

One of the first considerations for subject selection is defining the target popu-lation. Subject selection into a study may depend simply on self-reported use of a particular substance or follow stricter DSM-IV criteria for diagnoses of abuse or dependence.[iv] Abuse of multiple substances is common. As an example, the 2005 National Survey on Drug Use and Health reports strong associations of illicit drug use with cigarette smoking and alcohol use. In the past month among adolescents aged 12–17 years, illicit drug use occurred in 60% of those who drank 5 or more days or 5 or more drinks on the same occasion compared to 5% in non-drinkers.[45] Studies often target a primary substance of abuse and allow for secondary abuse or use of drugs or alcohol. Route of administration (i.e., oral, inhaled, smoked, injected) further contributes to sample diversity. Restriction of the study sample to a primary route or stratification by route of administration may be considered at randomization. With proper planning in study design, differences between strata may also be detected in statistical analysis. Frequency of substance use is also an important consideration in selecting an appropriate study population and target population. Frequency of use can lead to selection of very different samples, for example, the casual alcohol teetotaler is drawn from a different population than the 5 drink/day abuser.

In subject selection other characteristics leading to relapse are critical such as physical dependence and tolerance versus simply drug craving or a desire to "get high." Physical dependence in alcohol or opiate abusers may require detoxification prior to commencement of certain pharmacotherapies such as naltrexone for opi-ates, since it precipitates withdrawal. Drug craving or desire to "get high" as targets for treatment and relapse prevention might lead to selection of subjects in whom stress-related triggers drug seeking, or high baseline craving are sought. The Yale Cocaine Craving Scale, for example, can be used to select subjects with an intense desire for cocaine.[46] Alternatively, a broad-based scale such as the addiction severity index (ASI) is a useful tool in measuring the severity of problems associated with drug depen-dence and for selecting high or low severity patients for a particular intervention.[47]

The rates of depression, attention deficit hyperactivity disorder (ADHD), and anti-social personality disorders are significantly higher in stimulant abusers than com-munity controls.[48,49] Because psychiatric disorders may increase the risk for drug use (e.g., individuals may self-medicate to ease psychiatric symptomology), clinical tri-als need to address both the stimulant addiction and the comorbid disorder. Certain pharmacotherapies may be particularly useful for stimulant abusers with comorbid

---

[iv] Please refer to the *Diagnostic and Statistical Manual of Mental Disorders*, 4th edition – Text Revision (DSMIV-TR), or *The International Statistical Classification of Diseases and Related Health Problems*, 10th revision, published by the American Psychiatric Association and the World Health Organization, respectively, for current diagnostic criteria manuals in use.

psychopathology. For instance, treatment with antidepressants reduce depressive symptoms, cocaine use and craving in depressed cocaine addicts.[50,51] Also, methylphenidate has been reported to be effective in treating cocaine addicts with ADHD.[52] Unfortunately, psychiatric comorbidity has not yet shown a specific prognostic significance for behavioral therapies. Studies have used the Structured Clinical Interview for *Diagnostic and Statistical Manual of Mental Disorders* to assess current major psychiatric disorders at study screening.[53,54]

The subject's motivation for treatment may also affect retention and other aspects of treatment outcome. Social, legal, or physical stressors may play a role in motivating patients to enter a treatment program or a clinical trial; however, when the stressor passes, the subject may recall the positive aspects of their use and drop out of a trial. Involvement of family members, friends, employers, or significant others may encourage the participant to stay in the trial by offering an additional level of accountability.

General Design Issues

Research questions should be clearly stated and relate the research question to one or more of the following specific target areas for which pharmacotherapy can be beneficial: detoxification, maintenance treatment, reversal of acute overdose, maintenance of abstinence, prevention of relapse after a period of abstinence, reduction of craving, and normalization or stabilization of neurobiological dysfunction induced by chronic drug use. These target areas are not universally applicable to all classes of drugs of abuse, but the investigator needs to clearly identify the research question and the design best suited for that question. Study subject selection, study design choice, and determination of primary outcome measures will be influenced by the relationship between a specific research question and the specific medication effect to be studied.

Outpatient clinical trials remain the standard approach to assessing efficacy of a medication. While many principles for conducting randomized placebo controlled clinical trials in psychopharmacology apply to these studies, some specific considerations are relevant to outcome measures that are not found in other areas.[55] Urine toxicology is a most informative outcome that can be analyzed with both quantitative and qualitative approaches. Urine is typically obtained three times per week for maximum sensitivity to repeated stimulant or opiate use based on the duration that detectable metabolite levels remain after use. Analyses are most frequently done on the cocaine metabolite benzoylecognine. A cutoff score of 300 ng/ml, or any level above this, is considered an indication of cocaine use within the last 3 days. More complex analyses have been proposed using quantitative levels determined either directly with gas chromotography–mass spectroscopy or indirectly with semi-quantitation of immunoassays. This semi-quantitation can be combined with self-reported use and estimates can be made of new use of cocaine by comparing to urine levels obtained prior to the urine sample. With three times weekly toxicologies a heavy daily cocaine abuser can stop using cocaine for 2 or 3 days and yet still have a positive urine; as for example being positive on Monday and Wednesday when the last use was on Sunday.[56,57] While the goal of treatment is often complete abstinence, the sensitivity of these urine tests can be enhanced by these data manipulations. Thus, self-reported decreases in stimulant use may be important as a treatment outcome even when the goal may be abstinence initiation. Treatment retention is also critical in abstinence initiation, in order to keep the patient available for intervention.

In these outpatient studies relapse prevention is a conceptual outcome that follows abstinence initiation. Relapse as defined by recurrent use or dependence after "sustained abstinence" first requires a definition of sustained abstinence, particularly among the binge users of stimulants. These patients may use drugs weekly or even less often, in binges that last up to several days. While meeting DSM-IV criteria for stimulant dependence, during the periods of 5–10 days of non-use the patient can manifest reasonable psychosocial functioning. However, when after 5–10 days the patient returns to a binge of use, this return is not a relapse. Simple definitions of sustained abstinence can be defined as a period of cocaine- or amphetamine-free urines that lasts three, four or perhaps ten times longer than the typical inter-binge interval, but neurobiology may provide a better definition of this state of recovery. Neurobiology can be used to define abnormalities in brain function that need to be renormalized before the patient has fully entered recovery, and treatment outcome can then be focused on relapse prevention.

For current investigations an important prognostic stratification is evolving based on sustained abstinence before treatment entry rather than needing help in abstinence initiation. Patients who are abstinent during the 2–5 weeks prior to entering a medication trial and who exhibit other psychosocial indicators of treatment readiness have better treatment outcomes compared to patients who continue to use up to their entry into treatment.[58-60] Longer-term relapse prevention has also been an area where psychotherapy may synergize with pharmacotherapy.[61] For example, sustained abstinence with desipramine (Norpramin®) treatment for cocaine dependence was enhanced by CBT when examined at 6 and 12 month follow-up. Relapse was significantly higher after attaining abstinence with the medication alone than with both medication and the behavioral therapy.

Outcome measures such as retention in treatment and compliance with treatment programs are crucial to the success of clinical trials. Missing values or drop-outs may introduce bias and be related to the duration of treatment, the nature of the disease and the effectiveness, and/or toxicity of the study medication. Retention may be improved by incorporating behavioral interventions into substance abuse trials. Two types of intensive behavioral therapies, cognitive and contingency management (CM) therapies, are often effectively used to retain patients in clinical trials and positively affect abstinence.[62] These therapies can form the platform for any pharmacotherapy trial in order to engage the patient and facilitate more long-term changes including prevention of relapse.[63,64]

Using positive contingencies to initiate abstinence and prevent relapse has been quite successful for managing patients and improving treatment outcomes in individuals who abuse substances such as cocaine and amphetamines,[65-70] opiates,[71,72] and alcohol.[53,73,74] The goal of this approach has been to decrease behavior maintained by drug reinforcers and increase behavior maintained by non-drug reinforcers by presenting rewards contingent upon documented drug abstinence (positive contingencies) and withdrawing privileges contingent upon documented drug use (negative contingencies). There is, however, a significantly higher cost associated with the incentives group versus usual care group.[75]

CBT is also an efficacious intervention for the treatment of substance use disorders and offers an additional approach to engaging and retaining individuals in clinical trials. CBT has been examined in conjunction with pharmacotherapy to evaluate length of treatment, drug-free urinalyses, and reduction of alcohol and cocaine craving. In a recent study, CBT treated subjects remained in treatment longer than subjects who received both disulfiram (Antabuse®)/CBT and naltrexone (Revia®)/CBT.[76] In a study comparing CBT to CM, CM was efficacious during treatment application; where CM may be useful in engaging substance users, retaining them in treatment and helping them achieve abstinence, CBT has comparable longer-term outcomes.[77] Results of previous research also suggest that cognitive deficits predict low retention in outpatient CBT treatment programs for cocaine dependence.[78,79] Future studies should further examine the potential impact of differences in cognitive functioning on treatment outcomes.

Statistical Issues

Clinical trials evaluating the pharmacological treatment of drug abuse and dependence commonly suffer from disproportionately high sample attrition and treatment non-compliance. As a result, data sets often contain many gaps that can distort the characteristics of the underlying data. Since clinical trials often collect repeated measures, missing data presents problems in analyses of variance or endpoint analyses. The two statistical methods that are employed include the frequentist approach and the Bayesian approach. Though the use of Bayesian methods for analyzing data has increased in frequency in clinical research,[80-83] we will mainly focus on traditional methodologies for this discussion.[v]

It is impossible to address all questions with one trial, therefore identification of primary and secondary response variables should be chosen up front. Once response variables are chosen then the outcomes of treatment are defined for the intent of showing efficacy or safety.[84] The primary measures of pharmacological efficacy should be linked to the known or presumed effect of the investigational medication.

Sample size requirements vary based on the selected design; however, it should be reflective of the intent to show a clinically meaningful difference between an experimental treatment and the comparison group (which may be active placebo receiving standard therapy). Again, based on the significant drop-out rate anticipated in substance abuse trials, practical considerations for obtaining and retaining the necessary subjects should be made. As stated previously, use of behavioral therapies such as CM and CBT along with medical supervision work well in keeping difficult treatment populations engaged in therapy.

Interim analysis and data monitoring are important to discuss prior to starting any clinical investigation. Since the statistical model is based on the probability that the test statistic falls within a particular range, assuming that the null hypothesis is true is contingent upon how the experiment is done. Studies powered to detect an effect

---

[v] Please refer to McEvoy and Freudenreich, Issues in the design and conductance of clinical trials, in Volume 1, Psychiatric Disorders, for further discussion regarding statistical issues in clinical trial design.

using 100 subjects may therefore not be robust enough at an interim stage of analysis with accrual of only 50 subjects. Use of Bayesian methodology in clinical research allows for smaller, less expensive trials, and provides more leniency in evaluating outcomes as data accumulates compared to traditional methods that establish stopping rules in advance.[85] The *P*-value, a mainstay of inference, however, is lost with use of this method.

Two types of analyses need consideration: intent to treat (ITT) and efficacy analysis. The ITT includes all subjects regardless of treatment compliance, and is a conservative method for deciding that a difference exists between two groups. It may underestimate the magnitude of the effect for a new treatment, because it does not require that the patients have gotten an adequate dose of the treatment to have any efficacy.[86] An efficacy analysis includes only subjects who remained in the clinical trial a sufficient amount of time to allow them to have gotten some benefit from the treatment. For example, a medication that takes 3 weeks to reach therapeutic levels or a vaccination series that takes at least 2 months to attain therapeutic antibody levels would conduct efficacy analyses only on subjects who completed at least 3 weeks or 2 months of treatment, respectively.

A variety of models can be employed to assess outcome, dependent upon the type of data, the method of collection, and the intent of the measure. For example, differences between groups may be analyzed using chi-square test (categorical data) or Student's *t*-test or a non-parametric Mann–Whitney *U*-test (continuous data). Multivariate analysis of variance is useful in analyzing repeated measures, that is, consequences of specific treatments across several scales and time points. Analysis of co-variance (ANCOVA) may be used to examine primary outcome measures such as mean drinks per drinking day and percent days abstinent, using the baseline variables as covariates. Repeated measures, ANCOVA are useful in analyzing subjective measures such as craving; biological markers may undergo log-transformation to obtain values used in repeated measures ANCOVA.[53] Random regression models or hierarchical linear modeling (HLM) have been used to analyze urine toxicology in a cocaine vaccine trial. The HLM approach of modeling repeated measures allowed for unbalanced designs with missing data, intra-subject serial correlation, and unequal variance and co-variance structures over time.[86] The Wilcoxon rank sum test may be useful in analyzing a differential number of subjects within-groups who experience multiple relapse events, as was done in an alcohol dependence study of naltrexone and CBT compared to naltrexone and motivational enhancement.[53] Time-to-event analyses can also be of particular use and the Kaplan–Meier log-rank survival analysis, offers a useful time-to-event analysis of time to relapse.[72]

CONCLUSIONS

Drug addiction reflects the complex interactions of social, behavioral, genetic, and physiological facts. Some of these factors can be examined by manipulation in animals. Understanding the factors that promote the etiology of addiction is an important goal of animal research in this field; because many individuals exposed to psychoactive drugs do not move on to develop addiction. Despite the limitations inherent in the ability to translate animal research findings to humans, much can be learned and used to direct new approaches to the prevention and treatment of addiction. Application of human laboratory approaches, as either a precursor or supplement to traditional randomized, clinical trials, can help bridge the "bench to bedside" chasm. These human laboratory studies offer a specific translational approach to applying knowledge learned from animal research to human pharmacotherapy trials. Methods of subject retention will continue to offer added benefit for study adherence and ultimately robust statistical analyses. Finally, Bayesian modeling may reduce the numbers of patients needed for enrollment in early clinical trials testing new medications.

ACKNOWLEDGMENTS

Supported by National Institute on Drug Abuse grants K05 DA 0454 (TRK), P50-DA18197, and the Veterans Administration (VA) VISN 16 Mental Illness Research, Education and Clinical Center (MIRECC), and VA National Substance Use Disorders Quality Enhancement Research Initiative (QUERI).

REFERENCES

1. Gallup, G.G. and Suarez, S.D. (1985). Alternatives to the use of animals in psychological research. *Am Psychol*, 40:1104–1111.
2. Miller, N.E. (1985). The value of behavioral research on animals. *Am Psychol*, 40:423–440.
3. DeMontis, M.G., Co, C., Dworkin, S.I., and Smith, J.E. (1998). Modifications of dopamine D1 receptor complex in rats self-administering cocaine. *Eur J Pharmacol*, 362:9–15.
4. Dworkin, S.I., Mirkis, S., and Smith, J.E. (1995). Response-dependent versus response-independent presentation of cocaine: Differences in the lethal effects of the drug. *Psychopharmacology*, 117:262–266.
5. Hemby, S.E., Co, C., Koves, T.R., Smith, J.E., and Dworkin, S.I. (1997). Differences in extracellular dopamine concentrations in the nucleus accumbens during response-dependent and response-independent cocaine administration in the rat. *Psychopharmacology*, 133:7–16.
6. Self, D.W., McClenahan, A.W., Beitner-Johnson, D., Terwilliger, R.Z., and Nestler, E.J. (1995). Biochemical adaptations in the mesolimbic dopamine system in response to heroin self-administration. *Synapse*, 21:312–318.
7. Robinson, T.E. and Berridge, K.C. (1993). The neural basis of drug craving: An incentive sensitization theory of addiction. *Brain Res Rev*, 18:247–291.
8. Pickens, R., Meisch, R.A., and Thompson, T. (1978). Drug self-administration: An analysis of the reinforcing effects of drugs. In Iverson, L., Iverson, S., and Snyder, S. (eds.), *Handbook of Psychopharmacology*, Vol. 1. Plenum Press, New York, pp. 1–37.
9. Schuster, C.R. and Thompson, T. (1969). Self-administration of and behavioral dependence on drugs. *Annual Rev Pharmacol*, 9:483–502.
10. Collins, R.J., Weeks, J.R., Cooper, M.M., Good, P.I., and Russell, R.R. (1984). Prediction of abuse liability of drugs using IV self-administration by rats. *Psychopharmacology*, 82:6–13.
11. Deminiere, J.M., Piazza, P.V., LeMoal, M., and Simon, H. (1989). Experimental approach to individual vulnerability to psychostimulant addiction. *Neurosci Biobehav Rev*, 13:141–147.
12. Kosten, T.A., Miserendino, M.J.D., Haile, C.N., DeCaprio, J.L., Jatlow, P.I., and Nestler, E.J. (1997). Acquisition and maintenance of intravenous cocaine self-administration in Lewis and Fischer inbred rat strains. *Brain Res*, 778:418–429.

13. Kosten, T.A., Miserendino, M.J.D., and Kehoe, P. (2000). Enhanced acquisition of cocaine self-administration in adult rats with neonatal isolation stress experience. *Brain Res*, 875:44–50.

14. Hodos, W. (1961). Progessive ratio as a measure of reward. *Science*, 134:943–944.

15. Richardson, N.R. and Roberts, D.C. (1996). Progressive ratio schedules in drug self-administration studies in rats: A method to evaluate reinforcing efficacy. *J Neurosci Meth*, 66:1–11.

16. Rodefer, J.S. and Carroll, M.E. (1996). Progressive ratio and behavioral economic evaluation of the reinforcing efficacy of orally delivered phencyclidine and ethanol in monkeys: Effects of feeding conditions. *Psychopharmacology*, 128:265–273.

17. Mello, N.K. and Negus, S.S. (1996). Preclinical evaluation of pharmacotherapies for treatment of cocaine and opioid abuse using drug self-administration procedures. *Neuropsychopharmacology*, 14:375–424.

18. deWit, H. and Stewart, J. (1981). Reinstatement of cocaine-reinforced responding in the rat. *Psychopharmacology*, 75:134–143.

19. Shalev, U., Highfield, D., Yap, J., and Shaham, Y. (2000). Stress and relapse to drug seeking in rats: Studies on the generality of the effect. *Psychopharmacology*, 150:337–346.

20. Stewart, J. and deWit, H. (1987). Reinstatement of drug-taking behavior as a method of assessing incentive motivational properties of drugs. In Bozarth, M.A. (ed.), *Methods of assessing the reinforcing properties of abused drugs*. Springer, New York, pp. 211–227.

21. Bouton, M.E. and Swartzentruber, D. (1991). Sources of relapse after extinction in Pavlovian and instrumental learning. *Clin Psychol Rev*, 11:123–140.

22. Ahmed, S.H. and Koob, G.F. (1997). Cocaine- but not food-seeking behavior is reinstated by stress after extinction. *Psychopharmacology*, 132:289–295.

23. Erb, S., Shaham, Y., and Stewart, J. (1996). Stress reinstates cocaine-seeking behavior after prolonged extinction and a drug-free period. *Psychopharmacology*, 128:408–412.

24. Piazza, P.V. and LeMoal, M. (1998). The role of stress in drug self-administration. *Trends Pharmacol Sci*, 19:67–74.

25. Shaham, Y. and Stewart, J. (1995). Stress reinstates heroin-seeking in drug free animals: An effect mimicing heroin, not withdrawal. *Psychopharmacology*, 119:334–341.

26. Katz, J.L. and Higgins, S.T. (2003). The validity of the reinstatement model of craving and relapse to drug use. *Psychopharmacology*, 168:21–30.

27. Johanson, C.-E. and Fischman, M.M. (1989). The pharmacology of cocaine related to its abuse. *Pharmacol Rev*, 41:3–52.

28. Fischman, M.W., Schuster, C.R., Resnekov, I. *et al.* (1976). Cardiovascular and subject effects of intravenous cocaine administration in humans. *Arch Gen Psychiatry*, 10:535–546.

29. Fischman, M.W. and Johanson, C.E. (1998). Ethical and practical issues involved in behavioral pharmacology research that administers drugs of abuse to human volunteers. *Behav Pharmacol*, 9(7):479–498.

30. Foltin, R.W., Fischman, M.W., and Levin, F.R. (1995). Cardiovascular effects of cocaine in humans: Laboratory studies [Review]. *Drug Alcohol Depend*, 37(3):193–210.

31. Foltin, R.W. and Fischman, M.W. (1997). A laboratory model of cocaine withdrawal in humans: Intravenous cocaine. *Exp Clin Psychopharmacol*, 5(4):404–411.

32. Jaffe, J.H., Cascella, N.G., Kumor, K.M. *et al.* (1989). Cocaine-induced cocaine craving. *Psychopharmacology*, 97:59–64.

33. Fowler, J.S., Volkow, N.D., Ding, Y.S. *et al.* (1999). Positron emission tomography studies of dopamine-enhancing drugs [Review]. *J Clin Pharm* (Suppl):13S–16S.

34. Bartzokis, G., Beckson, M., Newton, T. *et al.* (1999). Selegiline effects on cocaine-induced changes in medial temporal lobe metabolism and subjective ratings of euphoria. *Neuropsychopharmacology*, 20(6):582–590.

35. Kosten, T.R., Markou, A., and Koob, G.F. (1998). Depression and stimulant dependence: Neurobiology and pharmacotherapy. *J Nerv Ment Dis*, 186:737–745.

36. Holman, B.L., Carvalo, P.A., Mendelson, J. *et al.* (1991). Brain perfusion is abnormal in cocaine dependent polydrug users: A study using Tc-99m-HMPAO and ASPECT. *J Nucl Med*, 32:1206-1210.

37. Holman, B.L., Garada, B., Johnson, K.A. *et al.* (1992). A comparison of brain perfusion SPECT in cocaine abuse and AIDS dementia complex. *J Nucl Med*, 33(7):1312-1315.

38. Holman, B.L., Mendelson, J., Garada, B. *et al.* (1993). Regional cerebral blood flow improves with treatment in chronic cocaine polydrug users. *J Nucl Med*, 34:723-727.

39. Mena, I., Giombetti, R., and Cody, C.K. (1990). Acute cerebral blood flow changes with cocaine intoxication [Abstract]. *Neurology*, 40(Suppl 1):179.

40. Miller, B.L., Mena, I., Giombetti, R. *et al.* (1992). Neuropsychiatric effects of cocaine: SPECT measurements. *J Addict Dis*, 11:47-58.

41. Strickland, A., Mena, I., Villanueva-Meyer, J. *et al.* (1991). Long-term effect of cocaine abuse on brain perfusion: Assessment with Xe-133 rCBF and Tc-99m-HMPAO. *J Nucl Med*, 32(5):1021.

42. Tumeh, S.S., Nagel, J.S., English, R.J. *et al.* (1990). Cerebral abnormalities in cocaine abusers: Demonstration by SPECT perfusion brain scintigraphy [Work in progress]. *Radiology*, 176:821-824.

43. Volkow, N.D., Mullani, N., Gould, L. *et al.* (1988). Cerebral blood flow in chronic cocaine users: A study with positron emission tomography. *Br J Psychiatry*, 152:641-648.

44. Malison, R.T., Wright, S., Sanacora, G., *et al.* (1998). Reductions in cocaine-induced craving following the selective D1 dopamine receptor agonist ABT-431 in human cocaine abusers [Abstract]. College for Problems of Drug Dependence Annual Meeting, Scottsdale, AZ:87.

45. Substance Abuse and Mental Health Services Administration (2006). National Survey on Drug Use and Health: National Findings, Office of Applied Studies, NSDUH Series H-30, DHHA Publication No. SMA 06-4194, Rockville, MD.

46. Gawin, F.H., Kleber, H.D., Rounsaville, B.J., Kosten, T.R., Jatlow, P.I., and Morgan, C. (1989). Desipramine facilitation of initial cocaine abstinence. *Arch Gen Psychiatry*, 46:117-121.

47. McLellan, A.T., Luborsky, L., Cacciola, J., and Griffith, J.E. (1985). New data from the addiction severity index: Reliability and validity in three centers. *J Nerv Ment Dis*, 173:412-423.

48. Rounsaville, B.J., Anton, S.F., Carroll, K.M. *et al.* (1991). Psychiatric diagnosis of treatment seeking cocaine abusers. *Arch Gen Psychiatry*, 18:43-51.

49. Weiss, R.D., Mirin, S.M., Michael, J.L. *et al.* (1986). Psychopathology in chronic cocaine abusers. *Am J Drug Alcohol Abuse*, 12:17-29.

50. Nunes, E.V., Quitkin, F.M., Brady, R. *et al.* (1991). Imipramine treatment of methadone maintenance patients with affective disorder and illicit drug use. *Am J Psychiatry*, 148:667-669.

51. Ziedonis, D.M. and Kosten, T.R. (1991). Depression as a prognostic factor for pharmacological treatment of cocaine dependence. *Psychopharmacol Bull*, 27:337-343.

52. Khantzian, E.J., Gawin, F., Kleber, H.D. *et al.* (1984). Methylphenidate (Ritalin) treatment of cocaine dependence - a preliminary report. *J Subst Abuse Treat*, 1:107-112.

53. Anton, R.F., moak, D.H., Latham, P. *et al.* (2005). Naltrexone combined with either cognitive behavioral or motivational enhancement therapy for alcohol dependence. *J Clin Psychopharmacol*, 25:349-357.

54. Chick, J., Anton, R., Checinski, K. *et al.* (2000). A multicentre, randomized, double-blind, placebo-controlled trial of naltresone in the treatment of alcohol dependence or abuse. *Alcohol Alcohol*, 35:578-593.

55. Blaine, J.D., Ling, W., Kosten, T.R. *et al.* (1994). Establishing the efficacy and safety of medications for the treatment of drug dependence and abuse: Methodological issues. In Prien, R. and Robinson, E. (eds.), *Clinical evaluation of psychotropic drugs: Principles and guidelines*. Raven Press, New York, pp. 593-623.

56. Preston, K.L., Silverman, K., Schuster, C.R. *et al.* (1997). Assessment of cocaine use with quantitative urinalysis and estimation of new uses. *Addiction*, 92(6):717-727.

57. Preston, K.L., Silverman, K., Schuster, C.R. *et al.* (1997). Use of quantitative urinalysis in monitoring cocaine use. *NIDA Research Monograph*, 75:253–264.
58. Margolin, A., Avants, S.K., and Kosten, T.R. (1995). Mazindol for relapse prevention to cocaine abuse in methadone-maintained patients. *Am J Drug Alcohol Abuse*, 21(4):469–481.
59. Avants, S.K., Margolin, A., and Kosten, T.R. (1996). Influence of treatment readiness on outcomes of two pharmacotherapy trials for cocaine abuse among methadone-maintained patients. *Psychol Addict Behav*, 10(3):147–156.
60. Preston, K.L., Silverman, K., Higgins, S.T. *et al.* (1998). Cocaine use early in treatment predicts outcome in a behavioral treatment program. *J Consult Clin Psychol*, 66(4):691–716.
61. Carroll, K.M., Rounsaville, B.J., Gordon, L.T. *et al.* (1994). Psychotherapy and pharmacotherapy for ambulatory cocaine abusers. *Arch Gen Psychiatry*, 51:177–187.
62. Crits-Christoph, P., Siqueland, L., Blaine, J. *et al.* (1999). Psychosocial treatments for cocaine dependence: National Institute on Drug Abuse Collaborative Cocaine Treatment Study. *Arch Gen Psychiatry*, 56:493–502.
63. Carroll, K.M. (1996). Relapse prevention as a psychosocial treatment approach: A review of controlled clinical trials. *Exp Clin Psychopharmacol*, 4:46–54.
64. Carroll, K.M. (1997). Manual-guided psychosocial treatment: A new virtual requirement for pharmacotherapy trials?. *Arch Gen Psychiatry*, 54:923–928.
65. Higgins, S.T., Budney, A.J., and Bickel, W.K. (1994). Applying behavioral concepts, principles to the treatment of cocaine dependence. *Drug Alcohol Depend*, 34:87–97.
66. Higgins, S.T., Badger, G.J., and Budney, A.J. (2000). Initial abstinence success in achieving longer term cocaine abstinence. *Exp Clin Psychopharmacol*, 8:377–386.
67. Higgins, S.T., Wong, C.J., Badger, G.J. *et al.* (2000). Contingent reinforcement increases cocaine abstinence during outpatient treatment and 1 year of follow-up. *J Consult Clin Psychol*, 68:64–72.
68. Petry, N.M., Peirce, J.M., Stitzer, M.L. *et al.* (2005). Effect of prize-based incentives on outcomes in stimulant abusers in outpatient psychosocial treatment programs. *Arch Gen Psychiatry*, 62:1148–1156.
69. Weinstock, J., Alessi, S.M., and Petry, N.M. (2007). Regardless of psychiatric severity the addition of contingency management to standard treatment improves retention and drug use outcomes. *Drug Alcohol Depend*, 87:288–296.
70. Schottenfeld, R.S., Chawarski, M.C., Pakes, J.R. *et al.* (2005). Methadone versus buprenorphine with contingency management or performance feedback for cocaine and opioid dependence. *Am J Psychiatry*, 162:340–349.
71. Poling, J., Oliveto, A., Petry, N. *et al.* (2006). Six-month trial of buproprion with contingency management for cocaine dependence in a methadone-maintained population. *Arch Gen Psychiatry*, 63:219–228.
72. Silverman, K., Higgins, S.T., Brooner, R.K. *et al.* (1996). Sustained cocaine abstinence in methadone maintenance patients through voucher-based reinforcement therapy. *Arch Gen Psychiatry*, 53:409–415.
73. Anton, R.F., Moak, D.H., Latham, P.K., Waid, L.R., Malcolm, R.J. *et al.* (2001). Posttreatment results of combining naltrexone with cognitive-behavior therapy for the treatment of alcoholism. *Psychopharmacology*, 21:72–77.
74. Balldin, J., Berglund, M., Borg, S., Mansson, M., Bendtsen, P. *et al.* (2003). A 6-month controlled naltrexone study: Combined effect with cognitive behavioral therapy in outpatient treatment of alcohol dependence. *Alcohol Clin Exp Res*, 27(7):1142–1149.
75. Olmstead, T.A., Sindelar, J.L., and Petry, N.M. (2007). Cost-effectiveness of prize-based incentives for stimulant abusers in outpatient psychosocial treatment programs. *Drug Alcohol Depend*, 87:175–182.

76. Grassi, M.C., Cioce, A.M., Giudici, F.D., Antonilli, L., and Nencini, P. (2007). Short-term efficacy of disulfiram or naltrexone in reducing positive urinalysis for both cocaine and cocaethylene in cocaine abusers: A pilot study. *Pharmacol Res*, 55(2):117–121.

77. Rawson, R.A., McCann, M.J., Fammino, F. *et al.* (2006). A comparison of contingency management and cognitive-behavioral approaches for stimulant-dependent individuals. *Addiction*, 101(2):267–274.

78. Aharonovich, E., Nunes, E., and Hasin, D. (2003). Cognitive impairment, retention and abstinence among cocaine abusers in cognitive-behavioral treatment. *Drug Alcohol Depend*, 71(2):207–211.

79. Aharonovich, E., Hasin, D.S., Brooks, A.C. *et al.* (2006). Cognitive deficits predict low treatment retention in cocaine dependent patients. *Drug Alcohol Depend*, 81(3):313–322.

80. Hennekens, C.H. *et al.* (2004). Additive benefits of pravastatin and aspirin to decrease risks of cardiovascular disease: Randomized and observational comparisons of secondary prevention trials and their meta-analysis. *Arch Intern Med*, 164:40–44.

81. Berry, S.M. and Berry, D.A. (2004). Accounting for multiplicities in assessing drug safety: A three-level hierarchical mixed model. *Biometrics*, 60:418–426.

82. Berry, S.M. *et al.* (2004). Bayesian survival analysis with nonproportional hazards: Meta-analysis of pravastatin–aspirin. *J Am Stat Assoc*, 99:36–44.

83. Inoue, L.Y.T., Thall, P., and Berry, D.A. (2002). Seamlessly expanding a randomized phase II trial to phase III. *Biometrics*, 58:264–272.

84. Chow, S. and Liu, J. (eds.) (2004). *Design and analysis of clinical trials: Concepts and methodologies*, 2nd edition. Wiley Interscience, pp. 93–118.

85. Berry, D.A. (2006). Bayesian clinical trials. *Nat Rev*, 5:27–36.

86. Lavori, P.W., Laska, E.M., and Uhlenhuth, E.H. (1994). Statistical issues for the clinical evaluation of psychotropic drugs. In Prien, R.F. and Robinson, D.F. (eds.), *Clinical Evaluation of Psychotropic Drugs: Principles and Guidelines*, Raven Press Ltd, New York, pp. 139–160.

The Role of Animal Models in Reward Deficit Disorders: Views from Academia

George F. Koob

Committee on the Neurobiology of Addictive Disorders,
The Scripps Research Institute, La Jolla, CA, USA

Animal and Translational Models for CNS Drug Discovery,
Vol. 3 of 3: Reward Deficit Disorders
Robert McArthur and Franco Borsini (eds), Academic Press, 2008

INTRODUCTION AND DEFINITIONS

Reward and motivation deficits form key elements of a number of psychiatric disorders, ranging from depression to drug addiction to schizophrenia. The most obvious example of the role of such deficits is in criteria no. 1 (depressed mood) and no. 2 (markedly diminished pleasure) of the *Diagnostic and Statistical Manual of Mental Disorders* criteria of depression and in what has been termed the motivational deficits associated with substance dependence (or addiction).[1,2] Robust animal models exist for reward and motivation deficits associated with psychiatric disorders and include brain stimulation reward thresholds, preference for a sweet solution, and progressive-ratio responding for food, among others. The use of the three aforementioned models will be explored in the present chapter in the context of depression and drug addiction, and evidence for face and construct validity of these models will be provided.

Motivation as a concept has many definitions. Donald Hebb argued that motivation is "stimulation that arouses activity of a particular kind",[3] and C.P. Richter argued that "spontaneous activity arises from certain underlying physiological origins and such 'internal' drives are reflected in the amount of general activity".[4] Dalbir Bindra defined motivation as a "rough label for the relatively persisting states that make an animal initiate and maintain actions leading to particular outcomes or goals".[5] A more behavioristic view is that motivation is "the property of energizing of behavior that is proportional to the amount and quality of the reinforcer".[6] A *reinforcer* can be defined operationally as "any event that increases the probability of a response." *Reward*, in contrast, has been variously described but refers generally to some additional emotional value such as pleasure.

All of these definitions point to certain common characteristics of our concepts of reward and motivation. Motivation is a state that varies with arousal and guides behavior in relationship to changes in the environment. The environment can be external (incentives) or internal (central motive states or drives), and such motivation or motivational states are not constant and vary over time. The concept of motivation was linked inextricably with hedonic, affective, or emotional states in the context of temporal dynamics by Solomon's "opponent process" theory of motivation. Solomon and Corbit[7] postulated that hedonic, affective, or emotional states, once initiated, are automatically modulated by the central nervous system with mechanisms that reduce the intensity of hedonic feelings. Solomon argued that there are affective or hedonic habituation (or tolerance) and affective or hedonic withdrawal (abstinence). He defined two processes: the *a-process* and the *b-process*. The *a-process* consists of either positive or negative hedonic responses and occurs shortly after presentation of a stimulus and correlates closely with the stimulus intensity, quality, and duration of the reinforcer, and shows tolerance. In contrast, the *b-process* appears after the *a-process* has terminated and is sluggish in onset, slow to build up to an asymptote, slow to decay, and gets larger with repeated exposure. Thus, the affective dynamics of opponent process theory generates new motives and new opportunities for reinforcing and energizing behavior.[8]

From a neurobehavioral perspective, it was hypothesized that in brain motivational systems the initial acute effect of an emotional stimulus or a drug is opposed or counteracted by homeostatic changes in brain systems. Certain systems in the brain were hypothesized to suppress or reduce all departures from hedonic neutrality.[7] This affect control system was conceptualized as a single negative feedback, or opponent, loop that opposes the stimulus-aroused affective state.[7,9,10] In opponent process theory, tolerance and dependence are inextricably linked,[7] and affective states – pleasant or aversive – are hypothesized to be automatically opposed by centrally mediated mechanisms that reduce the intensity of these affective states. In the context of drug dependence, Solomon argued that the first few self-administrations of an opiate drug produce a pattern of motivational changes resembling an acute positive hedonic response where the onset of the drug effect produces a euphoria that is the *a-process*, and this is followed by a decline in intensity. After the drug wears off, the *b-process* state emerges as an aversive craving state.

More recently, opponent process theory has been expanded into the domains of the neurocircuitry and neurobiology of stress, depression, and drug addiction from a physiological perspective. An allostatic model of the brain motivational systems has been proposed to explain the persistent changes in motivation that are associated with the negative affective states associated with depression[11] and the withdrawal/negative affect stage in addiction.[2] The allostatic model may generalize to other psychopathology associated with dysregulated motivational systems. For further details, the reader is referred to a number of reviews linking reward deficit states and depression and addiction.[2,11-14]

In this framework, addiction is conceptualized as a cycle of increasing dysregulation of brain reward systems that progressively increases, resulting in the compulsive use of drugs. Counteradaptive processes, such as opponent process, that are part of the normal homeostatic limitation of reward function fail to return within the normal homeostatic range and are hypothesized to form an allostatic state. The allostatic state is further hypothesized to be reflected in a chronic deviation of reward set point that is fueled not only by dysregulation of reward circuits *per se*, but also by recruitment of brain and hormonal stress responses.

Depression, or a major depressive episode, also is a chronic relapsing disorder characterized by severe mood disorders with a significant genetic component[15] and a long-hypothesized contribution from stress.[16] Major depressive episodes show a kindling-like effect; the episodes, if left untreated, become worse over time and more severe. Suicide is very highly represented in affective disorders, with 60–70% of severely depressed individuals having suicidal ideations and 10–15% ultimately attempting suicide.[17]

FACE AND CONSTRUCT VALIDITY OF ANIMAL MODELS OF REWARD DEFICITS IN PSYCHIATRIC ILLNESS

Animal models are critical for understanding the neuropharmacological mechanisms involved in the development of psychiatric disorders. While there are no complete animal models of addiction or depression, they do exist for many elements of the syndrome. An animal model can be viewed as an experimental paradigm developed for the purpose of studying a given phenomenon found in humans. The most relevant conceptualization of validity for animal models of addiction is the concept of *construct validity*.[18] Construct validity refers to the interpretability, "meaningfulness," or explanatory power of each animal model and incorporates most other measures of validity where multiple measures or dimensions are associated with conditions known to affect the construct.[19] A procedure has construct validity if there are statistical or deterministic propositions that relate constructs (e.g., reward) to observables (e.g., thresholds) derived from the procedure.[20]

An alternative conceptualization of construct validity is the requirement that models meet the construct of functional equivalence, defined as "assessing how controlling variables influence outcome in the model and the target disorders".[21] The most efficient process for evaluating functional equivalence has been argued to be through common experimental manipulations which should have similar effects in the animal model and the target disorder.[21] This process is very similar to the broad use of the construct *predictive validity* (see below). *Face validity* often is the starting point in animal models where animal syndromes are produced which resemble those found in humans but are limited by necessity.[22] *Reliability* refers to the stability and consistency with which the variable of interest can be measured. Reliability is achieved when, following objective repeated measurement of the variable, small within- and between-subject variability is noted, and the phenomenon is readily reproduced under similar circumstances (for review, see[23]). *Predictive validity* in the more narrow sense refers to the model's ability to lead to accurate predictions about the human phenomenon based on the response of the model system. Predictive validity is used most often in animal models of psychiatric disorders to refer to the ability of the model to identify pharmacological agents with potential therapeutic value in humans.[22,24] Others have argued that this type of predictive validity can be considered more explicitly as "*pharmacological isomorphism*," which is the use of clinically active standards as positive controls to validate a model/procedure.[25,26] However, when predictive validity is more broadly expanded to understand the physiological mechanism of action of psychiatric disorders, others have argued that it incorporates other types of validity (e.g., etiological, convergent or concurrent, discriminant) considered to be important for animal models and approaches the concept of construct validity.[27]

The present chapter describes animal models that have face validity and some degree of reliability and construct validity for reward deficits associated with depression and addiction. The two independent variables explored in this review are chronic mild stress as an animal model of depression and chronic administration of drugs sufficient to produce dependence (defined as the manifestation of motivational withdrawal). Three dependent variables that have been hypothesized to measure reward motivational deficits are explored and include sucrose consumption and/or preference, brain stimulation reward, and progressive-ratio responding for a natural reward.

FACE VALIDITY AND CONSTRUCT VALIDITY OF ANIMAL MODELS OF REWARD DEFICITS IN PSYCHIATRIC DISORDERS

Brain Stimulation Reward

Brain stimulation reward (commonly referred to also as intracranial self-stimulation) is a procedure where animals perform a response to electrically stimulate parts of the central nervous system. Discovered by Olds and Milner,[28] brain stimulation reward has long been hypothesized to mediate reward because it activates some of the same structures in the brain that mediate the rewarding effects of natural reinforcers[29] and the euphorigenic-like effects of drugs of abuse.[30] Brain stimulation reward has many advantages over natural rewards. It directly activates brain reward systems, bypassing much of the input side of the circuit. In addition, no motivational deficit states (food or water deprivation) are required. It also is exquisitely controllable with very small increments or decreases in reward value changing behavior systematically.

Several different procedures have been used to measure brain stimulation reward, including rate of responding in a simple operant situation with a given current intensity, place preferences, and so on, but the field has more or less settled on two procedures: the rate-frequency curve-shift procedure[31] and the discrete-trial, current-intensity procedure.[27,32] A large proportion of the studies exploring reward deficits in psychiatric illness have used one of these two procedures.

The rate-frequency procedure involves the generation of a stimulation-input response-output function and provides a frequency threshold measure.[31,33] The frequency of the stimulation (input) is varied, and the subject's response rate (output) is measured as a function of frequency. Rate-frequency curves are collected by allowing the rats to press a lever or turn a wheel manipulandum for an ascending or descending series of pulse-frequency stimuli delivered through an electrode in the medial forebrain bundle or other rewarding brain site. The rate-frequency function is a sigmoid curve that offers two measures. The locus of rise is a frequency threshold measure and is presumed to be a measure of intracranial self-stimulation reward threshold.[31] Locus of rise refers to the "location" (i.e., frequency) at which the function rises from zero to an arbitrary criterion level of performance. The most frequently used criterion is 50% of maximal rate. The behavioral maximum measure is the asymptotic maximal response rate, and changes in the maximum measure are hypothesized to reflect motor or performance effects.[34] The procedure has been validated and investigated extensively, and studies indicate that changes in the reward efficacy of the stimulation (i.e., intensity manipulations) shift the rate-frequency functions laterally, translating into large changes in the locus of rise value, but produce no alterations in the asymptote or in the shape of the function.[34] In contrast, performance manipulations (e.g., weight on the lever, curare, etc.), including changes in motivation (i.e., priming), alter the maximum measure value and the shape of the function.[33-35] Drugs of abuse such as cocaine or amphetamine shift the dose–response curve to the left[36] (Figure 4.1).

A second procedure that controls for rate of responding and nonspecific performance deficits is a discrete-trial procedure. The discrete-trial procedure is a modification of the classical psychophysical method of limits and provides a current-intensity threshold measure.[27,32] At the start of each trial, rats receive a noncontingent, experimenter-administered electrical stimulus. The subjects then have, for example, 7.5 s to turn the wheel manipulandum one-quarter of a rotation to obtain a contingent stimulus identical to the previously delivered noncontingent stimulus (positive response).[27] If responding does not occur within the 7.5 s after the delivery of the noncontingent stimulus (negative response), the trial is terminated and an intertrial interval follows. Stimulus intensities vary according to the psychophysical method of limits. The procedure provides two measures for each test session. The threshold value is defined as

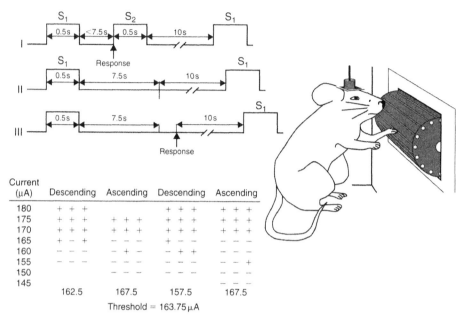

Current (μA)	Descending	Ascending	Descending	Ascending
180	+ + +		+ + +	+ + +
175	+ + +	+ + +	+ + +	+ + +
170	+ + +	+ + +	+ + +	+ + +
165	+ − +	− − −	+ − −	− − −
160	− − −	− + −	− + +	− − −
155	− − −	− − −	− − −	− − +
150		− − −	− − −	− − −
145				− − −
	162.5	167.5	157.5	167.5

Threshold = 163.75 μA

Figure 4.1 Intracranial self-stimulation threshold procedure. Panels I, II and III illustrate the timing of events during three hypothetical discrete trials. Panel I shows a trial during which the rat responded within the 7.5 s following the delivery of a noncontingent stimulus (positive response). Panel II shows a trial during which the animal did not respond (negative response). Panel III shows a trial during which the animal responded during the intertrial interval (negative response). For demonstration purposes, the intertrial interval was set at 10 sec. In reality, the interresponse interval had an average duration of 10 s and ranged from 7.5–12.5 s. The table depicts a hypothetical session and demonstrates how thresholds were defined for the four individual series. The threshold of the session is the mean of the four series' thresholds[a].
[a] Taken with permission from Markou and Koob.[27]

the midpoint in microamps between the current-intensity level at which the animal makes two or more positive responses out of the three stimulus presentations, and the level where the animal makes less than two positive responses at two consecutive intensities. Response latency is defined as the time in seconds that elapses between the delivery of the noncontingent electrical stimulus (end of the stimulus) and the animal's response on the wheel. Decreased thresholds are interpreted as an increase in reward value (or salience) of the stimulation, while increased thresholds reflect a decrease in reward value. Increases in response latency are interpreted as a motor/performance deficit.

In a systematic series of studies, the brain stimulation reward threshold procedure was shown to have construct validity.[37] The discrete-trial procedure provides four measures: current thresholds, response latency, extra responses, and time-out responses. The effects of a performance manipulation (variations in the force required to operate the manipulandum) and of a reward manipulation (variations in the train duration of the electrical stimulation) are evaluated on the four measures. Reward effects are reflected primarily in changes in thresholds, with no effect on any of the other three measures. Conversely, performance effects are reflected primarily in changes in response latency, extra responses, and time-out responses, with only a modest effect on thresholds.

Sucrose Consumption/Sucrose Preference

Sucrose is a highly reinforcing sweet substance in rodents, and rats will show a concentration-dependent increase in both consumption and preference for sucrose that forms an inverted U-shaped function.[38] On the ascending limb of the concentration-intake function, intake is monotonically related to preference.[38,39] Operant responding for sucrose shows a similar inverted U-shaped function.[40] Reduced preference for a sucrose solution in rats has been hypothesized to reflect a decreased sensitivity to reward and has been argued to be homologous with anhedonia.[41]

Sucrose consumption/preference typically is monitored by tracking, over repeated tests, a decrease in the consumption of and/or preference for a palatable, low-concentration (1–2%) sucrose solution. Testing often occurs in the animal's home cage. The animals are provided with two-bottle exposure to 1–2% sucrose versus water for 1 h.[42] Preference is determined by the formula: *% Preference = (Flavor intake/total intake) × 100.*

Construct validity for sucrose drinking as a valid measure of sensitivity to reward has been argued based on observations that treatments do not alter drinking of water in the choice situation.[43] Some studies have reported similar effects in a calorie-free test using saccharin.[42,44] Decreases in sucrose drinking have been observed in studies where the treatment to produce a reward deficit excluded periods of food and water deprivation.[45] However, others have questioned the validity and reliability of the measure in the context of weight loss in chronic mild stress procedures[46] (see below).

Progressive-Ratio Responding

Progressive-ratio responding has been used for over 40 years as a method for testing hypotheses of the relative strength of reinforcers.[47,48] The animal is required to emit a systematically increasing number of responses for each successive reinforcer. Thus, the number of responses in each successive ratio increases with a fixed increment. Using this index, early work provided validation that a progressive-ratio schedule was a sensitive and reliable measure of reinforcer value. The number or responses in the final ratio increased as a function of the salience of the reinforcer, and as the size of the ratio increment increased, the number of responses increased and the number of reinforcers obtained decreased sharply.[48] In addition, the number of responses in the final completed run of the progressive-ratio (i.e., break point) is sensitive to a number of motivational variables, including increasing the degree of food deprivation and the concentration of reinforcer, both of which increase the break point.[47]

A more recent evaluation of the progressive-ratio schedule has raised the hypothesis that progressive-ratio performance conforms to a microeconomic conservation model where there is a curvilinear relationship between economic supply and labor.[49]

In progressive-ratio studies, relative reinforcer value was defined in terms of changes in behavior from a balance point and allows for a definition of the relative value of a reinforcer based on changes in consumption across response costs. For example, when the maximally effective doses of cocaine and procaine were compared using the labor–supply approach, the relative reinforcing value of cocaine was confirmed as higher than that of procaine.[49] A fundamental difference between cocaine and procaine self-administration is that the consumption of cocaine is relatively resistant to increases in the "prices" (or labor) that effectively suppressed procaine consumption. The labor–supply approach also provided accurate mathematical descriptions of actual performance, lending further validity to the model.[49]

DRUG WITHDRAWAL MODEL OF DEPENDENCE

Drug addiction, also referred to diagnostically as substance dependence,[50] is a chronically relapsing brain disorder characterized by (i) compulsion to seek and take a given drug, (ii) loss of control in limiting intake, and (iii) emergence of a negative emotional state (e.g., dysphoria, anxiety, irritability) when the drug is unavailable or access to it is prevented, defined here as dependence.[51]

For this chapter, *addiction* and *alcoholism* will be held equivalent. Addiction and substance dependence (as currently defined by the *Diagnostic and Statistical Manual of Mental Disorders*, fourth edition) will be used to refer to a final stage of a usage process that moves from drug use to abuse to addiction. Drug addiction has been conceptualized as a disorder that progresses from impulsivity to compulsivity in a cycle that comprises three stages: *preoccupation/anticipation*, *binge/intoxication*, and *withdrawal/negative affect*.[51,52] As individuals move from an impulsive to a compulsive disorder, the drive for the drug-taking behavior shifts from positive to negative reinforcement. Impulsivity and compulsivity can coexist in different stages of the addiction cycle.

The withdrawal/negative affect stage defined above consists of key motivational elements such as chronic irritability, emotional pain, malaise, dysphoria, alexithymia, and loss of motivation for natural rewards. Different drugs produce different patterns of addiction with an emphasis on different components of the addiction cycle. Opioids are probably the classic drug of addiction and can be used here as an example. A pattern of drug-taking evolves, including intense intoxication (via the intravenous or smoking routes for heroin and oral or intravenous routes for opioid analgesics), the development of tolerance, escalation in intake, and profound dysphoria, physical discomfort, and somatic withdrawal signs during abstinence. Intense preoccupation with obtaining opioids (craving) develops that often precedes the somatic signs of withdrawal, and this preoccupation is linked not only to stimuli associated with obtaining the drug, but also to stimuli associated with withdrawal and internal and external states of stress. The drug must be obtained to avoid the severe dysphoria and discomfort during abstinence. Different patterns result from the use of psychostimulants, alcohol, nicotine, and marijuana, with some drugs producing more binge-like intake (psychostimulants), but as noted above, compulsive use, loss of control over limiting intake, and emergence of a negative emotional state are common elements of addiction to all drugs of abuse.

Brain Stimulation Reward As a Measure of the Reward Deficits Associated with Drug Dependence

Psychostimulant withdrawal has much face validity with the symptoms of a major depressive episode such that clinicians often cannot distinguish the two syndromes without a patient's history.[1] Such symptoms include the two most related to reward deficits: depressed mood and markedly diminished pleasure[53] (see above). These symptoms are most severe at the beginning of withdrawal but can persist into protracted abstinence. Also psychostimulant withdrawal has been proposed as an inducing condition in animal models of depression and has high predictive and construct validity[54] (see below).

As noted above, however, a common component of dependence on all drugs of abuse is the withdrawal/negative affect stage characterized by a negative emotional state. This negative emotional state varies from irritability to severe depressive symptoms depending on the drug of abuse, dose, and duration of addiction, but with some element of dysphoria manifested with all drugs of abuse. Most compelling for the present treatise, a series of studies have established that brain reward thresholds also are elevated during acute withdrawal from all major drugs of abuse, and thus intracranial self-stimulation thresholds have been used to assess changes in systems mediating reward and reinforcement processes during the course of drug dependence.

Acute administration of psychostimulant drugs lowers brain reward thresholds (i.e., increased reward) (for reviews, see[32,55]) (Table 4.1), and withdrawal from chronic administration of virtually all major drugs of abuse elevates thresholds (i.e., decreases reward)[37,56-62] (Figure 4.2).

Sucrose Intake/Preference As a Measure of the Reward Deficits Associated with Drug Dependence

There are few studies of sucrose consumption and preference as used in the chronic mild stress procedure in animal or human studies of acute withdrawal from drugs of abuse. Sucrose preference was not changed during withdrawal in adolescent rats exposed to ethanol vapor, even though locomotor activity was suppressed.[65] Sucrose consumption was decreased in rats during acute opiate abstinence following heroin self-administration.[66] However, others have shown enhanced taste reactivity for sucrose 24 h after 21 days of nicotine treatment.[67] No effect of precipitated opiate withdrawal was observed on sucrose (20%) consumption,[68] but conditioned taste aversions have been readily observed in precipitated opiate withdrawal.[69] Indeed, human studies have shown increases in preference for sweet solutions during protracted abstinence from both nicotine and ethanol,[70,71] an effect with nicotine that may have a metabolic component. Thus, sucrose consumption or preference as used in the chronic mild stress procedure has not been explored systematically as a dependent

Table 4.1 Effects of various drugs on intracranial self-stimulation reward threshold[a]

Lowers	No change	Raises
Morphine	Tetrahydrocannabinol	Haloperidol
6-Acetylmorphine	Lysergic acid diethylamide	Pimozide
Buprenorphine	Naloxone	Chlorpromazine
Nalbuphine[a]	Naltrexone	Imipramine
Methamphetamine	Cyclazocine	Atropine
Amfonelic acid	Ethylketocyclazocine	Scopolamine
Tripelennamine[b]	Nisoxetine	
Methylenedioxymethamphetamine	Apomorphine	
Heroin	U50,488	
Cocaine	Pentobarbital	
Pentazocine[a]	Procaine	
D-amphetamine		
Phencyclidine		
Bromocriptine		
Nicotine		
Ethanol[c]		

Note: The drugs listed are only those that have been tested in the laboratory of the original source of this work using the rate-independent threshold procedure. Some drugs that caused no change, especially pentobarbital, might lower the threshold under drug self-administration conditions, as observed with ethanol. Taken with permission from Kornetsky and Bain.[64]
[a] *Especially in combination with tripelennamine.*
[b] *Especially in combination with pentazocine or nalbuphine.*
[c] *Only under conditions of self-administration.[63]*

variable for reward deficits associated with drug withdrawal and may be confounded by metabolic and/or pronounced conditioned aversive effects.

Progressive-Ratio Responding As a Measure of Reward Deficits Associated with Drug Dependence

Early animal models of post-psychostimulant withdrawal have employed the use of brain reward thresholds[72] and psychomotor retardation as an index of withdrawal following a sustained regimen of amphetamine self-administration.[73] More recently, an animal model aimed at investigating the changes in motivation for natural reinforcers has been developed where rats withdrawing from a binge-like amphetamine treatment showed a reduction in progressive-ratio responding for a sweet solution.[74] Mildly food- and water-deprived rats were trained to work on a progressive-ratio schedule for

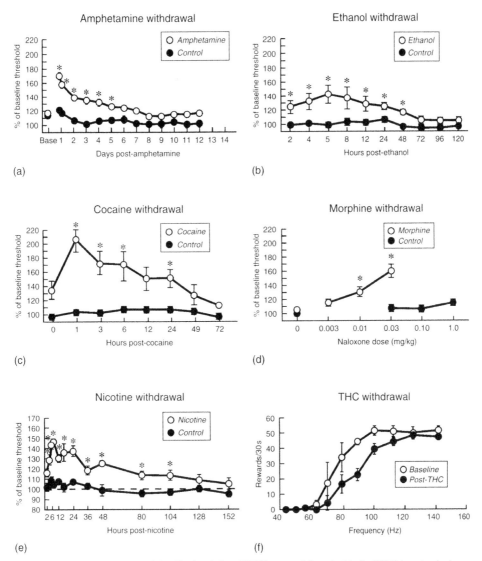

Figure 4.2 (a) Mean intracranial self-stimulation (ICSS) reward thresholds (±SEM) in rats during amphetamine withdrawal (10 mg/kg/day for 6 days). Data are expressed as a percentage of the mean of the last five baseline values prior to drug treatment. (b) Mean ICSS thresholds (±SEM) in rats during ethanol withdrawal (blood alcohol levels achieved: 197.29 mg%). Elevations in thresholds were time dependent. (c) Mean ICSS thresholds (±SEM) in rats during cocaine withdrawal 24 h following cessation of cocaine self-administration. (d) Mean ICSS thresholds (±SEM) in rats during naloxone-precipitated morphine withdrawal. The minimum dose of naloxone that elevated ICSS thresholds in the morphine group was 0.01 mg/kg. (e) Mean ICSS thresholds (±SEM) in rats during spontaneous nicotine withdrawal following surgical removal of osmotic minipumps delivering nicotine hydrogen tartrate (9 mg/kg/day) or saline. (f) Mean ICSS thresholds (±SEM) in rats during withdrawal from an acute 1.0 mg/kg dose of Δ^9-tetrahydrocannabinol (THC). Withdrawal significantly shifted the reward function to the right (indicating diminished reward). Figures taken with permission from Paterson et al.[62]; Schulteis et al.[59]; Markou and Koob[37]; Schulteis et al.[58]; data adapted with permission from Epping-Jordan et al.[60]; Gardner and Vorel.[61] Note that because different equipment systems and threshold procedures were used in the collection of the above data, direct comparisons among the magnitude of effects induced by these drugs cannot be made. Asterisks (*) indicate statistically significant differences from the saline control group ($P < 0.05$).

Table 4.2 Progressive-ratio responding for a nondrug reinforcer during drug withdrawal in rodents

Drug	Reinforcer	Progressive-ratio responding	Reference
Morphine	Sucrose	↓	79
Amphetamine	Food	↓	80
	Sucrose	↓	75
	Sucrose	↓	74
Methamphetamine	Sucrose	↓	76
Nicotine	Sucrose	↓	78

a 4% sucrose solution. Following a 4-day regimen of injections of increasing doses of D-amphetamine, break points for progressive-ratio responding were significantly reduced for up to 4 days of withdrawal.[74] Using the same model, the dopamine partial agonist terguride was shown to possess an agonistic-like profile, preventing the occurrence of post-amphetamine decreases in progressive-ratio responding,[75] thus reducing the duration and severity of methamphetamine withdrawal, an effect similar to that produced by the dopamine full agonist ropinirole (Requip®;[76] but see[77], for a negative result with psychostimulant withdrawal). Similar decreases in progressive-ratio responding during withdrawal have been observed with chronic nicotine and opiate administration[78,79] (Table 4.2). Thus, progressive-ratio responding as a dependent measure also shows reward deficits in animal models of acute withdrawal to dependence-inducing doses of drugs of abuse.

CHRONIC MILD STRESS AND DEPRESSION

In this model, rats are exposed to chronic mild stress consisting of 5–6 weeks of stressors such as food and water deprivation, continuous lighting, cage tilt, paired housing, soiled cage, exposure to reduced temperature, intermittent white noise, stroboscopic lighting, exposure to an empty bottle following water deprivation, restricted access to food, novel odors, and presence of a foreign object in the cage[42,81] (see Table 4.3 for a typical paradigm). The face validity of this paradigm involves the administration of chronic mild stress. Stress has long been implicated in the etiology of depression[82] and the decrease in reward which has been argued to reflect aspects of anhedonia, a central feature of depression.

Brain Stimulation Reward As a Measure of the Reward Deficits Associated with the Chronic Mild Stress Model of Depression

Chronic mild stress produces a number of behavioral and physiological abnormalities that have face validity for the symptoms of depression.[83] These include decreased

Table 4.3 Schedule of chronic mild stress procedures[a]

	Food Deprivation	Water Deprivation	Continuous Lighting	Cage Tilt	Group Housing	Soiled Cage	Cold Room	Intermittent White Noise	Strobo-scopic Lighting	Empty Water Bottle	Restricted Access to Food	Foreign Object In Cage
a.m. MON				10:00			15:00–15:30		10:00			
p.m. MON	15:00	15:00				17:00			12:00			
a.m. TUES			17:00			10:00		15:00				
p.m. TUES	15:00							18:00				
a.m. WED		11:00		10:00					10:00	10:00–11:00		
p.m. WED	17:00	17:00							17:00			17:00
a.m. THURS		10:00		10:00			10:00–10:30					10:00
p.m. THURS				17:00	17:00							
a.m. FRI								12:00			10:00–12:00	
p.m. FRI	12:00					10:00		17:00				
a.m. SAT	10:00	10:00										
p.m. SAT			17:00									
a.m. SUN	12:00	12:00	12:00	17:00								
p.m. SUN												

[a] Taken with permission from Yu et al.[81]

sexual and exploratory behavior, sleep abnormalities, immune and hypothalamic-pituitary-adrenal dysregulation, and hedonic deficits measured by sucrose consumption, place conditioning, and brain stimulation reward. Using brain stimulation reward, Jean-Luc Moreau and colleagues have consistently shown reliable hedonic deficits following chronic mild stress as measured by threshold changes in brain stimulation reward.[83–88] Dr. Moreau's use of chronic mild stress was adapted from Willner et al.[42] and consisted of 4 consecutive weeks, 5 days per week, of confinement to small cages, overnight illumination, food and water deprivation, restricted food, exposure to an empty water bottle, and soiled cage exposure. Brain stimulation reward thresholds were measured from electrodes implanted in the ventral tegmental area using a nose-poke rate-frequency function.[84] Elevations in reward thresholds were observed during the chronic mild stress procedure. Chronic administration of a number of antidepressant treatments and drug treatments, including desipramine (Norpramin®), mianserin (Tolvon®), tolcapone (Tasmar®), and moclobemide (Manerix®/Aurorix®), reversed the increase in reward thresholds produced by this chronic stress procedure[83–88] (Table 4.4).

Sucrose Intake/Preference As a Measure of the Reward Deficits Associated with the Chronic Mild Stress Model of Depression

Chronic sequential exposure to mild unpredictable stress has been found to decrease the consumption of palatable sweet solutions and in some cases preference for palatable sweet solutions[42,43,81,90] (Table 4.5). Seventy-two hours prior to the start of chronic mild stress, rats are given continuous-access 48h exposure to two bottles; one bottle containing 1% sucrose and the other tap water. Prior to testing for fluid consumption, animals are deprived of food and water for 23h. Testing is carried out in the home cage 6h into the light cycle. Sucrose preference is calculated as sucrose intake divided

Table 4.4 Pharmacological reversal of chronic mild stress-induced increases in thresholds for brain stimulation reward.

Antidepressant	Class	Reference
Tolcapone	Catechol-*O*-methyltransferase inhibitor	86
Moclobemide	Monoamine oxidase-A inhibitor	85
Desipramine	Tricyclic norepinephrine reuptake inhibitor	84
Mianserin	α_2 Noradrenergic receptor antagonist	87
Ro 60-0175	Serotonin-2C receptor agonist	89
Ro 60-0332	Serotonin-2C receptor agonist	89

Table 4.5 Effects of stress of varying types and intensities on sucrose/saccharin consumption and preference in rodents

| | Sucrose/Saccharin | | |
	Consumption/ Responding	Preference	Reference
Stress-induced depressive-like behavior			
Chronic mild stress	↓	↓	42
	↓	–	103
	↓	nt	98
	↓	nt	104
	↓	nt	105
	↓	nt	99
	↓	nt	106
	↓	–	107
	↓	↓	108
	–[a]	–[a]	109
	↓[b]	↓[b]	109
	–[c]	–[c]	109
	↓	nt	110
	↓	nt	111
	↓	–	112
	↓	↓	113
	↓ (sucrose)	nt	114
	– (saccharin)	nt	114
	↓	–	115
	nt	↓	102
	↓	↓	81
	↓	↓	116
	nt	↓ (single-housed)	117
	nt	– (pair-housed)	117

Table 4.5 (Continued)

| | Sucrose/Saccharin | | |
	Consumption/Responding	Preference	Reference
Social stress	↓	nt	91
	↓	↓	118
	nt	↓	100,101
	↓	↓	119
	↓	nt	120
Novelty stress	↓	↓	92
Forced swim stress	↓	↓	81
Non-stress-induced depressive-like behavior			
Interferon-α	↓	nt	93
Olfactory bulbectomy	nt	↓ (female)	96
Olfactory bulbectomy/ovariectomy	nt	↓ (female)	96
Reserpine	–	↓	121
Interleukin-1β	↓	nt	94
Prenatal cocaine (40 mg/kg) + nicotine (2.5 mg/kg)	↓	–	122
Interferon-γ	↓	↓	95
Serotonin transporter knockout (C57BL/6 background)	nt	–	123
Olfactory bulbectomy	–	↑	97
Vasopressin knockout (Brattleboro rats)	nt	↑	124
Dexamethasone	nt	↓	102

nt, not tested; –, no change.
[a] 6 weeks chronic mild stress.
[b] 3 weeks chronic mild stress + 24 h water deprivation.
[c] 3 weeks chronic mild stress + 5 h water deprivation.

by total intake.[42,81] Numerous studies have shown that chronic mild stress decreases sucrose consumption, and some also have shown a decrease in preference. In addition, the chronic mild stress-induced decrease in palatable solution intake has been replicated with social stress,[91] novelty stress,[92] forced swim stress,[81] and a variety of pharmacological agents hypothesized to induce depressive-like states, including interferon-α,[93] interleukin-1β,[94] and interferon-γ.[95] Not all depression models are effective,

Table 4.6 Pharmacological reversal of decreased stress-induced sucrose intake and preference

Stressor	Antidepressant	Class	Reference
Chronic mild stress	Desmethylimipramine	Tricyclic norepinephrine reuptake inhibitor	42
	Fluoxetine	Selective serotonin reuptake inhibitor	98,106,111
	Maprotiline	Tetracyclic monoamine reuptake inhibitor	98
	Quinpirole	Dopamine D_2/D_3 agonist	104
	Mianserin	α_2 Noradrenergic antagonist	99
	Carbamazepine	Anticonvulsant	106
	Lithium		106
	Ketoconazole	Antifungal ergosterol synthesis inhibitor	106
	Imipramine	Tricyclic monoamine reuptake inhibitor	106
	MMAI (5-methoxy-6-methyl-2-aminoindan)	Selective serotonin releaser	108
	p-methylthioamphetamine	Selective serotonin releaser	108
	Banxia-houpu decoction	Herbal medicine (pinellia tuber + magnolia bark + hoelen + perilla herb + ginger rhizome; mechanism unknown)	111
	Chlorimipramine	Selective serotonin reuptake inhibitor	116
	Paroxetine	Selective serotonin reuptake inhibitor	102
	Clomipramine	Selective serotonin reuptake inhibitor	81
Social stress	Citalopram	Selective serotonin reuptake inhibitor	101
	Imipramine	Tricyclic monoamine reuptake inhibitor	91
Novelty stress	Citalopram	Selective serotonin reuptake inhibitor	92

notably olfactory bulbectomy which produced decreases in preference only in female rats[96] or increases in preference.[97] Perhaps more impressive have been numerous studies showing that the decrease in consumption and/or preference was reversible by chronic but not acute treatment with antidepressants[42,81,98-102] (Table 4.6).

However, the chronic mild stress procedure combined with decreased sucrose intake and/or preference is not without controversy. In a series of studies, the hypothesis that weight loss *per se* accounts for the decrease in sucrose intake was proposed.

Intermittent food restriction (a component of chronic mild stress) would be predicted to lead to reduced caloric requirements for maintenance of stable body weight and thus account for the decrease in sucrose intake.[46,103,107] Possibly masking a true hedonic deficit, this interpretation has not held up in all studies. Several studies that involve other stressor and pharmacological challenges have led to decreases in sucrose consumption and/or preference. Also, a number of studies have shown that animals subjected to chronic mild stress do not show weight changes compared to nonstressed controls.[81] Regardless, because caloric restriction-induced weight changes at a minimum confound interpretation, it may be best to employ a regimen of stressors in chronic mild stress procedures that limits the confound of food restriction.

Progressive-Ratio Responding As a Measure of the Reward Deficits Associated with the Chronic Mild Stress Model of Depression

Curiously, only a few studies have explored progressive-ratio responding as a dependent variable with the chronic mild stress paradigm. Chronic mild stress had no effect on break points when responding under a progressive-ratio schedule for 1% or 7% sucrose, although the subjects did show the previously observed decrease in consumption of 1% sucrose.[125] This contrasts with a robust decrease in progressive-ratio responding observed during psychostimulant withdrawal[74] (see above). Similar results were observed with rats exposed to chronic mild stress that responded more for pellets containing 10% sucrose and with humans exposed to depressive musical mood induction who responded more for chocolate buttons leading the authors to argue that the progressive-ratio procedure measures craving rather than reward[126] (but see above regarding drug withdrawal and progressive-ratio responding).

Other stressors such as neonatal isolation increase progressive-ratio responding for food.[127] However, others have shown decreases in progressive-ratio responding for sucrose following early deprivation[128] or no effect for progressive-ratio responding for sucrose following maternal separation.[129] Thus, there is no consistent effect of chronic mild stress or developmental stressors on progressive-ratio responding for food or sucrose. The results suggest that the decreases in sucrose consumption and increases in brain stimulation reward thresholds observed with chronic mild stress or the decreases in progressive-ratio responding and brain stimulation reward observed with drug withdrawal may be reflecting different reward disorder deficits.

VALIDITY OF ANIMAL MODELS OF REWARD DEFICITS

Clearly, there are significant differences between the sensitivity, reliability, and construct validity of the three measures of reward deficits explored in this review. Brain stimulation reward provides a reliable and sensitive measure of a reward deficit in drug withdrawal consistent with reward deficits described in the human condition. It also has been validated with manipulation of both reward and performance variables.[37] Neuropharmacological validation of brain stimulation reward has shown that agents that decrease thresholds increase reward in humans (drugs of abuse), and agents that increase thresholds generally produce dysphoric responses in humans, lending some predictive, and thus construct, validity.

With much more limited data, progressive-ratio responding shows a similar pattern, with decreases in break point for natural rewards associated with drug withdrawal and chronic mild stress. Progressive-ratio responding has a long history of validation as a measure of motivation[48,49] and is sensitive to changes in reward value.[48] Using a microeconomic framework, progressive-ratio schedules conform to labor–supply relationships[49] and allow for a definition of relative value of a reinforcer.

However, sucrose consumption and/or preference, though used extensively, has much less reliability and validity. First, a number of studies have failed to see a change in sucrose consumption and/or preference (see Table 4.5). Second, there is a confound associated with weight loss in some studies, though others report decreases in consumption without weight loss. Third, decreases in sucrose consumption are not observed reliably during drug withdrawal, or indeed at all in the human condition. Fourth, the positive data largely reflect decreases in consumption, raising the issue of nonspecific effects more related to anergia, metabolic issues, or aversions. Twenty-six of 30 studies in Table 4.5 show decreases in sucrose consumption. Only 18 of 29 studies show decreases in sucrose preference. The data in fact indicate, to a large extent, that humans with major depressive episodes report increased "craving" for sweets, the direct opposite of rats exposed to chronic mild stress. However, sucrose consumption following chronic mild stress does have predictive validity for antidepressant treatments, presumably driving its popularity as the use of sucrose consumption/preferences has increased asymptotically in the literature. One of the other reasons for the extensive use of sucrose consumption/preference presumably is due to the ease of measurement as opposed to brain stimulation reward and progressive-ratio responding, which both require specialized operant equipment.

COMMON NEUROBIOLOGICAL SUBSTRATES FOR THE REWARD DEFICITS IN ADDICTION AND DEPRESSION

The neural substrates and neuropharmacological mechanisms for the reward deficits associated with affective disorders and withdrawal from drugs of abuse may have some common neurobiological substrates for reward dysfunction that can be measured with brain stimulation reward and progressive-ratio schedules.

Neurochemical Substrates in Dependence

The neuropharmacological effects of drug withdrawal involve disruption of the same neurochemical systems and neurocircuits implicated in the positive reinforcing effects of drugs of abuse. Repeated administration of psychostimulants produces an initial facilitation of dopamine and glutamate neurotransmission.[130,131] However, chronic administration leads to decreases in dopaminergic and glutamatergic neurotransmission in the nucleus accumbens during acute withdrawal, opposite responses of opioid receptor transduction mechanisms in the nucleus accumbens during opioid withdrawal, changes in γ-aminobutyric acid (GABA) neurotransmission during alcohol withdrawal, and differential regional changes in nicotinic acetylcholine receptor function during nicotine withdrawal. These decreases in reward neurotransmitter function have been hypothesized to contribute significantly to the negative motivational state associated with acute drug abstinence and may trigger long-term biochemical changes that contribute to the clinical syndrome of protracted abstinence and vulnerability to relapse.

Different neurochemical systems involved in stress modulation also may be engaged within the neurocircuitry of the brain stress systems in an attempt to overcome the chronic presence of the perturbing drug and to restore normal function despite the presence of drug. Both the hypothalamic-pituitary-adrenal axis and the brain stress system mediated by corticotropin-releasing factor (CRF) are dysregulated by chronic administration of drugs of abuse, with a common response of elevated adrenocorticotropic hormone, corticosterone, and CRF during acute withdrawal from all major drugs of abuse. Acute withdrawal from drugs of abuse also may increase the release of norepinephrine in the bed nucleus of the stria terminalis and decrease levels of neuropeptide Y (NPY) in the amygdala.

These results suggest not only a change in the function of neurotransmitters associated with the acute reinforcing effects of drugs of abuse during the development of dependence, such as dopamine, opioid peptides, serotonin and GABA, but also recruitment of the CRF and norepinephrine brain stress systems and dysregulation of the NPY brain antistress system. Additionally, activation of the brain stress systems may contribute to the negative motivational state associated with acute abstinence and to the vulnerability to stressors observed during protracted abstinence in humans.

Thus, acute withdrawal from drugs of abuse produces opponent process-like changes in reward neurotransmitters in specific elements of reward circuitry associated with the extended amygdala, as well as recruitment of brain stress systems that motivationally oppose the hedonic effects of drugs of abuse. Such changes in these brain systems associated with the development of motivational aspects of withdrawal are hypothesized to be a major source of potential allostatic changes that drive and maintain addiction. In this context, allostasis is defined as the process of achieving stability through change; an allostatic state is a state of chronic deviation of the regulatory system from its normal (homeostatic) operating level; and allostatic load is the cost to the brain and body of the deviation, accumulating over time, and reflecting in many cases pathological states and accumulation of damage. More specifically, allostasis from the drug addiction perspective is the process of maintaining apparent reward function stability through changes in reward and stress system neurocircuitry.[2] Decreases in the function of dopamine, serotonin, and opioid peptides are hypothesized to contribute to a shift in reward set point as well as recruitment of brain stress systems such as CRF (Figure 4.3). All of these changes are hypothesized to be focused on dysregulated function within the neurocircuitry of the basal forebrain macrostructure of the extended amygdala.

Neurochemical Substrates in Depression

The reward deficits associated with major depressive disorders also have been hypothesized to derive from dysfunction in the "dark side" of the reward system

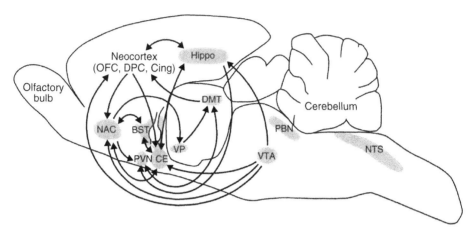

Figure 4.3   An anatomical circuit that may underlie the reward deficits associated with drug withdrawal. CE, central nucleus of the amygdala; BST, bed nucleus of the stria terminalis; LH, lateral hypothalamus; ZI, zona incerta; PBN, parabrachial nucleus; NTS, nucleus of the tractus solitarius; Hippo, hippocampus; OFC, orbitofrontal cortex; DPC, dorsolateral prefrontal cortex; Cing, cingulate cortex; DMT, dorsomedial thalamus; PVN, paraventricular nucleus; NAC, nucleus accumbens; VP, ventral pallidum[a].
[a] Adapted from a brain diagram from Schulkin et al.[11]

(i.e., brain stress systems).[11] Patients with major depressive episodes often are in a state of chronic expectation of negative outcomes, and loss of control over the environment would exaggerate normal coping responses associated with activation and preparation for stressors. Expectations of aversive events have been shown to induce anxiety and fear and activate stress hormones such as glucocorticoids. Glucocorticoids are hypothesized, in turn, to facilitate CRF expression in the amygdala. This CRF and norepinephrine in the extended amygdala are then hypothesized to mediate anticipated fearful and anxiety-producing events[132,133].[i] As a result, similar neurocircuitry has been hypothesized to mediate the reward deficits associated with major depressive episodes (Figure 4.4).

Extended Amygdala

The neuroanatomical entity termed the "extended amygdala" thus may represent not only a common anatomical substrate for the negative effects on reward function produced by stress that help drive compulsive drug administration and major depressive episodes. The extended amygdala is composed of the bed nucleus of the stria terminalis, the central nucleus of the amygdala, and a transition zone in the medial subregion

---

[i] For further discussion regarding stress systems in mood disorders such as depression and anxiety, please refer to Cryan et al., Developing more efficacious antidepressant medications: Improving and aligning preclinical and clinical assessment tools, or, Steckler et al., Developing novel anxiolytics: Improving preclinical detection and clinical assessment in Volume 1, Psychiatric Disorders, respectively.

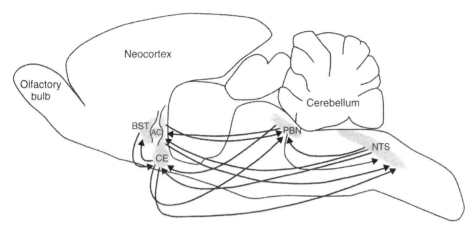

Figure 4.4 An anatomical circuit that may underlie states such as melancholic depression and adversity. BST, bed nucleus of the stria terminalis; AC, anterior commissure; CE, central nucleus of the amygdala; PBN, parabrachial nucleus; NTS, nucleus of the tractus solitarius[a].
[a] Taken with permission from Schulkin et al.[11]

of the nucleus accumbens (shell of the nucleus accumbens). Each of these regions has certain cytoarchitectural and circuitry similarities. The extended amygdala receives numerous afferents from limbic structures such as the basolateral amygdala and hippocampus, and sends efferents to the medial part of the ventral pallidum and lateral hypothalamus, thus further defining the specific brain areas that interface classical limbic (emotional) structures with the extrapyramidal motor system.

Allostasis in Reward Function

An allostatic view of motivation provides a framework for the development of reward psychopathology in a variety of domains. Allostasis originally was formulated as a hypothesis to explain the physiological basis for changes in patterns of human morbidity and mortality associated with modern life.[134] High blood pressure and other pathologies were linked to social disruption by a brain–body interaction. Using the arousal/stress continuum as their physiological framework, Peter Sterling and James Eyer argued that homeostasis was not adequate to explain such brain–body interactions, and the concept of allostasis had several unique characteristics that had more explanatory power. These characteristics include a continuous reevaluation of the organism's need and continuous readjustments to new set points, depending on demand. Allostasis can anticipate altered need, and the system can make adjustments in advance. Allostatic systems also were hypothesized to use past experience to anticipate demand.[134]

Extended to the domains of stress and the hypothalamic-pituitary-adrenal axis by Bruce McEwen[135,136] and anxiety disorders and central CRF by Jay Schulkin et al.,[11] the concept of allostatic load was introduced, which is the price the body pays to adapt to adverse psychosocial or physical situations.[136] Allostatic load represents either external demands, such as too much stress, or internal demands, such as inefficient operation of the stress hormone response system. Similar connections have been made between allostatic changes in brain stress systems and posttraumatic stress disorder and anxiety disorders.[11,137] A positive feedback interaction between glucocorticoids and CRF in the extended amygdala was hypothesized by Schulkin and colleagues[11] as a substrate for specific symptoms of depressed patients such as expectation of adversity or negative outcomes. They argued that the loss of predictability and loss of perceived control lead to perpetual anticipation, are mediated by CRF in the amygdala, and are modulated by glucocorticoids. They provided a neurobiological basis for the allostatic load and pathological arousal associated with melancholic depression. This early extension of the concept of allostasis clearly anticipated some of the observations associated with reward deficits in animal models of drug dependence.

Others similarly have argued that depression is an established outcome of stress (and indeed, an ongoing depressive episode itself can constitute a chronic stress) and fits an allostatic model.[138] The neuroendocrinology of human depression closely resembles that of chronic stress in the laboratory, including increased hypothalamic-pituitary-adrenal axis activity, reduced glucocorticoid feedback, and dysregulated diurnal rhythms or cortisol.[139] Consistent with the concept of allostatic load, there is a strong relationship between the number of depressive symptoms exhibited by subjects, premature mortality, and cardiovascular risk factors.[138] In addition, one can reasonably see how both developmental and genetic domains can modify allostatic load that may determine vulnerability to pathology.[136] The challenge for future research will be to explore how the neurochemical/neurocircuitry changes associated with drug addiction, including the dysregulation of brain stress systems, extend to other mood and motivational disorders.

TRANSLATIONAL VALUE OF THESE MODELS AND PROCEDURES FOR THE STUDY OF REWARD DEFICITS IN MAN

The present chapter has reviewed three dependent variables as measures of reward deficits in animal models using two independent procedures for inducing reward deficit-like effects in animals. Ignoring the circularity of reasoning here, it is obvious that withdrawal from drugs of abuse effectively increases reward thresholds for brain stimulation reward. These results convey some reciprocal construct validity in that these results suggest that withdrawal from drugs of abuse produces reward deficits, as observed in humans; and conversely, brain stimulation reward is a valid measure of brain reward. Elevations in brain stimulation reward with chronic mild stress is not reproducible in different laboratories,[140,141] casting doubt on the robustness of the independent variable, chronic mild stress. Progressive-ratio responding holds up equally well for withdrawal from drugs of abuse, but not so for chronic mild stress with limited studies in animals. From a translational perspective, human patients exposed to depressive mood induction actually showed increased progressive-ratio responding for chocolate rewards.[126] Similarly, sucrose preference holds up well for chronic mild stress, but not for withdrawal from drugs of abuse in animal models. Again in humans, there is an increase in preference for sweet solutions for both nicotine and ethanol.[70,71] Curiously, the model with the most predictive validity for existing antidepressant medications is the sucrose preference measure.

Several interesting conclusions *vis a vis* translational efforts that point to future studies can be derived from this work. First, it would be useful to have a human laboratory model of the reward measures isomorphic with brain stimulation reward. There are several hedonic scales currently in use: Positive and Negative Affect Scale (PANAS), Fawcett–Clark Pleasure Scale (FCPS), and Beck Depression Inventory-II (BDI-II).[142-144] However, a possible novel approach would be actually to have human subjects respond to hedonic stimuli as opposed to self-reporting on endogenous hedonic states.[ii] Second, measures in humans that reflect loss of hedonic function may be translated to the animal models. Some such measures include progressive-ratio responding for items without caloric value, anticipation of reward phenomena, and impaired reward responsiveness using a signal detection differential reinforcement approach.[145]

PROSPECTS FOR THE DEVELOPMENT OF NEW THERAPEUTICS OF REWARD DEFICIT DISORDERS

New therapeutics for reward deficit disorders will require validated measures of reward deficits. Outlined above are three measures that have some construct validity. How they translate to the human condition has been discussed above, but one further element of translation would be how they predict novel treatments for reward deficit disorders in humans. To date, most studies have explored only known antidepressants or functional derivatives and sucrose intake and preference. Exceptions include ketoconazole, a steroid synthesis inhibitor and an herbal medicine with an unknown mechanism. Clearly, novel drugs for novel targets that show selective effects on reward deficits will be an "out of the box" advance. Embedded also in this discourse is the observation that some of the brain hedonic systems, particularly those implicated in the "dark side"[146] may overlap between the reward deficits observed with drugs of abuse and depression. It is these targets that may provide new directions in the pursuit of novel therapeutics for reward deficit disorders.

SUMMARY AND CONCLUSIONS

In summary, from an "academic" perspective, there *are* reliable and validated measures of reward deficits associated with psychiatric disorders. Three measures were considered: brain stimulation reward, progressive-ratio, and sucrose consumption/preference. Brain stimulation reward and progressive-ratio, while used less than sucrose consumption/preference, have more reliability and construct validity to date. Both have been

---

[ii] Please refer to Gardener *et al.*, Issues in designing and conducting clinical trials for reward disorders: A clinical view, in this volume for further discussion regarding human testing paradigms and the use of self-report scales.

rigorously tested for responses to reward value (quantity of reinforcer) and have been shown to be sensitive. Both are resistant to performance confounds. Both generalize to different independent variables known to produce reward deficits in humans. Sucrose consumption appears to be a reliable predictor of antidepressant efficacy. However, the construct validity of sucrose consumption is limited. Sucrose consumption/preference responds to changes in sucrose concentration, but this variable is rarely tested in reward deficit studies. Sucrose consumption is sensitive to performance variables in the sense that taste aversions and loss of weight can confound interpretation. Finally, sucrose consumption does not generalize well to other independent variables, and indeed in the human situation there is evidence of the opposite effect where reward deficit disorders show increases in sweet solution craving. Thus, sucrose consumption/preference to date has excellent predictive validity in the narrow sense and is much more readily employed without sophisticated equipment, and as such probably makes an excellent initial screen for reward deficit disorders. However, progressive-ratio responding and brain stimulation reward would appear at the present time to be important confirmatory procedures that would lend construct validity to any reward deficit disorder hypothesis tested.

New efforts should be made on these fronts to advance the field. First, certain animal models should be used to create human laboratory measures of reward deficits. Second, human laboratory measures of reward deficits should be used to create novel animal models. Third, novel targets based on the common neurobiology of reward deficits should be explored with existing reward deficit models and models yet to be developed. Fourth, genetic models of reward deficits should be combined with environmental challenges to produce better independent variables related to depression.

ACKNOWLEDGMENTS

This is publication number 19124 from The Scripps Research Institute. The author would like to thank Mike Arends and Mellany Santos for their assistance with manuscript preparation.

REFERENCES

1. American Psychiatric Association (1994). *Diagnostic and Statistical Manual of Mental Disorders*, 4th edition. American Psychiatric Press, Washington, DC.
2. Koob, G.F. and Le Moal, M. (2001). Drug addiction, dysregulation of reward, and allostasis. *Neuropsychopharmacology*, 24:97–129.
3. Hebb, D.O. (1949). *Organization of behavior: A neuropsychological theory*. Wiley, New York.
4. Richter, C.P. (1927). Animal behavior and internal drives. *Q Rev Biol*, 2:307–343.
5. Bindra, D. (1976). *A theory of intelligent behavior*. Wiley, New York.
6. Kling, J.W. and Riggs, L.A. (1971). *Woodworth and Schlosberg's experimental psychology*, 3rd edition. Holt, Rinehart and Winston, New York.

7. Solomon, R.L. and Corbit, J.D. (1974). An opponent-process theory of motivation: 1. Temporal dynamics of affect. *Psychol Rev*, 81:119–145.

8. Solomon, R.L. (1980). The opponent-process theory of acquired motivation: The costs of pleasure and the benefits of pain. *Am Psychol*, 35:691–712.

9. Siegel, S. (1975). Evidence from rats that morphine tolerance is a learned response. *J Comp Physiol Psychol*, 89:498–506.

10. Poulos, C.X. and Cappell, H. (1991). Homeostatic theory of drug tolerance: A general model of physiological adaptation. *Psychol Rev*, 98:390–408.

11. Schulkin, J., McEwen, B.S., and Gold, P.W. (1994). Allostasis, amygdala, and anticipatory angst. *Neurosci Biobehav Rev*, 18:385–396.

12. Koob, G.F. and Le Moal, M. (2006). *Neurobiology of addiction*. Academic Press, London.

13. Markou, A., Kosten, T.R., and Koob, G.F. (1998). Neurobiological similarities in depression and drug dependence: A self-medication hypothesis. *Neuropsychopharmacology*, 18:135–174.

14. Kosten, T.R., Markou, A., and Koob, G.F. (1998). Depression and stimulant dependence: Neurobiology and pharmacotherapy. *J Nerv Ment Dis*, 186:737–745.

15. Levinson, D.F. (2006). The genetics of depression: A review. *Biol Psychiatry*, 60:84–92.

16. Wurtman, R.J. (2005). Genes, stress, and depression. *Metabolism*, 54(5 Suppl 1):6–19.

17. Moller, H.J. (2003). Suicide, suicidality and suicide prevention in affective disorders. *Acta Psychiatry Scand Suppl*, 418:73–80.

18. Ebel, R.L. (1961). Must all tests be valid? *Am Psychol*, 16:640–647.

19. Sayette, M.A., Shiffman, S., Tiffany, S.T., Niaura, R.S., Martin, C.S., and Shadel, W.G. (2000). The measurement of drug craving. *Addiction*, 95(Suppl 2):S189–S210.

20. Cronbach, L.J. and Meehl, P.E. (1955). Construct validity in psychological tests. *Psychol Bull*, 52:281–302.

21. Katz, J.L. and Higgins, S.T. (2003). The validity of the reinstatement model of craving and relapse to drug use. *Psychopharmacology*, 168:21–30. [erratum: 168:244].

22. McKinney, W.T. (1988). *Models of mental disorders: A new comparative psychiatry*. Plenum, New York.

23. Geyer, M.A. and Markou, A. (1995). Animal models of psychiatric disorders. In Bloom, F.E. and Kupfer, D.J. (eds.), *Psychopharmacology: The fourth generation of progress*. Raven Press, New York, pp. 787–798.

24. Willner, P. (1984). The validity of animal models of depression. *Psychopharmacology*, 83:1–16.

25. Matthysse, S. (1986). Animal models in psychiatric research. *Prog Brain Res*, 65:259–270.

26. Ellenbroek, B. and Cools, A.R. (1988). The Paw test: An animal model for neuroleptic drugs which fulfils the criteria for pharmacological isomorphism. *Life Sci*, 42:1205–1213.

27. Markou, A. and Koob, G.F. (1992). Construct validity of a self-stimulation threshold paradigm: Effects of reward and performance manipulations. *Physiol Behav*, 51:111–119.

28. Olds, J. and Milner, P. (1954). Positive reinforcement produced by electrical stimulation of septal area and other regions of rat brain. *J Comp Physiol Psychol*, 47:419–427.

29. Olds, M.E. and Forbes, J.C. (1981). Central basis of motivation: Self-stimulation studies. *Ann Dev Psychol*, 32:527–574.

30. Kornetsky, C. and Bain, G. (1992). Brain-stimulation reward: A model for the study of the rewarding effects of abused drugs. In Fracella, J. and Brown, R.M. (eds.), *Neurobiological approaches to brain–behavior interaction*. National Institute on Drug Abuse, Rockville, MD, pp. 73–93. (NIDA research monograph; Vol. 124).

31. Campbell, K.A., Evans, G., and Gallistel, C.R. (1985). A microcomputer-based method for physiologically interpretable measurement of the rewarding efficacy of brain stimulation. *Physiol Behav*, 35:395–403.

32. Kornetsky, C. and Esposito, R.U. (1979). Euphorigenic drugs: Effects on the reward pathways of the brain. *Fed Proc*, 38:2473–2476.

33. Miliaressis, E., Rompre, P.P., Laviolette, P., Philippe, L., and Coulombe, D. (1986). The curve-shift paradigm in self-stimulation. *Physiol Behav*, 37:85-91.

34. Edmonds, D.E. and Gallistel, C.R. (1974). Parametric analysis of brain stimulation reward in the rat: III. Effect of performance variables on the reward summation function. *J Comp Physiol Psychol*, 87:876-883.

35. Fouriezos, G., Bielajew, C., and Pagotto, W. (1990). Task difficulty increases thresholds of rewarding brain stimulation. *Behav Brain Res*, 37:1-7.

36. Bauco, P. and Wise, R.A. (1997). Synergistic effects of cocaine with lateral hypothalamic brain stimulation reward: Lack of tolerance or sensitization. *J Pharmacol Exp Ther*, 283:1160-1167.

37. Markou, A. and Koob, G.F. (1991). Post-cocaine anhedonia: An animal model of cocaine withdrawal. *Neuropsychopharmacology*, 4:17-26.

38. Muscat, R., Kyprianou, T., Osman, M., Phillips, G., and Willner, P. (1991). Sweetness-dependent facilitation of sucrose drinking by raclopride is unrelated to calorie content. *Pharmacol Biochem Behav*, 40:209-213.

39. Young, P.T. and Madsen, C.H., Jr. (1963). Individual isohedons in sucrose-sodium chloride and sucrose-saccharin gustatory areas. *J Comp Physiol Psychol*, 56:903-909.

40. Guttman, N. (1953). Operant conditioning, extinction, and periodic reinforcement in relation to concentration of sucrose used as reinforcing agent. *J Exp Psychol*, 46:213-224.

41. Willner, P., Muscat, R., and Papp, M. (1992). Chronic mild stress-induced anhedonia: A realistic animal model of depression. *Neurosci Biobehav Rev*, 16:525-534.

42. Willner, P., Towell, A., Sampson, D., Sophokleous, S., and Muscat, R. (1987). Reduction of sucrose preference by chronic unpredictable mild stress, and its restoration by a tricyclic antidepressant. *Psychopharmacology*, 93:358-364.

43. Muscat, R. and Willner, P. (1992). Suppression of sucrose drinking by chronic mild unpredictable stress: A methodological analysis. *Neurosci Biobehav Rev*, 16:507-517.

44. Ayensu, W.K., Pucilowski, O., Mason, G.A., Overstreet, D.H., Rezvani, A.H., and Janowsky, D.S. (1995). Effects of chronic mild stress on serum complement activity, saccharin preference, and corticosterone levels in Flinders lines of rats. *Physiol Behav*, 57:165-169.

45. Willner, P. (1997). Validity, reliability and utility of the chronic mild stress model of depression: A 10-year review and evaluation. *Psychopharmacology*, 134:319-329.

46. Reid, I., Forbes, N., Stewart, C., and Matthews, K. (1997). Chronic mild stress and depressive disorder: A useful new model? *Psychopharmacology*, 134:365-367. (discussion: 371-377).

47. Hodos, W. (1961). Progressive ratio as a measure of reward strength. *Science*, 134:943-944.

48. Hodos, W. and Kalman, G. (1963). Effects of increment size and reinforcer volume on progressive ratio performance. *J Exp Anal Behav*, 6:387-392.

49. Rowlett, J.K. (2000). A labor-supply analysis of cocaine self-administration under progressive-ratio schedules: Antecedents, methodologies, and perspectives. *Psychopharmacology*, 153:1-16.

50. American Psychiatric Association (2000). *Diagnostic and Statistical Manual of Mental Disorders*, 4th edition, text revision. American Psychiatric Press, Washington, DC.

51. Koob, G.F. and Le Moal, M. (1997). Drug abuse: Hedonic homeostatic dysregulation. *Science*, 278:52-58.

52. Koob, G.F. (2004). Allostatic view of motivation: Implications for psychopathology. In Bevins, R.A. and Bardo, M.T. (eds.), *Motivational Factors in the Etiology of Drug Abuse*. University of Nebraska Press, Lincoln, NE, pp. 1-18. (Nebraska symposium on motivation; Vol. 50).

53. Gawin, F.H. and Kleber, H.D. (1986). Abstinence symptomatology and psychiatric diagnosis in cocaine abusers: Clinical observations. *Arch Gen Psychiatry*, 43:107-113.

54. Barr, A.M. and Markou, A. (2005). Psychostimulant withdrawal as an inducing condition in animal models of depression. *Neurosci Biobehav Rev*, 29:675-706.

55. Stellar, J.R. and Rice, M.B. (1989). Pharmacological basis of intracranial self-stimulation reward. In Liebman, J.M. and Cooper, S.J. (eds.), *The neuropharmacological basis of reward*. Clarendon Press, Oxford, pp. 14-65. (Topics in experimental psychopharmacology; Vol. 1).

56. Kokkinidis, L. and McCarter, B.D. (1990). Postcocaine depression and sensitization of brain-stimulation reward: Analysis of reinforcement and performance effects. *Pharmacol Biochem Behav*, 36:463-471.

57. Leith, N.J. and Barrett, R.J. (1976). Amphetamine and the reward system: Evidence for tolerance and post-drug depression. *Psychopharmacologia*, 46:19-25.

58. Schulteis, G., Markou, A., Gold, L.H., Stinus, L., and Koob, G.F. (1994). Relative sensitivity to naloxone of multiple indices of opiate withdrawal: A quantitative dose-response analysis. *J Pharmacol Exp Ther*, 271:1391-1398.

59. Schulteis, G., Markou, A., Cole, M., and Koob, G.F. (1995). Decreased brain reward produced by ethanol withdrawal. *Proc Natl Acad Sci USA*, 92:5880-5884.

60. Epping-Jordan, M.P., Watkins, S.S., Koob, G.F., and Markou, A. (1998). Dramatic decreases in brain reward function during nicotine withdrawal. *Nature*, 393:76-79.

61. Gardner, E.L. and Vorel, S.R. (1998). Cannabinoid transmission and reward-related events. *Neurobiol Dis*, 5:502-533.

62. Paterson, N.E., Myers, C., and Markou, A. (2000). Effects of repeated withdrawal from continuous amphetamine administration on brain reward function in rats. *Psychopharmacology*, 152:440-446.

63. Moolten, M. and Kornetsky, C. (1990). Oral self-administration of ethanol and not experimenter-administered ethanol facilitates rewarding electrical brain stimulation. *Alcohol*, 7:221-225.

64. Kornetsky, C. and Bain, G. (1990). Brain-stimulation reward: A model for drug induced euphoria. In Adler, M.W. and Cowan, A. (eds.), *Testing and evaluation of drugs of abuse*. Wiley-Liss, New York, pp. 211-231. (Modern methods in pharmacology; Vol. 6).

65. Slawecki, C.J. and Roth, J. (2004). Comparison of the onset of hypoactivity and anxiety-like behavior during alcohol withdrawal in adolescent and adult rats. *Alcohol Clin Exp Res*, 28:598-607.

66. Hellemans, K.G., Shaham, Y., and Olmstead, M.C. (2002). Effects of acute and prolonged opiate abstinence on extinction behaviour in rats. *Can J Exp Psychol*, 56:241-252.

67. Parker, L.A. and Doucet, K. (1995). The effects of nicotine and nicotine withdrawal on taste reactivity. *Pharmacol Biochem Behav*, 52:125-129.

68. Schoenbaum, G.M., Martin, R.J., and Roane, D.S. (1989). Relationships between sustained sucrose-feeding and opioid tolerance and withdrawal. *Pharmacol Biochem Behav*, 34:911-914.

69. Pilcher, C.W. and Stolerman, I.P. (1976). Conditioned flavor aversions for assessing precipitated morphine abstinence in rats. *Pharmacol Biochem Behav*, 4:159-163.

70. Hatsukami, D., LaBounty, L., Hughes, J., and Laine, D. (1993). Effects of tobacco abstinence on food intake among cigarette smokers. *Health Psychol*, 12:499-502.

71. Yung, L., Gordis, E., and Holt, J. (1983). Dietary choices and likelihood of abstinence among alcoholic patients in an outpatient clinic. *Drug Alcohol Depend*, 12:355-362.

72. Wise, R.A. and Munn, E. (1995). Withdrawal from chronic amphetamine elevates baseline intracranial self-stimulation thresholds. *Psychopharmacology*, 117:130-136.

73. Pulvirenti, L. and Koob, G.F. (1993). Lisuride reduces psychomotor retardation during withdrawal from chronic intravenous amphetamine self-administration in rats. *Neuropsychopharmacology*, 8:213-218.

74. Barr, A.M. and Phillips, A.G. (1999). Withdrawal following repeated exposure to D-amphetamine decreases responding for a sucrose solution as measured by a progressive ratio schedule of reinforcement. *Psychopharmacology*, 141:99-106.

75. Orsini, C., Koob, G.F., and Pulvirenti, L. (2001). Dopamine partial agonist reverses amphetamine withdrawal in rats. *Neuropsychopharmacology*, 25:789–792.

76. Hoefer, M.E., Voskanian, S.J., Koob, G.F., and Pulvirenti, L. (2006). Effects of terguride, ropinirole, and acetyl-L-carnitine on methamphetamine withdrawal in the rat. *Pharmacol Biochem Behav*, 83:403–409.

77. Russig, H., Pezze, M.A., Nanz-Bahr, N.I., Pryce, C.R., Feldon, J., and Murphy, C.A. (2003). Amphetamine withdrawal does not produce a depressive-like state in rats as measured by three behavioral tests. *Behav Pharmacol*, 14:1–18.

78. LeSage, M.G., Burroughs, D., and Pentel, P.R. (2006). Effects of nicotine withdrawal on performance under a progressive-ratio schedule of sucrose pellet delivery in rats. *Pharmacol Biochem Behav*, 83:585–591.

79. Zhang, D., Zhou, X., Wang, X., Xiang, X., Chen, H., and Hao, W. (2007). Morphine withdrawal decreases responding reinforced by sucrose self-administration in progressive ratio. *Addiction Biol*, 12:152–157.

80. Schwabe, K. and Koch, M. (2007). Effects of aripiprazole on operant responding for a natural reward after psychostimulant withdrawal in rats. *Psychopharmacology*, 191:759–765.

81. Yu, J., Liu, Q., Wang, Y.Q., Wang, J., Li, X.Y., Cao, X.D. *et al.* (2007). Electroacupuncture combined with clomipramine enhances antidepressant effect in rodents. *Neurosci Lett*, 421:5–9.

82. Aneshensel, C.S. and Stone, J.D. (1982). Stress and depression: A test of the buffering model of social support. *Arch Gen Psychiatry*, 39:1392–1396.

83. Moreau, J.-L. (1997). Reliable monitoring of hedonic deficits in the chronic mild stress model of depression. *Psychopharmacology*, 134:357–358.

84. Moreau, J.-L., Jenck, F., Martin, J.R., Mortas, P., and Haefely, W.E. (1992). Antidepressant treatment prevents chronic unpredictable mild stress-induced anhedonia as assessed by ventral tegmentum self-stimulation behavior in rats. *Eur Neuropsychopharmacol*, 2:43–49.

85. Moreau, J.-L., Jenck, F., Martin, J.R., Mortas, P., and Haefely, W. (1993). Effects of moclobemide, a new generation reversible Mao-A inhibitor, in a novel animal model of depression. *Pharmacopsychiatry*, 26:30–33.

86. Moreau, J.-L., Borgulya, J., Jenck, F., and Martin, J.R. (1994a). Tolcapone: A potential new antidepressant detected in a novel animal model of depression. *Behav Pharmacol*, 5:344–350.

87. Moreau, J.-L., Bourson, A., Jenck, F., Martin, J.R., and Mortas, P. (1994b). Curative effects of the atypical antidepressant mianserin in the chronic mild stress-induced anhedonia model of depression. *J Psychiatry Neurosci*, 19:51–56.

88. Moreau, J.-L., Scherschlicht, R., Jenck, F., and Martin, J.R. (1995). Chronic mild stress-induced anhedonia model of depression: Sleep abnormalities and curative effects of electroshock treatment. *Behav Pharmacol*, 6:682–687.

89. Moreau, J.-L., Bos, M., Jenck, F., Martin, J.R., Mortas, P., and Wichmann, J. (1996). 5HT$_2$C receptor agonists exhibit antidepressant-like properties in the anhedonia model of depression in rats. *Eur Neuropsychopharmacol*, 6:169–175.

90. Katz, R.J. (1982). Animal model of depression: Pharmacological sensitivity of a hedonic deficit. *Pharmacol Biochem Behav*, 16:965–968.

91. Von Frijtag, J.C., Van den Bos, R., and Spruijt, B.M. (2002). Imipramine restores the long-term impairment of appetitive behavior in socially stressed rats. *Psychopharmacology*, 162:232–238.

92. Duncko, R., Schwendt, M., and Jezova, D. (2003). Altered glutamate receptor and corticoliberin gene expression in brain regions related to hedonic behavior in rats. *Pharmacol Biochem Behav*, 76:9–16.

93. Sammut, S., Goodall, G., and Muscat, R. (2001). Acute interferon-alpha administration modulates sucrose consumption in the rat. *Psychoneuroendocrinology*, 26:261–272.

94. Merali, Z., Brennan, K., Brau, P., and Anisman, H. (2003). Dissociating anorexia and anhedonia elicited by interleukin-1beta: Antidepressant and gender effects on responding for "free chow" and "earned" sucrose intake. *Psychopharmacology*, 165:413-418.
95. Kwant, A. and Sakic, B. (2004). Behavioral effects of infection with interferon-gamma adenovector. *Behav Brain Res*, 151:73-82.
96. Stock, H.S., Ford, K., and Wilson, M.A. (2000). Gender and gonadal hormone effects in the olfactory bulbectomy animal model of depression. *Pharmacol Biochem Behav*, 67:183-191.
97. Slattery, D.A., Markou, A., and Cryan, J.F. (2007). Evaluation of reward processes in an animal model of depression. *Psychopharmacology*, 190:555-568.
98. Muscat, R., Papp, M., and Willner, P. (1992). Reversal of stress-induced anhedonia by the atypical antidepressants, fluoxetine and maprotiline. *Psychopharmacology*, 109:433-438.
99. Cheeta, S., Broekkamp, C., and Willner, P. (1994). Stereospecific reversal of stress-induced anhedonia by mianserin and its (+)-enantiomer. *Psychopharmacology*, 116:523-528.
100. Rygula, R., Abumaria, N., Domenici, E., Hiemke, C., and Fuchs, E. (2006a). Effects of fluoxetine on behavioral deficits evoked by chronic social stress in rats. *Behav Brain Res*, 174:188-192.
101. Rygula, R., Abumaria, N., Flugge, G., Hiemke, C., Fuchs, E., Ruther, E. *et al.* (2006b). Citalopram counteracts depressive-like symptoms evoked by chronic social stress in rats. *Behav Pharmacol*, 17:19-29.
102. Casarotto, P.C. and Andreatini, R. (2007). Repeated paroxetine treatment reverses anhedonia induced in rats by chronic mild stress or dexamethasone. *Eur Neuropsychopharmacol*, 17:735-742.
103. Forbes, N.F., Stewart, C.A., Matthews, K., and Reid, I.C. (1996). Chronic mild stress and sucrose consumption: Validity as a model of depression. *Physiol Behav*, 60:1481-1484.
104. Papp, M., Willner, P., and Muscat, R. (1993). Behavioural sensitization to a dopamine agonist is associated with reversal of stress-induced anhedonia. *Psychopharmacology*, 110:159-164.
105. Papp, M. and Moryl, E. (1993). New evidence for the antidepressant activity of MK-801, a non-competitive antagonist of NMDA receptors. *Polish J Pharmacol*, 45:549-553.
106. Sluzewska, A. and Nowakowska, E. (1994). The effects of carbamazepine, lithium and ketoconazole in chronic mild stress model of depression in rats. *Behav Pharmacol*, 5(Suppl 1):86.
107. Matthews, K., Forbes, N., and Reid, I.C. (1995). Sucrose consumption as an hedonic measure following chronic unpredictable mild stress. *Physiol Behav*, 57:241-248.
108. Marona-Lewicka, D. and Nichols, D.E. (1997). The effect of selective serotonin releasing agents in the chronic mild stress model of depression in rats. *Stress*, 2:91-100.
109. Harris, R.B., Zhou, J., Youngblood, B.D., Smagin, G.N., and Ryan, D.H. (1997). Failure to change exploration or saccharin preference in rats exposed to chronic mild stress. *Physiol Behav*, 63:91-100.
110. Grippo, A.J., Beltz, T.G., and Johnson, A.K. (2003). Behavioral and cardiovascular changes in the chronic mild stress model of depression. *Physiol Behav*, 78:703-710.
111. Li, J.M., Kong, L.D., Wang, Y.M., Cheng, C.H., Zhang, W.Y., and Tan, W.Z. (2003). Behavioral and biochemical studies on chronic mild stress models in rats treated with a Chinese traditional prescription Banxia-houpu decoction. *Life Sci*, 74:55-73.
112. Konkle, A.T., Baker, S.L., Kentner, A.C., Barbagallo, L.S., Merali, Z., and Bielajew, C. (2003). Evaluation of the effects of chronic mild stressors on hedonic and physiological responses: Sex and strain compared. *Brain Res*, 992:227-238.
113. Strekalova, T., Spanagel, R., Bartsch, D., Henn, F.A., and Gass, P. (2004). Stress-induced anhedonia in mice is associated with deficits in forced swimming and exploration. *Neuropsychopharmacology*, 29:2007-2017.

114. Gronli, J., Murison, R., Fiske, E., Bjorvatn, B., Sorensen, E., Portas, C.M. *et al.* (2005). Effects of chronic mild stress on sexual behavior, locomotor activity and consumption of sucrose and saccharine solutions. *Physiol Behav*, 84:571–577.

115. Baker, S.L., Kentner, A.C., Konkle, A.T., Santa-Maria Barbagallo, L., and Bielajew, C. (2006). Behavioral and physiological effects of chronic mild stress in female rats. *Physiol Behav*, 87:314–322.

116. Yu, J., Li, X.Y., Cao, X.D., and Wu, G.C. (2006). Sucrose preference is restored by electro-acupuncture combined with chlorimipramine in the depression-model rats. *Acupunct Electrother Res*, 31:223–232.

117. Baker, S. and Bielajew, C. (2007). Influence of housing on the consequences of chronic mild stress in female rats. *Stress*, 10:283–293.

118. Rygula, R., Abumaria, N., Flugge, G., Fuchs, E., Ruther, E., and Havemann-Reinecke, U. (2005). Anhedonia and motivational deficits in rats: Impact of chronic social stress. *Behav Brain Res*, 162:127–134.

119. Grippo, A.J., Cushing, B.S., and Carter, C.S. (2007a). Depression-like behavior and stressor-induced neuroendocrine activation in female prairie voles exposed to chronic social isolation. *Psychosom Med*, 69:149–157.

120. Grippo, A.J., Gerena, D., Huang, J., Kumar, N., Shah, M., Ughreja, R. *et al.* (2007b). Social isolation induces behavioral and neuroendocrine disturbances relevant to depression in female and male prairie voles. *Psychoneuroendocrinology*, 32:966–980.

121. Skalisz, L.L., Beijamini, V., Joca, S.L., Vital, M.A., Da Cunha, C., and Andreatini, R. (2002). Evaluation of the face validity of reserpine administration as an animal model of depression: Parkinson's disease association. *Prog Neuropsychopharmacol Biol Psychiatry*, 26:879–883.

122. Sobrian, S.K., Marr, L., and Ressman, K. (2003). Prenatal cocaine and/or nicotine exposure produces depression and anxiety in aging rats. *Prog Neuropsychopharmacol Biol Psychiatry*, 27:501–518.

123. Kalueff, A.V., Gallagher, P.S., and Murphy, D.L. (2006). Are serotonin transporter knockout mice "depressed"? hypoactivity but no anhedonia. *Neuroreport*, 17:1347–1351.

124. Mlynarik, M., Zelena, D., Bagdy, G., Makara, G.B., and Jezova, D. (2007). Signs of attenuated depression-like behavior in vasopressin deficient Brattleboro rats. *Horm Behav*, 51:395–405.

125. Barr, A.M. and Phillips, A.G. (1998). Chronic mild stress has no effect on responding by rats for sucrose under a progressive ratio schedule. *Physiol Behav*, 64:591–597.

126. Willner, P., Benton, D., Brown, E., Cheeta, S., Davies, G., Morgan, J., and Morgan, M. (1998). "Depression" increases "craving" for sweet rewards in animal and human models of depression and craving. *Psychopharmacology*, 136:272–283.

127. Kosten, T.A., Zhang, X.Y., and Kehoe, P. (2006). Heightened cocaine and food self-administration in female rats with neonatal isolation experience. *Neuropsychopharmacology*, 31:70–76.

128. Ruedi-Bettschen, D., Pedersen, E.M., Feldon, J., and Pryce, C.R. (2005). Early deprivation under specific conditions leads to reduced interest in reward in adulthood in Wistar rats. *Behav Brain Res*, 156:297–310.

129. Shalev, U. and Kafkafi, N. (2002). Repeated maternal separation does not alter sucrose-reinforced and open-field behaviors. *Pharmacol Biochem Behav*, 73:115–122.

130. Ungless, M.A., Whistler, J.L., Malenka, R.C., and Bonci, A. (2001). Single cocaine exposure in vivo induces long-term potentiation in dopamine neurons. *Nature*, 411:583–587.

131. Vorel, S.R., Ashby, C.R., Jr., Paul, M., Liu, X., Hayes, R., Hagan, J.J. *et al.* (2002). Dopamine D3 receptor antagonism inhibits cocaine-seeking and cocaine-enhanced brain reward in rats. *J Neurosci*, 22:9595–9603.

132. Koob, G.F. (1999). Corticotropin-releasing factor, norepinephrine and stress. *Biol Psychiatry*, 46:1167–1180.

133. Heinrichs, S.C. and Koob, G.F. (2004). Corticotropin-releasing factor in brain: A role in activation, arousal, and affect regulation. *J Pharmacol Exp Ther*, 311:427–440.

134. Sterling, P. and Eyer, J. (1988). Allostasis: A new paradigm to explain arousal pathology. In Fisher, S. and Reason, J. (eds.), *Handbook of life stress, cognition and health*. John Wiley, Chichester, pp. 629–649.

135. McEwen, B.S. (1998). Stress, adaptation, and disease: Allostasis and allostatic load. In McCann, S.M., Lipton, J.M., Sternberg, E.M., Chrousos, G.P., Gold, P.W., and Smith, C.C. (eds.), *Neuroimmunomodulation: Molecular aspects, integrative systems, and clinical advances*. New York Academy of Sciences, New York, pp. 33–44. (Annals of the New York Academy of Sciences; Vol. 840).

136. McEwen, B.S. (2000). Allostasis and allostatic load: Implications for neuropsychopharmacology. *Neuropsychopharmacology*, 22:108–124.

137. Lindy, J.D. and Wilson, J.P. (2001). An allostatic approach to the psychodynamic understanding of PTSD. In Wilson, J.P., Friedman, M.J., and Lindy, J.D. (eds.), *Treating psychological trauma and PTSD*. Guilford Press, New York, pp. 125–138.

138. Carroll, B.J. (2002). Ageing, stress and the brain. In Chadwick, D.J. and Goode, J.A. (eds.), *Endocrine facets of ageing*. Wiley, New York, pp. 26–45. (Novartis Foundation Symposium; Vol. 242).

139. Checkley, S. (1996). The neuroendocrinology of depression and chronic stress. *Br Med Bull*, 52:597–617.

140. Lin, D., Bruijnzeel, A.W., Schmidt, P., and Markou, A. (2002). Exposure to chronic mild stress alters thresholds for lateral hypothalamic stimulation reward and subsequent responsiveness to amphetamine. *Neuroscience*, 114:925–933.

141. Nielsen, C.K., Arnt, J., and Sanchez, C. (2000). Intracranial self-stimulation and sucrose intake differ as hedonic measures following chronic mild stress: Interstrain and interindividual differences. *Behav Brain Res*, 107:21–33.

142. Watson, D., Clark, L.A., and Tellegen, A. (1988). Development and validation of brief measures of positive and negative affect: The PANAS scales. *J Pers Soc Psychol*, 54:1063–1070.

143. Fawcett, J., Clark, D.C., Scheftner, W.A., and Gibbons, R.D. (1983). Assessing anhedonia in psychiatric patients: The Pleasure Scale. *Arch Gen Psychiatry*, 40:79–84.

144. Beck, A.T., Ward, C.H., Mendelson, M., Mock, J., and Erbaugh, J. (1961). An inventory for measuring depression. *Arch Gen Psychiatry*, 4:561–571.

145. Pizzagalli, D.A., Jahn, A.L., and O'Shea, J.P. (2005). Toward an objective characterization of an anhedonic phenotype: A signal-detection approach. *Biol Psychiatry*, 57:319–327.

146. Koob, G.F. and Le Moal, M. (2005). Plasticity of reward neurocircuitry and the "dark side" of drug addiction. *Nat Neurosci*, 8:1442–1444.

CHAPTER 5

Pharmacotherapy of Alcohol Dependence: Improving Translation from the Bench to the Clinic

Hilary J. Little[1], David L. McKinzie[2], Beatrice Setnik[3], Megan J. Shram[3] and Edward M. Sellers[3,4]

[1]Departments of Basic Medical Sciences and Mental Health, St George's, University of London, Cranmer Terrace, London, UK
[2]Eli Lilly & Co., Lilly Research Laboratories, Lilly Corporate Center DC0510, Indianapolis, IN, USA
[3]DecisionLine Clinical Research Corporation, Toronto, ON, Canada
[4]Departments of Pharmacology, Medicine and Psychiatry, University of Toronto, ON, Canada

Animal and Translational Models for CNS Drug Discovery,
Vol. 3 of 3: Reward Deficit Disorders
Robert McArthur and Franco Borsini (eds). Academic Press, 2008

INTRODUCTION

Epidemiology/Clinical and Societal Impact

Alcohol abuse and dependence are the third highest risk to health in developing countries, and cause 3.2% of deaths worldwide.[1] The prevalence of alcohol abuse and dependence disorders are striking when compared to the incidence of other established mental illnesses. For instance, alcohol abuse and dependence affect nearly 9% of the US population.[2] Alcohol accounted for 6.1% of deaths, 10.7% disability adjusted life years and 12.3% of years of life lost in Europe during 2002.[3] Information from the World Health Organization shows that alcohol was the greatest contributor to neuropsychiatric disorders, which accounted for 32.2% of the disease burden.[3] This level of incidence is comparable to that observed for major depression and greater than the worldwide reported prevalence of schizophrenia. Alcohol dependence is clinically defined by the symptomatic criteria discussed below (see Table 5.1), but there are both similarities and differences between alcohol dependence and dependence on other drugs, and some important characteristics of alcohol dependence are not covered by the diagnostic criteria. Alcohol abuse is diagnosed if less than three of the DSM-IV-TR criteria[105] for alcohol dependence are met and patients have experienced legal infractions (e.g., driving while impaired) or interpersonal problems (e.g., loss of employment) as a result of heavy alcohol drinking.

A common feature of drugs of dependence is that they are reinforcing to the organism. This is defined behaviorally as the tendency to repeat the action that preceded the effects of the drug, and therefore involves behaviors such as drug seeking and drug taking. In addition the effects of drug-associated environments and drug priming (discussed below) are seen in dependence on a wide range of types of drug. There is now a considerable amount of experimental data indicating that there may be commonalities in the neuronal mechanisms of development of dependence on the different types of drug.[4,5] Mechanistic studies have for many years concentrated on dopamine transmission, because the mesolimbic dopamine pathways from the cell bodies in the ventral tegmentum are activated by drugs that cause dependence, including alcohol (with the exception of benzodiazepines). More recently, researchers have begun to explore pathways other than dopamine, which will likely lead to pharmacotherapeutic advances over existing treatments.[i] In accordance with this expansion from a dopamine-centric focus, there is growing recognition that addiction disorders are not simply driven by the rewarding attributes of the drug, and that maladaptive neuroadaptations occur with chronic abuse that subvert normal limbic and cortical function. Recent studies are examining a wider range of neuronal target sites, including many of the more recently discovered neurotransmitters that have previously been much neglected with regard to the development of alcohol dependence, such as ghrelin,[6,7] leptin[8,9] and neuropeptide Y.[10]

A comparison of the degree of dependence liability of alcohol with that of other types of drug is not simple. The addiction liability of drugs could be quantified in several ways, such as the amount or duration of use needed to cause dependence, or the level of difficulty of quitting. One important characteristic of human dependence on alcohol, however, is that it tends to develop slowly, and the diagnostic criteria are often not exhibited until 10 years or more after the first drink. In this it differs somewhat from other types of dependence-inducing drugs, for example nicotine causes dependence

---

[i] Please refer to Koob, The Role of animal models in reward deficit disorders: Views of academia; Heidbreder, Impulse and reward deficit disorders: Drug discovery and development, in this volume for further discussion regarding current behavioral theories of reward deficit disorders and advances in pharmacotherapeutic approaches.

much more rapidly.[ii] Despite this slow onset, drinking alcohol during the early years, particularly during adolescence, is extremely important; this is discussed in detail below. Another most notable characteristic is that the great majority of people who consume alcohol do not become dependent; alcohol dependence rates are around 5% for an adult population, but 80-90% consumes alcohol. Again, this differs from some other dependence-producing drugs, particularly nicotine.[11] These aspects might be thought to lead to the conclusion that alcohol dependence is a less serious problem, but the widespread legal use of alcohol throughout so many countries and the very large numbers of people who consume it regularly mean that alcohol dependence is one of the major world health problems. In those individuals that succumb to alcohol dependence, remission from the disease has proven challenging. For example, Zweben and Cisler[12] found a 70% relapse rate within 12 months. Thus, once alcohol dependence occurs, it remains a life-long struggle consisting of high rates of relapse.

There are certain characteristics of alcohol dependence that are of clinical importance, but are seen to a lesser extent with nicotine dependence, though some of these are also features of dependence on opiates or psychostimulant drugs. A large proportion of alcoholics suffer from other serious psychiatric disorders, many are malnourished, suffering in particular from thiamine deficiency, and have unhealthy diets. Fifty to eighty percent of alcoholics have cognitive deficits[13] and these deficits not only affect the quality of life of alcoholics and the amount of health care they need, but also have a detrimental effect on their ability to benefit from current treatment programs.[14-16] The effects of long-term alcohol consumption on memory also need to be considered in the context of animal models. Multidrug use is a problem with all types of dependence, including alcohol, but animal models of alcohol dependence have tended to avoid that particular complication.

Types of Alcoholism

Alcoholism is not a single entity and there have been many attempts to characterize the different types of the disorder based on time of onset of problem drinking, comorbid psychopathology, drinking patterns and other aspects, since the early studies of Jellinek.[17] The Type 1/Type 2 alcoholism distinction described by Cloninger[18] is widely known; Type 1 alcoholics are said to have more consistent alcohol intake, slow onset of the problem, more anxious personalities, low novelty-seeking and high harm avoidance. In contrast, Type 2 alcoholics show high intermittent, binge-like consumption of alcohol and early onset problem of alcohol drinking. Moreover, Type 2 alcoholics have been described as impulsive and antisocial, with high novelty-seeking and low harm avoidance. However, further examination of this two-class distinction has not fully supported its value and overlap between the two groups has been demonstrated (e.g.,[19]). Babor *et al.*[20] used a multidimensional analysis and described two main classes of alcohol dependence, Type A with milder dependence, less psychopathology, less family history of the problem, and fewer physical and

---

[ii] Please refer to Markou, Contribution of animal models and preclinical human studies to medication development for nicotine dependence, in this volume for further discussion of nicotine abuse.

social consequences and Type B with greater psychiatric comorbidity, more family alcohol problems, more polysubstance abuse and more severe dependency. Carpenter and Halsin,[19] however, found that 77% of a general population sample did not fit either of these two categories. Bau *et al.*[21] concluded that the most appropriate classification was into three groups and Windle and Scheidt identified four different subtypes. Epstein *et al.*[22] examined the value of four current classifications and concluded Type A/B was the most promising but there was heterogeneity within the groupings. Finally, a recent publication by Moss *et al.*[23] sampled respondents from the NESARC database and found five distinct alcohol-dependent subtypes that were distinguishable based on the following factors: family history, age of alcohol dependence onset, number of diagnostic criteria from the DSM-IV, presence of comorbid psychiatric disorders and presence of additional substance use disorders.

Important factors contributing to the above discrepancies among classification schemes are sample demographics, gender and whether the samples were from the community or patients who were in hospital for treatment of the problem. Many early studies were conducted on Veterans Affairs (VA) in patients who were predominantly male and Caucasian. Severity of dependence is also problematic for distinguishing subtypes of the disorder owing to its dependence on age and duration of alcohol drinking. Overall, little consistency has been found among these various classifications of subtypes of alcoholism, and, most importantly in the present context, none has yet been found to be useful in the determination of the optimum type of treatment. So far, no clear patterns have been found in attempts to match treatment outcome to a two-group classification. A large multicenter study comparing three types of psychosocial treatment found that only psychiatric severity showed a significant attribute by treatment interaction and depressed individuals were slower to benefit from treatment.[24] Consideration of the heterogeneity of alcohol dependence, however, may be of considerable importance in the development of animal models.

Diagnostic Features of Alcohol Dependence

The current situation for pharmacological treatment of alcohol dependence differs in one important respect from that for other psychiatric disorders; that only recently have medications been considered seriously within the range of treatments for alcohol dependence. Until relatively recently, the suggestion that drugs could be used to control or treat alcohol dependence was received by research funding bodies with skepticism and by pharmaceutical companies with disinterest. The introduction of naltrexone and acamprosate as alcohol dependence medications, despite their limited effectiveness, has gradually changed this dim view of pharmacotherapy. New medication development, however, has been hampered by animal models of alcohol dependence that have uncertain construct, discriminant and predictive validity for the human condition. For instance, the efficacy of selective serotonin (5-HT) reuptake inhibitors (SSRIs), such as fluoxetine (Prozac®), in rodent alcohol drinking paradigms provided impetus to examine SSRIs in several clinic trials. Despite some early indications of efficacy, the general conclusion, after many years of study, is that SSRIs are not robust medications for alcoholism.[25] Examples of such failures to translate preclinical efficacy to effective clinical outcomes have caused researchers to reevaluate the *status quo*

Table 5.1 DSM-IV diagnostic criteria for substance abuse as applied to alcohol

	Criterion
1	Tolerance, as defined by either (i) a need for markedly increased amounts of alcohol to achieve intoxication or the desired effect or (ii) markedly diminished effect with continued use of the same amount of alcohol
2	Withdrawal, as manifest by the characteristic withdrawal syndrome
3	Alcohol often taken in larger amounts or over a longer period than was intended
4	Persistent desire or unsuccessful efforts to cut down or control alcohol use
5	Great deal of time spent in activities necessary to obtain alcohol or recover from its effects
6	Important social, occupational or recreational activities are given up or reduced because of alcohol use
7	Alcohol use continued despite knowledge of having persistent or recurrent physical or psychological problem that is likely to have been caused or exacerbated by alcohol

Source: *Adapted from DSM-IV.*

of animal dependence models and to develop new approaches with a focus on specific symptoms of the disease rather than using rodent alcohol intake as a homolog to the human condition. Models of specific clinical symptoms, such as cue- or stress-induced craving, may yield better correspondence to the clinical situation providing clinicians use new medications in accord with the appropriate patient population, but models with more extensive parallels to the clinical disorder could be more useful.

Before discussing the different models of alcohol dependence, it is useful to consider what exactly the clinical disorder is that needs to be modeled. While the comparison between the clinical diagnostic criteria for alcohol dependence and the characteristics of the various animal models is useful, it is also necessary to evaluate how well the diagnostic criteria reflect the characteristics of alcohol dependence. Table 5.1 lists the DSM-IV diagnostic criteria for substance dependence as applied to alcohol. The relative importance of each diagnostic criterion to the clinical situation will be discussed, then aspects of alcoholism that are not covered by the diagnostic criteria. The sections describing each type of animal model will then examine the extent to which they parallel, or in many cases do not parallel, these criteria, what further is needed to provide a better reflection of the human situation and the importance or otherwise of the modeling of each criterion to the development of effective pharmacological treatments.

Clinical Diagnostic Criteria for Alcohol Dependence and Their Importance in the Development of Animal Models

The first DSM-IV criterion for dependence, tolerance to alcohol, is seen in those not dependent on alcohol, even in moderate drinkers. It can easily be demonstrated experimentally and includes both cellular and environmental components, as well as acute and extended tolerance. In early literature it was often suggested that tolerance and the withdrawal syndrome were both due to adaptive neuronal changes that first decreased the effects of alcohol, and then caused withdrawal hyperexcitability when the alcohol is removed. However, the dissociation between the two phenomena that can be produced experimentally,[26] as well as the multiple components of tolerance, suggests that this is a considerable oversimplification. Furthermore, tolerance to alcohol dissipates rapidly when alcohol intake ceases, showing that the neuronal mechanisms of tolerance are not closely related to those of alcohol dependence, since alcohol dependence, once established, is a very long-lasting syndrome. There is a general agreement among researchers that tolerance to alcohol is not a critical aspect of alcohol dependence, but the development of tolerance does precede physical dependence and may be a predictor of alcohol drinking history. The existence of tolerance can increase the amount of alcohol consumed, particularly by heavy drinkers prior to the development of full dependence. Development of tolerance has also been suggested to reduce the aversive effects of alcohol, revealing the reinforcing effects.[27] There is also some evidence of a link between the extent of development of acute tolerance to alcohol (i.e., tolerance developed during a single dose) and heavy drinking.[28] Family history positive subjects were more sensitive to the subjective effects of alcohol and developed acute tolerance while family history negative subjects did not exhibit the latter effect.[29] Thus, the preclinical study of tolerance has clinical utility in identification of risk factors (i.e., genetic) as well as modeling adaptive aspects of chronic alcohol exposure.

The acute alcohol withdrawal syndrome (Criterion 2) involves anxiety, tremors, convulsions, hallucinations, confusion, dysphoria, sympathetic hyperactivity and cortisol release, lasting 1–2 weeks in humans (8–12 h in rodents). As described by Victor and Adams[30] as long ago as 1953, the time courses of the different behavioral symptoms of alcohol withdrawal vary; the confusion ("delirium tremens") and hallucinations are exhibited later than the tremor and convulsive components. This suggests that the symptoms have different neuronal origins. The exact temporal relationships between the neuronal and behavioral changes have not been delineated, but decreased mesolimbic dopamine transmission considerably outlasts the behavioral hyperexcitability and is a feature of the protracted withdrawal syndrome, described below.[31] The behavioral aspects of the alcohol withdrawal syndrome can be prevented by benzodiazepines and by some anticonvulsant drugs, as described below, but the current treatments do not treat important clinical areas of unmet need (i.e., relapse prevention, cognitive deficits and psychiatric comorbidities). It is hoped that future, novel pharmacological approaches may be more effective in this respect.

Although withdrawal signs are included in the DSM-IV criteria, the relationship between this phenomenon and alcohol dependence was considered for many years to be less important than other aspects of alcohol dependence. The reasons for this were 2-fold. Unlike alcohol dependence, the acute alcohol withdrawal syndrome is transient, and after recovery from the symptoms, a large proportion of alcoholics relapse back into excess drinking (e.g.,[12]). Secondly, and perhaps more importantly, pharmacological prevention of the alcohol withdrawal syndrome (e.g., benzodiazepine administration) does not prevent the majority of alcoholics from returning to excessive alcohol use. These aspects suggest that the acute withdrawal syndrome is not related mechanistically to alcohol dependence. Over the last decade, however, research interest in alcohol withdrawal has been revitalized with the recognition of the effects of repeated alcohol detoxifications and the importance of what is now known as the "protracted withdrawal syndrome."

Repeated alcohol withdrawal is associated with higher levels of craving and relapse drinking.[32,33] It is known to cause "kindling," which is a progressive increase in neuronal hyperexcitability resulting in increased severity of withdrawal symptoms.[34-36] Repeated cycles of alcohol withdrawal cause greater neuronal damage than a single episode.[37-39] Greater neuronal degeneration and memory deficits are seen after cessation than during chronic alcohol consumption.[40-42] In addition, evidence from preclinical and clinical studies show that alcohol withdrawal plays a major causal role in the cognitive problems suffered by alcoholics.[38]

The protracted withdrawal syndrome in humans includes psychological disturbances, particularly anxiety and sleep disorders, that extend for weeks or months after cessation of alcohol drinking (e.g.,[43,44]). Strowig[45] identified depressed mood as the most frequent determinant of relapse during 12 months of abstinence, while a high level of irritability was noted by Brandt *et al.*[46] as preceding relapse episodes. Sleep disturbances have also been shown to predict relapse in abstinent alcoholics.[47,48] Anxiety levels during abstinence have also been shown to be a predictor of relapse drinking.[49] In 1982, Gorski and Miller[50] found that formulating the protracted withdrawal syndrome into a series of psychological symptoms gave good predictions of the probability of relapse drinking, and Miller and Harris[51] showed that "general demoralization" (i.e., depression, anxiety and anger) had predictive validity for relapse. In total, these studies indicate that changes in CNS function are still present long after the acute withdrawal phase and are likely key physiological markers of alcohol dependence severity. Although overlap of involved neurocircuitry is suspected to be high, specific patterns of neuroadaptation likely occur across different drugs of abuse. Recent studies have described behavioral changes in rodents that reflect these symptoms, such as anxiety-related behavior[52] and increased locomotor effects of psychostimulant drugs and nicotine.[53]

Turning to the other clinical diagnostic criteria, increased level of alcohol consumption after prolonged excess drinking (Criterion 3) is an important aspect of dependence diagnosis and could even be said to summarize the clinical problem. It is theoretically possible to demonstrate this behavior in animal models, but researchers have had some difficulties in producing this crucial behavior pattern in rodents, although some recent studies have shown increased alcohol intake and operant responding for alcohol. Criterion 4, the desire to reduce alcohol consumption and unsuccessful attempts to do so, is also of great importance clinically, but is far more difficult to parallel in animal models, because it involves the "intention" to carry out a behavior. Some models of "craving" for alcohol are described below, which address this problem but inevitably suffer from difficulties of interpretation and definition of the abstract construct of craving. There is also a complication that craving in alcoholics does not appear to correlate with relapse drinking,[46,54] suggesting that it may be an epiphenomenon rather than an essential component of dependence.

The amount of work that a human or animal is willing to do to obtain a drug is Criterion 5 for diagnosis and is an important clinical aspect. This is relatively easily measured in rodents, and is the basis of the operant models, which are described in detail below. The reduction of activities that are unrelated to drug taking (Criterion 6) is something that could potentially be measured in rodents, but this aspect has not so far attracted a lot of attention. A major component of dependence on any drug is the continuation of drug intake despite knowledge of the adverse consequences (Criterion 7). In the case of alcohol, this is an extremely important aspect clinically. Alcoholics continue to drink heavily despite pain from ulcers caused by alcohol or the knowledge that their liver function will be seriously, and often fatally, compromised. As for Criterion 6, this aspect could be included to a greater extent in behavioral models of alcohol dependence but only a few researchers have so far addressed this.

In summary, the tolerance and withdrawal diagnostic criteria are of less importance than originally believed, in either the clinical situation or the development of behavioral models of alcohol dependence, although the long-term consequences of alcohol withdrawal may be more fundamental than is currently realized. Although the "intentions" of rodents are not measurable, the amount of work a rodent is prepared to do to obtain alcohol is an integral part of many current animal models. Increased alcohol consumption after long-term drinking, giving up other pleasurable activities and continuation of alcohol consumption despite knowledge of adverse consequences are all of great clinical relevance that could be usefully incorporated to a much greater extent into the animal models of alcohol dependence suitable for testing potential pharmacological treatments.

Important Non-diagnostic Components of Alcohol Dependence

There are certain aspects of alcohol dependence that are not included in the above diagnostic criteria which are of great importance both in the disorder and in the development of animal models by which potential pharmacological treatments can be tested. These are craving, priming, the effects of drug-associated cues, cognitive deficits, the high incidence of comorbidity and adolescent drinking. Additionally, alcoholism is highly heritable in some patient populations, and environmental factors, such as life stress, are strong risk factors for development of alcohol addiction. We shall discuss these non-diagnostic factors in further detail.

Craving, or the overwhelming desire for alcohol, is experienced by a large proportion, if not all, alcoholics; clinical and laboratory findings have been reviewed by Sinha and O'Malley.[55] Priming is the phenomenon whereby intake of even a small amount of alcohol results in the longing to consume more. Whereas priming refers to brief contact with the actual reinforcer, such as alcohol, stimuli merely associated with access to the reinforcer have the ability to form second-order associations and, in turn, modulate behavior. Both priming and cue-induced reinstatement of alcohol seeking are currently popular preclinical models of alcoholism, as described below, and have been said to model some of the cognitive constructs of addiction, such as craving.[56] Similarly, exposure to a salient context in which drug exposure occurs has been shown to induce subjective reports of craving in alcoholics that correlate with activation of specific brain regions.[57,58] The conditioned place preference model is often used in rodents to assess associations formed between the rewarding (or aversive) properties of drugs and the context in which they are administered.

Comorbidity

Comorbid psychiatric symptomatology has a high incidence in alcohol-dependent individuals. Di Sclafani et al.[59] reported that over 85% of long-term abstinent alcoholics had a lifetime diagnosis of psychiatric disorder, compared with 50% of comparable non-alcoholic controls. The severity of alcohol dependence was found by Landa et al.[60] to be significantly related to comorbidity of psychiatric symptomatology, and Morgenstern et al.[61] showed that personality disorders were linked to more severe symptoms of alcohol dependence. Although it has been thought that comorbidity is associated with poorer prognosis, Mann et al.[62,63] found that accompanying psychiatric disorders did not correlate with relapse drinking, although in their sample of socially well integrated patients, 28% of the men and 65% of the women exhibited such problems. It has also been reported that psychopathology recorded during the first weeks of detoxification improves without specific treatment.[64] Depression is a major symptom in many alcoholics[65] and is associated with greater tendency to relapse drinking.[66] There is substantial evidence relating depressive symptoms and cognitive deficits[67,68] although the direction of a potential cause/effect relationship is not yet fully understood.

Recently the pharmacological treatment of patients with comorbid alcohol dependence and psychiatric disorders has been considered.[69] Although antidepressants are commonly prescribed to alcohol-dependent patients who complain of concomitant anxiety and depression, their effectiveness for the primary endpoint of increased abstinence or reduced heavy drinking is poor. Moreover, the most effective pharmacological treatments for reducing harmful alcohol drinking behavior (i.e., naltrexone and acamprosate) do not appear to demonstrate benefit for treating comorbid anxiety or depressive symptoms. Because psychiatric comorbidities are observed in the majority of alcoholic individuals, clearly next generation medications that jointly impact alcohol dependence and accompanying psychiatric symptoms will become the first-line treatments of choice.

Adolescent Drinking

The importance of adolescence with regard to drug dependence has been recognized over the last decade. Adolescent alcohol use is a significant risk factor for subsequent alcohol dependence, and emerging evidence suggests that adolescents respond differently from adults to alcohol.[70] Early onset of alcohol drinking is a strong predictor of subsequent dependence; those who drink alcohol earlier than 14 years of age have a 41% chance of developing dependence later in life.[70,71] This aspect is important not only with respect to early interventions and health services, but also to research using animal models.

Adolescence in humans is regarded as the transition phase between childhood and adulthood, lasting from approximately 12 to 18 years of age and including the periods prior to and after puberty.[70] Definitions of adolescence in rats and mice are not totally consistent, but it has been defined as between postnatal 28 and 42 days,[70] and also as from 7 to 10 days before puberty (about 40 days postnatal) to approximately 1 week after puberty.[72] During adolescence, maturation of brain regions takes place with changes particularly in the prefrontal cortex and hippocampus,[70,73] areas particularly damaged by long-term alcohol consumption.[74]

Alcohol appears to be less aversive in adolescent rodents than in adults, as described below, and has more effect on memory and learning during this developmental stage.[75] Osborne and Butler[76] reported greater cognitive deficits caused by chronic alcohol intake in rats during the adolescent phase. High dose "binge" alcohol treatments given at an early age have been shown to have negative effects on memory performance later in life.[77,78] Alcohol has been found to have greater effects on neuronal plasticity during the adolescent phase.[79,80] The effects of alcohol in depressing long-term potentiation (LTP) were found to be more pronounced in tissues from adolescent rats than those from adults.[79,80] In addition, high dose alcohol administration during adolescence enhanced the memory impairing effects of alcohol in adulthood.[81]

Cognitive Deficits

The cognitive functions particularly affected in alcoholics are learning new skills, visuospatial ability, executive function (planning, capacity for abstraction and effortful processing) and aspects of attention, whereas verbal skills are usually remain intact.[16,82-86] The cognitive deficits not only affect the quality of life of alcoholics and the amount of health care they need, but also have a detrimental effect on ability to benefit from treatment programs.[14,16] Long-term cessation of drinking requires the ability for psychosocial learning, but alcohol- and abstinence-related impairments in decision-making make it very difficult for alcoholics to benefit from their treatment regimes. Problems with executive function make extremely difficult the decision-making needed to effect changes in life-style and habits.

The specific memory deficits seen in alcoholics point to involvement of frontal cortex regions and this area is known to be particularly affected by prolonged alcohol consumption. Postmortem studies show anatomical changes in the frontal association cortex brains of alcoholics.[87,88] Similar alterations occur in experimental animals, but neuronal damage and cell loss are more pronounced in the hippocampus in animals than in humans.[38,89] Although alcohol has well-established acute amnesic effects, there is evidence that it is the cessation of drinking and subsequent neurotoxicity that cause the cognitive deficits during abstinence, rather than the presence of alcohol in the brain. Neurotoxicity clearly occurs, including cell loss, but prolonged changes in neuronal function without actual cell death have been demonstrated experimentally after cessation of prolonged alcohol consumption. Decreases in LTP are likely to underlie memory deficits, but only a small number of experimental studies have investigated this during the abstinence phase.[90,91]

Partial recovery of cognitive function occurs after long-term abstinence (months or years) from alcohol, but some residual memory problems do not recover. Although some of the functional impairment is due to permanent structural damage, the occurrence of gradual recovery suggests that preventative or remedial treatment could be developed once the mechanisms are understood. There is currently no effective treatment and few studies in animal models of alcohol dependence have included this aspect of alcohol dependence. The importance of memory deficits lies not only in the human cost, but is also directly relevant to animal models. If a rat in an operant cage cannot remember which lever results in which consequences, then it is likely to produce misleading results in experiments on reinforcement.

Stress

The "tension-reduction" hypothesis[126] suggests that the anxiolytic effect of ethanol is of considerable importance in contributing to excessive drinking in humans. However, several reviews have concluded that there is little evidence for a direct correlation.[92-94] Only limited support was obtained for the self-medication hypothesis in a volunteer study.[95] It appears that stress is involved in the longer-term aspects of alcohol dependence rather than having acute effects on consumption. Recent imaging and other studies have demonstrated the close similarities between the brain areas involved in relapse and those affected by stress, as reviewed by Sinha and Li.[58]

Clinical studies have shown a higher incidence of stressful major life events in alcohol dependence.[96] Employment in occupations that provide high strain and low control was found to be associated with an increased risk of alcohol abuse.[97] The relapse rate in abstinent alcoholics was higher in individuals who had experienced severe psychosocial stress.[98] In addition, posttraumatic stress disorder (PTSD) is often comorbid with alcohol and substance misuse.[99] Stressful experiences in early childhood, including physical and sexual abuse, are more common in alcoholics than in the normal population. A study of patients in three alcohol treatment units found that 54% of the women and 24% of the men had experienced sexual abuse or assault, mainly before the age of 16 years.[100] Langeland et al.[101] reported over 30% of a sample of treatment seeking alcoholics had experienced childhood physical or sexual abuse. A study of childhood traumatic experiences by Dom et al.[102] found that patients with an early onset of alcohol dependence had a high number and more severe childhood trauma than late-onset alcoholics. Stressful consequences of alcohol abuse clearly need to be distinguished from adverse experiences that occur prior to the development of dependence, and the picture is complicated by stress due to the experiences of growing up in a family with a history of alcohol dependence, and the problems of self-report, but the involvement of stress could be considered to a greater extent in animal models of alcohol dependence. The hypothalamo-pituitary adrenal (HPA) function appears to be involved in dependence on alcohol, as well as that on other drugs of abuse.[103-104,106]

CURRENT TREATMENT OF ALCOHOL DEPENDENCE

Current Food and Drug Administration (FDA)-approved medications for the treatment of alcohol dependence include disulfiram (Antabuse®), naltrexone (ReVia®), acamprosate (Campral®) and a depot formulation of naltrexone (Vivitrol®). Disulfiram blocks the normal metabolism of alcohol, resulting in accumulation of the toxic metabolite acetaldehyde. Disulfiram has been used for over 50 years in the treatment of alcohol dependence as an aversive agent to curb alcohol intake; however, owing to poor compliance and safety issues, disulfiram is little used in current clinical practice. Naltrexone is a µ-subtype preferring, pan opioid receptor antagonist that is believed to exert its therapeutic effects by blocking the rewarding effects of alcohol, an effect thought to be driven largely via µ-opioid receptor blockade. Naltrexone was first demonstrated to improve abstinence in alcohol-dependent patients by Volpicelli *et al.*[107] and was later approved by the US FDA in 1994 for use in alcohol dependence. Clinical use of naltrexone within the United States remains low (around 13% of alcohol-dependent patients), but is reported to be yet lower in most European and Asian countries.[108,109]

Acamprosate has been widely used to treat alcohol dependence in Europe since the early demonstration of its effectiveness against relapse drinking[110] and has been considered more useful than naltrexone.[109] A number of US clinical trials demonstrated its effectiveness in alcoholism[111-113] and acamprosate was approved by the FDA in the United States in 2004. Acamprosate's mechanism of action remains elusive.[114,115] Although it has functional glutamate antagonist properties, under certain conditions, acamprosate does not appear to directly bind to ionotropic (iGlu) or metabotropic (mGlu) glutamate receptors or glutamate transporters. Besides glutamatergic functioning, acamprosate also reduces calcium flux[116] and binds to a subunit of the dihydropyridine receptor. A recent clinical study evaluated acamprosate, naltrexone, and their combination in a large alcohol-dependent population, which determined that acamprosate did not differ from placebo, nor did it improve the modest, but significant, efficacy of naltrexone.[117] However, the study population differed from those in which acamprosate had previously been shown to be effective (i.e., alcoholics maintaining abstinence) and underscores the importance of patient segmentation and clinical study design. Acamprosate is the most commonly used pharmacotherapeutic treatment of alcohol dependence in Europe, while naltrexone remains the standard of care in the United States. A myriad of factors, not the least of which is modest efficacy, likely contribute to low usage of these pharmacotherapies by physicians and addiction specialists.[108]

Several non-approved medications have also shown some degree of efficacy in the treatment of alcohol dependence in clinical trials, including topiramate (Topamax®,[12,118-122]), baclofen (Kemstro®/Lioresal®,[123,124]) and serotonergic agents such as ondansetron (Zofran®,[125,127-129]). Of these, topiramate thus far appears to be the most promising, with a medium effect size in clinical trials.[130] Several clinical trials treating alcohol-dependent subjects, ranging from 6 to 14 weeks in duration, have demonstrated a significant reduction of the percentage of heavy drinking days following treatment of up to 300 mg of topiramate per day compared to placebo.[118-122] Other classes of drugs that have undergone clinical trials in alcohol-dependent individuals, but either have not been studied extensively enough or have not shown promising therapeutic effects, include dopamine receptor agonists (i.e., bromocriptine (Parlodel®) and 7-OH-DPAT),[132-134] *N*-methyl-D-aspartate (NMDA) receptor antagonists (i.e., memantine, Ebixa®),[135] SSRIs (i.e., citalopram (Celexa®/Cipramil®) and fluoxetine),[136-138] 5-HT$_2$ receptor antagonist (i.e., ritanserin (Tisterton®)).[139,140] Buspirone (Buspar®), a 5-HT$_{1A}$ partial receptor agonist, generally did not demonstrate improvement in non-comorbid alcoholics but showed some benefit in alcoholics with comorbid anxiety.[141,142] A summary of both approved and non-approved drugs used and tested for the treatment of alcoholism can be found in Table 5.2.

Table 5.2 Summary of US-approved and experimental drugs showing some degree of efficacy for the treatment of alcohol dependence

Drug	Mechanism of action	US drug approval date	Route of administration/ dosage	Treatment outcomes in clinical trials
Disulfiram	Acetaldehyde dehydrogenase inhibition	1951	Oral	May help prevent relapse in compliant patients but ineffective at promoting continuous abstinence or a delay in the resumption of drinking.[143] Effective in subjects who are highly motivated to comply with treatment.
Naltrexone	Competitive antagonism at μ- and κ-opioid receptors, and to a lesser extent at δ-opioid receptors	1994	Oral (50–100 mg/ day)	Found to reduce the risk of relapse in recently abstinent, alcohol-dependent individuals[144,145], lower percentage of drinking days and heavy drinking days compared to placebo in patients who maintained ≥80% treatment compliance[117,146,147]). Some studies have shown a significant reduction in craving scores compared to placebo.[148-151] Good clinical response observed in patients with a family history of alcohol dependence or strong cravings or urges for alcohol.[149,152,153]
Naltrexone depot (Vivitrol® (Alkermes Inc., Cambridge MA, USA), Naltrel® (Drug Abuse Sciences, Inc., Paris, France), Depotrex® (Biotek, Inc., Woburn, MA, USA)	Competitive antagonism at μ- and κ-opioid receptors, and to a lesser extent at δ-opioid receptors	Vivitrol® (2006)	Intra-muscular (150–300 mg/ month)	High dose of Vivitrol® (380 mg) shown to significantly reduce the percentage of heavy drinking days in male but not female alcohol-dependent subjects. No differences were observed between the lower dose (190 mg) and placebo.[154] Naltrel® (initial dose of 300 mg followed by monthly injections of 150 mg for 3 months) was shown to significantly increase mean number of cumulative abstinent days and time to first drink in subjects able to achieve at least 3 consecutive days of sobriety. The effects of gender on treatment outcome were not examined.[155] A higher dose of Naltrel® (300 mg) was shown to have improved outcomes on drinking including reduced number of drinks per day, heavy drinking days and proportion of drinking days.[156] A preliminary study with 206 mg of Depotrex® showed a significant reduction in the frequency of heavy drinking days.[157]

Drug	Mechanism	Year	Route (dose)	Notes
Acamprosate	Metabotropic glutamate receptor-5 modulator and *N*-methyl-D-aspartate (NMDA) antagonist	2004	Oral (1.6–3 g/day)	Early European studies have demonstrated increased continuous abstinence rates in alcohol-dependent individuals receiving acamprosate versus placebo,[62,63] and generally showed a small effect size for increasing the percentage of non-heavy drinking days[158] and reducing the relapse to heavy drinking.[159] The multisite COMBINE project in the United States failed to find any therapeutic benefit of acamprosate over placebo on any drinking outcome measures.[117] Based on the European studies, acamprosate appears to benefit alcohol-dependent individuals with increased levels of anxiety, physiological dependence, negative family history, late age of onset, and female gender.[160]
Topiramate	Alpha-amino-3-hydroxy-5-methylisoxazole-4-propionic acid and kainite glutamate receptor antagonist	NA*	Oral (200–300 mg/day)	In a 12-week double-blind randomized trial, topiramate (up to 300 mg/day) was shown to improve all drinking outcomes, decrease craving and improve the quality of life in alcohol-dependent individuals.[119,122] These results were in agreement with those from a 6-week trial of 76 heavy drinkers who were not seeking treatment and showed a significant decrease in the percentage of heavy drinking days following low and high dose topiramate (200 and 300 mg/day)[161] and for a larger 17 site trial (*N* = 371) showing that topiramate (up to 300 mg/day) showed improvements on all self-reported drinking outcomes, as well as GGT levels, compared to placebo.[121] Adverse events have been reported as mild to moderate with the most common being paresthesia, anorexia, difficulty with memory or concentration, and taste perversion.[162]
Baclofen	GABA_B receptor agonist	NA	Oral (up to 30 mg/day)	In a 4-week open label trial (*N* = 9), seven of nine alcohol-dependent men achieved abstinence, and the other two showed improvements on self-reported drinking outcomes and an overall *(continued)*

Table 5.2 (Continued)

Drug	Mechanism of action	US drug approval date	Route of administration/ dosage	Treatment outcomes in clinical trials
				decrease in craving during the study period.[123] Improvements in drinking outcomes, state anxiety scores and craving measures were also seen in a 4-week, randomized, placebo controlled double-blind trial with 39 alcohol-dependent patients. Baclofen was generally well tolerated and had no apparent abuse liability. Common adverse events consisted of nausea, vertigo, transient sleepiness and abdominal pain.[124]
Ondansetron	Serotonin-3 (5-HT$_3$) receptor antagonist	NA	Oral	A 6-week double-blind, placebo controlled study in 71 alcohol-dependent males (non-severe) demonstrated that 0.5 mg daily showed a trend toward reducing alcohol consumption, compared to a higher dose of 4 mg. *Post hoc* analysis, removing 11 subjects who consumed less than 10 drinks/day rendered the drinking outcome measures between 0.5 mg ondansetron and placebo significant ($P = 0.001$).[129] A larger ($N = 231$) randomized, double-blind trial, showed that ondansetron (1, 4 and 16 µg/kg b.i.d.) was superior to placebo for improving drinking outcome measures. This was limited to alcohol-dependent subjects of early onset or Type B-like subtype, but not the late-onset Type A-like subtype.[125] This outcome was supported by another 8-week trial of ondansetron (4 µg/kg b.i.d.) which showed significant improvements in drinking outcomes in Type B-like alcoholics compared to late-onset Type A alcoholics.[127]

Source: Adapted from[130,131]
*NA – Not approved by the US FDA for the indication of alcohol dependence.

HUMAN PHARMACOLOGY AND EXPERIMENTAL MEDICINE APPROACHES TO THE TREATMENT OF ALCOHOL DEPENDENCE

Review of Clinical Trial Design

A practical goal in the treatment of alcoholism is to reduce or eliminate clinical diagnostic symptoms of the disease. In defining clinical endpoints for an alcohol dependence trial, many measures have been used but can largely be segregated into measures of abstinence (e.g., total non-drinking days, days until relapse, etc.) and reduction in heavy drinking days. Historically and particularly in US clinical alcohol dependence trials, abstinence has been the key clinical endpoint for which success of a treatment was determined. Both naltrexone and acamprosate had to demonstrate an increase in abstinence rates for FDA approval and similar guidelines exist in Europe. The emphasis on obtaining abstinence as a determinant of clinical efficacy is based on the assumption that once an individual becomes alcohol dependent, controlled drinking is not an achievable goal on the road to recovery. This view has been endorsed by support groups such as Alcoholics Anonymous and generally espoused by the medical community.

While abstinence may indeed be a necessary goal for some individuals, many alcoholic patients do not want to become abstinent because it alienates them from many social activities where alcohol is a common feature. In these individuals, the ultimate treatment goal is controlled or social alcohol drinking. It is well established that alcohol-related problems linearly increases as a function of total alcohol intake. Thus, if an individual was able to reduce their alcohol drinking session from 12 drinks down to 4 drinks, a benefit would likely be seen in terms of a reduction in health, legal and social problems. The choice of primary clinical endpoint is not trivial and different medications may demonstrate different levels of efficacy depending on whether a clinical trial is designed to demonstrate abstinence or reduction in heavy drinking behavior. The demonstration of abstinence or reduction in heavy drinking behavior may be influenced by the type of clinical study design. Generally the assessment of drinking behavior can be divided into two types of study models. The first involves a scenario where subjects are abstinent for a defined period of time prior to the start of drug treatment, and the second in which subjects continue to drink prior to treatment initiation. In the former case, the design addresses whether the investigational drug is able to maintain a state of abstinence, whereas in the latter case, abstinence would have to be virtually drug induced and is therefore oftentimes difficult to demonstrate. In the former case, endpoints including abstinence and time to relapse (e.g., time to first drink, time to first episode of heavy drinking) are applicable measures in addition to those assessing drinking behavior. In the latter case, changes of drinking behavior (e.g., number of drinks per day, number of heavy drinking days) are more appropriate study outcomes.

For instance, although naltrexone was approved for the maintenance of abstinence, past studies have repeatedly shown statistical significance relative to placebo in a reduction of number of drinking days, number of heavy drinking days, time to heavy drinking and craving, and have not supported the maintenance of complete abstinence (review by[130]) For example, one of the earlier trials used to support the

FDA approval of naltrexone for the treatment of alcohol dependence (total $N = 70$) showed that naltrexone therapy for 12 weeks (50 mg/day) significantly reduced relapse rates (defined as drinking 5 or more drinks per day) in subjects treated with naltrexone (23%) versus placebo (54%).[107] Volpicelli *et al.* further commented that naltrexone treatment did not appear to prevent subjects from sampling alcohol, as 46% of subjects in the naltrexone-treated group admitted to having at least one alcoholic drink, while 57% of the placebo-treated group drank. This demonstrated that there was not a statistically significant difference in complete abstinence between the naltrexone- and placebo-treated groups.

The demonstration of abstinence has not been clear for other pharmacotherapies. Data from a meta-analysis indicated that acamprosate may be more effective in maintaining abstinence compared to naltrexone,[144] but this has not been consistently demonstrated within controlled studies comparing both drugs. For example, Rubio *et al.*[163] conducted a 3-month trial in 157 recently detoxified alcohol-dependent men with moderate dependence and treated them with naltrexone (50 mg/day) or acamprosate (1665–1998 mg/day). There was no difference between treatments in mean time to first drink (naltrexone 44 days, acamprosate 39 days) but the time to first relapse was 63 days (naltrexone) versus 42 days (acamprosate) ($P = 0.02$). At the end of 1 year, 41% receiving naltrexone and 17% receiving acamprosate had not relapsed ($P = 0.0009$). The cumulative number of days of abstinence was significantly greater, and the number of drinks consumed at one time and severity of craving were significantly less in the naltrexone group compared to the acamprosate group, as was the percentage of heavy drinking days ($P = 0.038$). It should be noted that in the Rubio *et al.*[163] study, as well as others,[164,165] "relapse" was defined as having consumed more than 5 drinks or 40 g of ethanol per day and did not follow a more conservative definition of consuming any amount of alcohol.

Other studies have defined relapse more precisely as a relapse to "heavy drinking" rather than any drinking[166,167] and have further defined this term according to gender. Heavy drinking has been generally classified as ≥ 5 drinks/day for men and ≥ 4 drinks/day for women,[131,166,168] but this definition has differed among studies. In other studies "heavy drinking" has been defined between ≥ 3 drinks per day (gender differences not specified)[163] to ≥ 6 drinks per day for males.[144,169,247] To further convolute matters, the definition of a standard "drink" has typically ranged between 8 g[163] and 12 g of ethanol.[170] To facilitate the interpretation of future clinical outcomes, standardized definitions of these terms including "relapse" and "abstinence" should be defined.

Additional factors, including severity of alcohol dependence and psychiatric comorbidities, have not been fully explored in clinical evaluation of potential medications for alcohol dependence. The mechanisms that alleviate both acute and protracted withdrawal states, such as the corticotropin releasing factor (CRF)1 receptor antagonists and GABAergic modulators, may exhibit optimal efficacy in more severe alcoholic patients. Alternatively, preclinical data indicate that mechanisms such as mGluR2/3 receptor agonists and mGluR5 receptor antagonists have anxiolytic- and antidepressant-like activity.[171] To date, most clinical studies for alcohol dependence exclude patients with comorbid anxiety and depression symptoms, and thus future clinical studies need to integrate, rather than exclude, such patient segments.

BEHAVIORAL MODELS OF ALCOHOL DEPENDENCE

At first sight, it might appear that alcohol dependence is easy to model in rodents because, unlike the majority of psychiatric disorders, the causative factor (i.e., drinking alcohol) is known. However, difficulties rapidly arise in the laboratory because the majority of rodents do not drink alcohol when given the opportunity and even if they do drink, they do not become dependent. The first of these problems has largely been overcome by gradual introduction of alcohol into the drinking fluid, addition of palatable flavors and development of genetic lines with high alcohol preference. The second difficulty has proved more difficult to resolve.

The discussion below has been confined to rodent models, since these offer considerable advantages over primates for drug testing. The evaluation has also been limited to the relevance of the models to alcohol dependence, rather than abuse or binge drinking, because dependence appears to be more likely to be responsive to pharmacological therapy.

Voluntary Drinking Choice Model

The voluntary drinking model, in which mice or rats are provided with a choice of two drinking fluid bottles containing tap water and dilute alcohol, has been very widely used but the interpretation of the results is not simple. Although it has the major advantage of close superficial resemblance to human alcohol consumption, there remains the problem that taste has a primary influence on the behavior. Most rodents do not appear to like the taste of alcohol, as evidenced by their facial reactions.[172] However, few humans enjoy their first taste of alcohol, and human introduction to alcohol is normally not in an unflavored form. Many studies have used rodents' liking for sweet flavors by introducing the alcohol mixed with sucrose. The sucrose concentration is gradually lowered ("sucrose-fading") and the rats or mice continue to consume the mixture and then go on to drink dilute alcohol without flavoring. The addition of flavor has obvious parallels in human drinking, and this would suggest that studies that used flavored mixtures throughout would be more appropriate models. However, the choice of flavor is difficult and has its own problems of interpretation. Rodents have not yet had the advantage of thousands of years of alcohol consumption to perfect their choice of flavoring for their alcohol. A further complication is the caloric value of alcohol that can influence its consumption by rodents.[173,174]

The concentration of alcohol chosen for rodent preference studies is crucial.[175] Most voluntary choice studies use alcohol concentrations of around 8% or 10% v/v, as these have been found to result in the highest g/kg alcohol intake. With concentrations of 5% v/v or lower, studies have shown little distinction between the patterns of consumption of the dilute alcohol and water, suggesting that rodents cannot discriminate between water and low alcohol concentrations. Concentrations higher than 15% or 20% v/v alcohol are not popular even with "alcohol-preferring" strains of mice or rats. Another variable is that of access, limited access paradigms such as 2 h/day[176,177] have been used that increase the amount of alcohol consumed and exhibit closer resemblance to the patterns of human alcohol drinking than 24 h consumption. A further complication with this type of model is that the choice has to be made between single housing rodents, which enables measurement of individual consumption levels but which is known to have adverse effects on rodents, and group housing in which individual alcohol drinking patterns cannot be distinguished without extensive technical arrangements.

Another aspect of voluntary drinking and other models is the dietary constituents. This is largely beyond the scope of this chapter, but Carrillo *et al.*[178] reported that intake of 9% v/v alcohol by rats was increased after 7 days on a diet containing 50% of calories as fat (75% as saturated fat, 25% unsaturated) compared with a diet containing 10% calories as fat. Pekkanen *et al.*[179] found that after 4 weeks on a diet containing 65% calories as unsaturated fat, rats had a significantly higher intake of alcohol during a 4-week, two-bottle, free choice period, compared with rats on diets containing 15% or 0.4% of calories from fat. Other important effects of high fat diets have been reported, including memory deficits[180,181] and reduced brain-derived neurotrophic factor content.[182] The fat content may not be the only aspect of diet to influence rodent intake of alcohol, since Tordoff *et al.*[183] demonstrated substantial differences in alcohol preference between mice fed on purified diets compared with standard cereal-based diets. In view of the high proportion of saturated fats in Western diets and the poor nutrition of many alcoholics, this aspect could usefully be examined in relation to models of alcohol dependence.

Chronic Alcohol Consumption, Alcohol Deprivation and Voluntary Alcohol Drinking

Prolonged alcohol drinking does not necessarily lead to increased voluntary consumption, even in high alcohol-preferring rodent strains, but this does depend on whether or not the chronic drinking is voluntary or involuntary and variations between laboratories have been reported. Early studies were able to demonstrate some increases. Veale and Myers[184] and Marfaing-Jallat and Le Magnen[185] demonstrated increased consumption of alcohol by rats after forced drinking and intragastric administration of alcohol, respectively, and Camarini and Hodge[186] were able to increase the alcohol intake of the alcohol-avoiding DBA/2J strain of mice by forced alcohol consumption. The importance of the access paradigms was demonstrated by Boyle *et al.*[187]

Behavior relevant to human alcoholism is seen in the deprivation models. In these models, mice or rats are given a choice of two or more bottles containing dilute alcohol or water, and then the alcohol is removed for periods ranging from a few days to a few weeks. When alcohol is reintroduced, the amount of alcohol consumed is considerably increased compared with that prior to the deprivation phase. When a single deprivation period is used, the increased alcohol consumption is transient, lasting only a few hours or a day or two. However, when repeated deprivation periods are applied, the elevated consumption is prolonged and greater in magnitude.[188,189] Four cycles of 2 weeks deprivation and 2 weeks alcohol choice increased the alcohol intake of three non-alcohol-preferring lines of rats, NP, LAD-1 and LAD-2,[190] showing that such alcohol exposure can overcome genetic dispositions not to choose to consume alcohol. Overall, these data not only show rodent behavior that is close to that of human alcoholics, but also raises again the strong possibility that neuronal changes during the absence of alcohol are responsible for the detrimental changes in behavior during the development of alcohol dependence.

A variation on the two-bottle choice model was that in which four bottles were provided, containing alcohol concentrations of 5%, 10% and 20% alcohol and water, to which group-housed rats had continuous access.[191] These authors reported steadily rising ethanol consumption over 9–18 months and a switch in preference to the higher ethanol concentrations, described as "uncontrolled intake." They suggested this to be a model of "behavioral dependence" on ethanol. This model also exhibited an increase in the levels of ethanol consumption after a forced abstinence of 9 months following 9 months of ethanol availability.[192] The duration of 9–18 months of ethanol drinking in this experimental design was suggested by the authors to be particularly useful as a model of the protracted time course of a lifetime of human drinking rather than the transient high level of consumption normally used in other models. Some difficulties were encountered, however, in reproducing this pattern of behavior.[193,194] Holter and Spanagel,[256] however, showed a similar pattern of increased alcohol consumption and choice of higher alcohol concentrations when rats were deprived of alcohol for 3 days every 4 weeks, after 8 weeks continuous drinking. Fachin-Scheit[195] reported that after 10 weeks free choice of alcohol 5% or 10% or water followed by one period deprivation some individual mice continued to prefer alcohol when it was adulterated with bitter-tasting quinine (described as "addicted") while others did not. Naltrexone did not reduce the alcohol preference in "addicted" mice but did so in the other mice.

Effects of aversive consequences on voluntary alcohol drinking

Alcohol itself produces a conditioned taste aversion, but the relationship between this and the development of dependence is uncertain.[196,197] Studies by Stewart *et al.*[198] demonstrated that rats in a line bred for high ethanol preference showed lower sensitivity to the aversive effects of ethanol; this difference was suggested to contribute to the difference in preference between the two lines of rats. Mutant knockout mice, lacking 5-HT$_{1B}$ receptors, were reported to develop conditioned taste aversion to alcohol, but not conditioned place preference.[199]

Pairing of free choice alcohol drinking with administration of the aversive LiCl has been found to considerably decrease voluntary alcohol consumption in rodents; McKinzie *et al.*[200] demonstrated a persistent avoidance of the previously highly preferred alcohol solution following pairing of LiCl injections with alcohol access in adolescent rats. Administration of disulfiram decreased voluntary alcohol intake by rats.[201] However, several recent examinations of the efficacy of disulfiram have concluded that there is little evidence for this drug being of value in maintaining abstinence in alcoholics,[168,202] although compliance is a problem.[203] The demonstration of reinforcing effects of acetaldehyde[204] also needs to be taken into consideration, as disulfiram causes accumulation of this substance.

Quinine has been used in some voluntary drinking studies to examine the persistence of alcohol consumption in spite of its bitter taste. Mormede *et al.*[205] found this substance reduced alcohol intake in high ethanol-preferring rats, indicating that the high preference of this breeding line had a large taste component.

Adolescence and Voluntary Alcohol Drinking

The voluntary consumption of alcohol varies greatly with age, and a pattern of high consumption levels early in life that decrease over the life span has been demonstrated in rodents. Intake of dilute alcohol has been found to be substantially higher in pre-weanling rats and during the early postnatal weeks compared with later in life and consumption of alcohol from an early age can increase the consumption at later times. McKinzie et al.[206,207] showed that 7-week continuous access to a free choice of alcohol from the age of 21 days followed by a deprivation period of 4 weeks resulted in considerably higher alcohol consumption and preference in a two-bottle choice over the first 3 days of testing. Siciliano and Smith[208] found higher levels of voluntary consumption in a two-bottle preference test by rats given alcohol as the sole drinking fluid from postnatal day 21 for 35 days. Alcohol intake orally and by the intragastric route during the pre-weaning period increased alcohol preference on postnatal day 12.[209] Juvenile rats were also willing to consume high concentrations of alcohol that are aversive to adult rodents[210] and alcohol-induced conditioned taste aversions appear to be lower in adolescent, compared with adult rats.[211] Furthermore, nursing from a dam which had received repeated injections of alcohol when pups were PD3-13 resulted in a long-lasting increase in alcohol consumption in the pups, when tested in adolescence.[212]

Effects of Stress on Voluntary Alcohol Drinking

Correlation of the levels of anxiety-related behavior (as measured in the elevated plus maze) with voluntary alcohol drinking in individual Wistar rats suggests that basal anxiety levels play an important part in regulation of alcohol consumption.[213] Some anxiolytic drugs decrease alcohol consumption in rodents (for example, buspirone,[215] but there have been reports of increased alcohol preference following the administration of other compounds (for example, diazepam[216]). Acute defeat can cause increases in anxiety-related behavior that are reduced by administration of alcohol,[214] providing further indication that alcohol can influence the expression of anxiety-related behaviors.

Stressful experiences can increase voluntary alcohol drinking, although this depends on the type of stress, the time interval studied and the strain of rodent. Several reports showed increases in animals with an innate low preference for alcohol.[217-219] The latter authors and Croft et al.[220] found that repeated stressful experiences, such as social defeat, increased voluntary alcohol drinking after a delay of 2-3 weeks. The consumption was not altered before this, although some studies have shown transient decreases in consumption following social defeat.[221] The less "natural" stress of footshock has been found to reduce alcohol intake, for example preventing the alcohol deprivation effect.[222] However, results from studies on adolescent rodents have demonstrated the importance of early alcohol drinking in the consequences of stress. Siegmund et al.[223] showed that repeated footshock stress increased alcohol consumption to a greater extent in adult rats that initiated alcohol drinking during adolescence, compared with those that did not have access to alcohol until they were adults. These authors also found that repeated swim stress had similar effects in both groups, but Fullgrabe et al.[224] found both repeated swim stress and footshock stress caused greater increases in alcohol consumption in rats that initiated alcohol consumption during adolescence compared to rats that acquired alcohol drinking during adulthood. Repeated swim stress increased the alcohol intake of unselected Wistar rats but not that of high alcohol-preferring lines (HAD, P and AA lines) while repeated footshock increased alcohol drinking in all rats tested with greatest effects in the HAD and P lines.[225]

Maternal separation can increase alcohol preference in adult life, with daily separations of 180 min during postnatal days 2–14 resulting in a range of behavioral changes, including increased anxiety-related behaviors and increased alcohol consumption, which were decreased by administration of paroxetine (Seroxat®/Paxil®[226]). This effect is apparent when the periods of maternal deprivation are prolonged, 180 or 360 min, and is not seen with periods of 15 min.[227-229] The latter periods of maternal deprivation are comparable with those normally encountered by rodent pups during maternal foraging. Interestingly, the effect has been reported to be confined to males.[228,230] Early weaning (postnatal day 16) of rats, however, resulted in lower alcohol consumption and preference.[231] Several studies have shown that social isolation rearing increases alcohol consumption by rats,[232,233] and that social isolation also has a similar effect in adult animals.[234]

Evidence suggests that alcohol consumption is directly affected by circulating glucocorticoid concentrations. Rats will self-administer corticosterone to achieve the plasma concentrations of this hormone in the range that occurs during stressful experiences (1–1.5 μM;[235]). Corticosterone administration increased voluntary alcohol drinking in rats, while lowering glucocorticoid concentrations with the glucocorticoid synthesis inhibitor, metyrapone, or by adrenalectomy, decreases ethanol intake.[236,237] Changes in voluntary alcohol drinking and operant self-administration of alcohol are also seen following manipulations of CRF, as described below.

In social groups, rodents will form a stable hierarchy of social rank, with one dominant animal and degrees of subordinate rank among the other group members. Subordinate rank induces emotional stress, as distinct from physical stress produced by aversive stimuli such as footshock, and is associated with functional changes in the CNS[238-240] as well as increased susceptibility to tumors, infections, gastric ulcers and cardiovascular problems.[241,242] Social subordination produces a chronic stress situation, and is thought to be analogous to the human experience of a long-term lack of control of one's environment. Consistent results from several laboratories show that subordinate rodents in a social hierarchy voluntarily consume considerably larger amounts of ethanol than dominant animals. This difference has been demonstrated in a range of social situations. In a "semi-natural" colony model, a sub-population of Long Evans rats showed subordinate patterns of behavior (when measured in a variety of ways) and high alcohol consumption.[243] Blanchard *et al.*[244] saw this pattern in subordinate male rats, and to a more variable extent, in females, in mixed-sex colonies. Hilakivi-Clarke and Lister[245] showed that in group-housed NIH Swiss mice, high alcohol consumption was seen in fight-stressed subordinate individuals (which had undergone defeat in resident/intruder tests) compared with dominant animals or non-fighting controls.

Effects of clinically effective and novel drugs on voluntary alcohol drinking

Naltrexone and acamprosate Two-bottle choice preference studies have demonstrated the effectiveness of acamprosate and naltrexone, but other drugs that have shown positive effects in this model have been less effective in humans. Naltrexone reduces alcohol consumption in continuous access and limited access schedules of voluntary drinking,[246] although some studies have shown a complex dose–response relationship[248] and the effect was not selective for alcohol, as consumption of sweet solutions was also reduced.[249] Examination of orofacial responses following alcohol drinking indicated that naltrexone reduced the aversive effects of voluntary alcohol drinking[250] and reduced conditioned taste aversion was shown by Froehlich et al.[251] Davidson and Amit[252] found naloxone (Narcan®), an opiate antagonist with similar properties to naltrexone, is effective in decreasing voluntary drinking only when it is given during periods of alcohol availability, suggesting that an alcohol/opiate antagonist interaction is necessary for the effect. This, however, is in contrast to results from some studies in which naltrexone is given during forced abstinence periods. Naltrexone administered to rats for the last 3 days of a 10-day deprivation period abolished the deprivation-induced increase in alcohol consumption.[253] In a limited access model with a 30-day deprivation period, the combination of naltrexone and abstinence reduced subsequent alcohol intake.[254] Holter and Spanagel[256] concluded from studies on the alcohol deprivation effect following long-term alcohol drinking by rats that a selective action on alcohol consumption in this model required low doses and intermittent administration. The effects of naltrexone on the alcohol deprivation effect do not appear to be specific, since the rise in saccharin consumption after deprivation was also reduced.[257] Naltrexone has not reduced drinking in all voluntary models, McMillen and Williams[258] found it had no effect in Fawn Hooded rats, However, since this strain reduced their alcohol consumption when offered an alternative highly palatable drink, it does not appear to be a good model in this context. Some studies showed that the effects of naltrexone did not last after its elimination,[259] but prolonged administration has been found to have effects beyond this time.[260] Phillips et al.[261] found naltrexone reduced alcohol consumption by high preferring C57BL/6J mice during acquisition of drinking but not once the consumption was established.

Early studies showed that acamprosate reduces voluntary alcohol drinking,[262] but tolerance to this effect may develop in alcohol-preferring rats.[263] Acamprosate also reduces the alcohol deprivation effect.[264] The latter effect was selective for alcohol, since the saccharin deprivation effect was not reduced.[257] Potentially important difference in adolescence was shown by Fullgrabe et al.[224] who demonstrated that while acamprosate reduced the deprivation effect in adult rats but not in animals that had comparable durations of alcohol consumption that began during the adolescent period. Additive effects of naltrexone and acamprosate on voluntary alcohol consumption were demonstrated by Kim et al.[265] and on the alcohol deprivation effect by Heyser et al.,[266] but were not seen in Wistar rats by Stromberg et al.[268]

Serotonergic agents   Many early drug studies implicated 5-HT (serotonin) in voluntary alcohol consumption[269-271] and the evidence was reviewed by Sellers et al.[272] and by Le et al.[273] SSRIs were found to reduce alcohol drinking in rodents[274,275] and ritanserin was found by some workers to decrease alcohol preference[276] (but see[277,278]). The 5-HT$_3$ receptor antagonists[206,207,279] and 5-HT$_{1A}$ receptor activation[280] also reduced alcohol drinking in rodents. The effects of 5-HT$_3$ antagonists, but not that of SSRIs, were selective for alcohol consumption.[269]

GABAergic agents   Baclofen and other drugs acting on GABA$_B$ receptors have been extensively investigated in rodent models of alcohol consumption. These studies have shown promising results, but there are also contradictions and reports of lack of specificity in the effects of this type of drug on operant self-administration of alcohol (see below). Both baclofen and another GABA$_B$ agonist, CGP44532, reduced alcohol consumption in the two-bottle choice test[281] and baclofen has also been found effective in reducing the ADE.[282,283] Colombo *et al.*[284] however reported that baclofen alone did not alter alcohol intake in Sardinian alcohol-preferring (sP) rats, but did reduce the increases in alcohol drinking induced by either morphine or a cannabinoid receptor agonist.

Novel agents under development   Of the newer compounds being developed, promising effects on voluntary drinking have been found among drugs acting at cannabinoid and glutamate receptors. Both CB-1 receptor antagonism and CB-1 receptor knockout reduced alcohol preference in mice and rats, while activation of this receptor increased voluntary alcohol consumption.[281,285-288] The alcohol deprivation effect on voluntary alcohol consumption has also shown to be decreased by administration of an antagonist at the CB-1 receptor site.[289,290] The CB-1 receptor antagonists do affect food intake, particularly when given a high doses[291] but the intake of sucrose may not be affected in a free choice paradigm.[281,287] High doses of these antagonists may also increase alcohol consumption, as reported in rats following repeated cycles of chronic alcohol treatment and withdrawal;[292] this effect may be dependent upon the time of antagonist administration. Using SR147778, these researchers found that while administration of the CB-1 receptor antagonist during chronic alcohol treatment caused a transient increase in alcohol consumption, giving the drug during the intervals when the free choice alcohol drinking was measured between alcohol treatments resulted in a decrease in alcohol intake.[293]

Drugs acting at glutamate receptors have received extensive attention with regard to alcohol dependence. Perhaps the most hopeful results have come from studies on metabotropic receptors (mGluR), but both competitive and non-competitive NMDA receptor antagonists, and a partial agonist, have been found to reduce alcohol preference in voluntary drinking studies and to reduce the ADE.[294-296] Selective antagonists at mGluR5 receptors reduce alcohol drinking and this effect was shown to involve PKCepsilon, since the effect was not seen in PKCepsilon knockout mice by Olive *et al.*[297] Backstrom *et al.*[298] found that the mGluR antagonist MPEP (2-methyl-6-(phenylethynyl)-pyridine), when given in repeated doses, had a greater effect on the increased alcohol drinking following deprivation than on the baseline consumption. Other metabotropic glutamate receptors also have potential, including the mGluR2/3 receptor agonist, LY404039, which reduces both maintenance and relapse drinking.[299]

Drugs affecting the HPA may prove of interest in reducing the effects of stress on alcohol consumption. The increased alcohol drinking produced by intermittent exposure to alcohol vapor was found to be decreased by infusion of the CRF receptor antagonist, D-Phe-CRF(12-41) into the amygdala, which did not have any effects on drinking in control rats previously unexposed to alcohol vapor.[300] Intracerebroventricular administration of CRF did not alter alcohol preference in C57/BL10 mice,[301] but caused a reduction in alcohol consumption by rats in limited access paradigms.[302,303] The latter effect is in agreement with the results of Palmer *et al.*[304] who showed that mice with overexpression of CRF voluntarily drank less than the wild-type animals and those of Olive *et al.*[305] who found CRF knockout mice consumed about twice as much alcohol as their wild-type controls. However, the situation appears more complex when the lack of conditioned place preference to alcohol in CRF-deficient mice and the prevention of reinstatement of responding for alcohol by CRF antagonists are considered. The Type 2 glucocorticoid antagonist, mifepristone (RU38486), was found to decrease voluntary alcohol consumption in mice and rats, while spironolactone, a Type 1 glucocorticoid receptor antagonist, had no effect.[301,306]

Genetic Aspects of Voluntary Drinking

There are considerable differences between strains of rats or mice in their tendency to consume alcohol voluntarily, and inbred rat and mouse strains with high and low preference for alcohol have been extensively studied with respect to the mechanisms of alcohol drinking behavior and the molecular biology of alcohol preference. This research has been well reviewed previously[307-311] so will not be examined in extensive detail here. The strain differences in alcohol preference have been stable over the last 40-50 years, and are consistent between laboratories.[312] Genetic correlations have been examined between different aspects of alcohol-related behaviors, low alcohol preference, high conditioned taste aversion response and high alcohol withdrawal severity are correlated.[313,314] Overstreet *et al.* have reported characteristics of a substrain of Fawn Hooded rats (FH/Wjd) that has high alcohol preference and also exhibits depression-related characteristics, including high basal corticosterone,[315] although the two aspects were under separate genetic control.[316] Both the effects of prior chronic alcohol administration[186] and the alcohol deprivation effect (described in the next section) have been found to vary between mouse strains.[317]

Conclusions About Voluntary Alcohol Drinking

Alcohol preference models reflect some of the diagnostic criteria for alcohol dependence. Tolerance and withdrawal signs can occur, although the latter are rarely studied. The deprivation schedules cause progressive increases in alcohol consumption that can be long lasting. These models do not, however, measure the amount of work the rodent is prepared to do to obtain the alcohol, and for this reason many authorities prefer the operant models described below. In addition, the lack of correlation between voluntary alcohol drinking and the rewarding effects of alcohol in different rodent strains, as measured by place conditioning, and intragastric operant self-administration illustrate the problems of the aversive taste of alcohol.[318] Do the preference models reflect the non-diagnostic characteristics of alcohol dependence? Stress-induced voluntary alcohol drinking has been demonstrated. Although results have not substantiated the tension-reduction hypothesis, the most important aspect is the slowly developing increase in voluntary alcohol consumption that is seen at later times. Genetic influences are well represented by this model but in the main these reflect genetic difference in willingness to consume alcohol, which is not the same as genetic difference in the liability to develop alcohol dependence.

Free choice preference studies may be said to model alcohol drinking, but not alcohol dependence, while the deprivation models more closely approach the clinical dependence situation. Both naltrexone and acamprosate are well established to reduce alcohol consumption in the free choice models, which parallel their capacity to reduce alcohol drinking in non-dependent individuals.[252,319] However, there are many examples of "false positive" results for drugs that reduced voluntary drinking but which have not been of value in the treatment of alcoholics.

Forced Alcohol Consumption

Rodents can be forced to consume alcohol by using dilute alcohol as the only drinking fluid, or adding it to a liquid diet or by inhalation. In each of these cases it is possible to achieve high levels of alcohol intake, but the animal does not have any choice about the intake. Such treatment schedules do not therefore have face validity with the human situation, but they have been of value not only in studies of the alcohol withdrawal syndrome, but also on the effects of prolonged intake on subsequent voluntary consumption or operant self-administration. Each of the methods of administration has disadvantages. Provision of alcohol as sole drinking fluid is only feasible with strains or breeding lines with a fairly high preference for alcohol, otherwise health problems ensue. Administration of alcohol via a liquid diet, such as that devised by Lieber and DeCarli,[320] requires pair feeding of controls with an isocaloric control diet because the rodents on the alcohol diet do not gain weight as fast as those on normal laboratory chow. This restriction of food intake could be said to be a disadvantage of this method of administration, but restricting the amount of food consumed actually increases the life span of rats considerably compared with unrestricted food intake[321] and *ad libitum* availability of food is the more "unnatural" situation. Inhalation avoids the problems stemming from rodents' dislike of the taste of alcohol, but is obviously a very different method of administration from the oral route and loses out in face validity. It also requires more complex equipment for administration and monitoring.

Effects of Clinically Effective and Novel Drugs on Alcohol Withdrawal

As noted above, it appears that important changes in neuronal function occur during the acute phase of alcohol withdrawal that are involved in subsequent behavioral problems including not only the progressive increase in severity of the withdrawal syndrome, but also in relapse drinking and cognitive deficits. This suggests that drug treatment during this time period might reduce the subsequent problems. Benzodiazepines are widely used clinically to decrease alcohol withdrawal symptoms, and experimentally these drugs reduce both the anxiogenic and the convulsive components of the alcohol withdrawal syndrome.[322] However, not only does benzodiazepine administration not reduce relapse drinking in alcoholics, but these drugs have their own dependence liability and do not prevent subsequent problems such as cognitive deficits and HPA dysfunction.[16,32,323,324] Chlormethiazole (Distraneurin®) is a frequently used alternative for the treatment of alcohol withdrawal symptoms, but despite its clinical use for many years there is no evidence that it prevents relapse drinking or the cognitive deficits. NMDA antagonists can decrease alcohol withdrawal signs,[325] but, depending on the treatment schedule, they can also increase both withdrawal severity and the withdrawal kindling[326-328] and so are unlikely to be of value therapeutically. MK-801 did not reduce memory loss after alcohol withdrawal,[329] although a protective effect was seen when this drug was given to rat pups during the withdrawal phase after neonatal alcohol administration.[78]

Of the two drugs in wide clinical use in alcoholics, naltrexone and acamprosate, the latter appears to have some effects on the alcohol withdrawal syndrome, although the primary effect of both compounds is reduction of alcohol consumption. Gewiss et al.[262] found reduced behavioral signs of alcohol withdrawal in rats given acamprosate and Spanagel et al.[330] found reduction of some, but not all, withdrawal signs, although in this study only a short (1 week) chronic alcohol treatment was given. When administered to alcoholics prior to drinking cessation, acamprosate can reduce subsequent sleep disturbances.[331]

Other compounds may offer more hope for prolonged beneficial effects. Gabapentin is a relatively new introduction in current use for chronic pain and anxiety. Its mechanism of action is complex, but likely involves modulation of alpha2delta subunit-containing calcium channels.[332,333] Experimentally and in humans it is effective in the acute alcohol withdrawal syndrome, both the anxiogenic and convulsant aspects,[334,335] but an initial report suggests that it does not affect craving for alcohol.[336] Effects on memory loss in alcoholics have not been investigated, but gabapentin improved memory retention in mice[337,338] and has neuroprotective effects.[333,339]

Topiramate has multiple sites of action, including ion channels and protein phosphorylation.[340] It is effective clinically in alcohol withdrawal[341] reducing anxiogenic and convulsive components.[342] It has a transient acute effect in reducing alcohol preference in mice[343] and is neuroprotective.[344-346] Effects on alcohol preference after intermittent abstinence in rodents have not been reported, but Johnson et al.[120] found improved quality of life and increased abstinence in alcoholics during a 12-week trial, Topiramate has also been reported to reduce craving and alcohol consumption.[119,347,348] A recent study by Johnson et al.[121] confirmed the efficacy of topiramate in a fully controlled 14-week trial of alcohol-dependent patients, demonstrating superiority over placebo treatment as evaluated with a variety of efficacy measures, although it was associated with many adverse events such as paresthesia, taste and eating abnormalities, and disruptions of cognition.

Carbamazepine is well established as an anticonvulsant, but recent evidence suggests that it may have other properties relevant to alcohol dependence. It is effective against the alcohol withdrawal syndrome in humans,[349] including anxiety,[350] but has been reported not to be neuroprotective.[351,352] The latter authors also found reduced relapse drinking, as did Mueller et al.[353] Carbamazepine has an acute action in decreasing alcohol preference and alcohol tolerance, suggesting it has an effect on adaptive changes.[354] Improved verbal memory performance was reported during withdrawal in alcoholics given carbamazepine compared to those receiving chlormethiazole, though this could have been due to the adverse effects associated with chlormethiazole rather than beneficial actions of carbamazepine.[349]

Alcohol Place Conditioning

Conditioned place preference is considered to measure the motivational effects of rewarding and aversive stimuli,[355] including drugs of abuse, food or sexual activity.[356-358] Place conditioning involves pairing of drug experience with a characteristic location so that the drug treatment serves as the unconditioned stimulus and the environmental cues the conditioned stimulus.[355] With repeated pairing the location therefore acquires motivational properties that can elicit either approach or aversion. The behavior patterns appear to follow classical conditioning. Place preference has been suggested to be a model of drug reward rather than reinforcement,[358] but this distinction is often unclear in the literature. Preference for a location can be induced by palatable food (e.g.,[356]) or sexual activity[357] as well as drug effects. One advantage of conditioned place preference over other animal models is that the measurements are made in the absence of alcohol, thus avoiding problems of sedation or other motor effects that would alter activity. The administration of alcohol via the intraperitoneal or intragastric routes avoids the complications of the taste of alcohol, but raises questions about the relevance of results to human oral alcohol consumption.

Pairing of drug effect and location can be via a range of schedules. This aspect will be described only briefly here, but there are several excellent reviews.[359-361] Important variations include the number of compartments, either just two, one drug associated and one vehicle associated, or three with a central neutral compartment. Preference for one or other environment prior to experiencing any drug effects is incorporated by some researchers by pairing of drug effects with the least preferred location. This is thought to increase the extent of drug-induced preference but may involve reduction of the aversive properties of the less preferred location. There are complications of interpretation with all conditioned place preference studies, for example the role of novelty seeking, and direct comparison with human preference for locations is debatable, as discussed by Bardo and Bevins.[359]

Most place conditioning drug effects follow the patterns that might be predicted from human experience, but the effects of alcohol are more problematical and there are major species differences. Although place preference is relatively easy to establish in mice, rats more commonly exhibit place aversion (see below). Even in mice, the effects are complex and dependent on the mouse strain as well the experimental schedule. Cunningham *et al.* have examined the latter aspect in detail and found the timing of the alcohol administration and place testing to be crucial. Intragastric administration of alcohol produced preference in mice tested after a 5-min delay, while before this time an aversive effect was seen[362] and administration of alcohol after exposure to the conditioning environment produced place aversion.[363] The extent of alcohol-induced preference was also inversely related to the duration of the conditioning trials.[364] An interesting parallel with humans was demonstrated by Morse *et al.*[365] who found rats showed an aversion to the place in which the "hangover" phase following alcohol administration was experienced.

Most authors have found that rats exhibit place aversion rather than preference after conditioning with alcohol injections.[128,366,367] Despite ingesting considerable amounts of alcohol when given the opportunity, rats previously trained via food reward to drink 8% alcohol exhibited aversion to the location in which they consumed the alcohol.[368] Comparison of place conditioning in the P and NP lines of rats showed that while both showed place aversion, the alcohol-preferring P rats exhibited less aversive conditioning than the non-alcohol-preferring NP rat line.[198] Place preference to alcohol in rats has, however, been seen in rats after longer conditioning periods than for other drugs,[369,370] when low alcohol doses were used,[371] after stress[372-376] and also with standard schedules[377] and in young rats.[378] Importantly, it has also recently been demonstrated after intracerebroventricular infusions of alcohol, suggesting ingestive effects are responsible for the place aversion.[379]

Changes in Alcohol Place Conditioning After Chronic Treatment and Reinstatement of Conditioned Place Preference

There has been little examination of the effects of prolonged alcohol intake on alcohol-induced place conditioning in mice. Previous intake of alcohol however may enable establishment of place preference in rats. As long ago as 1985, Reid et al.[255] reported that after 26 days consumption of a 6% alcohol solution, place preference was seen in rats only after a few conditioning trials. Bienkowski et al.[369,380] found repeated injections of alcohol prior to conditioning led to development of alcohol place preference, but their demonstration that repeated saline injections also had a similar effect suggests that this was not due to pharmacological actions of the alcohol. Gauvin and Holloway[381] found previous experience of alcohol-induced place preference when this was not apparent without such experience, while Ciccocioppo et al.[382] found the alcohol dose range producing place preference increased in Sardinian alcohol-preferring rats after prior administration via indwelling intragastric catheters, though not via intraperitoneal injection or by intragastric gavage. The effects of repeated withdrawal from alcohol on alcohol place conditioning in rats do not appear to have been studied.

Reinstatement of place conditioning to morphine after extinction of the response and priming with morphine injections has been demonstrated,[383] but few studies appear to have been done on the corresponding behavior patterns with alcohol. Reinstatement of alcohol-conditioned place preference after extinction can, however, be demonstrated,[384] although this appears to have been little studied compared with reinstatement of operant responding for alcohol (see below).

It is possible that the difficulties encountered in rodents in establishing a clear effect of chronic alcohol treatment in increasing alcohol-conditioned place preference is due to complications of alcohol pharmacokinetics and difficulties in finding the most appropriate experimental protocols, rather than to it being a poor model for alcohol dependence. In zebra fish, a conditioned place preference can be established to alcohol and that persists for at least 3 weeks after cessation of conditioning. It also persists in the face of adverse consequences.[385]

Effects of Aversive Consequences on Alcohol Place Conditioning

LiCl causes gastric malaise, and place aversion can be produced by its administration to rodents. Devaluation of the motivation for a food reward is possible by administration of LiCl after the place conditioning has been established.[386] Although knowledge of the adverse consequences of alcohol drinking in humans applies over a longer time scale, if the shorter lifetime and memory span of rodents are taken into account, a reasonable parallel can be drawn and this methodology could be said to model DSM Criterion 7. Examination of effects of such consequences on alcohol place conditioning is only just beginning. A recent conference abstract reported that alcohol place preference in mice was resistant to devaluation by pairing of alcohol with LiCl administration,[387] suggesting an important difference from food conditioning.

Adolescence and Alcohol Place Conditioning

Few studies have examined the effects of age on alcohol place conditioning, but Philpot *et al.*[378] found that very young rats, in contrast to adult rats, do exhibit a place preference. At postnatal day 25, place preference was seen after conditioning to a low, 0.2 g/kg, alcohol dose and on postnatal day 45, 0.5 or 1 g/kg alcohol conditioning doses resulted in place preference. Place aversion was seen to higher doses and by day 60 there was place aversion with all doses.

Stress and Alcohol Place Conditioning

The small number of studies on the effects of stress on alcohol place conditioning does not appear to have included the effects of early stressful experiences, even though maternal deprivation has been found to increase morphine-conditioned place preference in adult life.[388] A single session of tail shock did not alter conditioned place preference to alcohol commenced 24 h later[377] even though such experience increased morphine place conditioning. Either intermittent footshock or social defeat blocked the place aversion to alcohol in rats, although place preference was not observed.[389]

A series of publications on conditioned fear stress and alcohol place conditioning have reported some interesting and potentially important results.[372-376] They showed when rats were given intraperitoneal injections of a low dose, 0.3 g/kg, of alcohol immediately after exposures to a compartment in which they had previously experienced footshock, they developed a place preference over 12 days for that location. The effect was prevented by naloxone and by the δ-opioid receptor antagonist naltrindole, but not by a κ-receptor antagonist, by D_1 and D_2 dopamine receptor antagonists and by ondansetron, when these were given during the alcohol conditioning. Such place conditioning would involve anxiety-related effects but the effects of alcohol were not reproduced when diazepam was substituted for the alcohol.

Effects of Clinically Effective and Novel Drugs on Alcohol Place Conditioning

A variety of approved and novel compounds have been tested in the conditioned place paradigm to examine their effects on the rewarding properties of alcohol and to identify the mechanisms underlying these effects. Opiate antagonists reduce alcohol-conditioned place preference. Cunningham *et al.*[390] showed a 10 mg/kg dose of naloxone had little effect on expression of conditioned place preference but facilitated its extinction, and it also enhanced place aversion caused by alcohol. A reduction in alcohol place conditioning with naltrexone was demonstrated by Middaugh and Bandy[391] at low doses (0.6 and 1 mg/g), but not at a higher dose (3 mg/kg). Kuzmin *et al.*[384] found naloxone at 1 mg/kg was very effective in preventing both expression of alcohol-conditioned place preference and reinstatement of such conditioning after extinction, but when given alone it caused place aversion. Administration of a non-selective opiate antagonist into the ventral tegmental area reduced alcohol-induced place preference in mice.[392] M-opiate receptor knockout mice exhibited less alcohol-conditioned place preference than the wild type.[393]

Acamprosate reduced the development of alcohol-induced place conditioning, and also that of cocaine.[395] When given alone, it did not produce either place conditioning or place aversion. This compound also reduced behavior on the elevated plus maze conditioned by prior administration of alcohol, an effect not seen with naltrexone.[396] Environmental tolerance to alcohol was also reduced by acamprosate.[397]

Few studies have examined the effects of modification of 5-HT transmission on place conditioning to alcohol. Risinger and Boyce[398] found that a specific 5-HT_{1A} antagonist (pindobind-5HT_{1A}) enhanced conditioned place preference to alcohol in mice. Mianserin, a 5-HT_2 receptor antagonist, also increased place preference conditioning to alcohol, without exhibiting any place conditioning when give alone.[399] Antagonists of dopamine D_1, D_2 or D_3 receptors did not alter place preference to alcohol in mice[400] and dopamine D_3 knockout mice did not differ in this respect from the wild type.[401] However, an enhancement of alcohol-conditioned place preference by the dopamine D_3 receptor antagonist, U99194A, which was devoid of place conditioning when given alone suggested a possible involvement of these receptors,[398,402] and D_2 receptor knockout mice showed no evidence of place conditioning to alcohol in a schedule that reliably produced this effect in wild-type and heterozygous animals.[403]

Although potentiation of GABA transmission has been suggested to be involved in the rewarding actions of alcohol, this action of alcohol is seen only in certain neurons[404] and GABA receptor antagonists increase, rather than decrease, alcohol-induced place preference.[405] Studies on knockout mice showed no difference in alcohol place preference with GABA receptor $\alpha1$ subunit deletion while $\beta2$ subunit knockouts had slightly less place preference.[406] The $GABA_B$ receptor agonist, baclofen, given systemically did not alter the acquisition of conditioned place preference to alcohol or conditioned taste avoidance induced by alcohol,[407] in contrast to the effects of this compound on voluntary drinking and operant self-administration of alcohol. However, infusion of baclofen into the ventral tegmental area did reduce alcohol-conditioned place preference,[392] suggesting that $GABA_B$ receptor in this brain region may be involved in the reinforcing effects of alcohol, but there may be actions at other sites that affect the consequences of systemic administration.

Expression of alcohol-place conditioning in mice can be blocked by an mGluR1a antagonist and by an mGluR5 antagonist;[408] however, McGeehan and Olive[395] did not find effects of an mGluR5 antagonist in the acquisition of alcohol-conditioned place preference. The non-competitive NMDA receptor antagonist, neramexane, reduces both acquisition and expression of alcohol-induced conditioned place preference;[409] but pretreatment with the non-competitive NMDA receptor antagonists, MK-801 or ketamine, and NMDA2R antagonists does not alter acquisition of alcohol-induced place preference[409] and pretreatment with the channel blockers, MK-801 or ketamine, and NMDA2R antagonists does not alter acquisition of alcohol place preference in mice.[410] The competitive antagonist CGP37849 prevented acquisition of both place preference and place aversion caused by alcohol,[410] suggesting effects on place conditioning processes rather than the rewarding properties of alcohol.

Both cannabinoid CB-1 receptor antagonism and CB-1 receptor knockout reduced alcohol-conditioned place preference in mice.[285,286,411-413] The CB-1 knockout mice did not exhibit any differences in cocaine-conditioned place preference, so the effect appears to be specific for alcohol, and there were no differences in sensitivity to the actions of alcohol on anxiety-related behavior on the elevated plus maze. Topiramate, an anticonvulsant that has received attention owing to its reported effects in decreasing alcohol craving and relapse drinking, does not alter acquisition of alcohol-conditioned place preference in mice,[414] indicating that topiramate may not influence the rewarding effects of alcohol.

CRF-deficient mice did not exhibit any place preference conditioning to alcohol.[305] Since CRF has been reported to decrease alcohol drinking while CRF antagonists reduce reinstatement of responding for alcohol, it is not clear how this result should be interpreted. Studies on the effects of CRF receptor antagonists on place conditioning to alcohol would be of considerable interest in this context.

Genetic Aspects of Alcohol-Induced Place Conditioning

Apart from the species differences, there are also strain differences within alcohol place conditioning, although these have not always correlated with other aspects such as voluntary alcohol consumption or operant self-administration. The high drinking short-term selected (STDRHI) mouse line exhibited greater place preference than that corresponding low drinking line (STDRLO).[415] Alcohol was more effective in producing place preference in high alcohol-preferring mice than in low alcohol-preferring mice.[416] Chester *et al.*[417] found that mice selectively bred for high sensitivity to the alcohol withdrawal syndrome exhibited conditioned place preference to alcohol while the parallel bred line with low withdrawal sensitivity did not show such conditioning.

Conclusions on Alcohol Place Conditioning

Comparison of conditioned place preference with clinical diagnostic criteria shows that DSM Criteria 1, 2 and 4–7 are not reflected by this model. The difficulties of showing place preference in rats might at first sight something of a barrier to inclusion of place conditioning in simple test batteries, but this may actually be extremely important in our development of better models of alcohol dependence. The demonstration of conditioned place preference in very young rats and the increase in preference in rats when combined with conditioned fear stress suggest parallels with the human situation. The recent demonstration of devaluation with LiCl and studies on the influence of chronic alcohol treatment suggest that place conditioning may be a better model of dependence than it appears at first sight. Reports on these aspects are so far few in numbers and could with considerable advantage be extended. This model has not been extensively used in drug testing, although both naltrexone and acamprosate have been reported to decrease the effects of alcohol.

Operant Self-administration

Operant self-administration of alcohol has been extensively studied and reported, so this section will be primarily confined to important aspects of the experimental schedules, the effects of chronic alcohol consumption, the influences of stress and devaluation and the most significant drug effects. The seminal work of Samson *et al.* has characterized operant self-administration of alcohol. It has been suggested that alcohol is a weaker reinforcer than drugs such as psychostimulants since it is more difficult to demonstrate high levels of operant self-administration, but this is, at least in part, due to complications of taste and pharmacokinetics of alcohol. Introduction of alcohol by mixing with sucrose enabled the demonstration that alcohol mixtures supported greater responding than higher concentrations of sucrose and that the alcohol played a major role in the reinforcement.[418,419] Concentration effects could be demonstrated and progressive ratio schedules established.[419,420] Studies involving separation of responding for alcohol from alcohol consumption demonstrated that alcohol can exhibit strong reinforcing properties in this model.[421,422] Removal of a secondary reinforcer can increase responding for alcohol. Podlesnik *et al.*[423] trained rats to respond for alcohol by lever pressing, then discontinued the alcohol and trained the animals to respond for food pellets by chain pulling. Cessation of the food reward resulted in resumption of lever pressing even though no alcohol was available.

Changes in operant self-administration of alcohol after chronic alcohol treatment and after deprivation

Increased alcohol self-administration after prolonged chronic alcohol consumption in rodents would parallel the many years of excess drinking that are involved in the development of alcohol dependence in humans. Despite this, not very many published studies have demonstrated this behavior so far and there have been negative reports (e.g.,[424]). In many rodent strains and treatment schedules prior long-term chronic alcohol intake does not increase operant self-administration. The parameters measured are also important, for example Files *et al.*[425] found increases only in the g/kg alcohol consumed per drinking bout after operant presentation of alcohol via sipper tube after 5 weeks voluntary alcohol drinking. Training in operant self-administration prior to the chronic alcohol treatment may be an important factor in demonstrating increased responding for alcohol. It is also possible that memory deficits caused by chronic alcohol consumption could affect measurements of operant responding. In the main such studies have concentrated on the early withdrawal phase but some have examined longer time intervals after cessation of chronic alcohol consumption which is more relevant to the human situation. Schulteis *et al.*[426] demonstrated increased lever responding for alcohol when tested using continuous reinforcement at 8h after cessation of 17 days of alcohol consumption via a liquid diet. Importantly, there was no corresponding increase in responding for saccharin. Roberts *et al.*[427] also found increased responding on a continuous reinforcement schedule during a 12-h withdrawal period after 2 weeks chronic alcohol inhalation. Withdrawal from 9 days consumption of an alcohol liquid diet increased the break point for self-administration of ethanol on a progressive ratio schedule at 24h after the alcohol withdrawal, suggesting increased reinforcing properties of alcohol.[428] O'Dell *et al.*[429] found that intermittent exposure to alcohol vapor increased responding for alcohol at 2 and 8h after withdrawal, while continuous exposure to the same duration of vapor only had a small effect. Self-administration of alcohol into the posterior ventral tegmental area was also found to be increased by chronic consumption.[430] Availability of a 24-h free choice between alcohol and water for 8 weeks in the alcohol-preferring P rat line resulted in a higher level of self-infusion than that in alcohol-naive rats.

Other studies have examined longer durations of abstinence. McKinzie *et al.*[206,207] found that after 6 or 7 weeks of operant responding for alcohol by alcohol-preferring P rats, there was a transient increase in responding only on the first test day after 2 weeks of abstinence. Roberts *et al.*[431,432] reported increased operant self-administration of alcohol by rats 4–8 weeks after withdrawal from 2 weeks of chronic alcohol treatment via inhalation. Later studies from this group have shown effects of CRF antagonism on operant self-administration,[433,434] as described below. The pattern of behavior is not confined to rats. Middaugh *et al.*[435] demonstrated operant self-administration of alcohol in C57/BL6 mice and showed enhanced responding after a 2-week enforced abstinence following 2 weeks of overnight access to alcohol. Repeated cycles of deprivation and operant access to alcohol (15%) result in increased responding for alcohol and 1.5- or 2-fold higher break points in progressive ratio responding by two high alcohol drinking (HAD) rat lines.[436] Deprivation periods can also increase the reinforcing effects of alcohol infused into the posterior ventral tegmental area.[430] Eight months of continuous access to alcohol increased the level of alcohol self-infusion and this was greater in rats that were given repeated deprivation periods during the chronic treatment than in the animals that had continuous access to the alcohol.

Reinstatement of operant responding for alcohol after extinction of the response has been rather more extensively examined than the effects of abstinence after chronic alcohol consumption without extinction. The value of studying reinstatement of self-administration of drugs has recently been reviewed by Epstein *et al.*[438] and the deprivation model by McBride *et al.*[439] Reinstatement of operant responding to alcohol after extinction of the response can be induced by non-contingent presentation of the dipper-containing alcohol,[440–442] by priming with an injection of alcohol and by application of acute stress, such as footshock.[443] Exposure to alcohol-related cues or stressful experiences or administration of priming doses of alcohol result in resumption of previously extinguished responding for alcohol by both rats and mice.[435] However self-administration of a small amount of alcohol was found by Samson and Chappell[444] to have little effect in reinstating responding for the drug. Responding for intravenous alcohol can also be reinstated by cues or priming.[445] The effect of an abstinence phase in increasing responding for alcohol after extinction was found to be greater with 56 days abstinence than after 28 days, suggesting a progressive development of the relevant neuronal changes.[446]

There are difficulties with the reinstatement after extinction model in the comparison with the human situation, even apart from the fact that such extinction, that is, non-provision of alcohol previously supplied in response to lever pressing or other operant response, is a situation that rarely occurs for humans; the cognitive processes that would be involved in humans are substantially different from those in rodents and the behavior patterns after experimental extinction in humans did not follow those predicted from animal studies.[447] Alcohol-related cues, priming or stressful experiences cause relapses back to excess drinking in alcohol-dependent humans but have considerably less effect in those who are not alcohol dependent. Drug treatments aimed at decreasing relapse drinking are intended for use in alcoholics in which external stimuli that provoke relapse will have very different neurobiological effects than in non-dependent individuals.

Operant self-administration of alcohol during adolescence

Only a small number of studies have been made on the effects of adolescent alcohol drinking on operant self-administration of alcohol but it appears from these that consumption of alcohol from an early age increases the propensity of rats to work to obtain alcohol later in life. Siciliano and Smith[208] found higher levels of responding for alcohol in rats given 5% or 10% alcohol as their sole drinking fluid from PD21 to PD105. Rodd-Henricks et al.[448] examined the effects of free choice access to alcohol during adolescence (postnatal days 30–60) and in adulthood (over 135 days of age). The animals with previous experience of alcohol drinking during adolescence showed faster acquisition of responding for alcohol, greater resistance to extinction of the responding and more prolonged high alcohol responding during reinstatement than alcohol-naive animals, but this effect was not produced by alcohol exposure during adult life.

Effects of Aversive Consequences on Operant Self-administration of Alcohol

Punishment procedures compare the effectiveness of a standard punisher, such as administration of LiCl, in suppressing lever pressing by rats for alcohol, with the effects of punishment of responding for a natural reinforcer, a sucrose solution. Meachum[449] demonstrated that if lever pressing by rats for a natural reward was followed by LiCl-induced malaise, lever pressing progressively declined across successive sessions. Responding for a cocaine reward, however, was found to be unaffected by LiCl devaluation.[450] If alcohol self-administration by rodents is particularly persistent in the face of adverse consequences, then lever pressing reinforced by alcohol delivery will be more resistant to suppression by punishment than responding reinforced by a natural reward, even if the alcohol does not maintain a higher baseline level of self-administration.

Dickinson et al.[451] showed that rats will respond for alcohol and food reward when the presentation of these is contingent on responding on two different levers. In this study, the effect of a devaluation procedure on responding for alcohol and food reward was examined. After operant training, an aversion was conditioned to either the alcohol or the food pellets by pairing consumption with the gastric malaise induced by administration of lithium chloride, thus "devaluing" one or other reinforcer. When subsequently tested, the rats pressed less on the pellet lever if the pellets, rather than the ethanol, had been devalued by the aversion conditioning. By contrast, performance on the ethanol lever was unaffected by the devaluation procedure. The level of responding for the food reward prior to devaluation in this study was approximately twice that for the alcohol reward. These results showed that it was more difficult to devalue the reinforcement produced by alcohol, as the rats persisted in responding for the presentation of this substance after its devaluation by LiCl administration, in contrast to the reduction in responding for the food reward seen after devaluation of that reward. Samson et al.,[452] however, demonstrated reduced operant responding by rats for alcohol when LiCl was paired with alcohol, when measurements were made on the responding following extinction that is considered a model of "alcohol seeking." After 3 days with 10% alcohol as the sole drinking fluid followed by sucrose-fading training in operant self-administration of alcohol, an extinction test was carried out. Rats were then given alcohol gavage and LiCl injections, either paired or unpaired for 6 days. When a post-devaluation test was run, rats given both paired and unpaired LiCl showed decreased levels of lever pressing, but the group that had paired alcohol and LiCl exhibited significantly lower levels than those given unpaired administration. The paired group also showed lower alcohol consumption during the first post-devaluation consummatory session. As the authors indicated, a major difference between the experimental procedures in these studies and those of Dickinson *et al.*[451] was that in the latter case the rats were self-administering alcohol during the devaluation procedure while during the studies of Samson *et al.*[452] alcohol was not available for self-administration during this time. More extensive study of this aspect of alcohol self-administration would clarify the situation and would be of great value in the development of better animal models of alcohol dependence.

Effects of Stress on Operant Self-administration of Alcohol

Studies on the effects of stressful experiences on operant self-administration of alcohol have primarily concentrated on the reinstatement of responding after extinction, but there has been some investigation of stress effects in the absence of a reinstatement protocol. Isolation rearing of rats during postnatal days 21–111 resulted in higher rates of responding for alcohol in the sucrose-fading procedure carried out at the end of this time, compared with social rearing when two rats were housed per cage. A single experience of social defeat resulted in a transient reduction of operant responding for alcohol, delayed for 1–2 days; the responding then returned to pre-stress levels over the subsequent few days.[453]

Le *et al.*[443,454] found that intermittent footshock stress potently reinstates responding for alcohol after the response had been extinguished. This is a pattern similar to that seen in responding for heroin and cocaine, but these authors found that responding for sucrose, in contrast, was not reinstated by the footshock. The effect of footshock does not appear to involve glucocorticoid action, since adrenalectomy followed by maintenance of basal corticosterone levels does not prevent the behavioral changes. Interactions between stress and alcohol cues on alcohol responding were investigated by Liu and Weiss.[455] A combination of stress and alcohol cues had greater effects on the reinstatement of responding for alcohol than either condition alone. Importantly, both the single and interactive effects of these procedures were more effective in rats which become alcohol dependent by 12 days of alcohol vapor inhalation than in animals that had been housed similarly without alcohol inhalation.

Effects of clinically effective and novel drugs on operant self-administration of alcohol

Hyytia and Sinclair[456] showed that naltrexone reduced operant responding of rats for alcohol and found the effect developed slowly. Later studies confirmed this effect, but it was not selective for alcohol since responding for sucrose is also decreased.[457] A selective effect of naltrexone was seen by Holter and Spanagel[256] only when an intermittent alcohol exposure schedule and low doses of the drug were used. Reduction of reinstatement of responding for alcohol after naltrexone has been demonstrated by Bienkowski *et al.*[440,441] and Burattini *et al.*[458] The former authors showed single doses of the opiate antagonist increased extinction and cue-induced responding while its repeated administration led to progressive reduction in standard responding.

Heidbreder et al.[459] however, found neither naltrexone nor acamprosate prevented reinstatement of alcohol responding after extinction. Cue-induced responding for alcohol was shown by Katner et al.[460] to be reduced by naltrexone. Gonzales and Weiss[461] demonstrated that naltrexone suppressed the rise in dopamine levels in the nucleus accumbens following operant responding for alcohol, but not the increase in dopamine in this brain region seen during the anticipation period.

Acamprosate reduces operant self-administration of alcohol, but repeated injection of this drug results in tolerance to this effect.[263] Similarly, Heyser et al.[462] showed that acamprosate has limited effects on operant responding for alcohol, but does prevent the alcohol deprivation effect-induced increase in responding. Acamprosate also reduces context-induced responding for alcohol.[463] In contrast to the effects of naltrexone, this effect was found to be selective for alcohol since no change in operant responding for sucrose was seen at doses that reduced responding for alcohol.[464]

Heyser et al.[462] showed acamprosate not only had some effects on operant responding for alcohol, but also prevented the deprivation-induced increase in responding. Responding for alcohol elicited by exposure to alcohol-related environmental stimuli was found by Bachteler et al.[463] to be reduced by acamprosate. In contrast to the effects of naltrexone, this effect was found to be selective for alcohol as no change in operant responding for sucrose was seen at doses that reduced responding for alcohol.[464] Acamprosate reduced operant self-administration of alcohol by alcohol-preferring Fawn Hooded and iP rats, but repeated injection of this drug resulted in tolerance to this effect.[263]

SSRIs have been reported to reduce operant self-administration of alcohol and Le et al.[465] demonstrated decreased reinstatement of responding following extinction. Ginsberg and Lamb,[466] however, found that decreases in responding are not selective for alcohol and vary considerably between individual animals. 5-HT$_3$ antagonists have also been reported to reduce operant responding for alcohol,[467,468] without affecting responding for water.[467] In addition, fluoxetine and the 5-HT$_3$ receptor antagonist, ondansetron, were found by Le et al.[469] to reduce the effects of footshock in reinstating operant responding for alcohol. Activation of 5-HT$_{1B}$ receptors[470,471] and antagonism of 5-HT$_{2C}$ receptors[472] also reduced the operant self-administration of alcohol.

Blockade of cannabinoid CB-1 receptors has consistently been found to reduce operant self-administration of alcohol[473] and extinction of responding for alcohol was shown to be facilitated by a CB-1 receptor antagonist, SR147778.[289] Rates of responding for palatable foods and for normal food have also been found to be reduced by CB-1 receptor antagonists.[474] Reinstatement of responding induced by presentation of alcohol-associated cues was shown to be decreased by the CB-1 receptor antagonist, SR14716A, as well as the break points for alcohol in progressive ratio responding, but in addition this compound reduced the corresponding responding for sucrose.[475] The effects of this antagonist on alcohol responding were found to be greater in an alcohol-preferring line than in heterogeneous Wistar stock animals.[476] Conversely, the cannabinoid agonist, WIN55,212-2, given during a deprivation period caused extended increased responding after the deprivation.[477] Reinstatement of responding for alcohol induced by nicotine was reduced by the CB-1 receptor antagonist rimonabant,[478] but reinstatement of responding for alcohol induced by acute stress was unaffected by SR14716A.[475]

A number of reports have shown that mGluR5 receptor antagonists reduce operant responding for alcohol. Backstrom *et al.*[298] demonstrated reduction of reinstatement of alcohol responding with MPEP, although at the highest dose this drug also reduced responding on the inactive lever. The increased responding following repeated deprivation periods was also reduced by MPEP.[479] Another mGluR5 receptor antagonist, MTEP (2-methyl-1,3-thiazole-4-yl)ethynyl-pyridine), reduced both responding for alcohol and its consumption by C57BL/6 strain mice. Hodge *et al.*[480] also found reduction of operant responding for alcohol with MPEP, but saw no effects of the mGluR1 antagonist CPCCOEt or the mGluR2/3 antagonist LY341495. The potential importance of the mGLuR2/3 receptor was recently demonstrated by Rodd *et al.*[299] who found that the mGluR2/3 agonist, LY404039, reduced the alcohol deprivation effect on operant responding for alcohol, but did not alter maintenance responding. Other glutamate receptor antagonists have also been found to reduce operant responding for alcohol, including the AMPA antagonist, GYKI 52466.[481]

Decreases in operant responding for alcohol have been reported following administration of baclofen, but there is some uncertainty about the specificity of this effect. Anstrom *et al.*[482] found a dose-dependent reduction in responding by rats for both alcohol and sucrose with baclofen. Reductions in the responding of mice for alcohol were reported by Besheer *et al.*[483] after baclofen administration, but these authors also reported that the sedative effects of alcohol were increased by this compound and this could have affected the responding. A reduction of extinction responding for alcohol by baclofen was suggested to indicate an effect on the motivation to consume alcohol.[282] However, baclofen was also found to reduce operant responding for sucrose.[484] In contrast to previous reports, Czachowski *et al.*[485] did not find reductions in responses for alcohol following doses of baclofen that were found by others to have this effect, while responding for sucrose was reduced. These authors found a reduction in the "alcohol-seeking" responding for alcohol but only after 3 days administration of the highest dose of baclofen tested (3 mg/kg). When consumption of the alcohol and sucrose following presentation of these rewards was examined, the alcohol intake was increased and that of sucrose decreased. Possible beneficial effects of other forms of modulation of the $GABA_B$ receptor were indicated by the specific reduction of responding for alcohol, but not that for sucrose, after administration of the positive allosteric modulator, GS39783.[486]

Operant responding for alcohol after extinction of responses has been consistently shown to be decreased by administration of CRF1 receptor antagonists. The non-selective antagonist, D-phe-CRF, and the selective CRF1 receptor blocker, CP-154,526, reduced the reinstatement of responding induced by intermittent footshock.[454] In agreement with these results, CRF itself has the opposite effect. Le *et al.*[487] found that administration of CRF into the median raphé nucleus resulted in reinstatement of responding for alcohol.

Differences in drug effects on alcohol self-administration in animals previously treated chronically with alcohol compared with alcohol-naive animals are likely to be crucially important not only for the development of effective treatments, but also for our understanding of the biological mechanisms. The increased operant responding seen at 2 h and at 3–5 weeks after withdrawal from chronic treatment was prevented by central administration of the CRF antagonist D-phe-CRF.[433] Local administration of D-phe-CRF(12-41) identified the site of this effect as the central nucleus of the amygdala,[488] and systemic administration of selective antagonists showed the receptor involved to be of the CRF1 subtype.[489] Involvement of the CRF1 receptor in increased responding for alcohol after cessation of chronic alcohol treatment was also demonstrated by Chu *et al.*[490] who showed the CRF1 antagonist antalarmin prevented this change in mice and, importantly, that the increased in responding for alcohol was not seen in CRF knockout mice. The CRF2 receptor, however, appears to play a role in operant responding for alcohol that may be opposite to that of the CRF1 receptor, since the selective CRF2 agonist, urocortin 3, prevented the increase in responding for alcohol after withdrawal from chronic alcohol consumption.[434,489]

Walker and colleagues[379,491] demonstrated that a lower dose of baclofen decreased responding for alcohol in rats that had undergone a 1-month intermittent chronic alcohol vapor treatment than those effective in parallel control animals. De Fonseca *et al.*[492] found the CB-1 receptor antagonist, SR14716A, reduced operant responding for alcohol only in rats previously made dependent by exposure to alcohol vapor, without effects on food responding. Antagonism of NPY-Y2 receptors reduced by approximately half the responding for alcohol by Wistar rats previously given long-term intermittent exposure to alcohol vapor, but the effective dose of the antagonist, BIIE0246, had no effect on responding in rats that had not undergone the chronic treatment.[493]

Genetic Aspects of Operant Self-administration of Alcohol

Rodent lines with high alcohol preference for alcohol do not necessarily also exhibit higher responding for alcohol than their low alcohol-preferring counterparts. Risinger *et al.*[494] demonstrated greater operant self-administration of alcohol by the high alcohol-preferring C57Bl/6 strain than the lower preferring DBA strain. Samson *et al.*[495] examined operant responding for alcohol among several selected lines and concluded that 62% of genetic variance in operant self-administration resulted from genes selected for voluntary drinking. The P, HAD and NP lines all acquired alcohol self-administration and only the LAD line did not maintain responding after initiation. The ANA line self-administered less alcohol than the AA line. Among the high alcohol-preferring rats, Czachowski and Samson[496] found the P rat line responded more for both alcohol and sucrose than the HAD line. The Sardinian alcohol-preferring and -non-preferring lines did, however, show a similar pattern in operant self-administration as the sNP animals showed minimal operant self-administration of alcohol after the sucrose-fading procedure while the sP rats were able to acquire the response.

Conclusions on the Operant Self-administration Model

Operant responding for alcohol has face validity with respect to the willingness of rats to perform work to obtain alcohol. Most of the experimental schedules do not reflect the clinical diagnostic criteria detailed in Table 5.1, with the exception of Criterion 5 ("great deal of time spent in activities necessary to obtain alcohol") and the non-diagnostic aspects of priming and effects of alcohol-related cues and the effects of stress. The model however has the potential to examine important but currently neglected aspects, particularly giving up other activities (Criterion 6) and continued alcohol use despite knowledge of adverse effects (Criterion 7) as well as the potential influence of early stress and commencement of alcohol consumption at an early age. The alcohol intake is low compared with voluntary or forced drinking but some evidence of withdrawal signs has been reported (e.g.,[497]). Standard instrumental and consummatory procedures for studying alcohol self-administration by animals do not measure the persistence of drug taking in the face of adverse consequences and therefore fail to assess dependence as defined by DSM Criterion 7; the effects of devaluation have not been widely studied and more studies in this area would be useful. Chronic alcohol treatment does increase responding for alcohol, but again this has not been widely studied and the possible influence of cognitive deficits has been little examined. In contrast, there has been extensive work on the effects of deprivation periods and robust increases in responding for alcohol are seen. Important effects of alcohol consumption during adolescence on operant self-administration in adult life have been reported that will be of value in future drug testing. Stress can produce reinstatement of responding for alcohol, but long-term consequences of stress and effects of early stressful experiences need to be further examined to parallel the human situation. Operant models of alcohol self-administration have been suggested to model craving, as they can demonstrate "drug seeking" and conditioning of such behavior. However, craving is defined as the "uncontrollable desire" for alcohol and although many of these operant behavior schedules measure apparent "desire" on the part of the animals but few examine the "uncontrollable" nature of such desire. The demonstration of alcohol seeking in the face of adverse consequences or the availability of other rewards would more closely resembles the human situation of alcohol dependence. The drug treatments effective in humans, naltrexone and acamprosate, decrease responding for alcohol but compounds affecting 5-HT transmission also showed positive effects, in contrast to their lack of benefit in humans (see above/below). Perhaps the most important pharmacological results in relation to alcohol dependence are those that have demonstrated drug effects only after long-term alcohol intake.

Genetics of Alcohol Dependence

Genetic and environmental factors are both known to contribute to the expression of alcohol dependence, and the risk of developing this disorder is also dependent upon their interaction.[498] Environment appears to play a more influential role in alcohol-taking behaviors during adolescence (for review, see[499]), but evidence from familial, twin and adoption studies indicate that the influence of genetics increases with age, with alcohol dependence exhibiting a strong heritability of 40–60%.[500-504] Similar to other psychiatric disorders, alcohol dependence is multigenic and comprises a heterogeneous population. Owing to its complex nature, the determination of genetic risk factors for alcohol dependence has been and remains challenging.

A number of strategies have been employed to explore the genetics of alcohol dependence including animal models, and more recently with the completion of the human genome project, association and linkage studies. The following subsections will describe two approaches commonly used in alcohol genetics research. The first subsection will focus on familial alcoholism and the animal models developed based on selection for high and low alcohol preference. The second subsection will examine more recent advances in human genetics research and the genetically engineered animal models that have been developed to identify the role of specific genes and their products in alcohol-related behaviors. The effects of currently approved medications and other pharmacological agents for the treatment of alcohol dependence have not been extensively compared across human genotypes or assessed in the many animal models, but in trying to demonstrate translation from bench to clinic, these will be described where possible.

Family History of Alcohol Dependence

Individuals with a positive family history of alcohol dependence (FHP, e.g., Type 2/B) have a 4- to 10-fold risk of developing alcohol dependence compared to individuals with a negative family history of alcohol dependence (FHN;[498]) and their responses to alcohol have been of widespread interest in the search for potential markers. In particular, a low level of response (e.g., body sway) to the acute effects of alcohol in FHP individuals has been repeatedly demonstrated by Schuckit[505-508] and is considered an important predictor of risk for alcohol dependence (cf. FHN more sensitive:[509-511] no difference:[512,513]).

Considering the strong heritable component of alcohol dependence, much effort has been placed in selectively breeding rodent models to examine the heritability of physiological and behavioral differences in the response to alcohol. Several rodent lines have been selected for differences in alcohol withdrawal severity (withdrawal seizure prone and resistant mice;[514]), sensitivity to hypnotic effects of alcohol (long sleep and short sleep mice;[515]) and alcohol-induced hypothermia (HOT and COLD mice;[516]), etc., but the majority have been selected over multiple generations for high and low alcohol preference/consumption (see Table 5.3).[255,267,394,437] A brief review of the literature shows that, similar to human alcoholics, there is wide variation in the response to alcohol even following similar selection pressures. It is beyond the scope of the present chapter to review the numerous differences across selectively bred strains, but reference to the motor impairing effects of alcohol is presented here. As observed in FHP individuals, rodent lines bred for high alcohol preference have been shown to be less sensitive to the motor impairing effects of alcohol compared to their low alcohol-preferring counterparts,[517-520] but other lines have shown no difference,[521,522] or even greater sensi-tivity to alcohol compared to low alcohol-preferring rats[522-526] but (see[527]). Line differences have also been observed in procedures measuring alcohol reward and reinforcement, including conditioned place preference (e.g.,[198,416,417]) and operant self-administration (e.g.,[497,498]), but not all high alcohol-preferring lines demonstrate enhanced sensitivity to the rewarding and reinforcing effects of alcohol compared to low alcohol-preferring lines (e.g.,[528,529]). These observations underscore the idea that one measure is insufficient as a sole predictor of risk for susceptibility to alcohol, and that other behaviors (e.g., tolerance, reinstatement) need to be assessed. In so far as tolerance development, there are a number of studies indicating that selective breeding for high alcohol preference is associated with greater tolerance to acute alcohol (e.g., P versus NP rats and Wistar rats;[530]).

Family History and Treatment Outcomes

A comparison of the effects of approved and potential pharmacological treatments on alcohol-related behaviors in FHP and FHN humans are presented in Table 5.3. As can be seen from the available clinical studies, there are a number of interactions between

Table 5.3 A comparison of the effects of approved and potential pharmacological treatments on alcohol-related behaviors in FHP and FHN humans, and in genetic rodent models of alcohol dependence

Human		Model	
Family history		**Phenotype selection**	
Positive (FHP)	*Negative †FHN*	*High alcohol preference*	*Low alcohol preference*
■ Type 2/Type B ■ Early onset of alcohol dependence (<25) ■ Male predominance ■ History of sociopathic activity	■ Type 1/Type A ■ Later onset of dependence (>25) ■ Greater anxiety	■ P rats and replicate lines ■ HAD rats and replicate lines ■ AA rats ■ UChB rats ■ HARF rats ■ C57 mice	■ NP rats and replicate lines ■ LAD rats and replicate lines ■ ANA rats ■ UChA rats ■ LARF rats ■ DBA mice
Disulfiram			
■ N/A		■ Disulfiram or cyanimide decrease alcohol consumption in alcohol-preferring, but not non-preferring lines[531,532]	
Naltrexone			
■ Naltrexone induces greater reduction in subjective effects of alcohol (euphoria, sedation), heavy drinking and number of drinks consumed in FHP versus FHN.[149,164,533]		■ Naltrexone and naloxone reduce alcohol consumption and reinstatement in alcohol-preferring lines[534-536] ■ μ-Antagonism reduces responding for alcohol more potently in alcohol-preferring rats compared to non-selected rats[537]	
Acamprosate			
■ Greater efficacy of acamprosate in FHN versus FHP[160]		■ Acamprosate reduces responding, consumption and motivational salience of alcohol-related cues in alcohol-preferring rats[263,536]	
Serotonergic agents			
■ SSRIs may improve outcome in late, but not early onset alcoholics[538-541] ■ Ondansetron reduces drinking in early, but not late-onset alcoholics[125,127-129]		■ Fluoxetine reduces palatable alcohol consumption in non-preferring rats[542] ■ 5-HT$_3$ antagonism reduces responding for alcohol, but no consistent effects on reinstatement in alcohol-preferring rats[543]	
Dopaminergic agents			
■ N/A		■ Raclopride reduces alcohol seeking in alcohol-preferring rats[536]	
GABAergic agents			
■ N/A		■ Baclofen reduces alcohol drinking, withdrawal severity, reinstatement in alcohol-preferring rats[281,544]	

family history of alcohol dependence and the effectiveness of pharmacological treatments. For example, naltrexone may be more effective in FHP alcoholics, whereas acamprosate may be more beneficial in FHN alcoholics. There has also been limited assessment of the efficacy of pharmacological agents in rodent lines selected for high and low alcohol preference. The preclinical literature supports the clinical findings for naltrexone in that alcohol-preferring rats may be more responsive to μ-opioid receptor antagonism on alcohol consumption and reinstatement of alcohol seeking compared to non-preferring and non-selected strains. Acamprosate also reduces operant responding for alcohol in alcohol-preferring rats; however, this has not been assessed in non-preferring rats making it difficult to compare with the clinical studies. The table provides a very brief overview of the effects of pharmacological treatments on alcohol-taking behaviors; however, it must be noted that not all rat lines have been assessed and the reader is referred to the primary literature for greater detail.

Emerging evidence indicates that the presence/absence of a family history of alcohol dependence can influence treatment outcomes, but further research into the genes involved in specific features of alcohol dependence is needed to tailor and individualize pharmacological treatments. Association and linkage studies are providing further insight into the contribution of individual genes in alcohol dependence. In combination with preclinical research with genetically engineered strains, these will facilitate identification of more selective subtypes of alcoholics and potentially, development of pharmacological agents targeted for the associated behavioral phenotypes.

Genes Associated with Risk of Alcohol Dependence

The contribution of genes in phenotypic expression is becoming an increasingly important area of study in a variety of diseases and research has been particularly fruitful when studying genes of large effect. It becomes much more challenging in polygenic disorders such as alcohol dependence in which each gene is likely only to have a small effect on the phenotype. In order to identify possible genes associated with alcohol dependence, geneticists have relied on linkage and association studies.

Briefly, linkage studies examine patterns of DNA sharing among family members that match phenotypic patterns. Thus, if a particular locus influences a specific trait, then relatives who demonstrate a similar phenotype should also share more genetic material (concordant) compared to relatives who do not show the phenotype of interest. This approach uses a genome-wide scan and genotyping of a few hundred markers and does not require *a priori* knowledge of the genes involved in the phenotype. Furthermore, the alleles common in one set of concordant relatives do not have to be the same as those shared by another set of concordant relatives. In contrast, association studies examine the frequency with which certain alleles are present in samples of affected (case) and unaffected (control) individuals drawn from the population. Previous knowledge of the neurobiology of alcohol dependence is commonly used to guide the selection of candidate genes in these types of studies.

The following sections will describe current knowledge of genes associated with alcohol dependence in humans and the advancements being made in genetic animal models. Table 5.4 compares the genetic variations associated with alcohol dependence in humans with the effect of genetic selection on alcohol consumption in rodent models.

Table 5.4 Genotype and alcohol dependence. Comparison of the genetic variations associated with alcohol dependence in humans with the effect of genetic selection on alcohol consumption in rodent models

Human	Model
Genetic variant	**Genotype selection**

Alcohol metabolism

*ADH1B*2* (rapid alcohol oxidation)
- Associated with 3-fold protection from alcohol dependence
- Up to 90% prevalence in Asian population; ~5% of European population[545]

*ADH1B*3 and ADH1C*1* (rapid alcohol oxidation)
- 1B*3 exhibits higher prevalence in African American, Native American[547–549]
- 1C*1 exhibits 50% prevalence in European population, up to 90% in some Asian populations[545,548–551]
- Associated with protection from alcohol dependence, but inconsistent[552]

*ALDH2*2* (slow/inactive acetaldehyde oxidation)
- Up to 50% prevalence in Asian populations[545]
- Homozygotism associated with 10-fold protection from alcohol dependence[554]

UChA/B rats
- ALDH2 activity higher in alcohol-preferring UChB compared with non-alcohol-preferring UChA[531,546]

C57/DBA mice
- Alcohol-avoiding DBA mice show higher acetaldehyde levels following alcohol consumption compared to alcohol-preferring C57 mice[553]

Opioid system

OPRM1 (μ-opioid receptor 118G variant: Asn → Asp substitution)
- Associated with 3-fold increase in opioid binding[555], but see [556,557]
- Associated with higher subjective feelings of intoxication, stimulation, sedation and happiness compared with 118A variant[558,559]

μ-opioid receptor knockout (KO) mice
- Reduced alcohol preference, consumption, operant self-administration compared to wild type (WT[393,431,432])
- Effects depend upon line and sex[393,560]

κ-opioid receptor KO mice
- Reduced alcohol consumption compared to WT[561]

δ-opioid receptor KO mice
- Enhanced alcohol preference, operant self-administration, but similar intake compared to WT[562]

β-endorphin deficient mice
- Enhanced alcohol consumption compared to WT[563,564]

(continued)

Table 5.4 (Continued)

Human — Genetic variant	Model — Genotype selection
Dopaminergic system	
DRD2 (TaqIA polymorphism-A1)	*D1 or D2 receptor KO mice*
▪ A1 allele associated with higher prevalence of alcohol dependence compared to A2 allele[565-567]	▪ Reduced alcohol consumption compared to WT[412,413,569-572]
▪ A1 allele associated with reduced D2 receptors numbers[568]	*D3 receptor KO mice*
DRD4 (highly polymorphic 48 bp VNTR)	▪ Similar alcohol preference, operant self-administration, compared to WT[401]
▪ Long (7-repeat) allele associated with stronger urge to drink following alcohol challenge compare to shorter (2- and 4-repeat) alleles[573]	*D4 receptor KO mice*
	▪ Similar alcohol consumption to WT[574]
	DAT KO mice
	▪ Enhanced alcohol consumption compared to WT, but may be sex- and line-dependent[575], but see[576]
Serotonergic system	
5-HTTLPR (serotonin transporter (5-HTT) protein promoter polymorphism)	*5-HTT KO mice*
▪ SS associated with lower expression of 5-HTT sites and 5-HT reuptake, greater prevalence of alcohol dependence and more severe alcohol withdrawal[577-580] but (see[581])	▪ Reduced alcohol consumption compared to WT[586,587]
▪ LL associated with greater 5-HT reuptake, low level of response (LR) to alcohol and early onset of dependence[582-584], but (see[581,585])	*5-HT$_{1B}$ receptor KO mice*
	▪ Enhanced alcohol consumption and operant self-administration compared to WT[589,590]
5-HT$_{1B}$ receptor gene (861G > C polymorphism)	*5-HT$_3$ overexpressing transgenic mice*
▪ Associated with antisocial alcohol dependence and alcohol dependence with inactive ALDH2 (Lappalainen et al., 1998;[588]	▪ Reduced alcohol consumption compared to WT[591]

GABAergic system

GABRA1 (α1 subunit variant)
- May be associated with alcohol dependence, blackouts, age at first drink, LR[592] but (see[593])

GABRA2 (α2 subunit single nucleotide polymorphism (SNP))
- May be associated with alcohol dependence in adults[594-597], but not children/adolescents[592]
- GG/AA more sensitive to acute effects of alcohol versus heterozygotes[598]

GABRA3 (α3 subunit dinucleotide repeat polymorphism)
- May be associated with alcohol dependence[593]

GABRA6 (α6 subunit SNP)
- May be associated with alcohol dependence and low LR to alcohol challenge[582,601,602]

GABRB1 (β1 subunit tetranucleotide repeat polymorphism)
- May be associated with alcohol dependence[603]

GABRB2 (β2 subunit SNP)
- May be associated with alcohol dependence[601]

GABRB3 (β3 subunit dinucleotide repeat polymorphism)
- May be associated with alcohol dependence[604]

GABRG3 (γ3 subunit SNP)
- May be associated with alcohol dependence[605] but (see[606])

α1 KO
- Reduced alcohol drinking, operant self-administration compared to WT (June *et al.*, 2006;[406])

α2 KO
- Reduced alcohol drinking, withdrawal severity compared to WT[599]

α5 KO
- Reduced alcohol consumption and withdrawal severity compared to WT[599,600]

α6 KO
- Alcohol consumption not assessed

β2 KO
- Similar alcohol consumption, possibly greater withdrawal severity[406]

γ2 KO
- Alcohol consumption not assessed

Glutamatergic system

EAAT2 (glutamate transporter silent G603A nucleotide exchange)
- A603 allele associated with antisocial alcoholism[607]

GRIK3 (kainate receptor Ser310Ala polymorphism)
- Ser310 allele associated with delirium tremens history, but not withdrawal seizures[609]

Per2 Brdm1 null mutants
- Clock gene associated with decreased levels of EAAT1, enhanced glutamate levels
- Enhanced alcohol consumption, operant self-administration compared to WT[608], no difference in reinstatement (Zghoul *et al.*, 2007)

GluR1 KO
- Similar alcohol consumption compared to WT[610]

Genetic Variation in Alcohol Metabolism

Thus far, only two genes have convincingly been associated with the risk of alcohol dependence, ALDH2 and ADH1, which encode aldehyde dehydrogenase (ALDH) and alcohol dehydrogenase (ADH), enzymes involved in the metabolism of acetaldehyde and alcohol, respectively. Both mutations lead to accumulation of acetaldehyde, which limits the use of alcohol as it is associated with nausea, flushing, headache, dysphoria and cardiovascular effects. This aversive reaction is the basis behind the use of disulfiram, an inhibitor of ALDH, in promoting abstinence. Although it is likely that acetaldehyde accumulation limits alcohol use, other factors or genes may also be influencing this effect,[552] particularly considering the evidence that acetaldehyde itself can be reinforcing.[611]

Genetic Variation in the Opioid System

The opioid system and the μ-opioid receptor in particular have been implicated in the modulation of alcohol reward. Recently, variants of the μ-receptor gene, OPRM1 (amino acid substitution, asparagine (Asn) to aspartic acid (Asp)), have been identified in humans. In terms of genetic variation in treatment response, individuals with the Asp variant exhibit greater activation of the hypothalamic-pituitary-adrenal axis in response to naltrexone[612] and are more sensitive to naltrexone-induced reductions in alcohol-associated high and craving compared with their Asn variant counterparts.[558,613] Asp carriers may also be less likely to return to heavy drinking while taking naltrexone; however, there does not appear to be any difference in terms of overall relapse.[614]

Genetic animal models have also been used to study the role of the opioid system in alcohol-related behaviors. Interestingly, naltrexone administration reduces alcohol consumption in β-endorphin-deficient mice, but this paradoxical effect has been attributed to compensatory developmental mechanisms in these genetically engineered mice.[615]

Genetic Variation in the Dopamine System

The dopamine system has been of great interest in addiction research, including that of alcohol. Dopamine is associated with reward, reinforcement and motivation, and all drugs of abuse tested thus far have demonstrated dopamine releasing activity in the mesocorticolimbic system.[616] Considering the focus on dopamine's role in drug addiction, many genetic studies have concentrated on genes involved in dopamine neurotransmission. Thus far, three genetic variants within the dopamine system have been associated with alcohol dependence.

With respect to treatment response, Lawford et al.[617] reported that alcoholics with the TaqIA1 allele polymorphism of the DRD2 gene (A1/A1 and A1/A2 genotypes) experience greater decreases in anxiety and craving following bromocriptine, a dopamine D_2 receptor agonist than those without this allele. This small study indicated that dopamine-targeted treatment may be beneficial in subtypes of alcoholics who demonstrate this allelic variant at the DRD2 gene, although more recent work has shown that dopamine receptor sensitivity does not vary by DRD2 polymorphism.[618] Thus, treatments targeting the dopaminergic system may potentially have a different impact on alcoholics with these genetic polymorphisms, although further verification is needed.

Several studies have examined the role of the gene encoding the dopamine transporter (DAT1) in alcohol dependence since the DAT is important in regulating dopamine neurotransmission. These studies have largely been negative,[619-621] although there is some evidence that the A9 allele is associated with alcohol dependence and delirium tremens.[624] No comparisons have yet been made in terms of treatment response by genotype.

The gene encoding the dopamine D_4 receptor contains a highly polymorphic 48 base pair variable number of tandem repeats (48 bp VNTR) sequence in exon III. The evidence supporting a role of the D_4 polymorphism in alcohol dependence is limited;[623] however, the inconsistent association between the VNTR polymorphism of the DRD4 gene and alcohol dependence may be related to the observation that this polymorphism has also been associated with novelty seeking and attention-deficit hyperactivity disorder, which both predispose to substance use and misuse.[624-626] [iii] Nonetheless, there is evidence that naltrexone may have a greater beneficial effect on heavy drinking in individuals with the long allele compared with those carrying the shorter alleles[627] but (see[628]).

Several knockout mouse lines have been created to study the role of specific dopamine receptors and transporters in behavior. Results of studies indicating reduced alcohol consumption should be interpreted with caution, however, given that some of these models also show reduced consumption of water and food, highlighting the importance of dopamine in motivated behaviors, and that the changes in response are not limited to alcohol. Further work is needed to evaluate potential dopamine-targeted agents in these models.

Genetic Variation in the Serotonergic System

The role of serotonin in alcohol-related behaviors has been extensively studied. Of particular interest thus far in genetics research of alcohol dependence are variants of the serotonin transporter (5-HTT) 5-HTTLPR gene and the serotonin 1B receptor (5-HT_{1B}) gene. Thus far, no direct comparisons across genotype have been made with respect to treatment outcome; however, early onset alcoholics, who may have serotonergic deficiency, are more responsive to ondansetron, a 5-HT_3 receptor antagonist. This apparent paradoxical effect may be related to alcohol-induced changes in 5-HTT function, but speculation is beyond the scope of this chapter (see[130]).

A number of knockout models and mutants have been developed to examine the role of specific 5-HT receptors in alcohol-related behaviors. Although reduced serotonergic activity is associated with higher alcohol consumption, it is also implicated in impulsivity, aggression, anxiety, etc., all factors thought to modulate the development of alcohol dependence.

Genetic Variation in the GABAergic System

The role of GABA in alcohol's effects and alcohol dependence has been an area of intense research. Alcohol facilitates GABA-A receptor function and there is substantial

---

[iii] See Tannock *et al.*, "Towards a biological understanding ADHD of and the discovery of novel therapeutic approaches", in Volume 1, Psychiatric Disorders for further discussion of the DRD4 gene and involvement in ADHD.

evidence of neuroadaptations at this receptor that are thought to underlie alcohol tolerance, dependence and withdrawal symptoms.[629] Variation at the GABRA2 gene may influence the response to low doses of alcohol,[630] but the α_2 receptor subunit is also important in anxiety, judgment and impulsivity, factors with known involvement in alcohol use and misuse. Other polymorphisms in genes encoding subunits of the GABA-A receptor (e.g., GABRA1, GABRA6, GABRB2 and GABRG2) have been identified; however, most linkage and association studies have not been replicated[629,631] and the functional significance of some of these subtypes in the response to alcohol and/ or susceptibility to alcohol dependence remains unknown. Knockout mouse models have demonstrated the importance of specific functional GABA-A receptor subunits in alcohol consumption, as well as alcohol-induced locomotor activation, sedation and withdrawal (see[307,310]), but replication and viable knockout have been difficult to achieve. Thus far, the influence of genotype on treatment response has not yet been assessed in either human alcoholics or in genetic animal models.

Genetic Variation in the Glutamatergic System

There is increasing evidence of glutamate's importance in alcohol's effects and its withdrawal. Some genetic variation has been observed to influence the risk of alcohol dependence, but so far, there have not been many studies. There also do not appear to be any studies examining treatment outcome by genotypic variation in this neurotransmitter system. Acamprosate, which is currently marketed for reducing craving and relapse in alcohol dependence, is thought to act partly by attenuating the hyperglutamatergic state associated with chronic alcohol exposure (see above). Mice with a PerBrdm1 mutation typically have higher glutamate levels in their brain due to reduced glutamate transporter availability. These mice demonstrate enhanced alcohol consumption that can be attenuated by acamprosate administration, which also reduces glutamate levels.[608]

Genes Influencing Relapse

Alcohol dependence is a chronic relapsing disorder in which many neuroadaptations occur that may interact with genotype, and genes identified in animal models of acute or subchronic alcohol effects may not confer a significant risk as would those identified following chronic alcohol exposure and the behaviors associated with alcohol dependence (e.g., tolerance, craving, relapse). As is seen with other drugs of abuse, craving and relapse are the most persistent and difficult symptoms of alcohol dependence. Several animal models have been developed to address these features (e.g., reinstatement of drug seeking). Medications approved for the treatment of alcohol dependence, as well as drugs currently in development, have been shown to block reinstatement of alcohol seeking, supporting their predictive validity. However, these procedures have not yet been implemented in genetically engineered animals. The lack of reinstatement models in genetically engineered rodents has largely been due to difficulties in implementing these complex behavioral paradigms, including the hyperactivity and inattention of mice, and the technical challenges of developing genetically engineered rats. Thus, specific genes underlying risk of relapse have not been identified in animal models and this may partly explain why drugs to prevent relapse and craving have only met with limited success in humans. With greater knowledge of drug effects in genetically driven alcohol-related behaviors, more selective drug development will be enabled.

In sum, there is increasing awareness of the genetic contribution to alcohol dependence and there have been significant advancements in genetic models to understand the neurobiological response to alcohol and how this may influence susceptibility to alcohol dependence. There has been little work, however, in terms of assessing the effectiveness of potential pharmacological treatments. Additional models are needed to identify genes associated with the chronic, relapsing aspect of alcohol dependence in order to enhance the predictive validity of animal models when testing new pharmacological agents.

CORRESPONDENCE OF EFFICACY OF ESTABLISHED AND NOVEL TREATMENTS FOR ALCOHOL DEPENDENCE IN ANIMALS AND HUMANS

One of the primary goals of preclinical research in the treatment of alcohol dependence is to test the efficacy of currently available drugs and novel pharmacological agents. The predictive validity of these behavioral models has been challenged/disputed since several compounds that have proven promising in reducing alcohol consumption and reinstatement of alcohol seeking (i.e., relapse) in these rodent models, but have failed to demonstrate efficacy in clinical trials. As previously alluded to with respect to clinical trial design, the lack of translation may be related to the endpoints chosen. In animal models of self-administration, reductions in responding for alcohol, alcohol consumption and reinstatement of alcohol seeking are the most common dependent variables used in assessing the effects of pharmacological treatments. In contrast, to demonstrate therapeutic efficacy in humans, abstinence must be used as an endpoint (e.g., number of non-drinking days, days until relapse). Thus, it may not be surprising that findings from the laboratory do not extend into the clinic. That being said, however, if one compares laboratory and clinical data more closely, one may begin to see more similarities than differences, particularly if one shifts the goal of treatment to reductions in drinking, rather than complete abstinence. Examples using naltrexone and acamprosate are provided here to illustrate this point.

Naltrexone has been shown repeatedly to reduce the number of heavy drinking days, time to heavy drinking (relapse) and craving.[107,117,144-151] Although an increase in total abstinent days has been reported, abstinence has typically been defined as abstaining from heavy drinking, complete abstinence, that is, not having consumed a single drink for the duration of the trial period, has not been demonstrated in any clinical trial. In rodent models, naltrexone or other opioid receptor antagonists can reduce operant self-administration and consumption of alcohol, but also does not completely block alcohol consumption.[246,456] Thus, in terms of correlating preclinical and clinical findings, naltrexone administration suppresses drinking, but does not result in complete abstinence in either situation. With respect to relapse, naltrexone can increase the time to heavy drinking in alcohol-dependent subjects[107,144,145] and reduces reinstatement of alcohol seeking triggered by re-exposure to alcohol and associated cues in rodents.[440,441,459,461,467] Naltrexone also reduces cue and alcohol priming-induced craving in alcohol-dependent subjects[632,633] and has been shown to abolish the ADE, a rodent model of craving.[253,254] The preclinical evidence indicates that the effects of naltrexone may not be selective for alcohol-related behaviors and at times show a complex dose–response relationship, which may provide clues in terms of issues of compliance and appropriate dose selection, respectively. Overall, the findings obtained with naltrexone in the laboratory are mainly consistent with the clinical findings when similar endpoints are used.

Acamprosate also appears to demonstrate consistent effects in rodent models and clinical studies. Acamprosate can reduce the percentage of heavy drinking days and craving in alcohol-dependent subjects, whereas its effects on cumulative abstinence have shown conflicting results.[144,163] Similarly, acamprosate reduces the ADE-induced increases in responding for alcohol, as well as reinstatement of alcohol seeking in rodents.[264,463] Unlike naltrexone, acamprosate's effects appear to be selective for alcohol, but it has limited effects in reducing alcohol consumption, under both operant and voluntary consumption conditions.[262,263,463]

As in clinical studies, the conditions and measures used in the laboratory can have a significant impact on the outcome, and not all experiments with naltrexone and acamprosate have demonstrated positive effects in the preclinical models (e.g.,[460]). With respect to predicting efficacy in the clinical setting, it might be crucial to demonstrate a strong signal in the preclinical models, that is, in reducing alcohol consumption, craving *and* reinstatement, as well as demonstrate selectivity for alcohol-related behaviors. Alternatively, combination therapy, involving agents that target different aspects of alcohol dependence and via different mechanisms, may be more effective. For example, naltrexone and acamprosate have additive effects in animal models of craving and reinstatement[265-266] but (see[268]) and this has also been shown in alcohol-dependent subjects[634], but not replicated in the COMBINE study.[117]

CONSENSUS STATEMENT REGARDING ANIMAL MODELS OF ALCOHOL DEPENDENCE

The above discussion shows that the main animal models in current use for testing drugs for the treatment of alcohol dependence, voluntary drinking and operant self-administration have demonstrated effects of the treatments in current use, but have also shown positive results with compounds that have not proved of value in humans. In particular, the SSRIs gave positive results in all three types of animal model described above, but did not prove of clinical benefit, suggesting that a re-assessment of the models of alcohol dependence would be of benefit. It is important to consider the different situations in which drugs may be of benefit, the treatment of heavy drinking without attempts at abstinence is a very different situation from maintenance of abstinence by motivated, detoxified alcoholics. Of the current treatments, the effects of opioid antagonists in decreasing the rewarding effects of alcohol would be expected to be of more use in the former situation while the antirelapse actions of acamprosate are of more value in the latter patients. It is also of importance in future animal models of dependence to take more account of the prolonged time course and progressive stages of development of alcohol dependence. Treatment during acute withdrawal phases could have prolonged beneficial effects on the later consequences of stopping drinking.

It is also crucial to take into account the other important factors of early initiation of alcohol consumption, comorbid depression, stressful experiences and the complications of cognitive deficits. Each of the stages of dependence development, initiation, which normally takes place during adolescence or even earlier, then excess consumption over many years interspersed with periods of abstinence composed of acute and protracted withdrawal phases have different neurochemical bases. Drugs to alter alcohol consumption in dependent animals are likely to act via different target sites than those which decrease alcohol consumption in animals that are not dependent, owing to changes in neuronal systems that control consumption. Although pharmacological treatment to affect responses to the initial consumption of alcohol in humans is not feasible, it may be possible to prevent the longer-term consequences of early alcohol consumption once the mechanisms of the neuronal changes are understood. The studies on the deprivation effect take into account the repeated abstinence periods, but the prolonged development of alcohol dependence in humans suggests it might be of advantage to examine drug effects in animals that have consumed alcohol over a comparable proportion of their life span. The duration of excess consumption before the clinical symptoms of alcohol dependence appear is extensive, frequently 10–15 years, that would be approximately equivalent to 6 months in rodents. The possibility of drug treatment during the acute withdrawal phase to alter the long-term consequences not only relapse drinking, but also the cognitive deficits would benefit from further investigation. Cognitive deficits caused by chronic alcohol consumption not only form a potential target for drug actions that would have major clinical benefit, but may cause unrecognized complications during behavioral testing following such intake.

The effects of stressful experiences need to be studied on a long-term basis. There is now considerable information about changes in both neuronal function and behavior in adult rodents caused by early adverse experiences that could be used in the study of the development of alcohol dependence. In addition, whilst some of the clinical diagnostic criteria for alcohol dependence are not of great value with regard to animal models, Criterion 7, persistent of intake despite knowledge of adverse consequences is crucial to our understanding of dependence and has been largely neglected in the models. Knowledge in rodents of the consequences of consumption of a drug can only be implied, but persistence of alcohol consumption or responding for alcohol following experiences that can be regarded as adverse or decreased consumption of another reward, such as palatable food, can be measured. The quantity of alcohol consumed also needs more consideration. Owing to the high rates of metabolism in rodents, comparisons need to be made of blood or, where possible, brain concentrations, not by dose. In the majority of animal models such levels are nearer to, or even below, those seen in social drinkers rather than alcoholics. The duration of the potential effects of stress also requires consideration, in accordance with the different life spans of humans and rodents, as described above.

A large proportion of alcoholics started drinking alcohol at a very early age are clinically depressed, have memory problems particularly in executive function, have a family history of alcohol problems, have experienced early abuse or neglect and have undergone many years of high levels of drinking interspersed with periods of abstinence.

Many are malnourished and the great majority in Western countries has a high fat diet. Most rodents in models of alcohol dependence commence alcohol consumption as adults, are not "depressed," have not experienced early stress, have a balanced diet, have only consumed alcohol for short periods and have not drunk sufficient alcohol to cause cognitive deficits. Although the family history aspect is addressed by the genetic studies on animal models, it would be of advantage to include more of these factors in our test methods.

REFERENCES

1. Alcohol Concern Factsheet (2003) 37:i–vi.
2. Grant, B.F., Dawson, D.A., Stinson, F.S., Chou, S.P., Dufour, M.C., and Pickering, R.P. (2004). The 12-month prevalence and trends in DSM-IV alcohol abuse and dependence: United States, 1991–1992 and 2001–2002. *Drug Alcohol Depend*, 74:223–234.
3. Rehm, J., Taylor, B., and Patra, J. (2006). Volume of alcohol consumption, patterns of drinking and burden of disease in the European region (2002). *Addiction*, 101:1086–1095.
4. Dichiara, G. (1997). Alcohol and dopamine. *Alcohol Health Res World*, 21(2):108–114.
5. Pulvirenti, L. and Diana, M. (2001). Drug dependence as a disorder of neural plasticity: Focus on dopamine and glutamate. *Rev Neurosci*, 12(2):141–158.
6. Addolorato, G., Capristo, E., Leggio, L., Ferrulli, A., Abenavoli, L., Malandrino, N., Farnetti, S., Domenicali, M., D'Angelo, C., Vonghia, L., Mirijello, A., Cardone, S., and Gasbarrini, G. (2006). Relationship between ghrelin levels, alcohol craving, and nutritional status in current alcoholic patients. *Alcohol Clin Res Exp*, 30(11):1933–1937.
7. Kraus, T., Schanze, A., Groschl, M., Bayerlein, K., Hillemacher, T., Reulbach, U., Kornhuber, J., and Bleich, S. (2005). Ghrelin levels are increased in alcoholism. *Alcohol Clin Exp Res*, 29(12):2154–2157.
8. Hillemacher, T., Kraus, T., Rauh, J., Weiss, J., Schanze, A., Frieling, H., Wilhelm, J., Heberlein, A., Groschl, M., Sperling, W., Kornhuber, J., and Bleich, S. (2007). Role of appetite-regulating peptides in alcohol craving: An analysis in respect to subtypes and different consumption patterns in alcoholism. *Alcohol Clin Exp Res*, 31(6):950–954.
9. Kiefer, F., Jahn, H., Otte, C., Demiralay, C., Wolf, K., and Wiedemann, K. (2005). Increasing leptin precedes craving and relapse during pharmacological abstinence maintenance treatment of alcoholism. *J Psychiatry Res*, 39(5):545–551.
10. Fox, H.C., Hong, K.I., Siedlarz, K.M., Bergquist, K.L., Hauger, R.L., and Sinha, R. (2007). Suppressed NPY levels in 4-week abstinent alcoholics: Association with alcohol withdrawal and alcohol craving ratings. *Alcohol Clin Exp Res*, 31(6 Suppl S):110A–110A.
11. Hughes, J.R. (2001). Why does smoking so often produce dependence? A somewhat different view. *Tobac Cont*, 10(1):62–64.
12. Zweben, A. and Cisler, R.A. (2003). Clinical and methodological utility of a composite outcome measure for alcohol treatment research. *Alcohol Clin Exp Res*, 27(10):1680–1685.
13. Bates, M.E., Bowdon, S.C., and Barry, D. (2002). Neurocognitive impairment associated with alcohol use disorders: Implications for treatment. *Exp Clin Psychopharmacol*, 10:193–212.
14. Bowdon, S.C., Crews, F.T., Bates, M.E., Fals-Stewart, W., and Ambrose, M.L. (2001). Neurotoxicity and neurocognitive impairments with alcohol and drug-use disorders: Potential roles in addiction and recovery. *Alcohol Clin Exp Res*, 25:317–321.
15. Davies, S.J.C., Pandit, S.A., Feeney, A., Stevenson, B.J., Kerwin, R.W., Nutt, D.J., Marshall, E.J., and Boddington, S. (2005). Is there cognitive impairment in clinically "healthy" abstinent alcohol dependence? *Alcohol Alcohol*, 40:498–503.

16. Ihara, H., Berrios, G.E., and London, M. (2000). Group and case study of the dysexecutive syndrome in alcoholism without amnesia. *J Neurol Neurosurg Psychiatry*, 68:731-737.

17. Jellinek, E.M. (1960). Alcoholism: A genus and some of its species. *Can Med Assoc J*, 83:1341-1345.

18. Cloninger, C.R. (1987). Neurogenic adaptive mechanisms in alcoholism. *Science*, 236:410-416.

19. Carpenter, K.M. and Halsin, D.S. (2001). Reliability and discriminant validity of the Type I/II and Type A/B alcoholic subtype classifications in untreated problem drinkers: A test of the Apollonian/Dionysian hypothesis. *Drug Alcohol Depend*, 63:51-67.

20. Babor, T.F., Dolinsky, Z.S., Meyer, R.E., Hesselbrock, M., Hofmann, M., and Tennen, H. (1992). Types of alcoholics – concurrent and predictive-validity of some common classification schemes. *Br J Addiction*, 87(10):1415-1431.

21. Bau, C.H.D., Spode, A., Ponson, A.C., Elisa, E.p., Garcia, C.E.D., Costa, F.t., and Hutz, M.H. (2001). Heterogeneity in early onset alcoholism suggests a third group of alcoholics. *Alcohol*, 23:9-13.

22. Epstein, E.E., Labouvie, E., mccrady, B.S., Jensen, N.K., and Hayaki, J. (2002). A multidimensional study of alcohol subtypes: Classification and overlap of unidimensional and multidimensional typologies. *Addiction*, 97:1041-1053.

23. Moss, H.B., Chen, C.M., and Yi, H.Y. (2007). Subtypes of alcohol dependence in a nationally representative sample. *Drug Alcohol Depend*, 91(2-3):149-158.

24. Allen, J., Anton, R.F., Babor, T.F., Carbonari, J., Carroll, K.M., Connors, G.J., Cooney, N.L., Del Boca, F.K., diclemente, C.C., Donovan, D., Kadden, R.M., Litt, M., Longabaugh, R., Mattson, M., Miller, W.R., Randall, C.L., Rounsaville, B.J., Rychtarik, R.G., Stout, R.L., Tonigan, J.S., Wirtz, P.W., and Zweben, A. (1997). Project MATCH secondary a priori hypotheses. *Addiction*, 92(12):1671-1698.

25. Johnson, B.A. (2003). The role of serotonergic agents as treatments for alcoholism. *Drugs Today*, 39:665-672.

26. Ritzmann, R.F. and Tabakoff, B. (1976). Dissociation of alcohol tolerance and dependence. *Nature*, 263(5576):418-420.

27. Gauvin, D.V., Baird, T.J., and Briscoe, R.J. (2000). Differential development of behavioral tolerance and the subsequent hedonic effects of alcohol in AA and ANA rats. *Psychopharmacology*, 151(4):335-343.

28. Ramchandani, V.A., Flury, L., Morzorati, S.L., Kareken, D., Blekher, T., Foroud, T., Li, T.K., and O'Connor, S. (2002). Recent drinking history: Association with family history of alcoholism and the acute response to alcohol during a 60 mg% clamp. *J Stud Alcohol*, 63(6):734-744.

29. Morzorati, S.L., Ramchandani, V.A., Flury, L., Li, T.K., O'Connor, S. (2002). Self-reported subjective perception of intoxication reflects family history of alcoholism when breath alcohol levels are constant. *Alcohol Clin Exp Res* 26:1299-1306.

30. Victor, M. and Adams, K.D. (1953). The effect of alcohol on the nervous system. *Res Publ Assoc Res Nerv Ment Dis*, 32:526-573.

31. Diana, M., Brodie, M., Muntoni, A., Puddu, M.C., Pillolla, G., Steffensen, S., Spiga, S., and Little, H.J. (2003). Enduring effects of chronic ethanol in the CNS: Basis for alcoholism. *Alcohol Clin Exp Res*, 27:354-361.

32. Duka, T., Townsend, J.M., Collier, K., and Stephens, D.N. (2003a). Impairment in cognitive functions after multiple detoxifications in alcoholic inpatients. *Alcohol Clin Exp Res*, 27:1563-1572.

33. Malcolm, R., Roberts, J.S., Wang, W., Myrick, H., and Anton, R.F. (2000). Multiple previous detoxifications are associated with less responsive treatment and heavier drinking during an index outpatient detoxification. *Alcohol*, 22(3):159-164.

34. Adinoff, B., Oneill, H.K., and Ballenger, J.C. (1995). Alcohol-withdrawal and limbic kindling a hypothesis of relapse. *Am J Addiction*, 4(1):5-17.

35. Becker, H.C. (1998). Kindling in alcohol withdrawal. *Alcohol Health Res World*, 22(1):25–33.

36. Post, R.M. (1992). Transduction of psychosocial stress into the neurobiology of recurrent affective disorder. *Am J Psychiatry*, 149:999–1010.

37. Bond, N.W. (1979). Impairment of shuttlebox avoidance learning following repeated alcohol withdrawal episodes in rats. *Pharmacol Biochem Behav*, 11:589–591.

38. Fadda, F. and Rossetti, Z.L. (1998). Chronic ethanol consumption: From neuroadaptation to neurodegeneration. *Progr Neurobiol*, 55:1–47.

39. Lundqvist, C., Volk, B., Knoth, R., and Alling, C. (1994). Long-term effects of intermittent versus continuous ethanol exposure on hippocampal synapses of the rat. *Acta Neuropathol*, 87:242–249.

40. Cadete-Leite, (1990). Structural changes in the hippocampal formation and frontal cortex after long-term alcohol consumption and withdrawal in the rat. In Normal Palmer, T. (ed.), *Alcoholism: A Molecular Perspective*. Plenum Press, New York and London.

41. Lukoyanov, N.V., Madeira, M.D., and Paulabarbosa, M.M. (1999). Behavioral and neuroanatomical consequences of chronic ethanol intake and withdrawal. *Physiol Behav*, 66:337–346.

42. Steigerwald, E.S. and Miller, M.W. (1997). Performance by adult rats in sensory-mediated radial arm maze tasks is not impaired and may be transiently enhanced by chronic exposure to ethanol. *Alcohol Clin Exp Res*, 21(9):1553–1559.

43. Aston-Jones, G. and Harris, G.C. (2004). Brain substrates for increased drug seeking during protracted withdrawal. *Neuropharmacology*, 47:167–179.

44. Begleiter, H. and Porjesz, B. (1999). What is inherited in the predisposition toward alcoholism? A proposed model. *Alcohol Clin Exp Res*, 23(7):1125–1135.

45. Strowig (2000). Relapse determinants reported by men treated for alcohol addiction – the prominence of depressed mood. *J Subst Abuse Treat*, 19:469–474.

46. Brandt, A.M.H., Thorburn, D., Hiltunen, A.J., and Borg, S. (1999). Prediction of single episodes of drinking during the treatment of alcohol-dependent patients. *Alcohol*, 18(1):35–42.

47. Brower, K.J., Aldrich, M.S., and Hall, J.M. (1998). Polysomnographic and subjective sleep predictors of alcoholic relapse. *Alcohol Clin Exp Res*, 22:1864–1871.

48. Gillin, J.C., Smith, T.L., Irwin, M., Butters, N., Demodena, A., and Schuckit, M. (1994). Increased pressure for rapid eye-movement sleep at time of hospital admission predicts relapse in nondepressed patients with primary alcoholism at 3-month follow-up. *Arch Gen Psychiatry*, 51:189–197.

49. Sloan, T.B., Roache, J.D., and Johnson, B.A. (2003). The role of anxiety in predicting drinking behaviour. *Alcohol. Alcohol*, 38:360–363.

50. Gorski, and Miller, (1982). *Counselling for Relapse Prevention*. Herald-House-Independence Press, Independence, MO.

51. Miller, W.R. and Harris, R.J. (2000). A simple scale of Gorski's warning signs for relapse. *J Stud Alcohol*, 61:759–765.

52. Breese, G.R., Overstreet, D.H., Knapp, D.J., and Navarro, M. (2005). Prior multiple ethanol withdrawals enhance stress-induced anxiety-like behavior: Inhibition by CRF1- and benzodiazepine-receptor antagonists and a 5-HT1A-receptor agonist. *Neuropsychopharmacology*, 30(9):1662–1669.

53. Manley, S. and Little, H.J. (1997). Enhancement of amphetamine- and cocaine-induced locomotor activity after chronic ethanol administration. *J Pharmacol Exp Ther*, 281:1330–1339.

54. Rohsenow, D.J. and Monti, P.M. (1999). Does urge to drink predict relapse after treatment? *Alcohol Res Health*, 23(3):225–232.

55. Sinha, R. and O'Malley, S.S. (1999). Craving for alcohol: Findings from the clinic and the laboratory. *Alcohol Alcohol*, 34(2):223–230.

56. Littleton, J.M. (2000). Can craving be modeled in animals? The relapse prevention perspective. *Addiction*, 95:S83–S90.

57. Myrick, H., Anton, R.F., and Li, X.B. (2004). Differential brain activity in alcoholics and social drinkers to alcohol cues: Relationship to craving. *Neuropsychopharmacology*, 29(2):393–402.

58. Sinha, R. and Li, C.S.R. (2007). Imaging stress- and cue-induced drug and alcohol craving: Association with relapse and clinical implications. *Drug Alcohol Rev*, 26(1):25–31.

59. Di Sclafani, V., Finn, P., and Fein, G. (2007). Psychiatric comorbidity in long-term abstinent alcoholic individuals. *Alcohol Clin Exp Res*, 31(5):795–803.

60. Landa, N., Fernandez-Montalvo, J., Lopez-Goni, J.J., and Lorea, I. (2006). Psychopathological comorbidity in alcoholism: A descriptive study. *Int J Clin Health Psychol*, 6(2):253–269.

61. Morgenstern, J., Langenbucher, J., Labouvie, E., and Miller, k.J. (1997). The comorbidity of alcoholism and personality disorders in a clinical population: Prevalence rates and relation to alcohol typology variables. *J Abnorm Psychol*, 106(1):74–84.

62. Mann, K., Lehert, P., and Morgan, M.Y. (2004). The efficacy of acamprosate in the maintenance of abstinence in alcohol-dependent individuals: Results of a meta-analysis. *Alcohol Clin Exp Res*, 28:51–68.

63. Mann, K., Hintz, T., and Jung, M. (2004). Does psychiatric comorbidity in alcohol-dependent patients affect treatment outcome? *Eur Arch Psychiatry Clin Neurosci*, 254(3):172–181.

64. Wetterling, T. and Junghanns, K. (2000). Psychopathology of alcoholics during withdrawal and early abstinence. *Eur Psychiatry*, 15(8):483–488.

65. Petrakis, I., Ralevski, E., Nich, C., Levinson, C., Carroll, K., Poling, J., and Rounsaville, B.VA VISN I MIRECC Study Group. (2007). Naltrexone and disulfiram in patients with alcohol dependence and current depression. *J Clin Psychopharmacol*, 27(2):160–165.

66. Modesto-Lowe, V. and Kranzler, H.R. (1999). Diagnosis and treatment of alcohol-dependent patients with comorbid psychiatric disorders. *Alcohol Res Health*, 23(2):144–149.

67. Elderkin-Thompson, V., Mintz, J., and Haroon, E. (2007). Executive dysfunction and memory in older patients with major and minor depression. *Arch Clin Neuropsychol*, 22(2):259.

68. Sachs, G., Schaffer, M., and Winklbaur, B. (2007). Cognitive deficits in bipolar disorder. *Neuropsychiatrie*, 21(2):93–101.

69. Goldstein, B.I. and Levitt, A.J. (2006). A gender-focused perspective on health service utilization in comorbid bipolar I disorder and alcohol use disorders: Results from the national epidemiologic survey on alcohol and related conditions. *J Clin Psychiatry*, 67(6):925–932.

70. Spear, L.P. (2000). The adolescent brain and age-related behavioral manifestations. *Neurosci Biobehav Rev*, 24:417–463.

71. Grant, B.F., Stinson, F.S., and Harford, T.C. (2001). Age at onset of alcohol use and DSM-IV alcohol abuse and dependence: A 12-year follow-up. *J Subst Abuse*, 13(4):493–504.

72. Laviola, G., Macri, S., Morley-Fletcher, S., and Adriani, W. (2003). Risk-taking behavior in adolescent mice: Psychobiological determinants and early epigenic influence. *Neurosci Biobehav Rev*, 27:19–31.

73. Monti, P.M., Miranda, R., and Nixon, K. (2005). Adolescence: Booze, brains, and behavior. *Alcohol Clin Exp Res*, 29(2):207–220.

74. Harper, C. (1998). The neuropathology of alcohol-specific brain damage, or does alcohol damage the brain? *J Neuropathol Exp Neurol*, 57(2):101–110.

75. Brown, S.A. and Tapert, S.S. (2004). Adolescence and the trajectory of alcohol use: Basic to clinical studies. *Ann NY Acad Sci*, 1021:234–244.

76. Osborne, G.L. and Butler, A.C. (1983). Enduring effects of periadolescent alcohol exposure on passive avoidance performance in rats. *Physiol Psych*, 11:205–208.

77. Girard, T.A., Xig, H.C., Ward, G.R., and Wainwright, P.E. (2000). Early postnatal ethanol exposure has longterm effects on the performance of male rats in a delayed matching-to-place task in the water maze. *Alcohol Clin Exp Res*, 24:300–306.

78. Thomas, J.D., Weinert, S.P., Sharif, S., and Riley, E.P. (1997). MK801 administration during ethanol withdrawal in neonatal rat pups attenuates ethanol-induced behavioral deficits. *Alc Clin Exp Res*, 21:1218–1225.

79. Pyapali, G.K., Turner, D.A., Wilson, W.A., and Swartzwelder, H.A. (1999). Age and dose-dependent effects of ethanol on the induction of hippocampal long term potentization. *Alcohol*, 19:107–111.

80. Swartzwelder, H.S., Wilson, W.A., and Tayyeb, M.I. (1995). Age-dependent inhibition of long term potentiation by ethanol in immature versus mature hippocampus. *Alcohol Clin Exp Res*, 19:320–323.

81. White, A.M., Ghia, A.J., Levin, E.D., and Schartzwelder, H.C. (2000). Binge pattern ethanol exposure in adolescent and adult rats: Differential impact on subsequent responsiveness to alcohol. *Alcohol Clin Exp Res*, 24:1251–1256.

82. Ciesielski, K.T., Waldorf, A.V., and Jung, R.E. (1995). Anterior brain deficits in chronic alcoholism – cause or effect? *J Nerv Ment Diss*, 183:756–761.

83. Di Sclafani, V., Ezekiel, F., Meyerhof, D.J., Mackay, S., Dillon, W.p., Weiner, M.w., and Fein, G. (1995). Brain atrophy and cognitive function in older abstinent alcoholic men. *Alcohol Clin Exp Res*, 19:1121–1126.

84. Sullivan, E.V., Rosenbloom, M.J., and Pfefferbaum, A. (2000). Pattern of motor and cognitive deficits in detoxified alcoholic men. *Alcohol Clin Exp Res*, 24(5):611–621.

85. Sullivan, E.V., Fama, R., Rosenbloom, M.J., and Pfefferbaum, A. (2002). A profile of neuropsychological deficits in alcoholic women. *Neuropsychology*, 16:74–83.

86. Uekermann, J., Daum, I., Schlebusch, P., Wiebel, B., and Trenckmann, U. (2003). Depression and cognitive functioning in alcoholism. *Addiction*, 98:1521–1529.

87. Harper, C., Dixon, G., Sheedy, D., and Garrick, T. (2003). Neuropathological alterations in alcoholic brains. Studies arising from the New South Wales Tissue Resource Centre. *Progr Neuro Psychopharmacol Biol Psychiatry*, 27:951–961.

88. Kril, J.J., Halliday, G.M., Svoboda, M.D., and Cartright, H. (1997). The cerebral cortex is damaged in chronic alcoholics. *Neuroscience*, 79:983–998.

89. Irle, E. and Markowitsch, H.J. (1983). Widespread neuroanatomical damage and learning deficits following chronic alcohol consumption or vitamin B1 (thiamine) deficiency in rats. *Behav Brain Res*, 9:277–294.

90. Durand, D. and Carlen, P.L. (1984). Decreased neuronal inhibition in vitro after long-term administration of ethanol. *Science*, 224:1359–1361.

91. Roberto, M., Nelson, T.E., Ur, C.L., and Gruol, D.L. (2002). Long-term potentiation in the rat hippocampus is reversibly depressed by chronic ethanol exposure. *J Neurophys*, 87:2385–2397.

92. Goldenberg, I.M., Mueller, T., Fierman, E.J., Gordon, A., Pratt, L., Cox, K., Park, T., Lavori, P., Goisman, R.M., and Keller, M.B. (1995). Specificity of substance use in anxiety-disordered subjects. *Compr Psychiatry*, 36(5):319–328.

93. Nunes, E.V., Mcgrath, P.J., and Quitkin, F.M. (1995). Treating anxiety in patients with alcoholism. *J Clin Psychiatry*, 56(Suppl 2):3–9.

94. Schuckit, M.A. and Hesselbrock, V. (1994). Alcohol dependence and anxiety disorders – what is the relationship? *Am J Psychiatry*, 151(12):1723–1734.

95. Chutuape, M.A.D. and Dewit, H. (1995). Preferences for ethanol and diazepam in anxious individuals – an evaluation of the self-medication hypothesis. *Psychopharmacology*, 121(1):91–103.

96. Gorman, D.M. and Brown, G.W. (1992). Recent developments in life-event research and their relevance for the study of addictions. *Br J Addiction*, 87:837–849.

97. Crum, R.M., Muntaner, C., and Eaton, W.W. (1995). Occupational stress and the risk of alcohol-abuse and dependence. *Alcohol Clin Exp Res*, 19(3):647–655.

98. Brown, S.A., Vik, P.W., Patterson, T., and Grant, I. (1995). Stress, vulnerability and adult alcohol relapse. *J Stud Alcohol*, 56:538-545.

99. Kessler, R.C., Crum, R.M., and Warner, L.A. (1997). Lifetime co-occurrence of DSM-III-R alcohol abuse and dependence with other psychiatric disorders in the national comorbidity survey. *Arch Gen Psychiatry*, 54(4):313-321.

100. Moncrieff, J., Drummond, D.C., Candy, B., Chechinski, K., and Farmer, R. (1996). Sexual abuse in people with alcohol problems: A study of the prevalence of sexual abuse and its relationship to alcohol drinking behavior. *Br J Psychiatry*, 169:355-360.

101. Langeland, W., Draijer, N., and van den Brink, W. (2004). Psychiatric comorbidity in treatment-seeking alcoholics: The role of childhood trauma and perceived parental dysfunction. *Alcohol Clin Exp Res*, 28(3):441-447.

102. Dom, G., De Wilde, B., Hulstijn, W. *et al.* (2007). Traumatic experiences and posttraumatic stress disorders: Differences between treatment-seeking early- and late-onset alcoholic patients. *Compr Psychiatry*, 48(2):178-185.

103. Piazza, P.V. and Le Moal, M. (1996). Pathophysiological basis of vulnerability to drug abuse: Role of an interaction between stress, glucocorticoids, and dopaminergic neuron. *Ann Rev Pharmacol Toxicol*, 36:359-378.

104. Piazza, P.V. and Le Moal, M. (1997). Glucocorticoids as a biological substrate of reward: Physiological and pathological implications. *Brain Res Rev*, 25:359-372.

105. Diagnostic and Statistical Manual IV TR (2000) American Psychiatric Association, *Arlington*.

106. Piazza, P.V. and Le Moal, M. (1998). The role of stress in drug self-administration. *Trends Pharmacol Sci*, 19(2):67-74.

107. Volpicelli, J.R., Alterman, A.I., Hayashida, M., and O'Brien, C.P. (1992). Naltrexone in the treatment of alcohol dependence. *Arch Gen Psychiatry*, 49:876-880.

108. Mark, T., Kranzler, H.R., and Song, X. (2003). Understanding US addiction physicians' low rate of naltrexone prescription. *Drug Alcohol Depend*, 71:219-228.

109. Soyka, M. and Chick, J. (2003). Use of acamprosate and opioid antagonists in the treatment of alcohol dependence: A European perspective. *Am J Addiction*, 12:S69-S80.

110. Lhuintre, J.P., Daoust, M., and Moore, N.D. (1990). Acamprosate appears to decrease alcohol intake in weaned alcoholics. *Alcohol Alcohol*, 25:213-222.

111. Chick, J. (1995). Acamprosate as an aid in the treatment of alcoholism. *Alcohol Alcohol*, 30:784-787.

112. Mann, K. (1996). The pharmacological treatment of alcohol dependence: Needs and possibilities. *Alcohol Alcohol*, 31:55-58.

113. Whitworth, A.B., Fischer, F., Lesch, O.M., Nimmerrichter, A., Oberbauer, H., Platz, T., Potgieter, A., Walter, H., and Fleischhacker, W.W. (1996). Comparison of acamprosate and placebo in long-term treatment of alcohol dependence. *Lancet*, 347:1438-1442.

114. Littleton, J.M. (1995). Acamprosate in alcohol dependence: How does it work. *Addiction*, 90:1179-1188.

115. Zornoza, T., Cano, M.J., Polache, A., and Granero, L. (2003). Pharmacology of acamprosate: An overview. *CNS Drug Rev*, 9:359-374.

116. Allgaier, C., Franke, H., Sobottka, H., and Scheibler, P. (2000). Acamprosate inhibits Ca^{2+} influx mediated NMDA by receptors and voltage-sensitive Ca^{2+} channels in cultured rat mesencephalic neurons. *N Schmied Arch Pharmacol*, 362(4-5):440-443.

117. Anton, R.F., O'Malley, S.S., Ciraulo, D.A., Cisler, R.A., Couper, D., Donovan, D.M., Gastfriend, D.R., Hosking, J.D., Johnson, B.A., Locastro, J.S., Longabaugh, R., Mason, B.J., Mattson, M.E., Miller, W.R., Pettinati, H.M., Randall, C.L., Swift, R., Weiss, R.D., Williams, L.D., and Zweben, A. (2006). COMBINE Study Research Group. Combined pharmacotherapies

and behavioral interventions for alcohol dependence: The COMBINE study: A randomized controlled trial. *J Am Med Assoc*, 295:2003–2017.

118. Conigrave, K.M., Degenhardt, L.J., Whitfield, J.B., Saunders, J.B., Helander, A., Tabakoff, B. *et al.* (2002). CDT, GGT, and ASTG as markers of alcohol use: The WHO/ISBRA collaborative project. *Alcohol Clin Exp Res*, 26:332–339.

119. Johnson, B.A., Ait-Daoud, N., Bowden, C.L., Diclemente, C.C., Roache, J.D., Lawson, K., Javors, M.A., and Ma, J.Z. (2003). Oral topiramate for treatment of alcohol dependence: A randomised controlled trial. *Lancet*, 361(9370):1677–1685.

120. Johnson, B.A., Ait-Daoud, N., Akhtar, F.Z., and Ma, J.Z. (2004). Oral topiramate reduces the consequences of drinking and improves the quality of life of alcohol-dependent individuals – a randomized controlled trial. *Arch Gen Psychiatry*, 61:905–912.

121. Johnson, B.A., Rosenthal, N., Capece, J.A., Wiegand, F., Mao, L., Beyers, K., McKay, A., Ait-Daoud, N., Anton, R.F., Ciraulo, D.A., Kranzler, H.R., Mann, K., O'Malley, S.S., and Swift, R.M. Topiramate for Alcoholism Advisory Board, Topiramate for Alcoholism Study Group. (2007). Topiramate for treating alcohol dependence: A randomized controlled trial. *J Am Med Assoc*, 298:1641–1651.

122. Ma, J.Z., Ait-Daoud, N., and Johnson, B.A. (2006). Topiramate reduces the harm of excessive drinking: implications for public health and primary care. *Addiction*, 101:1561–1568.

123. Addolorato, G., Caputo, F., Capristo, E., Colombo, G., Gessa, G.L., and Gasbarrini, G. (2000). Ability of baclofen in reducing alcohol craving and intake. II: Preliminary clinical evidence. *Alcohol Clin Exp Res*, 24:67–71.

124. Addolorato, G., Caputo, F., Capristo, E., Domenicali, M., Bernardi, M., Janiri, L. *et al.* (2002). Baclofen efficacy in reducing alcohol craving and intake: A preliminary double-blind randomized controlled study. *Alcohol*, 37:504–508.

125. Johnson, B.A., Roache, J.K., Javors, M.A., Declemente, C.C., Cloninger, C.R., Prihoda, T.J. *et al.* (2000). Ondansetron for reduction of drinking among biologically predisposed alcoholic patients: A randomized controlled trial. *J Am Med Assoc*, 284:963–971.

126. Conger, J.J. (1956). Reinforcement theory and the dynamics of alcoholism. *Quart J Stud Alcohol*, 18:296–305.

127. Kranzler, H.R., Pierucci-Lagha, A., Feinn, R., and Hernandez-Avila, C. (2003). Effects of ondansetron in early- versus late-onset alcoholics: A prospective, open-label study. *Alcohol Clin Exp Res*, 27:1150–1155.

128. Van der Kooy, D., O'Shaughnessy, M., Mucha, R.F. and Kalant, H. (1983). Motivational properties of ethanol in naive rats as studied by place conditioning. *Pharmacol Biochem Behav*, 19:441–445.

129. Sellers, E.m., Toneatto, T., Romach, M.K., Somer, G.R., Sobell, L.C., and Sobell, M.B. (1994). Clinical efficacy of the 5-HT3 antagonist ondansetron in alcohol abuse and dependence. *Alcohol Clin Exp Res*, 18:879–885.

130. Johnson, B.A. (2008). Update on neuropharmacological treatments for alcoholism: Scientific basis and clinical findings. *Biochem Pharmacol*, 75:34–56.

131. Hersh, D., Van Kirk, J.R. and Kranzler, H.R, (1998). Naltrexone treatment of comorbid alcohol and cocaine use disorders. *Psychopharmacology*, 139:44–52.

132. Dongier, M., Vachon, L., and Schwartz, G. (1991). Bromocriptine in the treatment of alcohol dependence. *Alcohol Clin Exp Res*, 15:970–977.

133. Naranjo, C.A., George, S.R., and Bremner, K.E. (1992). Novel neuropharmacological treatments of alcohol dependence. *Clin Neuropharmacol*, 15(Suppl 1, Pt A):74A–75A.

134. Powell, B.J., Campbell, J.L., Landon, J.F., Liskow, B.I., Thoas, h.m., Nickel, E.J. *et al.* (1995). A double-blind, placebo-controlled study of nortiptyline and bromocriptine in male alcoholics subyped by comorbid psychiatric disorders. *Alcohol Clin Exp Res*, 19:462–468.

135. Evans, S.m., Levin, F.R., Brooks, D.J., and Garawi, F. (2007). A pilot double-blind treatment trial for memantine for alcohol dependence. *Alcohol Clin Exp Res*, 31:775-782.

136. Kabel, D.I. and Petty, F. (1996). A placebo-controlled, double-blind study of fluoxetine in severe alcohol dependence: Adjunctive pharmacotherapy during and after inpatient treatment. *Alcohol Clin Exp Res*, 20:780-784.

137. Kranzler, H.R., Burleson, J.A., Korner, P., Del Boca, F.K., Bohn, M.J., Brown, J. *et al.* (1995). Placebo-controlled trial of fluoxetine as an adjunct to relapse prevention in alcoholics. *Am J Psychiatry*, 152:391-397.

138. Naranjo, C.A., Bremner, K.E., and Lanctot, K.L. (1995). Effects of cilaopram and a brief psycho-social intervention on alcohol intake, dependence and problems. *Addiction*, 90:87-99.

139. Johnson, B.A., Jasinski, D.R., Galloway, G.P., Kranzler, H., Weinreib, R., Anton, F.R. *et al.* (1996). Ritanserin in the treatment of alcohol dependence - a multi-center clinical trial. *Psychopharmacology*, 128:206-215.

140. Ugedo, L., Grenhoff, J., and Svensson, T.H. (1989). Ritanserin, a 5-HT2 receptor antagonist, activates midbrain dopamine neurons by blocking serotonergic inhibition. *Psychopharmacolgy*, 98:45-50.

141. Bruno, F. (1989). Buspirone in the treatment of alcoholic patients. *Psychopathology*, 22(Suppl 1):49-59.

142. Malec, T.S., Malec, E.A., and Dongier, M. (1996). Efficacy of buspirone in alcohol dependence: A review. *Alcohol Clin Exp Res*, 20:853-858.

143. Fuller, R.K., Branchey, L., Brightwell, D.R., Derman, R.M., Emrick, C.D., Iber, F.L. *et al.* (1986). Disulfira treatment of alcoholism. A veterans administration cooperative study. *J Am Med Assoc*, 256:1449-1455.

144. Bouza, C., Angeles, M., Munoz, A., and Amate, J.M. (2004). Efficacy and safety of naltrexone and acamprosate in the treatment of alcohol dependence: A systematic review. *Addiction*, 99:811-828.

145. Srisurapanont, M., Jarusuraisin, N. (2005). Opioid antagonists for alcohol dependence. *Cochrane Database Syst Rev* 1:CD001867.

146. Litten, R.X., Allen, J., and Fertig, J. (1991). Pharmacotherapies for alcohol problems: A review of research with focus on developments since. *Alcohol Clin Exp Res*, 20:859-876.

147. Litten, R.Z. and Allen, J.P. (1998). Advances in development of medications for alcoholism treatment. *Psychopharmacology*, 139:20-33.

148. Salloum, I.M., Cornellus, J.R., Thase, M.E., Daley, D.C., Kirisci, L., and Spotts, C. (1998). Naltrexone utility in depressed alcoholics. *Psychopharmacol Bull*, 34(1):111-115.

149. Monterosso, J.R., Flannery, B.A., Pettinati, H.M., Oslin, D.W., Rukstalis, M., O'Brien, C.P., and Volpicelli, J.R. (2001). Predicting treatment response to naltrexone: The influence of craving and family history. *Am J Addiction*, 10:258-268.

150. Huang, M.C., Chen, C.H., Yu, J.M., and Chen, C.C. (2005). A double-blind, placebo-controlled study of naltrexone in the treatment of alcohol dependence in Taiwan. *Addiction Biol*, 10:289-292.

151. Pertakis, I.L., Poling, J., Levinson, C., Nich, C., Carroll, K., and Rounsavill, B. VA New England VISN I MIRECC Study Group. (2005). Naltrexone and disulfiram in patients with alcohol dependence and comorbid psychiatric disorders. *Biol Psychiatry*, 57(10):1128-1137.

152. Jaffe, A.J., Rounsaville, B., Chang, G., Schottenfield, R.S., Meyer, R.E., and O'Malley, S.S. (1996). Naltrexone, relapse prevention, and supportive therapy with alcoholics: An analysis of patient treatment matching. *J Consult Clin Psychol*, 64:1044-1053.

153. King, A.C., Schluger, J., Gunduz, M., Borg, L., Perret, G., Ho, A. *et al.* (2002). Hypothalamic-pituitary-adrenocortical (HPA) axis response and biotransformation oral naltrexone: preliminary examination of relationship to family history of alcoholism. *Neuropsychopharmacology*, 26:778-788.

154. Garbutt, J.C., Kranzler, H.R., O'Malley, S.S., Gastriend, D.R., Pettinati, H.M., Silverman, B.L. *et al.* (2005). Efficacy and tolerability of long-acting injectable naltrexone for alcohol dependence: A randomized controlled trial. *J Am Med Assoc*, 293:1617–1625.

155. Kranzler, H.R., Wesson, D.R., and Billot, L.Drug Abuse Sciences Naltrexone Depot Study Group (2004). Naltrexone depot for treatment of alcohol dependence: A multicenter, randomized, placebo-controlled clinical trial. *Alchol Clin Exp Res*, 28:1050–1059.

156. Galloway, G.P., Koch, M., Cello, R., and Smith, D.E. (2005). Pharmacokinetics, safety, and tolerability of a depot formulation of naltrexone in alcoholics: An open-label trial. *BMC Psychiatry*, 5:18.

157. Kranzler, H.R., Modesto-Lowe, V., and Nuwayser, E.S. (1998). Sustained-release naltrexone for alcoholism treatment: A preliminary study. *Alcohol Clin Exp Res*, 22:1074–1079.

158. Kranzler, H.R. and Van Kirk, J. (2001). Efficacy of naltrexone and acamprosate for alcoholism treatment: A meta-analysis. *Alcohol Clin Exp Res*, 25:1335–1341.

159. Chick, J., Lehert, P., Landron, F., and Plinius Major Society, (2003). Does acamprosate improve reduction of drinking as well as aiding abstinence? *J Psychopharmacol*, 17:397–402.

160. Verheul, R., Lehert, P., Geerlings, P.J., Koeter, M.W., and van den, B.W. (2005). Predictors of acamprosate efficacy: Results from a pooled analysis of seven European trials including 1485 alcohol-dependent patients. *Psychopharmacology (Berl)*, 178:167–173.

161. Miranda, R., Monti, P., Swift, R., Mackillop, J., Tidey, J., Gwaltney, C., *et al.* (2006). Effects of topiramate on alcohol cue reactivity and the subjective effects of drinking, in Poster presentation at the *45th Annual Meeting of the American College of Neuropsychopharmacology*.

162. Biton, V., Edwards, K.R., Montouris, G.D., Sackellares, J.C., Harden, C.L., Kamin, M. *et al.* (2001). Topiramate titration and tolerability. *Ann Pharacother*, 35:173–179.

163. Rubio, G., Jimenez-Arriero, M.A., Ponce, G., and Palomo, T. (2001). Naltrexone versus acamprosate: One year follow-up of alcohol dependence treatment. *Alcohol Alcohol*, 36(5):419–425.

164. King, A.C., Volpicelli, J.R., Frazer, A., and O'Brien, C.P. (1997). Effect of naltrexone on subjective alcohol response in subjects at high and low risk for future alcohol dependence. *Psychopharmacology (Berl)*, 129:15–22.

165. Oslin, D.W., Pettinati, H., and Volpicelli, J.R. (2002). Alcoholism treatment adherence: Older age predicts better adherence and drinking outcomes. *Am J Geriatr Psychiatry*, 10(6):740–747.

166. Anton, R.F., Moak, D.H., Latham, P., Waid, R.L., Myrick, H., Voronin, K., Thevos, A., and Woolson, W. (2005). Naltrexone combined with either cognitive behavioural or motivational enhancement therapy for alcohol dependence. *J Clin Psychopharmacol*, 25(4):349–357.

167. Romach, M.K., Sellers, E.M., Somer, G.R., Landry, M., Cunningham, G.M., Jovey, R.D., McKay, C., Boislard, J., Mercier, C., Pepin, J.M., Perreault, J., Lernire, E., Baker, R.P., Campbell, W., and Ryan, D. (2002). Naltrexone in the treatment of alcohol dependence: A Canadian trial. *Can J Clin Pharmacol*, 9(3):130–136.

168. Kranzler, H.R. (2000). Pharmacotherapy of alcoholism: Gaps in knowledge and opportunities for research. *Alcohol Alcohol*, 35(6):537–547.

169. Killeen, T.K., Brady, K.T., Gold, P.B., Simson, K.N., Faldowski, R.A., Tyson, C., and Anton, R.F. (2004). Effectiveness of naltrexone in a community treatment program. *Alcohol Clin Exp Res*, 28(11):1710–1717.

170. Gastpar, M., Bonnet, U., Boning, J., Mann, K., Schmidt, L.G., Soyka, M., Wetterling, T., Kielstein, V., Lariola, D., Croop, R. (2002). Lack of efficacy of naltrexone in the prevention of alcohol relapse: results from a German multicenter study. *J Clin Psychopharmacol*. 592–598.

171. Mutel, V. (2002). Therapeutic potential of non-competitive, subtype-selective metabotropic glutamate receptor ligands. *Expert Opin Ther Patents*, 12(12):1845–1852.

172. Kiefer, S.W., Hill, K.G., and Kaczmarek, H.J. (1998). Taste reactivity to alcohol and basic tastes in outbred mice. *Alcohol Clin Exp Res*, 22(5):1146–1151.

173. Forsander, O.A. (1994). Hypothesis – factors involved in the mechanisms regulating food-intake affect alcohol-consumption. *Alcohol Alcohol*, 29(5):503–512.

174. Richardson, A., Rumsey, R.D.E., and Read, N.W. (1984). The effect of ethanol on the normal food intake and eating behavior of the rat. *Physiol Behav*, 48:845–847.

175. Myers, R.D. (1996). Experimental alcoholism: Worldwide need to standardize animal models. *Alcohol*, 13:109–110.

176. Nowak, K.L., McKinzie, D.L., McBride, W.J., and Murphy, J.M. (1999). Patterns of ethanol and saccharin intake in P rats under limited access conditions. *Alcohol*, 19:85–96.

177. Sinclair, J.D., Hyytia, P., and Nurmi, M. (1992). The limited access paradigm: Description of one method. *Alcohol*, 9:441–444.

178. Carrillo, C.A., Leibowitz, S.F., Karatayev, O., and Hoebel, B.G. (2004). A high-fat meal or injection of lipids stimulates ethanol intake. *Alcohol*, 34:197–202.

179. Pekkanen, L., Eriksson, K., and Sihvonen, M.-L. (1978). Dietarily-induced changes in voluntary ethanol-consumption and ethanol-metabolism in rat. *Br J Nutr*, 40:103–113.

180. Greenwood, C.E. and Winocur, G. (1996). Cognitive impairment in rats fed high-fat diets: A specific effect of saturated fatty-acid intake. *Behav Neurosci*, 110:451–459.

181. Greenwood, C.E. and Winocur, G. (2001). Glucose treatment reduces memory deficits in young adult rats fed high-fat diets. *Neurobiol Learn Mem*, 75:179–189.

182. Molteni, R., Barnard, R.J., Ying, Z., Roberts, C.K., and Gomez-Pinilla, F. (2002). A high-fat, refined sugar diet reduces hippocampal brain-derived neurotrophic factor, neuronal plasticity, and learning. *Neuroscience*, 112:803–814.

183. Tordoff, M.G., Pilchak, D.M., Williams, J.A., McDaniel, A.H., and Bachmanov, A.A. (2002). The maintenance diets of C57BL/6J and 129X1/svj mice influence their taste solution preferences: Implications for large-scale phenotyping projects. *J Nutr*, 132(8):2288–2297.

184. Veale, W. and Myers, R.D. (1969). Increased alcohol preference in rats following repeated exposures to alcohol. *Psychopharmacologia*, 15:361–372.

185. Marfaing-Jallat, P. and Le Magnen, J. (1982). Induction of high voluntary ethanol intake in dependent rats. *Pharmacol Biochem Behav*, 17:609–612.

186. Camarini, R. and Hodge, C.W. (2004). Ethanol preexposure increases ethanol self-administration in C57BL/6J and DBA/2J mice. *Pharmacol Biochem Behav*, 79(4):623–632.

187. Boyle, A.E.L., Smith, B.R., Spivak, K., and Amit, Z. (1994). Voluntary ethanol consumption in rats: The importance of the exposure paradigm in determining final intake outcome. *Behav Pharmacol*:501–512.

188. Koros, E., Piasecki, J., Kostowski, W., and Bienkowski, P. (1999a). Development of alcohol deprivation effect in rats: Lack of correlation with saccharin drinking and locomotor activity. *Alcohol Alcohol*, 34(4):542–550.

189. Spanagel, R. and Holter, S.M. (1999). Long term alcohol self-administration with repeated alcohol deprivation phases: And animal model of alcoholism? *Alcohol Alcohol*, 34:231–243.

190. Bell, R.L., Rodd, Z.A., Boutwell, C.L., Hsu, C.C., Lumeng, L., Murphy, J.M., Li, T.K., and McBride, W.J. (2004). Effects of long-term episodic access to ethanol on the expression of an alcohol deprivation effect in low alcohol-consuming rats. *Alcohol Clin Exp Res*, 28(12):1867–1874.

191. Wolffgramm, J. and Heyne, A. (1995). From controlled intake to loss of control: The irreversible development of drug addiction in the rat. *Behav Br Res*, 70:77–94.

192. Wolffgramm, J. (1991). Social behaviour, dominance, and social deprivation of rats determine drug choice. *Pharmacol Biochem Behav*, 38:389–399.

193. Holt, J.D.S., Watson, W.P., and Little, H.J. (2001). Studies on a model of long term alcohol drinking. *Behav Brain Res*, 123:193–200.

194. Terenina-Rigaldie, E., Jones, B.C., and Mormede, P. (2004). The high-ethanol preferring rat as a model to study the shift between alcohol abuse and dependence. *Eur J Pharmacol*, 504(3):199–206.

195. Fachin-Scheit, D.J., Ribeiro, A.F., Pigatto, G., Goeldner, F.O., and de Lacerda, R.B. (2006). Development of a mouse model of ethanol addiction: Naltrexone efficacy in reducing consumption but not craving. *J Neural Transm*, 113(9):1305–1321.

196. Cannon, D.S., Leeka, J.K., and Block, A.K. (1994). Ethanol self-administration patterns and taste-aversion learning across inbred rat strains. *Pharmacol Biochem Behav*, 47(4):795–802.

197. Kiefer, S.W. (1995). Alcohol, palatability, and taste reactivity. *Neurosci Biobehav Rev*, 19(1):133–141.

198. Stewart, R.B., Murphy, J.M., McBride, W.J., Lumeng, L., and Li, T.K. (1996). Place conditioning with alcohol in alcohol-preferring and -nonpreferring rats. *Pharmacol Biochem Behav*, 53:487–491.

199. Risinger, F.O., Bormann, N.M., and Oakes, R.A. (1996). Reduced sensitivity to ethanol reward, but not ethanol aversion, in mice lacking 5-HT1B receptors. *Alcohol Clin Exp Res*, 20(8):1401–1405.

200. McKinzie, D.L., Eha, R., Murphy, J.M., McBride, W.J., Lumeng, L., and Li, T.K. (1996). Effects of taste aversion training on the acquisition of alcohol drinking in adolescent P and HAD rat lines. *Alcohol Clin Exp Res*, 20(4):682–687.

201. Garver, E., Ross, A.D., Tu, G.C., Cao, Q.N., Zhou, F., and Israel, Y. (2000). Paradigm to test a drug-induced aversion to ethanol. *Alcohol Alcohol*, 35(5):435–438.

202. Hughes, J.C. and Cook, C.C.H. (1997). The efficacy of disulfiram: A review of outcome studies. *Addiction*, 92(4):381–395.

203. Suh, J.J., Pettinati, H.M., Kampman, K.M., and O'Brien, C.P. (2006). The status of disulfiram – a half of a century later. *J Clin Psychopharmacol*, 26(3):290–302.

204. Rodd-Henricks, Z.A., Melendez, R.I., Zaffaroni, A., Goldstein, A., McBride, W.J., and Li, T.K. (2002a). The reinforcing effects of acetaldehyde in the posterior ventral tegmental area of alcohol-preferring rats. *Pharmacol Biochem Behav*, 72(1–2):55–64.

205. Mormede, P., Colas, A., and Jones, B.C. (2004). High ethanol preferring rats fail to show dependence following short- or long-term ethanol exposure. *Alcohol Alcohol*, 39(3):183–189.

206. McKinzie, D.L., Eha, R., Cox, R., Stewart, R.B., Dyr, W., Murphy, J.M., McBride, W.J., Lumeng, L., and Li, T.-K. (1998). Serotonin-3 receptor antagonism of alcohol drinking intake: Effects of drinking conditions. *Alcohol*, 15:291–298.

207. McKinzie, D.L., Nowak, K.L., Yorger, L., McBride, W.J., Murphy, J.M., Lumeng, L., and Li, T.K. (1998). The alcohol deprivation effect in the alcohol-preferring P rat under free-drinking and operant access conditions. *Alcohol Clin Exp Res*, 22(5):1170–1176.

208. Siciliano, D. and Smith, R.F. (2001). Periadolescent alcohol alters adult behavioral characteristics in the rat. *Physiol Behav*, 74:637–643.

209. Sanders, S. and Spear, N.E. (2007). Ethanol acceptance is high during early infancy and becomes still higher after previous ethanol ingestion. *Alcohol Clin Exp Res*, 31:1148–1158.

210. Truxell, E.M., Molina, J.C., and Spear, N.E. (2007). Ethanol intake in the juvenile, adolescent, and adult rat: Effects of age and prior exposure to ethanol. *Alcohol Clin Exp Res*, 31(5):755–765.

211. Schramm-Sapyta, N.L., Morris, R.W., and Kuhn, C.M. (2006). Adolescent rats are protected from the conditioned aversive properties of cocaine and lithium chloride. *Pharmacol Biochem Behav*, 84(2):344–352.

212. Pepino, M.Y., Abate, P., Spear, N.E., and Molina, J.C. (2004). Heightened ethanol intake in infant and adolescent rats after nursing experiences with an ethanol-intoxicated dam. *Alcohol Clin Exp Res*, 28(6):895-905.

213. Spanagel, R., Montkowski, A., Allingham, K., Stohr, T., Shoaib, M., Holsboer, F., and Landgraf, R. (1995). Anxiety: A potential predictor of vulnerability to the initiation of ethanol self administration in rats. *Psychopharmacology*, 122(4):369-373.

214. Tornatzky, W. and Miczek, K.A. (1995). Alcohol, anxiolytics and social stress in rats. *Psychopharmacology*, 121(1):135-144.

215. Hedlund, L. and Wahlstrom, G. (1996). Buspirone as an inhibitor of voluntary ethanol intake in male rats. *Alcohol Alcohol*, 31(2):149-156.

216. Deutsch, J.A. and Walton, N.Y. (1977). Diazepam maintenance of alcohol preference during alcohol withdrawal. *Science*, 198(4314):307-309.

217. Little, H.J., Butterworth, A.R., O'Callaghan, M.J., Wilson, J., Cole, J., and Watson, W.P. (1999). Low alcohol preference among the "high alcohol preference" C57 strain mice; preference increased by saline injections. *Psychopharmacology*, 147:182-189.

218. Rockman, G.E., Hall, A.M., Markert, L., Glavin, G.B., and Pare, W.P. (1987). Ethanol-stress interaction – immediate versus delayed-effects of ethanol and handling on stress responses of ethanol-consuming rats. *Alcohol*, 4(5):391-394.

219. Volpicelli, J.R., Davis, M.A., and Olgin, J.E. (1986). Naltrexone blocks the post-shock increase of ethanol-consumption. *Life Sci*, 38(9):841-884.

220. Croft, A.P., Brooks, S.P., Cole, J., and Little, H.J. (2005). Social defeat increases alcohol preference of C57BL/10 strain mice; effect prevented by a CCKB antagonist. *Psychopharmacology*, 183:163-170.

221. Van Erp, A.M.M., Tachi, N., and Miczek, K.A. (2001). Short or continuous social stress: Suppression of continuously available ethanol intake in subordinate rats. *Behav Pharmacol*, 12(5):335-342.

222. Dayas, C.V., Martin-Fardon, R., Thorsell, A., and Weiss, F. (2004). Chronic footshock, but not a physiological stressor, suppresses the alcohol deprivation effect in dependent rats. *Alcohol Alcohol*, 39(3):190-196.

223. Siegmund, S., Vengeliene, V., Singer, M.V., and Spanagel, R. (2005). Influence of age at drinking onset on long-term ethanol self-administration with deprivation and stress phases. *Alcohol Clin Exp Research*, 29(7):1139-1145.

224. Fullgrabe, M.W., Vengeliene, V., and Spanagel, R. (2007). Influence of age at drinking onset on the alcohol deprivation effect and stress-induced drinking in female rats. *Pharmacol Biochem Behav*, 86(2):320-326.

225. Vengeliene, V., Siegmund, S., Singer, M.V., Sinclair, J.D., Li, T.K., and Spanagel, R. (2003). A comparative study on alcohol-preferring rat lines: Effects of deprivation and stress phases on voluntary alcohol intake. *Alcohol Clin Exp Res*, 27(7):1048-1054.

226. Huot, R.L., Thrivikraman, K.V., Meaney, M.J., and Plotsky, P.M. (2001). Development of adult ethanol preference and anxiety as a consequence of neonatal maternal separation in Long Evans rats and reversal with antidepressant treatment. *Psychopharmacology*, 158(4):366-373.

227. Gustafsson, L. and Nylander, I. (2006). Time-dependent alterations in ethanol intake in male Wistar rats exposed to short and prolonged daily maternal separation in a 4-bottle free-choice paradigm. *Alcohol Clin Exp Res*, 30(12):2008-2016.

228. Roman, E. and Nylander, I. (2005). The impact of emotional stress early in life on adult voluntary ethanol intake-results of maternal separation in rats. *Stress*, 8(3):157-174.

229. Roman, E., Ploj, K., and Nylander, I. (2004). Maternal separation has no effect on voluntary ethanol intake in female Wistar rats. *Alcohol*, 33(1):31-39.

230. Gustafsson, L., Ploj, K., and Nylander, I. (2005). Effects of maternal separation on voluntary ethanol intake and brain peptide systems in female Wistar rats. *Pharmacol Biochem Behav*, 81(3):506–516.

231. Fahlke, C., Hard, E., and Eriksson, C.J.P. (1997). Effects of early weaning and social isolation on subsequent alcohol intake in rats. *Alcohol*, 14(2):175–180.

232. Ehlers, C.L., Walker, B.M., Pian, J.P., Roth, J.L., and Slawecki, C.J. (2007). Increased alcohol drinking in isolate-housed alcohol-preferring rats. *Behav Neurosci*, 121(1):111–119.

233. Hall, F.S., Huang, S., Fong, G.W., Pert, A., and Linnoila, M. (1998). Effects of isolation-rearing on voluntary consumption of ethanol, sucrose and saccharin solutions in Fawn Hooded and Wistar rats. *Psychopharmacology*, 139(3):210–216.

234. Nunez, M.J., Riveiro, P., Becerra, M.A., De Miguel, S., Quintans, M.R., Nunez, L.A., Legazpi, M.P., Mayan, J.M., Rey-Mendez, M., Varela, M., and Freire-Garabal, M. (1999). Effects of alprazolam on the free-choice ethanol consumption induced by isolation stress in aged rats. *Life Sci*, 64(20):PL213–PL217.

235. Piazza, P.V., Deroche, V., Deminiere, J.M., Maccari, S., Lemoal, M., and Simon, H. (1993). Corticosterone in the range of stress-induced levels possesses reinforcing properties – implications for sensation-seeking behaviors. *Proc Natl Acad Sci USA*, 90(24):11738–11742.

236. Fahlke, C., Engel, J.A., Eriksson, C.J.P., Hard, E., and Soderpalm, B. (1994a). Involvement of corticosterone in the modulation of ethanol-consumption in the rat. *Alcohol*, 11(3):195–202.

237. Fahlke, C., Hard, E., Thomasson, R., Engel, J.A., and Hansen, S. (1994b). Metyrapone-induced suppression of corticosterone synthesis reduces ethanol-consumption in high-preferring rats. *Pharmacol Biochem Behav*, 48(4):977–981.

238. Blanchard, D.C., Sakai, R.R., Mcewen, B., Weiss, S.M., and Blanchard, R.J. (1993). Subordination stress – behavioral, brain, and neuroendocrine correlates. *Behav Brain Res*, 58(1–2):113–121.

239. Blanchard, R.J., McKittrick, C.R., and Blanchard, D.C. (2001). Animal models of social stress: Effects on behavior and brain neurochemical systems. *Physiol Behav*, 73(3):261–271.

240. Miczek, K.A., Covington, H.E., Nikulina, E.A., and Hammer, R.P. (2004). Aggression and defeat: Persistent effects on cocaine self-administration and gene expression in peptidergic and aminergic mesocorticolimbic circuits. *Neurosci Biobehav Rev*, 27(8):787–802.

241. Barnett, S.A., Hocking, W.E., Munro, K.M.H., and Walker, K.Z. (1975). Socially induced renal pathology of captive wild rats (*Rattus villosissimus*). *Aggressive Behavior*, 1(2):123–133.

242. Raab, A. and Oswald, R. (1980). Coping with social-conflict – impact on the activity of tyrosine-hydroxylase in the limbic system and in the adrenals. *Physiol Behav*, 24(2):387–394.

243. Ellison, G. (1987). Stress and alcohol intake – the socio-pharmacological approach. *Physiol Behav*, 40(3):387–392.

244. Blanchard, R.J., Hori, K., Tom, P., and Blanchard, D.C. (1987). Social-structure and ethanol-consumption in the laboratory rat. *Pharmacol Biochem Behav*, 28(4):437–442.

245. Hilakivi-Clarke, L. and Lister, R.G. (1992). Social-status and voluntary alcohol-consumption in mice – interaction with stress. *Psychopharmacology*, 108(3):276–282.

246. Le, A.D., Poulos, C.X., Quan, B., and Chow, S. (1993). The effects of selective blockade of delta and mu opiate receptors on ethanol-consumption by C57Bl/6 mice in a restricted access paradigm. *Brain Res*, 630(1–2):330–332.

247. Krystal, J.H., Cramer, J.A., Krol, W.F., Kirk, G.F., and Rosenheck, R.A. (2001). Naltrexone in the treatment of alcohol dependence. *N Engl J Med*, 345:1734–1739.

248. Davidson, D. and Amit, Z. (1997a). Naltrexone blocks acquisition of voluntary ethanol intake in rats. *Alcohol Clin Exp Res*, 21(4):677–683.

249. Biggs, T.A.G. and Myers, R.D. (1998). Naltrexone and amperozide modify chocolate and saccharin drinking in high alcohol-preferring P rats. *Pharmacol Biochem Behav*, 60(2):407–413.

250. Hill, K.G. and Kiefer, S.W. (1997). Naltrexone treatment increases the aversiveness of alcohol for outbred rats. *Alcohol Clin Exp Res*, 21(4):637–641.

251. Froehlich, J.C., Badia-Elder, N.E., Zink, R.W., McCullough, D.E., and Portoghese, P.S. (1998). Contribution of the opioid system to alcohol aversion and alcohol drinking behavior. *J Pharmacol Exp Ther*, 287(1):284–292.

252. Davidson, D. and Amit, Z. (1996). Effects of naloxone on limited-access ethanol drinking in rats. *Alcohol Clin Exp Res*, 20(4):664–669.

253. Kuzmin, A., Kreek, M.J., Bakalkin, G., and Liljequist, S. (2007). The nociceptin/orphanin FQ receptor agonist Ro 64-6198 reduces alcohol self-administration and prevents relapse-like alcohol drinking. *Neuropsychopharmacology*, 32(4):902–910.

254. Reid, L.D., Gardell, L.R., Chattopadhyay, S., and Hubbell, C.L. (1996). Periodic naltrexone and propensity to take alcoholic beverage. *Alcohol Clin Exp Res*, 20(8):1329–1334.

255. Eriksson, K. (1968). Genetic selection for voluntary alcohol consumption in the albino rat. *Science*, 159:739–741.

256. Holter, S.M. and Spanagel, R. (1999). Effects of opiate antagonist treatment on the alcohol deprivation effect in long-term ethanol-experienced rats. *Psychopharmacology*, 145(4):360–369.

257. Zakharova, E., Malyshkin, A., Kashkin, V., Neznanova, O., Sukhotina, I., Danysz, W., and Bespalov, A. (2004). The NMDA receptor channel blocker memantine and opioid receptor antagonist naltrexone inhibit the saccharin deprivation effect in rats. *Behav Pharmacol*, 15(4):273–278.

258. McMillen, B.A. and Williams, H.L. (1995). Volitional consumption of ethanol by Fawn-Hooded rats – effects of alternative solutions and drug treatments. *Alcohol*, 12(4):345–350.

259. Davidson, D. and Amit, Z. (1997a). Effect of ethanol drinking and naltrexone on subsequent drinking in rats. *Alcohol*, 14(6):581–584.

260. Stromberg, M.F., Volpicelli, J.R., and O'Brien, C.P. (1998). Effects of naltrexone administered repeatedly across 30 or 60 days on ethanol consumption using a limited access procedure in the rat. *Alcohol Clin Exp Res*, 22(9):2186–2191.

261. Phillips, T.J., Wenger, C.D., and Dorow, J.D. (1997). Naltrexone effects on ethanol drinking acquisition and on established ethanol consumption in C57BL/6J mice. *Alcohol Clin Exp Res*, 21:691–702.

262. Gewiss, M., Heidbreder, C., Opsomer, L., Durbin, P., and De, W.P. (1991). Acamprosate and diazepam differentially modulate alcohol-induced behavioural and cortical alterations in rats following chronic inhalation of ethanol vapour. *Alcohol Alcohol*, 26:129–137.

263. Cowen, M.S., Adams, C., Kraehenbuehl, T., Vengeliene, V., and Lawrence, A.J. (2005). The acute anti-craving effect of acamprosate in alcohol-preferring rats is associated with modulation of the mesolimbic dopamine system. *Addiction Biol*, 10:233–242.

264. Spanagel, R. and Holter, S.M. (2000). Pharmacological validation of a new animal model of alcoholism. *J Neural Transm*, 107(6):669–680.

265. Kim, S.G., Han, B.D., Park, J.M., Kim, M.J., and Stromberg, M.F. (2004). Effect of the combination of naltrexone and acamprosate on alcohol intake in mice. *Psychiatry Clin Neurosci*, 58:30–36.

266. Heyser, C.J., Moc, K., and Koob, G.F. (2003). Effects of naltrexone alone and in combination with acamprosate on the alcohol deprivation effect in rats. *Neuropsychopharmacology*, 28:1463–1471.

267. Mardones, J. and Segovia-Riquelme, N. (1983). Thirty-two years of selection of rats by ethanol preference: UChA and UChB strains. *Neurobehav Toxicol Teratol*, 2:171–178.

268. Stromberg, M.F., Mackler, S.A., Volpicelli, J.R., and O'Brien, C.P. (2001). Effect of acamprosate and naltrexone, alone or in combination, on ethanol consumption. *Alcohol*, 23(2):109–116.

269. Higgins, G.A., Tomkins, D.M., Fletcher, P.J., and Sellers, E.M. (1992). Effect of drugs influencing 5-HT function on ethanol drinking and feeding-behavior in rats – studies using a drinkometer system. *Neurosci Biobehav Rev*, 16(4):535–552.

270. McBride, W.J., Murphy, J.M., Yoshimoto, K., Lumeng, L., and Li, T.K. (1993). Serotonin mechanisms in alcohol-drinking behavior. *Drug Dev Res*, 30(3):170–177.

271. Murphy, J.M., McBride, W.J., Lumeng, L., and Li, T.K. (1988). Effects of serotonin and dopamine agents on ethanol intake of alcohol-preferring P-rats. *Alcohol Clin Exp Res*, 12(2):306.

272. Sellers, E.M., Higgins, G.A., and Sobell, M.B. (1992). 5-HT and alcohol-abuse. *Trends Pharmacol Sci*, 13(2):69–75.

273. Le, A.D., Tomkins, D.M., and Sellers, E.M. (1996). Use of serotonin (5-HT) and opiate-based drugs in the pharmacotherapy of alcohol dependence: An overview of the preclinical data. *Alcohol Alcohol*, 31(Suppl 1):27–32.

274. Koe, B.K. (1990). Preclinical pharmacology of sertraline – a potent and specific inhibitor of serotonin reuptake. *J Clin Psychiatry*, 51(Suppl B):13–17.

275. Myers, R.D. and Quarfordt, S.D. (1991). Alcohol drinking attenuated by sertraline in rats with 6-OHDA or 5,7-DHT lesions of N. accumbens – a caloric response. *Pharmacol Biochem Behav*, 40(4):923–928.

276. Meert, T.F. and Janssen, P.A.J. (1991). Ritanserin, a new therapeutic approach for drug-abuse. 1: Effects on alcohol. *Drug Dev Res*, 24(3):235–249.

277. Myers, R.D. and Lankford, M.F. (1993). Failure of the 5-HT2 receptor antagonist, ritanserin, to alter preference for alcohol in drinking rats. *Pharmacol Biochem Behav*, 45(1):233–237.

278. Panocka, I., Ciccocioppo, R., Pompei, P., and Massi, M. (1993). 5-HT(2) receptor antagonists do not reduce ethanol preference in sardinian alcohol-preferring (sP) rats. *Pharmacol Biochem Behavior*, 46(4):853–856.

279. Fadda, F., Garau, B., Marchei, F., Colombo, G., and Gessa, G.L. (1991). MDL 72222, a selective 5-HT3 receptor antagonist, suppresses voluntary ethanol-consumption in alcohol-preferring rats. *Alcohol Alcohol*, 26(2):107–110.

280. Kostowski, W. and Dyr, W. (1992). Effects of 5-HT-1A receptor agonists on ethanol preference in the rat. *Alcohol*, 9(4):283–286.

281. Colombo, G., Serra, S., Brunetti, G., Atzori, G., Pani, M., Vacca, G., Addolorato, G., Froestl, W., Carai, M.A., and Gessa, G.L. (2002). The GABA(B) receptor agonists baclofen and CGP 44532 prevent acquisition of alcohol drinking behaviour in alcohol-preferring rats. *Alcohol Alcohol*, 37:499–503.

282. Colombo, G., Addolorato, G., Agabio, R., Carai, M.A.M., Pibiri, F., Serra, S., Vacca, G., and Gessa, G.L. (2004a). Role of GABA(B) receptor in alcohol dependence: Reducing effect of baclofen on alcohol intake and alcohol motivational properties in rats and amelioration of alcohol withdrawal syndrome and alcohol craving in human alcoholics. *Neurotoxicity Research*, 6(5):403–414.

283. Colombo, G., Serra, S., Vacca, G., Carai, M.A.M., and Gessa, G.L. (2006). Baclofen-induced suppression of alcohol deprivation effect in Sardinian alcohol-preferring (sP) rats exposed to different alcohol concentrations. *Eur J Pharmacol*, 550(1–3):123–126.

284. Colombo, G., Serra, S., Vacca, G., Gessa, G.L., and Carai, M.A.M. (2004b). Suppression by baclofen of the stimulation of alcohol intake induced by morphine and WIN 55,212-2 in alcohol-preferring rats. *Eur J Pharmacol*, 492(2–3):189–193.

285. Basavarajappa, B.S. (2007a). Neuropharmacology of the endocannabinoid signaling system: Molecular mechanisms, biological actions and synaptic plasticity. *Curr Neuropharmacol*, 5(2):81–97.

286. Basavarajappa, B.S. (2007b). The endocannabinoid signaling system: A potential target for next-generation therapeutics for alcoholism. *Mini Reviews in Medicinal Chemistry*, 7(8):769–779.

287. Colombo, G., Serra, S., Brunetti, G., Gomez, R., Melis, S., Vacca, G., Carai, M.A.M., and Gessa, G.L. (2002a). Stimulation of voluntary ethanol intake by cannabinoid receptor agonists in ethanol-preferring sP rats. *Psychopharmacology*, 159(2):181–187.

288. Hungund, B.L., Szakall, I., Adam, A., Basavarajappa, B.S., and Vadasz, C. (2003). Cannabinoid CB1 receptor knockout mice exhibit markedly reduced voluntary alcohol consumption and lack alcohol-induced dopamine release in the nucleus accumbens. *J Neurochem*, 84(4):698–704.

289. Gessa, G.L., Serra, S., Vacca, G., Carai, M.A.M., and Colombo, G. (2005). Suppressing effect of the cannabinoid CB1 receptor antagonist, SR147778, on alcohol intake and motivational properties of alcohol in alcohol-preferring sP rats. *Alcohol Alcohol*, 40(1):46–53.

290. Serra, S., Brunetti, G., Pani, M., Vacca, G., Carai, M.A.M., Gessa, G.L., and Colombo, G. (2002). Blockade by the cannabinoid CB1 receptor antagonist, SR 141716, of alcohol deprivation effect in alcohol-preferring rats. *Eur J Pharmacol*, 443(1–3):95–97.

291. Rowland, N.E., Mukherjee, M., and Robertson, K. (2001). Effects of the cannabinoid receptor antagonist SR 141716, alone and in combination with dexfenfluramine or naloxone, on food intake in rats. *Psychopharmacology*, 159(1):111–116.

292. Lallemand, F., Soubrie, P., and De Witte, P. (2004). Effects of CB1 cannabinoid receptor blockade on ethanol preference after chronic alcohol administration combined with repeated re-exposures and withdrawals. *Alcohol Alcohol*, 39(6):486–492.

293. Lallemand, F. and De Witte, P. (2006). SR147778, a CB1 cannabinoid receptor antagonist, suppresses ethanol preference in chronically alcoholized Wistar rats. *Alcohol*, 39(3):125–134.

294. McMillen, B.A., Crawford, M.S., Kulers, C.M., and Williams, H.L. (2005). Effects of NMDA glutamate receptor antagonist drugs on the volitional consumption of ethanol by a genetic drinking rat. *Brain Res Bull*, 64(3):279–284.

295. Stromberg, M.F., Volpicelli, J.R., O'Brien, C.P., and Mackler, S.A. (1999). The NMDA receptor partial agonist, 1-aminocyclopropanecarboxylic acid (ACPC), reduces ethanol consumption in the rat. *Pharmacol Biochem Behav*, 64(3):585–590.

296. Vengeliene, V., Bachteler, D., Danysz, W., and Spanagel, R. (2005). The role of the NMDA receptor in alcohol relapse: A pharmacological mapping study using the alcohol deprivation effect. *Neuropharmacology*, 48(6):822–829.

297. Olive, M.F., McGeehan, A.J., Kinder, J.R., McMahon, T., Hodge, C.W., Janak, P.H., and Messing, R.O. (2005). The mGluR5 antagonist 6-methyl-2-(phenylethynyl)pyridine decreases ethanol consumption via a protein kinase C epsilon-dependent mechanism. *Mol Pharmacol*, 67(2):349–355.

298. Backstrom, P., Bachteler, D., Koch, S., Hyytia, P., and Spanagel, R. (2004). mGluR5 antagonist MPEP reduces ethanol-seeking and relapse behavior. *Neuropsychopharmacology*, 29(5):921–928.

299. Rodd, Z.A., McKinzie, D.L., Bell, R.L., McQueen, V.K., Murphy, J.M., Schoepp, D.D., and McBride, W.J. (2006). The metabotropic glutamate 2/3 receptor agonist LY404039 reduces alcohol-seeking but not alcohol self-administration in alcohol-preferring (P) rats. *Behav Brain Res*, 171(2):207–215.

300. Finn, D.A., Snelling, C., Fretwell, A.M., Tanchuck, M.A., Underwood, L., Cole, M., Crabbe, J.C., and Roberts, A.J. (2007). Increased drinking during withdrawal from intermittent ethanol exposure is blocked by the CRF receptor antagonist D-Phe-CRF(12-41). *Alcohol Clin Exp Res*, 31(6):939–949.

301. O'Callaghan, M.J., Croft, A.P., Jacquot, C., and Little, H.J. (2005). The hypothalamopituitary-adrenal axis and alcohol preference. *Brain Res Bull*, 68:171–178.

302. Bell, S.M., Reynolds, J.G., Thiele, T.E., Gan, J., Figlewicz, D.P., and Woods, S.C. (1998). Effects of third intracerebroventricular injections of corticotropin releasing factor (CRF) on ethanol drinking and food intake. *Psychopharmacology*, 139:128-135.

303. Thorsell, A., Slawecki, C.J., and Ehlers, C.L. (2005). Effects of neuropeptide Y and corticotropin-releasing factor on ethanol intake in Wistar rats: Interaction with chronic ethanol exposure. *Behav Brain Res*, 161(1):133-140.

304. Palmer, A.A., Sharpe, A.L., Burkhart-Kasch, S., McKinnon, C.S., Coste, S.C., Stenzel-Poore, M.P., and Phillips, T.J. (2004). Corticotropin-releasing factor overexpression decreases ethanol drinking and increases sensitivity to the sedative effects of ethanol. *Psychopharmacology*, 176(3-4):386-397.

305. Olive, M.F., Mehmert, K.K., Koenig, H.N., Camarini, R., Kim, J.A., Nannini, M.A., Ou, C.J., and Hodge, C.W. (2003). A role for corticotropin releasing factor (CRF) in ethanol consumption, sensitivity and reward as revealed by CRF-deficient mice. *Psychopharmacology*, 165:181-187.

306. Koenig, H.N. and Olive, M.F. (2004). The glucocorticoid receptor antagonist mifepristone reduces ethanol intake in rats under limited access conditions. *Psychoneuroendocrinology*, 29(8):999-1003.

307. Crabbe, J.C., Phillips, T.J., Harris, R.A., Arends, M.A., and Koob, G.F. (2006). Alcohol-related genes: Contributions from studies with genetically engineered mice. *Addiction Biol*, 11:195-269.

308. Crabbe, J.C., Phillips, T.J., Buck, K.J., Cunningham, C.L., and Belknap, J.K. (1999). Identifying genes for alcohol and drug sensitivity: Recent progress and future directions. *Trends Neuroscie*, 22(4):173-179.

309. Crabbe, J.C. (2002). Alcohol and genetics: New models. *Am J Med Genet*, 114(8):969-974.

310. Crabbe, J.C., Metten, P., Ponomarev, I., Prescott, C.A., and Wahlsten, D. (2006). Effects of genetic and procedural variation on measurement of alcohol sensitivity in mouse inbred strains. *Behav Genet*, 36(4):536-552.

311. Overstreet, D.H., Rezvani, A.H., and Parsian, A. (1999). Behavioural features of alcohol-preferring rats: Focus on inbred strains. *Alcohol Alcohol*, 34(3):378-385.

312. Wahlsten, D., Bachmanov, A., Finn, D.A., and Crabbe, J.C. (2006). Stability of inbred mouse strain differences in behavior and brain size between laboratories and across decades. *Proc Natl Acad Sci USA*, 103(44):16364-16369.

313. Broadbent, J., Muccino, K.J., and Cunningham, C.L. (2002). Ethanol-induced conditioned taste aversion in 15 inbred mouse strains. *Behav Neurosci*, 116(1):138-148.

314. Metten, P. and Crabbe, J.C. (2005). Alcohol withdrawal severity in inbred mouse (Mus musculus) strains. *Behav Neurosci*, 119(4):911-925.

315. Overstreet, D.H., Rezvani, A.H., Djouma, E., Parsian, A., and Lawrence, A.J. (2007). Depressive-like behavior and high alcohol drinking co-occur in the FH/Wjd rat but appear to be under independent genetic control. *Neurosci Biobehav Rev*, 31(1):103-114.

316. Rezvani, A.H., Parsian, A., and Overstreet, D.H. (2002). The Fawn-Hooded (FH/Wjd) rat: A genetic animal model of comorbid depression and alcoholism. *Psychiatry Genet*, 12(1):1-16.

317. Khisti, R.T., Wolstenholme, J., Shelton, K.L., and Miles, M.F. (2006). Characterization of the ethanol-deprivation effect in substrains of C57BL/6 mice. *Alcohol*, 40(2):119-126.

318. Fidler, T.L., Struthers, A.M., Powers, M.S., Cunningham, C.L. (2007). Hybrid C57BL/6JX mice do not self-infuse more ethanol than do C57BL/6J mice. *Alc Clin Exp Res* 31:90A.

319. De Wit, H., Svenson, J., and York, A. (1999). Non-specific effect of naltrexone on ethanol consumption in social drinkers. *Psychopharmacology*, 146(1):33-41.

320. Lieber, C.S. and DeCarli, L.M. (1989). Liquid diet technique of ethanol administration: 1989 update. *Alcohol Alcohol*, 24:197-211.

321. Vasselli, J.R., Weindruch, R., Heymsfield, S.B., Pi-Sunyer, F.X., Boozer, C.N., Yi, N., Wang, C.X., Pietrobelli, A., and Allison, D.B. (2005). Intentional weight loss reduces mortality rate in a rodent model of dietary obesity. *Obes Res*, 13(4):693-702.

322. Watson, W.P. and Little, H.J. (2002). Selectivity of the effects of dihydropyridine calcium channel antagonists against the alcohol withdrawal syndrome. *Brain Res*, 930:111-122.

323. Duka, T., Gentry, R., Ripley, T.L., Borlokova, G., Stephens, D.N., Veatch, L.M., Becker, H.C., and Crews, F.T. (2003). Consequences of multiple withdrawals from alcohol. *AlcoholClin Exp Res*, 28:233-246.

324. Noel, X., Van der Linden, M., Schmidt, N., Sferrazza, R., Hanak, C., Le Bon, O., De Mol, J., Kornreich, C., Pelc, I., and Verbanck, P. (2001). Supervisory attentional system in nonamnesic alcoholic men. *Arch Gen Psychiatry*, 58:1152-1158.

325. Grant, K.A., Valverius, P., Hudspith, M., and Tabakoff, B. (1990). Ethanol withdrawal seizures and the NMDA receptor complex. *Eur J Pharmacol*, 176:289-296.

326. Ripley, T.L. and Little, H.J. (1995). Effects on ethanol withdrawal hyperexcitability of chronic treatment with a competitive *N*-methyl-D-aspartate receptor antagonist. *J Pharmacol Exp Ther*, 272:112-118.

327. Ripley, T.L., Dunworth, S.J., and Stephens, D.N. (2002). Effect of CGP39551 administration on the kindling of ethanol-withdrawal seizures. *Psychopharmacology*, 163(2):157-165.

328. Veatch, L.M. and Becker, H.C. (2005). Lorazepam and MK-801 effects on behavioral and electrographic indices of alcohol withdrawal sensitization. *Brain Res*, 1065(1-2):92-106.

329. Lukoyanov, N.V. and Paula-Barbosa, M.M. (2001). Memantine, but not dizocilpine, ameliorates cognitive deficits in adult rats withdrawn from chronic ingestion of alcohol. *Neurosci Lett*, 309(1):45-48.

330. Spanagel, R., Putzke, J., Stefferl, A., Schobitz, B., and Zieglgansberger, W. (1996). Acamprosate and alcohol. 2: Effects on alcohol withdrawal in the rat. *Eur J Pharmacol*, 305(1-3):45-50.

331. Staner, L., Boeijinga, P., Danel, T., Gendre, I., Muzet, M., Landron, F., and Luthringer, R. (2006). Effects of acamprosate on sleep during alcohol withdrawal: A double-blind placebo-controlled polysomnographic study in alcohol-dependent subjects. *Alcohol Clin Exp Res*, 30(9):1492-1499.

332. Dooley, D.J., Taylor, C.P., Donevan, S., and Feltner, D. (2007). Ca^{2+} channel alpha(2)delta ligands: Novel modulators of neurotransmission. *Trends Pharmacol Sci*, 28(2):75-82.

333. Taylor, C.P., Gee, N.S., Su, T.Z., Kocsis, J.D., Welty, D.F., Brown, J.P., Dooley, D.J., Boden, P., and Singh, L. (1998). Summary of mechanistic hypotheses of gabapentin pharmacology. *Epil Res*, 29(3):233-249.

334. Mariani, J.J., Rosenthal, R.N., Tross, S., Singh, P., and Anand, O.P. (2006). A randomized, open-label, controlled trial of gabapentin and phenobarbital in the treatment of alcohol withdrawal. *Am J Addiction*, 15:76-84.

335. Watson, W.P., Robinson, E., and Little, H.J. (1997). Gabapentin prevents the anxiogenic as well as convulsive aspects of the ethanol withdrawal syndrome. *Neuropharmacology*, 36:1369-1375.

336. Bisaga, A. and Evans, S.M. (2006). The acute effects of gabapentin in combination with alcohol in heavy drinkers. *Drug Alc Dep*, 83:25-32.

337. Blake, M.G., Boccia, M.M., Acosta, G.B., and Baratti, C.M. (2004). Posttraining administration of pentylenetetrazol dissociates gabapentin effects on memory consolidation from that on memory retrieval process in mice. *Neurosci Lett*, 368:211-215.

338. Boccia, M.M., Acosta, G.B., and Baratti, C.M. (2001). Memory improving actions of gabapentin in mice: Possible involvement of central muscarinic cholinergic mechanism. *Neurosci Lett*, 311:153-156.

339. Baydas, G., Sonkaya, E., Tuzcu, M., Yasar, A., and Donder, E. (2005). Novel role for gabapentin in neuroprotection of central nervous system in streptozotocine-induced diabetic rats. *Acta Pharmacol Sin*, 26:417-422.

340. Shank, R.P., Gardocki, J.F., Streeter, A.J., and Maryanoff, B.E. (2000). An overview of the preclinical aspects of topiramate: Pharmacology, pharmacokinetics, and mechanism of action. *Epilepsia*, 41(Suppl 1):S3-S9.

341. Moak, D.H. (2004). Assessing the efficacy of medical treatments for alcohol use disorders. *Expert Opin Pharmacother*, 5:2075–2089.

342. Cagetti, E., Baicy, K.J., and Olsen, R.W. (2004). Topiramate attenuates withdrawal signs after chronic intermittent ethanol in rats. *Neuroreport*, 15:207–210.

343. Nguyen, S.A., Malcolm, R., and Middaugh, L.D. (2007). Topiramate reduces ethanol consumption by C57BL/6 mice. *Synapse*, 61:150–156.

344. Kudin, A.P., Debska-Vielhaber, G., Vielhaber, S., Elger, C.E., and Kunz, W.S. (2004). The mechanism of neuroprotection by topiramate in an animal model of epilepsy. *Epilepsia*, 45:1478–1487.

345. Noh, M.R., Kim, S.K., Sun, W., Park, S.K., Choi, H.C., Lim, J.H., Kim, I.H., Kim, H.J., Kim, H., and Eun, B.L. (2006). Neuroprotective effect of topiramate on hypoxic ischemic brain injury in neonatal rats. *Exp Neurol*, 201:470–478.

346. Sfaello, I., Baud, O., Arzimanoglou, A., and Gressens, P. (2005). Topiramate prevents excitotoxic damage in the newborn rodent brain. *Neurobiol Dis*, 20:837–848.

347. Knapp, C.M., Saird-Segal, O., Richardson, M.A., Afshar, M., Piechniczek-Buczek, J., Koplow, J., and Ciraulo, D.A. (2006). Effects of topiramate on alcohol use and cognitive function in alcohol dependence. *Alc Clin Exp Res*, 30(Suppl S):107A.

348. Rubio, G., Ponce, G., Jimenez-Arriero, M.A., Palomo, T., Manzanares, J., and Ferre, F. (2004). Effects of topiramate in the treatment of alcohol dependence. *Pharmacopsychiatry*, 37:37–40.

349. Seifert, J., Peters, E., Jahn, K., Metzner, C., Ohlmeier, M., Wildt, B.T., Emrich, H.M., and Schneider, U. (2004). Treatment of alcohol withdrawal: Chlormethiazole vs. carbamazepine and the effect on memory performance – a pilot study. *Addiction Biol*, 9:43–51.

350. Malcolm, R., Myrick, H., Roberts, J., Wang, W., and Anton, R.F. (2002). The differential effects of medication on mood, sleep disturbance, and work ability in outpatient alcohol detoxification. *Am J Addiction*, 11:141–150.

351. Ambrosio, A.F., Silva, A.P., Araujo, I., Malva, J.O., Soares-da-Silva, P., Carvalho, A.P., and Carvalho, C.M. (2000). Neurotoxic/neuroprotective profile of carbamazepine, oxcarbazepine and two new putative antiepileptic drugs, BIA 2-093 and BIA 2-024. *Eur J Pharmacol*, 406:191–201.

352. Pitkanen, A., Tuunanen, J., and Halonen, T. (1996). Vigabatrin and carbamazepine have different efficacies in the prevention of status epilepticus induced neuronal damage in the hippocampus and amygdale. *Epil Res*, 24:29–45.

353. Mueller, T.I., Stout, R.L., Rudden, S., Brown, R.A., Gordon, A., Solomon, D.A., and Recupero, P.R. (1997). A double-blind, placebo-controlled pilot study of carbamazepine for the treatment of alcohol dependence. *Alcohol Clin Exp Res*, 21:86–92.

354. Strzelec, J.S. and Czarnecka, E. (2001). Influence of clonazepam and carbamazepine on alcohol withdrawal syndrome, preference and development of tolerance to ethanol in rats. *Pol J Pharmacol*, 53:117–124.

355. Tzschentke, T.M. (2007). Measuring reward with the conditioned place preference (CPP) paradigm: Update of the last decade. *Addiction Biol*, 12:227–462.

356. Agmo, A. and Marroquin, E. (1997). Role of gustatory and postingestive actions of sweeteners in the generation of positive affect as evaluated by place preference conditioning. *Appetite*, 29(3):269–289.

357. Oldenburger, W.P., Everitt, B.J., and Dejonge, F.H. (1992). Conditioned place preference induced by sexual interaction in female rats. *Horm Behav*, 26(2):214–228.

358. White, N. (1989). Reward or reinforcement: What's the difference? *Neurosci Biobehav Rev*, 13:181–186.

359. Bardo, M.T. and Bevins, R.A. (2000). Conditioned place preference: What does it add to our preclinical understanding of drug reward? *Psychopharmacology*, 153(1):31–43.

360. Schechter, M.D. and Calcagnetti, D.J. (1998). Continued trends in the conditioned place preference literature from 1992 to 1996, inclusive, with a cross-indexed bibliography. *Neurosci Biobehav Rev*, 22(6):827–846.

361. Tzschentke, T.M. (1998). Measuring reward with the conditioned place preference paradigm: A comprehensive review of drug effects, recent progress and new issues. *Prog Neurobiol*, 56:613–672.

362. Cunningham, C.L., Clemans, J.M., and Fidler, T.L. (2002). Injection timing determines whether intragastric ethanol produces conditioned place preference or aversion in mice. *Pharmacol Biochem Behav*, 72(3):659–668.

363. Hill, K.G., Ryabinin, A.E., and Cunningham, C.L. (2007). FOS expression induced by an ethanol-paired conditioned stimulus. *Pharmacol Biochem Behav*, 87(2):208–221.

364. Cunningham, C.L. and Prather, L.K. (1992). Conditioning trial duration affects ethanol-induced conditioned place preference in mice. *Anim Learn Behav*, 20(2):187–194.

365. Morse, A.C., Schulteis, G., Holloway, F.A., and Koob, G.F. (2000). Conditioned place aversion to the "hangover" phase of acute ethanol administration in the rat. *Alcohol*, 22(1):19–24.

366. Cunningham, C.L. (1981). Spatial aversion conditioning with ethanol. *Pharmacol Biochem Behav*, 14(2):263–264.

367. Cunningham, C.L., Niehus, J.S., and Noble, D. (1993). Species difference in sensitivity to ethanols hedonic effects. *Alcohol*, 10(2):97–102.

368. Stewart, R.B. and Grupp, L.A. (1986). An investigation of the interaction between the reinforcing properties of food and ethanol using the place preference paradigm. *Prog Neuropsychopharmacol*, 5:609–613.

369. Bienkowski, P., Kuka, P., Piasecki, J., and Kostowski, Q. (1996). Low doses of ethanol induce conditioned place preference in rats after repeated exposures to ethanol or saline injections. *Alcohol Alcohol*, 31:547–553.

370. Bozarth, M.A. (1990). Evidence for the rewarding effects of ethanol using the conditioned place preference method. *Pharmacol Biochem Behav*, 35(2):485–487.

371. Biala, G. and Kotlinska, J. (1999). Blockade of the acquisition of ethanol-induced conditioned place preference by *N*-methyl-D-aspartate receptor antagonists. *Alcohol Alcohol*, 34(2):175–182.

372. Matsuzawa, S., Suzuki, T., and Misawa, M. (1998a). Conditioned fear stress induces ethanol-associated place preference in rats. *Eur J Pharmacol*, 341(2–3):127–130.

373. Matsuzawa, S., Suzuki, T., Misawa, M., and Nagase, H. (1998b). Involvement of mu- and delta-opioid receptors in the ethanol-associated place preference in rats exposed to foot shock stress. *Brain Res*, 803(1–2):169–177.

374. Matsuzawa, S., Suzuki, T., Misawa, M., and Nagase, H. (1999a). Different roles of mu-, delta- and kappa-opioid receptors in ethanol-associated place preference in rats exposed to conditioned fear stress. *Eur J Pharmacol*, 368(1):9–16.

375. Matsuzawa, S., Suzuki, T., Misawa, M., and Nagase, H. (1999b). Involvement of dopamine D-1 and D-2 receptors in the ethanol-associated place preference in rats exposed to conditioned fear stress. *Brain Res*, 835:298–305.

376. Matsuzawa, S. and Suzuki, M. (2000). Ethanol, but not the anxiolytic drugs buspirone and diazepam, produces a conditioned place preference in rats exposed to conditioned fear stress. *Pharmacol Biochem Behav*, 65(2):281–288.

377. Der-Avakian, A., Bland, S.T., Rozeske, R.R., Tamblyn, J.P., Hutchinson, M.R., and Watkins, L.R. (2007). The effects of a single exposure to uncontrollable stress on the subsequent conditioned place preference responses to oxycodone, cocaine, and ethanol in rats. *Psychopharmacology*, 191(4):909–917.

378. Philpot, R.M., Badanich, K.A., and Kirstein, C.L. (2003). Place conditioning: Age-related changes in the rewarding and aversive effects of alcohol. *Alcohol Clin Exp Res*, 27(4):593–599.

379. Walker, B.M. and Ettenberg, A. (2007). Intracerebroventricular ethanol-induced conditioned place preferences are prevented by fluphenazine infusions into the nucleus accumbens of rats. *Behav Neurosci*, 121(2):401–410.

380. Bienkowski, P., Kuka, P., and Kostowski, Q. (1995). Conditioned place preference after prolonged exposure to ethanol. *Pol J Pharmacol*, 47:185–187.

381. Gauvin, D.V. and Holloway, F.A. (1992). Historical factors in the development of etoh-conditioned place preference. *Alcohol*, 9(1):1–7.

382. Ciccocioppo, R., Panocka, I., Froldi, R., Quitadamo, E., and Massi, M. (1999). Ethanol induces conditioned place preference in genetically selected alcohol-preferring rats. *Psychopharmacology*, 141(3):235–241.

383. Mueller, D., Perdikaris, D., and Stewart, J. (2002). Persistence and drug-induced reinstatement of a morphine-induced conditioned place preference. *Behav Brain Res*, 136(2):389–397.

384. Kuzmin, A., Sandin, J., Terenius, L., and Ogren, S.O. (2003). Acquisition, expression, and reinstatement of ethanol-induced conditioned place preference in mice: Effects of opioid receptor-like 1 receptor agonists and naloxone. *J Pharmacol Exp Ther*, 304(1):310–318.

385. Kily, L., Cowe, Y., Michael-Titus, A., Brennan, C. (2007). A conditioned place preference assay for modelling addiction in zebrafish. *Br J Pharmacol* (in press).

386. Perks, S.M. and Clifton, P.G. (1997). Reinforcer revaluation and conditioned place preference. *Physiol Behav*, 61(1):1–5.

387. Cunningham, C.L., Young, E.A. (2007) Ethanol conditioned place preference is resistant to ethanol devaluation by lithium chloride. *Alc Clin Exp Res*, 31:150A.

388. Vazquez, V., Weiss, S., Giros, B., Martres, M.P., and Dauge, V. (2007). Maternal deprivation and handling modify the effect of the dopamine D3 receptor agonist, BP 897 on morphine-conditioned place preference in rats. *Psychopharmacology*, 193(4):475–486.

389. Funk, D., Vohra, S., and Le, A.D. (2004). Influence of stressors on the rewarding effects of alcohol in Wistar rats: Studies with alcohol deprivation and place conditioning. *Psychopharmacology*, 176(1):82–87.

390. Cunningham, C.L., Henderson, C.M., and Bormann, N.M. (1998). Extinction of ethanol-induced conditioned place preference and conditioned place aversion: Effects of naloxone. *Psychopharmacology*, 139:62–70.

391. Middaugh, L.D. and Bandy, A.L.E. (2000). Naltrexone effects on ethanol consumption and response to ethanol conditioned cues in C57BL/6 mice. *Psychopharmacology*, 151(4):321–327.

392. Bechtholt, A.J. and Cunningham, C.L. (2005). Ethanol-induced conditioned place preference is expressed through a ventral tegmental area dependent mechanism. *Behav Neurosci*, 119(1):213–223.

393. Hall, F.S., Sora, I., and Uhl, G.R. (2001). Ethanol consumption and reward are decreased in mu-opiate receptor knockout mice. *Psychopharmacology*, 154(1):43–49.

394. Colombo, G. (1997). ESBRA-Nordmann 1996 award lecture: Ethanol drinking behaviour in Sardinian alcohol-preferring rats. *Alcohol Alcohol*, 32:443–453.

395. McGeehan, A.J. and Olive, M.F. (2003). The anti-relapse compound acamprosate inhibits the development of a conditioned place preference to ethanol and cocaine but not morphine. *Br J Pharmacol*, 138(1):9–12.

396. Cole, J., Littleton, J.M., and Little, H.J. (2000). Acamprosate, but not naltrexone, inhibits conditioned abstinence behaviour associated with repeated ethanol administration and exposure to a plus maze. *Psychopharmacology*, 147:403–411.

397. Quertemont, E., Brabant, C., and De Witte, P. (2002). Acamprosate reduces context-dependent ethanol effects. *Psychopharmacology*, 164(1):10–18.

398. Risinger, F.O. and Boyce, J.M. (2002). 5-HT1A receptor blockade and the motivational profile of ethanol. *Life Sci*, 71(6):707–715.

399. Risinger, F.O. and Oakes, R.A. (1996). Mianserin enhancement of ethanol-induced conditioned place preference. *Behav Pharmacol*, 7(3):294–298.

400. Dickinson, S.D., Lee, E.L., Rindal, K., and Cunningham, C.L. (2003). Lack of effect of dopamine receptor blockade on expression of ethanol-induced conditioned place preference in mice. *Psychopharmacology*, 165(3):238–244.

401. Boyce-Rustay, J.M. and Risinger, F.O. (2003). Dopamine D3 receptor knockout mice and the motivational effects of ethanol. *Pharmacol Biochem Behav*, 75(2):373–379.

402. Boyce, J.M. and Risinger, F.O. (2000). Enhancement of ethanol reward by dopamine D3 receptor blockade. *Brain Res*, 880(1–2):202–206.

403. Cunningham, C.L., Howard, M.A., Gill, S.J., Rubinstein, M., Low, M.J., and Grandy, D.K. (2000). Ethanol-conditioned place preference is reduced in dopamine D2 receptor-deficient mice. *Pharmacol Biochem Behav*, 67(4):693–699.

404. Aguayo, L.G., Pancetti, F.C., Klein, R.L., and Harris, R.A. (1994). Differential effects of gabaergic ligands in mouse and rat hippocampal neurons. *Brain Res*, 647:97–105.

405. Chester, J.A. and Cunningham, C.L. (1999a). GABAA recepors modulate ethanol-induced conditioned place preference and taste aversion in mice. *Psychopharmacology*, 144:363–372.

406. Blednov, Y.A., Walker, D., Alva, H., Creech, K., Findlay, G., and Harris, R.A. (2003). GABAA receptor alpha 1 and beta 2 subunit null mutant mice: Behavioral responses to ethanol. *J Pharmacol Exp Ther*, 305:854–863.

407. Chester, J.A. and Cunningham, C.L. (1999b). Baclofen alters ethanol-stimulated activity but not conditioned place preference or taste aversion in mice. *Pharmacol Biochem Behav*, 63:325–331.

408. Lominac, K.D., Kapasova, Z., Hannun, R.A., Patterson, C., Middaugh, L.D., and Szumlinski, K.K. (2006). Behavioral and neurochemical interactions between Group 1 mGluR antagonists and ethanol: Potential insight into their anti-addictive properties. *Drug Alcohol Depend*, 85(2):142–156.

409. Kotlinska, J., Biala, G., Rafalski, P., Bochenski, M., and Danysz, W. (2004). Effect of neramexane on ethanol dependence and reinforcement. *Eur J Pharmacol*, 503:95–98.

410. Boyce-Rustay, J.M. and Cunningham, C.L. (2004). The role of NMDA receptor binding sites in ethanol place conditioning. *Behav Neurosci*, 118(4):822–834.

411. Houchi, H., Babovic, D., Pierrefiche, O., Ledent, C., Daoust, M., and Naassila, M.L. (2005). CB1 receptor knockout mice display reduced ethanol-induced conditioned place preference and increased striatal dopamine D2 receptors. *Neuropsychopharmacology*, 30(2):339–349.

412. Thanos, P.K., Dimitrakakis, E.S., Rice, O., Gifford, A., and Volkow, N.D. (2005). Ethanol self-administration and ethanol conditioned place preference are reduced in mice lacking cannabinoid CB1 receptors. *Behav Brain Res*, 164(2):206–213.

413. Thanos, P.K., Rivera, S.N., Weaver, K., Grandy, D.K., Rubinstein, M., Umegaki, H., Wang, G.J., Hitzemann, R., and Volkow, N.D. (2005). Dopamine D2R DNA transfer in dopamine D2 receptor-deficient mice: Effects on ethanol drinking. *Life Sci*, 77:130–139.

414. Gremel, C.M., Gabriel, K.I., and Cunningham, C.L. (2006). Topiramate does not affect the acquisition or expression of ethanol conditioned place preference in DBA/2J or C57BL/6J mice. *Alcohol Clin Exp Res*, 30(5):783–790.

415. Phillips, T.J., Broadbent, J., Burkhart-Kasch, S., Henderson, C., Wenger, C.D., McMullin, C., McKinnon, C.S., and Cunningham, C.L. (2005). Genetic correlational analyses of ethanol reward and aversion phenotypes in short-term selected mouse lines bred for ethanol drinking or ethanol-induced conditioned taste aversion. *Behav Neurosci*, 119(4):892–910.

416. Grahame, N.J., Chester, J.A., Rodd-Henricks, K., Li, T.K., and Lumeng, L. (2001). Alcohol place preference conditioning in high- and low-alcohol preferring selected lines of mice. *Pharmacol Biochem Behav*, 68:805–814.

417. Chester, J.A., Risinger, F.O., and Cunningham, C.L. (1998). Ethanol reward and aversion in mice bred for sensitivity to ethanol withdrawal. *Alcohol Clin Exp Res*, 22(2):468–473.

418. Files, F.J., Samson, H.H., and Brice, G.T. (1995). Sucrose, ethanol, and sucrose/ethanol reinforced responding under variable-interval schedules of reinforcement. *Alcohol Clin Exp Res*, 19(5):1271–1278.

419. Slawecki, C.J. and Samson, H.H. (1997). Changes in oral ethanol self-administration patterns resulting from ethanol concentration manipulations. *Alcohol Clin Exp Res*, 21(6):1144–1149.

420. Czachowski, C.L. and Samson, H.H. (1999). Breakpoint determination and ethanol self-administration using an across-session progressive ratio procedure in the rat. *Alcohol Clin Exp Res*, 23(10):1580–1586.

421. Samson, H.H., Czachowski, C.L., and Slawecki, C.J. (2000). A new assessment of the ability of oral ethanol to function as a reinforcing stimulus. *Alcohol Clin Exp Res*, 24(6):766–773.

422. Samson, H.H., Czachowski, C.L., Chappell, A., and Legg, B. (2003). Measuring the appetitive strength of ethanol: Use of an extinction trial procedure. *Alcohol*, 31(1–2):77–86.

423. Podlesnik, C.A., Jimenez-Gomez, C., and Shahan, T.A. (2006). Resurgence of alcohol seeking produced by discontinuing non-drug reinforcement as an animal model of drug relapse. *Behav Pharmacol*, 17(4):369–374.

424. Koros, E., Kostowski, W., and Bienkowski, P. (1999b). Operant responding for ethanol in rats with a long-term history of free-choice ethanol drinking. *Alcohol Alcohol*, 34(5):685–689.

425. Files, F.J., Samson, H.H., and Denning, C.E. (2000). Effects of prior ethanol exposure on ethanol self-administration in a continuous access situation using retractable drinking tubes. *Alcohol*, 21:97–102.

426. Schulteis, G., Hyytia, P., Heinrichs, S.C., and Koob, G.F. (1996). Effects of chronic ethanol exposure on oral self-administration of ethanol or saccharin by Wistar rats. *Alcohol Clin Exp Res*, 20(1):164–171.

427. Roberts, A.J., Cole, M., and Koob, G.F. (1996). Intra-amygdala muscimol decreases operant ethanol self-administration in dependent rats. *Alcohol Clin Exp Res*, 20(7):1289–1298.

428. Brown, G., Jackson, A., and Stephens, D.N. (1998). Effects of repeated withdrawal from chronic ethanol on oral self-administration of ethanol on a progressive ratio schedule. *Behav Pharmacol*, 9(2):149–161.

429. O'Dell, L.E., Roberts, A.J., Smith, R.T., and Koob, G.F. (2004). Enhanced alcohol self-administration after intermittent versus continuous alcohol vapor exposure. *Alcohol Clin Exp Res*, 28(11):1676–1682.

430. Rodd, Z.A., Bell, R.L., McQueen, V.K., Davids, M.R., Hsu, C.C., Murphy, J.M., Li, T.K., Lumeng, L., and McBride, W.J. (2005a). Prolonged increase in the sensitivity of the posterior ventral tegmental area to the reinforcing effects of ethanol following repeated exposure to cycles of ethanol access and deprivation. *J Pharmacol Exp Ther*, 315(2):648–657.

431. Roberts, A.J., Heyser, C.J., Cole, M., Griffin, P., and Koob, G.F. (2000). Excessive ethanol drinking following a history of dependence: Animal model of allostasis. *Neuropsychopharmacology*, 22(6):581–594.

432. Roberts, A.J., McDonald, J.S., Heyser, C.J., Kieffer, B.L., Matthes, H.W., Koob, G.F., and Gold, L.H. (2000). mu-Opioid receptor knockout mice do not self-administer alcohol. *J Pharmacol Exp Ther*, 293:1002–1008.

433. Valdez, G.R., Roberts, A.J., Chan, K., Davis, H., Brennan, M., Zorrilla, E.P., and Koob, G.F. (2002). Increased ethanol self-administration and anxiety-like behavior during acute ethanol withdrawal and protracted abstinence: Regulation by corticotropin-releasing factor. *Alcohol Clin Exp Res*, 26(10):1494–1501.

434. Valdez, G.R., Sabino, V., and Koob, G.F. (2004). Increased anxiety-like behavior and ethanol self-administration in dependent rats: Reversal via corticotropin-releasing factor-2 receptor activation. *Alcohol Clin Exp Res*, 28(6):865–872.

435. Middaugh, L.D., Lee, A.M., and Bandy, A.L.E. (2000). Ethanol reinforcement in nondeprived mice: Effects of abstinence and naltrexone. *Alcohol Clin Exp Res*, 24(8):1172–1179.

436. Oster, S.M., Toalston, J.E., Kuc, K.A., Pommer, T.J., Murphy, J.M., Lumeng, L., Bell, R.L., McBride, W.J., and Rodd, Z.A. (2006). Effects of multiple alcohol deprivations on operant ethanol self-administration by high-alcohol-drinking replicate rat lines. *Alcohol*, 38(3):155–164.

437. Lê, A.D., Israel, Y., Juzytsch, W., Quan, B. and Harding, S. (2001). Genetic selection for high and low alcohol consumption in a limited-access paradigm. *Alcohol Clin Exp Res*, 25:1613–1620.

438. Epstein, D.H., Preston, K.L., Stewart, J., and Shaham, Y. (2006). Toward a model of drug relapse: An assessment of the validity of the reinstatement procedure. *Psychopharmacology*, 189(1):1–16.

439. McBride, W.J., Le, A.D., and Noronha, A. (2002). Central nervous system mechanisms in alcohol relapse. *Alcohol Clin Exp Res*, 26(2):280–286.

440. Bienkowski, P., Kostowski, W., and Koros, E. (1999). Ethanol-reinforced behaviour in the rat: Effects of naltrexone. *Eur J Pharmacol*, 374:321–327.

441. Bienkowski, P., Kostowski, W., and Koros, E. (1999). The role of drug-paired stimuli in extinction and reinstatement of ethanol-seeking behaviour in the rat. *Eur J Pharmacol*, 374(3):315–319.

442. Bienkowski, P., Koros, E., Kostowski, W., and Bogucka-Bonikowska, A. (2000). Reinstatement of ethanol seeking in rats: Behavioral analysis. *Pharmacol Biochem Behav*, 66(1):123–128.

443. Le, A.D., Quan, B., Juzytch, W., Fletcher, P.J., Joharchi, N., and Shaham, Y. (1998). Reinstatement of alcohol-seeking by priming injections of alcohol and exposure to stress in rats. *Psychopharmacology*, 135(2):169–174.

444. Samson, H.H. and Chappell, A. (2002). Reinstatement of ethanol seeking responding after ethanol self-administration. *Alcohol*, 26(2):95–101.

445. Gass, J.T. and Olive, M.F. (2007). Reinstatement of ethanol-seeking behavior following intravenous self-administration in Wistar rats. *Alcohol Clin Exp Res*, 31(9):1441–1445.

446. Bienkowski, P., Rogowski, A., Korkosz, A., Mierzejewski, P., Radwanska, K., Kaczmarek, L., Bogucka-Bonikowska, A., and Kostowski, W. (2004). Time-dependent changes in alcohol-seeking behaviour during abstinence. *Eur Neuropsychopharmacol*, 14(5):355–360.

447. Stasiewicz, P.R., Brandon, T.H., and Bradizza, C.M. (2007). Effects of extinction context and retrieval cues on renewal of alcohol-cue reactivity among alcohol-dependent outpatients. *Psychol Addict Behav*, 21(2):244–248.

448. Rodd-Henricks, Z.A., Bell, R.L., Kuc, K.A., Murphy, J.M., McBride, W.J., Lumeng, L., and Li, T.K. (2002b). Effects of ethanol exposure on subsequent acquisition and extinction of ethanol self-administration and expression of alcohol-seeking behavior in adult alcohol-preferring (P) rats. I: Periadolescent exposure. *Alcohol Clin Exp Res*, 26(11):1632–1641.

449. Meachum, C.L. (1988). Toxicosis affects instrumental behavior in rats. *Q J Exp Psychol B*, 40(3):209–228.

450. Miles, F.J., Everitt, B.J., and Dickinson, A. (2003). Oral cocaine seeking by rats: Action or habit? *Behav Neurosci*, 117(5):927–938.

451. Dickinson, A., Wood, N., and Smith, J.W. (2002). Alcohol seeking by rats: Action or habit? *Q J Exp Psychol B*, 55(4):331–348.

452. Samson, H.H., Cunningham, C.L., Czachowski, C.L., Chappell, A., Legg, B., and Shannon, E. (2004). Devaluation of ethanol reinforcement. *Alcohol*, 32(3):203–212.

453. Van Erp, A.M.M. and Miczek, K.A. (2001). Persistent suppression of ethanol self-administration by brief social stress in rats and increased startle response as index of withdrawal. *Physiol Behav*, 73(3):301–311.

454. Le, A.D., Harding, S., Juzytsch, W., Watchus, J., Shalev, U., and Shaham, Y. (2000). The role of corticotrophin-releasing factor in stress-induced relapse to alcohol-seeking behavior in rats. *Psychopharmacology*, 150(3):317–324.

455. Liu, X. and Weiss, F. (2002). Additive effect of stress and drug cues on reinstatement of ethanol seeking: Exacerbation by history of dependence and role of concurrent activation of corticotropin-releasing factor and opioid mechanisms. *J Neuroscie*, 22(18):7856–7861.

456. Hyytia, P. and Sinclair, J.D. (1993). Responding for oral ethanol after naloxone treatment by alcohol-preferring AA rats. *Alcohol Clin Exp Res*, 17(3):631–636.

457. Sharpe, A.L. and Samson, H.H. (2003). Ethanol and sucrose self-administration components: Effects of drinking history. *Alcohol*, 29(1):31–38.

458. Burattini, C., Gill, T.M., Aicardi, G., and Janak, P.H. (2006). The ethanol self-administration context as a reinstatement cue: Acute effects of naltrexone. *Neuroscience*, 139:877–887.

459. Heidbreder, C.A., Andreoli, M., Marcon, C., Hutcheson, D.M., Gardner, E.L., and Ashby, C.R., Jr. (2007). Evidence for the role of dopamine D3 receptors in oral operant alcohol self-administration and reinstatement of alcohol-seeking behavior in mice. *Addiction Biol*, 12:35–50.

460. Katner, S.N., Magalong, J.G., and Weiss, F. (1999). Reinstatement of alcohol-seeking behavior by drug-associated discriminative stimuli after prolonged extinction in the rat. *Neuropsychopharmacology*, 20:471–479.

461. Gonzales, R.A. and Weiss, F. (1998). Suppression of ethanol-reinforced behavior by naltrexone is associated with attenuation of the ethanol-induced increase in dialysate dopamine levels in the nucleus accumbens. *J Neurosci*, 18(24):10663–10671.

462. Heyser, C.J., Schulteis, G., Durbin, P., and Koob, G.F. (1998). Chronic acamprosate eliminates the alcohol deprivation effect while having limited effects on baseline responding for ethanol in rats. *Neuropsychopharmacology*, 18:125–133.

463. Bachteler, D., Economidou, D., Danysz, W., Ciccocioppo, R., and Spanagel, R. (2005). The effects of acamprosate and neramexane on cue-induced reinstatement of ethanol-seeking behavior in rat. *Neuropsychopharmacology*, 30(6):1104–1110.

464. Czachowski, C.L., Legg, B.H., and Samson, H.H. (2001). Effects of acamprosate on ethanol-seeking and self-administration in the rat. *Alcohol Clin Exp Res*, 25(3):344–350.

465. Le, A.D., Poulos, C.X., Harding, S., Watchus, J., Juzytsch, W., and Shaham, Y. (1999). Effects of naltrexone and fluoxetine on alcohol self-administration and reinstatement of alcohol seeking induced by priming injections of alcohol and exposure to stress. *Neuropsychopharmacology*, 21(3):435–444.

466. Ginsburg, B.C. and Lamb, R.J. (2006). Fluvoxamine effects on concurrent ethanol- and food-maintained behaviors. *Exp Clin Psychopharmacol*, 14(4):483–492.

467. Hodge, C.W., Samson, H.H., Lewis, R.S., and Erickson, H.L. (1993). Specific decreases in ethanol- but not water-reinforced responding produced by the 5-HT3 antagonist ICS 205-930. *Alcohol*, 10(3):191–196.

468. McKinzie, D.L., McBride, W.J., Murphy, J.M., Lumeng, L., and Li, T.K. (2000). Effects of MDL 72222, a serotonin(3) antagonist, on operant responding for ethanol by alcohol-preferring P rats. *Alcohol Clin Exp Res*, 24(10):1500–1504.

469. Le, A.D., Funk, D., Harding, S., Juzytsch, W., Fletcher, P.J., and Shaham, Y. (2006). Effects of dexfenfluramine and 5-HT3 receptor antagonists on stress-induced reinstatement of alcohol seeking in rats. *Psychopharmacology*, 186(1):82–92.

470. Tomkins, D.M. and O'Neill, M.F. (2000). Effect of 5-HT1B receptor ligands on self-administration of ethanol in an operant procedure in rats. *Pharmacol Biochem Behav*, 66(1):129–136.

471. Wilson, A.W., Neill, J.C., and Costall, B. (1998). An investigation into the effects of 5-HT agonists and receptor antagonists on ethanol self-administration in the rat. *Alcohol*, 16(3):249–270.

472. Tomkins, D.M., Joharchi, N., Tampakeras, M., Martin, J.R., Wichmann, J., and Higgins, G.A. (2002). An investigation of the role of 5-HT2C receptors in modifying ethanol self-administration behaviour. *Pharmacol Biochem Behav*, 71(4):735–744.

473. Gallate, J.E., Mallet, P.E., and McGregor, I.S. (2004). Combined low dose treatment with opioid and cannabinoid receptor antagonists synergistically reduces the motivation to consume alcohol in rats. *Psychopharmacology*, 173(1–2):210–216.

474. Freedland, C.S., Poston, J.S., and Porrino, L.J. (2000). Effects of SR141716A, a central cannabinoid receptor antagonist, on food-maintained responding. *Pharmacol Biochem Behav*, 67(2):265–270.

475. Economidou, D., Mattioli, L., Cifani, C., Perfumi, M., Massi, M., Cuomo, V., Trabace, L., and Ciccocioppo, R. (2006). Effect of the cannabinoid CB1 receptor antagonist SR-141716A on ethanol self-administration and ethanol-seeking behaviour in rats. *Psychopharmacology*, 183(4):394–403.

476. Cippitelli, A., Bilbao, A., Hansson, A.C., del Arco, I., Sommer, W., Heilig, M., Massi, M., Bermudez-Silva, F.J., Navarro, M., Ciccocioppo, R., and de Fonseca, F.R. (2005). Cannabinoid CB1 receptor antagonism reduces conditioned reinstatement of ethanol-seeking behavior in rats. *Eur J Neurosci*, 21(8):2243–2251.

477. Lopez-Moreno, J.A., Gonzalez-Cuevas, C., de Fonseca, F.R., and Navarro, M. (2004). Long-lasting increase of alcohol relapse by the cannabinoid receptor agonist WIN 55,212-2 during alcohol deprivation. *J Neurosci*, 24(38):8245–8252.

478. Lopez-Moreno, J.A., Gonzalez-Cuevas, G., and Navarro, M. (2007). The CB1 cannabinoid receptor antagonist rimonabant chronically prevents the nicotine-induced relapse to alcohol. *Neurobiol Dis*, 25(2):274–283.

479. Schroeder, J.P., Overstreet, D.H., and Hodge, C.W. (2005). The mGluR5 antagonist MPEP decreases operant ethanol self-administration during maintenance and after repeated alcohol deprivations in alcohol-preferring (P) rats. *Psychopharmacology*, 179(1):262–270.

480. Hodge, C.W., Miles, M.F., Sharko, A.C., Stevenson, R.A., Hillmann, J.R., Lepoutre, V., Besheer, J., and Schroeder, J.P. (2006). The mGluR5 antagonist MPEP selectively inhibits the onset and maintenance of ethanol self-administration in C57BL/6J mice. *Psychopharmacology*, 183(4):429–438.

481. Sanchis-Segura, C., Borchardt, T., Vengeliene, V., Zghoul, T., Bachteler, D., Gass, P., Sprengel, R., and Spanagel, R. (2006). Involvement of the AMPA receptor glur-C subunit in alcohol-seeking behavior and relapse. *J Neurosci*, 26(4):1231–1238.

482. Anstrom, K.K., Cromwell, H.C., Markowski, T., and Woodward, D.J. (2003). Effect of baclofen on alcohol and sucrose self-administration in rats. *Alcohol Clin Res Exp*, 27(6):900–908.

483. Besheer, J., Lepoutre, V., and Hodge, C.W. (2004). GABA(B) receptor agonists reduce operant ethanol self-administration and enhance ethanol sedation in C57BL/6J mice. *Psychopharmacology*, 174(3):358–366.

484. Maccioni, P., Serra, S., Vacca, G., Orru, A., Pes, D., Agabio, R., Addolorato, G., Carai, M.A.M., Gessa, G.L., and Colombo, G. (2005). Baclofen-induced reduction of alcohol reinforcement in alcohol-preferring rats. *Alcohol*, 36(3):161–168.

485. Czachowski, C.L., Legg, B.H., and Stansfield, K.H. (2006). Ethanol and sucrose seeking and consumption following repeated administration of the GABA(B) agonist baclofen in rats. *Alcohol Clin Exp Res*, 30(5):812–818.

486. Maccioni, P., Pes, D., Orru, A., Froestl, W., Gessa, G.L., Carai, M.A.M., and Colombo, G. (2007). Reducing effect of the positive allosteric modulator of the GABA(B) receptor, GS39783, on alcohol self-administration in alcohol-preferring rats. *Psychopharmacology*, 193(2):171–178.

487. Le, A.D., Harding, S., Juzytsch, W., Fletcher, P.J., and Shaham, Y. (2002). The role of corticotropin-releasing factor in the median raphe nucleus in relapse to alcohol. *J Neurosci*, 22(18):7844-7849.

488. Funk, C.K., O'Dell, L.E., Crawford, E.F., and Koob, G.F. (2006). Corticotropin-releasing factor within the central nucleus of the amygdala mediates enhanced ethanol self-administration in withdrawn, ethanol-dependent rats. *J Neurosci*, 26(44):11324-11332.

489. Funk, C.K., Zorrilla, E.P., Lee, M.J., Rice, K.C., and Koob, G.F. (2007). Corticotropin-releasing factor 1 antagonists selectively reduce ethanol self-administration in ethanol-dependent rats. *Biol Psychiatry*, 61(1):78-86.

490. Chu, K., Koob, G.F., Cole, M., Zorrilla, E.P., and Roberts, A.J. (2007). Dependence-induced increases in ethanol self-administration in mice are blocked by the CRF receptor antagonist antalarmin and by CRF receptor knockout. *Pharmacol Biochem Behav*, 86(4):813-821.

491. Walker, B.M. and Koob, G.F. (2007). The gamma-aminobutyric acid-B receptor agonist baclofen attenuates responding for ethanol in ethanol-dependent rats. *Alcohol Clin Exp Res*, 31(1):11-18.

492. De Fonseca, F.R., Roberts, A.J., Bilbao, A., Koob, G.F., and Navarro, M. (1999). Cannabinoid receptor antagonist SR141716A decreases operant ethanol self administration in rats exposed to ethanol-vapor chambers. *Acta Pharmacol Sin*, 20(12):1109-1114.

493. Rimondini, R., Thorsell, A., and Heilig, M. (2005). Suppression of ethanol self-administration by the neuropeptide Y (NPY) Y2 receptor antagonist BIIE0246: Evidence for sensitization in rats with a history of dependence. *Neurosci Lett*, 375(2):129-133.

494. Risinger, F.O., Brown, M.M., Doan, A.M., and Oakes, R.A. (1998). Mouse strain differences in oral operant ethanol reinforcement under continuous access conditions. *Alcohol Clin Exp Res*, 22:677-684.

495. Samson, H.H., Files, F.J., Denning, C., and Marvin, S. (1998). Comparison of alcohol-preferring and nonpreferring selectively bred rat lines. I: Ethanol initiation and limited access operant self-administration. *Alcohol Clin Exp Res*, 22(9):2133-2146.

496. Czachowski, C.L. and Samson, H.H. (2002). Ethanol- and sucrose-reinforced appetitive and consummatory responding in HAD1, HAD2, and P rats. *Alcohol Clin Exp Res*, 26(11):1653-1661.

497. Samson, H.H., Files, F.J., and Denning, C. (1999). Chronic ethanol self-administration in a continuous-access operant situation: The use of a sucrose/ethanol solution to increase daily ethanol intake. *Alcohol*, 19(2):151-155.

498. Enoch, M.A. (2006). Genetic and environmental influences on the development of alcoholism: Resilience vs. risk. *Ann NY Acad Sci*, 1094:193-201.

499. Pagan, J.L., Rose, R.J., Viken, R.J., Pulkkinen, L., Kaprio, J., and Dick, D.M. (2006). Genetic and environmental influences on stages of alcohol use across adolescence and into young adulthood. *Behav Genet*, 36:483-497.

500. Cloninger, C.R., Bohman, A., and Sigvardsson, S. (1981). Inheritance of alcohol abuse: Cross-fostering of adopted men. *Arch Gen Psychiatry*, 38:861-868.

501. Cloninger, C.R., Bohman, M., and Sigvardsson, S. (1981). Inheritance of alcohol abuse: Cross-fostering analysis of adopted men. *Arch Gen Psychiatry*, 38:861-868.

502. Dick, D.M. and Bierut, L.J. (2006). The genetics of alcohol dependence. *Curr Psychiatry Rep*, 8:151-157.

503. Heath, A.C., Bucholz, K.K., Madden, P.A., Dinwiddie, S.H., Slutske, W.S., Bierut, L.J., Statham, D.J., Dunne, M.P., Whitfield, J.B., and Martin, N.G. (1997). Genetic and environmental contributions to alcohol dependence risk in a national twin sample: Consistency of findings in women and men. *Psychol Med*, 27:1381-1396.

504. Schuckit, M.A. (1998). Biological, psychological and environmental predictors of the alcoholism risk: a longitudinal study. *J Stud Alcohol*, 59:485-494.

505. Schuckit, M.A. (1980). Self-rating of alcohol intoxication by young men with and without family histories of alcoholism. *J Stud Alcohol*, 41:242–249.

506. Schuckit, M.A. (1985). Ethanol-induced changes in body sway in men at high alcoholism risk. *Arch Gen Psychiatry*, 42:375–379.

507. Schuckit, M.A. (1988). Reactions to alcohol in sons of alcoholics and controls. *Alcohol Clin Exp Res*, 12:465–470.

508. Schuckit, M.A. (1994). Low level of response to alcohol as a predictor of future alcoholism. *Am J Psychiatry*, 151:184–189.

509. Kaplan, R.F., Hesselbrock, V.M., O'Connor, S., and Depalma, N. (1988). Behavioral and EEG responses to alcohol in nonalcoholic men with a family history of alcoholism. *Prog Neuropsychopharmacol Biol Psychiatry*, 12:873–885.

510. Nagoshi, C.T. and Wilson, J.R. (1988). One-month repeatability of emotional responses to alcohol. *Alcohol Clin Exp Res*, 12:691–697.

511. Newlin, D.B. and Thomson, J.B. (1990). Alcohol challenge with sons of alcoholics: A critical review and analysis. *Psychol Bull*, 108:383–402.

512. Lipscomb, T.R. and Nathan, P.E. (1980). Blood alcohol level discrimination. The effects of family history of alcoholism, drinking pattern, and tolerance. *Arch Gen Psychiatry*, 37:571–576.

513. O'Malley, S.S. and Maisto, A. (1985). Effects of family drinking history and expectancies on responses to alcohol in men. *J Stud Alcohol*, 46:289–297.

514. Crabbe, J.C., Kosobud, A., and Young, E.R. (1983). Genetic selection for ethanol withdrawal severity: Differences in replicate mouse lines. *Life Sci*, 33:955–962.

515. Goldstein, D.B. and Kakihana, R. (1975). Alcohol withdrawal reactions in mouse strains selectively bred for long or short sleep times. *Life Sci*, 17:981–985.

516. Crabbe, J.C., Feller, D.J., and Dorow, J.S. (1989). Sensitivity and tolerance to ethanol-induced hypothermia in genetically selected mice. *J Pharmacol Exp Ther*, 249:456–461.

517. Kurtz, D.L., Stewart, R.B., Zweifel, M., Li, T.K., and Froehlich, J.C. (1996). Genetic differences in tolerance and sensitization to the sedative/hypnotic effects of alcohol. *Pharmacol Biochem Behav*, 53:585–591.

518. Lumeng, L., Waller, M.B., McBride, W.J., and Li, T.K. (1982). Different sensitivities to ethanol in alcohol-preferring and -nonpreferring rats. *Pharmacol Biochem Behav*, 16:125–130.

519. Stewart, R.B., Kurtz, D.L., Zweifel, M., Li, T.K., and Froehlich, J.C. (1992). Differences in the hypothermic response to ethanol in rats selectively bred for oral ethanol preference and nonpreference. *Psychopharmacology (Berl)*, 106:169–174.

520. Tampier, L. and Quintanilla, M.E. (2002). Effect of a dose of ethanol on acute tolerance and ethanol consumption in alcohol drinker(UChB) and non-drinker (UChA) rats. *Addiction Biol*, 7:279–284.

521. Le, A.D. and Kiianmaa, K. (1988). Characteristics of ethanol tolerance in alcohol drinking (AA) and alcohol avoiding (ANA) rats. *Psychopharmacology (Berl)*, 94:479–483.

522. Suwaki, H., Kalant, H., Higuchi, S., Crabbe, J.C., Ohkuma, S., Katsura, M., Yoshimura, M., Stewart, R.C., Li, T.K., and Weiss, F. (2001). Recent research on alcohol tolerance and dependence. *Alcohol Clin Exp Res*, 25:189S–196S.

523. Colombo, G., Agabio, R., Carai, M.A., Lobina, C., Pani, M., Reali, R., Addolorato, G., and Gessa, G.L. (2000a). Ability of baclofen in reducing alcohol intake and withdrawal severity. I: Preclinical evidence. *Alcohol Clin Exp Res*, 24:58–66.

524. Colombo, G., Agabio, R., Carai, M.A., Lobina, C., Pani, M., Reali, R., Vacca, G., and Gessa, G.L. (2000b). Different sensitivity to ethanol in alcohol-preferring sP and -nonpreferring snp rats. *Alcohol Clin Exp Res*, 24:1603–1608.

525. Moore, J.A. and Kakihana, R. (1978). Ethanol-induced hypothermia in mice: Influence of genotype on development of tolerance. *Life Sci*, 23:2331–2337.

526. Shram, M.J., Bahroos, M., Beleskey, J.I., Tampakeras, M., Le, A.D., and Tomkins, D.M. (2004). Motor impairing effects of ethanol and diazepam in rats selectively bred for high and low ethanol consumption in a limited-access paradigm. *Alcohol Clin Exp Res*, 28:1814–1821.

527. Froehlich, J.C. and Wand, G.S. (1997). Adenylyl cyclase signal transduction and alcohol-induced sedation. *Pharmacol Biochem Behav*, 58:1021–1030.

528. Cunningham, C.L., Niehus, D.R., Malott, D.H., and Prather, L.K. (1992). Genetic differences in the rewarding and activating effects of morphine and ethanol. *Psychopharmacology (Berl)*, 107:385–393.

529. Ritz, M.C., Garcia, J.M., Protz, D., Rael, A.M., and George, F.R. (1994). Ethanol-reinforced behavior in P, NP, HAD and LAD rats: Differential genetic regulation of reinforcement and motivation. *Behav Pharmacol*, 5:521–531.

530. Gatto, G.J., Murphy, J.M., Waller, M.B., McBride, W.J., Lumeng, L., and Li, T.K. (1987). Persistence of tolerance to a single dose of ethanol in the selectively-bred alcohol-preferring P rat. *Pharmacol Biochem Behav*, 28:105–110.

531. Mardones, J., Contreras, S., and Segovia-Riquelme, N. (1988). A method for recognizing specific effects on ethanol intake by experimental animals. *Alcohol*, 5:15–19.

532. He, X.X., Nebert, D.W., Vasiliou, V., Zhu, H., and Shertzer, H.G. (1997). Genetic differences in alcohol drinking preference between inbred strains of mice. *Pharmacogenetics*, 7:223–233.

533. Krishnan-Sarin, S., Krystal, J.H., Shi, J., Pittman, B., and O'Malley, S.S. (2007). Family history of alcoholism influences naltrexone-induced reduction in alcohol drinking. *Biol Psychiatry*, 62:694–697.

534. Froehlich, J.C., Harts, J., Lumeng, L., and Li, T.K. (1990). Naloxone attenuates voluntary ethanol intake in rats selectively bred for high ethanol preference. *Pharmacol Biochem Behav*, 35:385–390.

535. Badia-Elder, N.E., Mosemiller, A.K., Elder, R.L., and Froehlich, J.C. (1999). Naloxone retards the expression of a genetic predisposition toward alcohol drinking. *Psychopharmacology (Berl)*, 144:205–212.

536. Rodd, Z.A., Bell, R.L., Sable, H.J., Murphy, J.M., and McBride, W.J. (2004). Recent advances in animal models of alcohol craving and relapse. *Pharmacol Biochem Behav*, 79:439–450.

537. Froehlich, J., O'Malley, S., Hyytia, P., Davidson, D., and Farren, C. (2003). Preclinical and clinical studies on naltrexone: What have they taught each other? *Alcohol Clin Exp Res*, 27:533–539.

538. Kranzler, H.R., Burleson, J.A., Brown, J., and Babor, T.F. (1996). Fluoxetine treatment seems to reduce the beneficial effects of cognitive–behavioral therapy in type B alcoholics. *Alcohol Clin Exp Res*, 20:1534–1541.

539. Pettinati, H.M., Volpicelli, J.R., Kranzler, H.R., Luck, G., Rukstalis, M.R., and Cnaan, A. (2000). Sertraline treatment for alcohol dependence: Interactive effects of medication and alcoholic subtype. *Alcohol Clin Exp Res*, 24:1041–1049.

540. Chick, J., Aschauer, H., and Hornik, K. (2004). Efficacy of fluvoxamine in preventing relapse in alcohol dependence: A one-year, double-blind, placebo-controlled multicentre study with analysis by typology. *Drug Alcohol Depend*, 74:61–70.

541. Dundon, W., Lynch, K.G., Pettinati, H.M., and Lipkin, C. (2004). Treatment outcomes in type A and B alcohol dependence 6 months after serotonergic pharmacotherapy. *Alcohol Clin Exp Res*, 28:1065–1073.

542. Gatto, G.J., Murphy, J.M., McBride, W.J., Lumeng, L., and Li, T.K. (1990). Effects of fluoxetine and desipramine on palatability-induced ethanol consumption in the alcohol-nonpreferring (NP) line of rats. *Alcohol*, 7:531–536.

543. Rodd-Henricks, Z.A., McKinzie, D.L., Edmundson, V.E., Dagon, C.L., Murphy, J.M., McBride, W.J., Lumeng, L., and Li, T.K. (2000). Effects of 5-HT(3) receptor antagonists on

daily alcohol intake under acquisition, maintenance, and relapse conditions in alcohol-preferring (P) rats. *Alcohol*, 21:73–85.

544. Colombo, G., Serra, S., Brunetti, G., Vacca, G., Carai, M.A., and Gessa, G.L. (2003). Suppression by baclofen of alcohol deprivation effect in Sardinian alcohol-preferring (sP) rats. *Drug Alcohol Depend*, 70:105–108.

545. Goedde, H.W., Agarwal, D.P., Fritze, G., Meier-Tackmann, D., Singh, S., Beckmann, G., Bhatia, K., Chen, L.Z., Fang, B., and Lisker, R. (1992). Distribution of ADH2 and ALDH2 genotypes in different populations. *Hum Genet*, 88:344–346.

546. Quintanilla, M.E. and Tampier, L. (2003). Brain mitochondrial aldehyde dehydrogenase: Relation to acetaldehyde aversion in low-alcohol-drinking (UChA) and high-alcohol-drinking (UChB) rats. *Addiction Biol*, 8:387–397.

547. Wall, T.L., Garcia-Andrade, C., Thomasson, H.R., Carr, L.G., and Ehlers, C.L. (1997). Alcohol dehydrogenase polymorphisms in Native Americans: Identification of the ADH2*3 allele. *Alcohol Alcohol*, 32:129–132.

548. Osier, M.V., Pakstis, A.J., Goldman, D., Edenberg, H.J., Kidd, J.R., and Kidd, K.K. (2002a). A proline–threonine substitution in codon 351 of ADH1C is common in Native Americans. *Alcohol Clin Exp Res*, 26:1759–1763.

549. Osier, M.V., Pakstis, A.J., Soodyall, H., Comas, D., Goldman, D., Odunsi, A., Okonofua, F., Parnas, J., Schulz, L.O., Bertranpetit, J., Bonne-Tamir, B., Lu, R.B., Kidd, J.R., and Kidd, K.K. (2002b). A global perspective on genetic variation at the ADH genes reveals unusual patterns of linkage disequilibrium and diversity. *Am J Hum Genet*, 71:84–99.

550. Agarwal, D.P. and Goedde, H.W. (1992). Pharmacogenetics of alcohol metabolism and alcoholism. *Pharmacogenetics*, 2:48–62.

551. Shen, Y.C., Fan, J.H., Edenberg, H.J., Li, T.K., Cui, Y.H., Wang, Y.F., Tian, C.H., Zhou, C.F., Zhou, R.L., Wang, J., Zhao, Z.L., and Xia, G.Y. (1997). Polymorphism of ADH and ALDH genes among four ethnic groups in China and effects upon the risk for alcoholism. *Alcohol Clin Exp Res*, 21:1272–1277.

552. Quertemont, E. (2004). Genetic polymorphism in ethanol metabolism: Acetaldehyde contribution to alcohol abuse and alcoholism. *Mol Psychiatry*, 9:570–581.

553. Eriksson, C.J., Atkinson, N., Petersen, D.R., and Deitrich, R.A. (1984). Blood and liver acetaldehyde concentrations during ethanol oxidation in C57 and DBA mice. *Biochem Pharmacol*, 33:2213–2216.

554. Thomasson, H.R., Crabb, D.W., Edenberg, H.J., Li, T.K., Hwu, H.G., Chen, C.C., Yeh, E.K., and Yin, S.J. (1994). Low frequency of the ADH2*2 allele among Atayal natives of Taiwan with alcohol use disorders. *Alcohol Clin Exp Res*, 18:640–643.

555. Bond, C., Iaforge, K.S., Tian, M., Melia, D., Zhang, S., Borg, L., Gong, J., Schluger, J., Strong, J.A., Leal, S.M., Tischfield, J.A., Kreek, M.J., and Yu, L. (1998). Single-nucleotide polymorphism in the human mu opioid receptor gene alters beta-endorphin binding and activity: Possible implications for opiate addiction. *Proc Natl Acad Sci USA*, 95:9608–9613.

556. Beyer, A., Koch, T., Schroder, H., Schulz, S., and Hollt, V. (2004). Effect of the A118G polymorphism on binding affinity, potency and agonist-mediated endocytosis, desensitization, and resensitization of the human mu-opioid receptor. *J Neurochem*, 89:553–560.

557. Kroslak, T., Laforge, K.S., Gianotti, R.J., Ho, A., Nielsen, D.A., and Kreek, M.J. (2007). The single nucleotide polymorphism A118G alters functional properties of the human mu opioid receptor. *J Neurochem*, 103:77–87.

558. Ray, L.A. and Hutchison, K.E. (2004). A polymorphism of the mu-opioid receptor gene (OPRM1) and sensitivity to the effects of alcohol in humans. *Alcohol Clin Exp Res*, 28:1789–1795.

559. Bart, G., Kreek, M.J., Ott, J., Laforge, K.S., Proudnikov, D., Pollak, L., and Heilig, M. (2005). Increased attributable risk related to a functional mu-opioid receptor gene polymorphism

in association with alcohol dependence in central Sweden. *Neuropsychopharmacology*, 30:417–422.

560. Becker, A., Grecksch, G., Kraus, J., Loh, H.H., Schroeder, H., and Hollt, V. (2002). Rewarding effects of ethanol and cocaine in mu opioid receptor-deficient mice. *N Schmied Arch Pharmacol*, 365:296–302.

561. Kovacs, K.M., Szakall, I., O'Brien, D., Wang, R., Vinod, K.Y., Saito, M., Simonin, F., Kieffer, B.L., and Vadasz, C. (2005). Decreased oral self-administration of alcohol in kappa-opioid receptor knock-out mice. *Alcohol Clin Exp Res*, 29:730–738.

562. Roberts, A.J., Gold, L.H., Polis, I., McDonald, J.S., Filliol, D., Kieffer, B.L., and Koob, G.F. (2001). Increased ethanol self-administration in delta-opioid receptor knockout mice. *Alcohol Clin Exp Res*, 25:1249–1256.

563. Grisel, J.E., Mogil, J.S., Grahame, N.J., Rubinstein, M., Belknap, J.K., Crabbe, J.C., and Low, M.J. (1999). Ethanol oral self-administration is increased in mutant mice with decreased beta-endorphin expression. *Brain Res*, 835:62–67.

564. Hayward, M.D., Hansen, S.T., Pintar, J.E., and Low, M.J. (2004). Operant self-administration of ethanol in C57BL/6 mice lacking beta-endorphin and enkephalin. *Pharmacol Biochem Behav*, 79:171–181.

565. Berggren, U., Fahlke, C., Aronsson, E., Karanti, A., Eriksson, M., Blennow, K., Thelle, D., Zetterberg, H., and Balldin, J. (2006). The Taqi DRD2 A1 allele is associated with alcohol-dependence although its effect size is small. *Alcohol Alcohol*, 41:479–485.

566. Munafo, M.R., Johnstone, E.C., Welsh, K.I., and Walton, R.T. (2005). Association between the DRD2 gene Taq1A (C32806T) polymorphism and alcohol consumption in social drinkers. *Pharmacogenomics J*, 5:96–101.

567. Smith, L., Watson, M., Gates, S., Ball, D., Foxcroft, D. (2007). Meta-Analysis of the Association of the Taq1A Polymorphism with the Risk of Alcohol Dependency: A huge Gene-Disease Association Review. *Am J Epidemiol*.

568. Noble, E.P., Blum, K., Ritchie, T., Montgomery, A., and Sheridan, P.J. (1991). Allelic association of the D2 dopamine receptor gene with receptor-binding characteristics in alcoholism. *Arch Gen Psychiatry*, 48:648–654.

569. El-Ghundi, M., George, S.R., Drago, J., Fletcher, P.J., Fan, T., Nguyen, T., Liu, C., Sibley, D.R., Westphal, H., and O'Dowd, B.F. (1998). Disruption of dopamine D1 receptor gene expression attenuates alcohol-seeking behavior. *Eur J Pharmacol*, 353:149–158.

570. Phillips, T.J., Brown, K.J., Burkhart-Kasch, S., Wenger, C.D., Kelly, M.A., Rubinstein, M., Grandy, D.K., and Low, M.J. (1998). Alcohol preference and sensitivity are markedly reduced in mice lacking dopamine D2 receptors. *Nat Neurosci*, 1:610–615.

571. Risinger, F.O., Freeman, P.A., Rubinstein, M., Low, M.J., and Grandy, D.K. (2000). Lack of operant ethanol self-administration in dopamine D2 receptor knockout mice. *Psychopharmacology (Berl)*, 152:343–350.

572. Palmer, A.A., Low, M.J., Grandy, D.K., and Phillips, T.J. (2003). Effects of a Drd2 deletion mutation on ethanol-induced locomotor stimulation and sensitization suggest a role for epistasis. *Behav Genet*, 33:311–324.

573. Hutchison, K.E., McGeary, J., Smolen, A., Bryan, A., and Swift, R.M. (2002). The DRD4 VNTR polymorphism moderates craving after alcohol consumption. *Health Psychol*, 21:139–146.

574. Falzone, T.L., Gelman, D.M., Young, J.I., Grandy, D.K., Low, M.J., and Rubinstein, M. (2002). Absence of dopamine D4 receptors results in enhanced reactivity to unconditioned, but not conditioned, fear. *Eur J Neurosci*, 15:158–164.

575. Hall, F.S., Sora, I., and Uhl, G.R. (2003). Sex-dependent modulation of ethanol consumption in vesicular monoamine transporter 2 (VMAT2) and dopamine transporter (DAT) knockout mice. *Neuropsychopharmacology*, 28:620–628.

576. Savelieva, K.V., Caudle, W.M., Findlay, G.S., Caron, M.G., and Miller, G.W. (2002). Decreased ethanol preference and consumption in dopamine transporter female knock-out mice. *Alcohol Clin Exp Res*, 26:758-764.

577. Heils, A., Teufel, A., Petri, S., Stober, G., Riederer, P., Bengel, D., and Lesch, K.P. (1996). Allelic variation of human serotonin transporter gene expression. *J Neurochem*, 66:2621-2624.

578. Bonnet-Brilhault, F., Laurent, C., Thibaut, F., Campion, D., Chavand, O., Samolyk, D., Martinez, M., Petit, M., and Mallet, J. (1997). Serotonin transporter gene polymorphism and schizophrenia: An association study. *Biol Psychiatry*, 42:634-636.

579. Sander, T., Harms, H., Lesch, K.P., Dufeu, P., Kuhn, S., Hoehe, M., Rommelspacher, H., and Schmidt, L.G. (1997). Association analysis of a regulatory variation of the serotonin transporter gene with severe alcohol dependence. *Alcohol Clin Exp Res*, 21:1356-1359.

580. Little, K.Y., McLaughlin, D.P., Zhang, L., Livermore, C.S., Dalack, G.W., McFinton, P.R., Delproposto, Z.S., Hill, E., Cassin, B.J., Watson, S.J., and Cook, E.H. (1998). Cocaine, ethanol, and genotype effects on human midbrain serotonin transporter binding sites and mRNA levels. *Am J Psychiatry*, 155:207-213.

581. Edenberg, H.J., Reynolds, J., Koller, D.L., Begleiter, H., Bucholz, K.K., Conneally, P.M., Crowe, R., Goate, A., Hesselbrock, V., Li, T.K., Nurnberger, J.I., Jr., Porjesz, B., Reich, T., Rice, J.P., Schuckit, M., Tischfield, J.A., and Foroud, T. (1998). A family-based analysis of whether the functional promoter alleles of the serotonin transporter gene HTT affect the risk for alcohol dependence. *Alcohol Clin Exp Res*, 22:1080-1085.

582. Schuckit, M.A., Mazzanti, C., Smith, T.L., Ahmed, U., Radel, M., Iwata, N., and Goldman, D. (1999). Selective genotyping for the role of 5-HT2A, 5-HT2C, and GABA alpha 6 receptors and the serotonin transporter in the level of response to alcohol: A pilot study. *Biol Psychiatry*, 45:647-651.

583. Ishiguro, H., Saito, T., Akazawa, S., Mitushio, H., Tada, K., Enomoto, M., Mifune, H., Toru, M., Shibuya, H., and Arinami, T. (1999). Association between drinking-related antisocial behavior and a polymorphism in the serotonin transporter gene in a Japanese population. *Alcohol Clin Exp Res*, 23:1281-1284.

584. Hinckers, A.S., Laucht, M., Schmidt, M.H., Mann, K.F., Schumann, G., Schuckit, M.A., and Heinz, A. (2006). Low level of response to alcohol as associated with serotonin transporter genotype and high alcohol intake in adolescents. *Biol Psychiatry*, 60:282-287.

585. Sander, T., Harms, H., Dufeu, P., Kuhn, S., Hoehe, M., Lesch, K.P., Rommelspacher, H., and Schmidt, L.G. (1998). Serotonin transporter gene variants in alcohol-dependent subjects with dissocial personality disorder. *Biol Psychiatry*, 43:908-912.

586. Kelai, S., Aissi, F., Lesch, K.P., Cohen-Salmon, C., Hamon, M., and Lanfumey, L. (2003). Alcohol intake after serotonin transporter inactivation in mice. *Alcohol Alcohol*, 38:386-389.

587. Boyce-Rustay, J.M., Wiedholz, L.M., Millstein, R.A., Carroll, J., Murphy, D.L., Daws, L.C., and Holmes, A. (2006). Ethanol-related behaviors in serotonin transporter knockout mice. *Alcohol Clin Exp Res*, 30:1957-1965.

588. Hasegawa, Y., Higuchi, S., Matsushita, S., and Miyaoka, H. (2002). Association of a polymorphism of the serotonin 1B receptor gene and alcohol dependence with inactive aldehyde dehydrogenase-2. *J Neural Transm*, 109:513-521.

589. Crabbe, J.C., Phillips, T.J., Feller, D.J., Hen, R., Wenger, C.D., Lessov, C.N., and Schafer, G.L. (1996). Elevated alcohol consumption in null mutant mice lacking 5-HT1B serotonin receptors. *Nat Genet*, 14:98-101.

590. Risinger, F.O., Doan, A.M., and Vickrey, A.C. (1999). Oral operant ethanol self-administration in 5-HT1B knockout mice. *Behav Brain Res*, 102:211-215.

591. Engel, S.R., Lyons, C.R., and Allan, A.M. (1998). 5-HT3 receptor over-expression decreases ethanol self administration in transgenic mice. *Psychopharmacology (Berl)*, 140:243-248.

592. Dick, D.M., Bierut, L., Hinrichs, A., Fox, L., Bucholz, K.K., Kramer, J., Kuperman, S., Hesselbrock, V., Schuckit, M., Almasy, L., Tischfield, J., Porjesz, B., Begleiter, H., Nurnberger, J., Jr., Xuei, X., Edenberg, H.J., and Foroud, T. (2006). The role of GABRA2 in risk for conduct disorder and alcohol and drug dependence across developmental stages. *Behav Genet*, 36:577-590.

593. Parsian, A. and Cloninger, C.R. (1997). Human GABAA receptor alpha 1 and alpha 3 subunits genes and alcoholism. *Alcohol Clin Exp Res*, 21:430-433.

594. Soyka, M., Preuss, U.W., Hesselbrock, V., Zill, P., Koller, G., and Bondy, B. (2008). GABA-A2 receptor subunit gene (GABRA2) polymorphisms and risk for alcohol dependence. *J Psychiatry Res*, 42:184-191.

595. Edenberg, H.J., Dick, D.M., Xuei, X., Tian, H., Almasy, L., Bauer, L.O., Crowe, R.R., Goate, A., Hesselbrock, V., Jones, K., Kwon, J., Li, T.K., Nurnberger, J.I., Jr., O'Connor, S.J., Reich, T., Rice, J., Schuckit, M.A., Porjesz, B., Foroud, T., and Begleiter, H. (2004). Variations in GABRA2, encoding the alpha 2 subunit of the GABA(A) receptor, are associated with alcohol dependence and with brain oscillations. *Am J Hum Genet*, 74:705-714.

596. Covault, J., Gelernter, J., Hesselbrock, V., Nellissery, M., and Kranzler, H.R. (2004). Allelic and haplotypic association of GABRA2 with alcohol dependence. *Am J Med Genet B Neuropsychiatry Genet*, 129:104-109.

597. Lappalainen, J., Krupitsky, E., Remizov, M., Pchelina, S., Taraskina, A., Zvartau, E., Somberg, L.K., Covault, J., Kranzler, H.R., Krystal, J.H., and Gelernter, J. (2005). Association between alcoholism and gamma-amino butyric acid alpha2 receptor subtype in a Russian population. *Alcohol Clin Exp Res*, 29:493-498.

598. Haughey, H.M., Ray, L.A., Finan, P., Villanueva, R., Niculescu, M., Hutchison, K.E. (2007). The human GABA(A) receptor alpha2 gene moderates the acute effects of alcohol and brain mrna expression. *Genes Brain Behav*.

599. Boehm, S.L., Ponomarev, I., Jennings, A.W., Whiting, P.J., Rosahl, T.W., Garrett, E.M., Blednov, Y.A., and Harris, R.A. (2004). Gamma-aminobutyric acid A receptor subunit mutant mice: New perspectives on alcohol actions. *Biochem Pharmacol*, 68:1581-1602.

600. Stephens, D.N., Pistovcakova, J., Worthing, L., Atack, J.R., and Dawson, G.R. (2005). Role of GABAA alpha5-containing receptors in ethanol reward: the effects of targeted gene deletion, and a selective inverse agonist. *Eur J Pharmacol*, 526:240-250.

601. Loh, E.W., Smith, I., Murray, R., McLaughlin, M., McNulty, S., and Ball, D. (1999). Association between variants at the GABAAbeta2, GABAAalpha6 and GABAAgamma2 gene cluster and alcohol dependence in a Scottish population. *Mol Psychiatry*, 4:539-544.

602. Radel, M., Vallejo, R.L., Iwata, N., Aragon, R., Long, J.C., Virkkunen, M., and Goldman, D. (2005). Haplotype-based localization of an alcohol dependence gene to the 5q34 γ-aminobutyric acid type A gene cluster. *Arch Gen Psychiatry*, 62:47-55.

603. Parsian, A. and Zhang, Z.H. (1999). Human chromosomes 11p15 and 4p12 and alcohol dependence: Possible association with the GABRB1 gene. *Am J Med Genet*, 88:533-538.

604. Noble, E.P., Zhang, X., Ritchie, T., Lawford, B.R., Grosser, S.C., Young, R.M., and Sparkes, R.S. (1998). D2 dopamine receptor and GABA(A) receptor beta3 subunit genes and alcoholism. *Psychiatry Res*, 81:133-147.

605. Dick, D.M., Edenberg, H.J., Xuei, X., Goate, A., Kuperman, S., Schuckit, M., Crowe, R., Smith, T.L., Porjesz, B., Begleiter, H., and Foroud, T. (2004). Association of GABRG3 with alcohol dependence. *Alcohol Clin Exp Res*, 28:4-9.

606. Hsu, Y.P., Seow, S.V., Loh, E.W., Wang, Y.C., Chen, C.C., Yu, J.M., and Cheng, A.T. (1998). Search for mutations near the alternatively spliced 8-amino-acid exon in the GABAA receptor gamma 2 subunit gene and lack of allelic association with alcoholism among four aboriginal groups and Han Chinese in Taiwan. *Brain Res Mol Brain Res*, 56:284-286.

607. Sander, T., Ostapowicz, A., Samochowiec, J., Smolka, M., Winterer, G., and Schmidt, L.G. (2000). Genetic variation of the glutamate transporter EAAT2 gene and vulnerability to alcohol dependence. *Psychiatry Genet*, 10:103–107.

608. Spanagel, R., Pendyala, G., Abarca, C., Zghoul, T., Sanchis-Segura, C., Magnone, M.C., Lascorz, J., Depner, M., Holzberg, D., Soyka, M., Schreiber, S., Matsuda, F., Lathrop, M., Schumann, G., and Albrecht, U. (2005). The clock gene Per2 influences the glutamatergic system and modulates alcohol consumption. *Nat Med*, 11:35–42.

609. Preuss, U.W., Zill, P., Koller, G., Bondy, B., Hesselbrock, V., and Soyka, M. (2006). Ionotropic glutamate receptor gene GRIK3 SER310ALA functional polymorphism is related to delirium tremens in alcoholics. *Pharmacogenom J*, 6:34–41.

610. Cowen, M.S., Schroff, K.C., Gass, P., Sprengel, R., and Spanagel, R. (2003). Neurobehavioral effects of alcohol in AMPA receptor subunit (glur1) deficient mice. *Neuropharmacology*, 45:325–333.

611. Amit, Z. and Smith, B.R. (1985). A multi-dimensional examination of the positive reinforcing properties of acetaldehyde. *Alcohol*, 2:367–370.

612. Wand, G., McCaul, M.E., Gotjen, D., Reynolds, J., and Lee, S. (2001). Confirmation that offspring from families with alcohol-dependent individuals have greater hypothalamic-pituitary-adrenal axis activation induced by naloxone compared with offspring without a family history of alcohol dependence. *Alcohol Clin Exp Res*, 25:1134–1139.

613. Ray, L.A. and Hutchison, K.E. (2007). Effects of naltrexone on alcohol sensitivity and genetic moderators of medication response: A double-blind placebo-controlled study. *Arch Gen Psychiatry*, 64:1069–1077.

614. Oslin, D.W., Berrettini, W., Kranzler, H.R., Pettinati, H., Gelernter, J., Volpicelli, J.R., and O'Brien, C.P. (2003). A functional polymorphism of the mu-opioid receptor gene is associated with naltrexone response in alcohol-dependent patients. *Neuropsychopharmacology*, 28:1546–1552.

615. Grahame, N.J., Mosemiller, A.K., Low, M.J., and Froehlich, J.C. (2000). Naltrexone and alcohol drinking in mice lacking beta-endorphin by site-directed mutagenesis. *Pharmacol Biochem Behav*, 67:759–766.

616. Di, C.G. and Imperato, A. (1988). Drugs abused by humans preferentially increase synaptic dopamine concentrations in the mesolimbic system of freely moving rats. *Proc Natl Acad Sci USA*, 85:5274–5278.

617. Lawford, B.R., Young, R.M., Rowell, J.A., Qualichefski, J., Fletcher, B.H., Syndulko, K., Ritchie, T., and Noble, E.P. (1995). Bromocriptine in the treatment of alcoholics with the D2 dopamine receptor A1 allele. *Nat Med*, 1:337–341.

618. Wiesbeck, G.A., Dursteler-Macfarland, K.M., Wurst, F.M., Walter, M., Petitjean, S., Muller, S., Wodarz, N., and Boning, J. (2006). No association of dopamine receptor sensitivity in vivo with genetic predisposition for alcoholism and DRD2/DRD3 gene polymorphisms in alcohol dependence. *Addiction Biol*, 11:72–75.

619. Chen, W.J., Chen, C.H., Huang, J., Hsu, Y.P., Seow, S.V., Chen, C.C., and Cheng, A.T. (2001). Genetic polymorphisms of the promoter region of dopamine D2 receptor and dopamine transporter genes and alcoholism among four aboriginal groups and Han Chinese in Taiwan. *Psychiatry Genet*, 11:187–195.

620. Foley, P.F., Loh, E.W., Innes, D.J., Williams, S.M., Tannenberg, A.E., Harper, C.G., and Dodd, P.R. (2004). Association studies of neurotransmitter gene polymorphisms in alcoholic Caucasians. *Ann NY Acad Sci*, 1025:39–46.

621. Franke, P., Schwab, S.G., Knapp, M., Gansicke, M., Delmo, C., Zill, P., Trixler, M., Lichtermann, D., Hallmayer, J., Wildenauer, D.B., and Maier, W. (1999). DAT1 gene polymorphism in alcoholism: A family-based association study. *Biol Psychiatry*, 45:652–654.

622. Kohnke, M.D., Batra, A., Kolb, W., Kohnke, A.M., Lutz, U., Schick, S., and Gaertner, I. (2005). Association of the dopamine transporter gene with alcoholism. *Alcohol Alcohol*, 40:339-342.

623. Tyndale, R.F. (2003). Genetics of alcohol and tobacco use in humans. *Ann Med*, 35:94-121.

624. Bailey, J.N., Breidenthal, S.E., Jorgensen, M.J., McCracken, J.T., and Fairbanks, L.A. (2007). The association of DRD4 and novelty seeking is found in a nonhuman primate model. *Psychiatry Genet*, 17:23-27.

625. Lahoste, G.J., Swanson, J.M., Wigal, S.B., Glabe, C., Wigal, T., King, N., and Kennedy, J.L. (1996). Dopamine D4 receptor gene polymorphism is associated with attention deficit hyperactivity disorder. *Mol Psychiatry*, 1:121-124.

626. Laucht, M., Becker, K., Blomeyer, D., and Schmidt, M.H. (2007). Novelty seeking involved in mediating the association between the dopamine D4 receptor gene exon III polymorphism and heavy drinking in male adolescents: Results from a high-risk community sample. *Biol Psychiatry*, 61:87-92.

627. Tidey, J.W., Monti, P.M., Rohsenow, D.J., Gwaltney, C.J., Miranda, R., Jr., McGeary, J.E., MacKillop, J., Swift, R.M., Abrams, D.B., Shiffman, S., and Paty, J.A. (2008). Moderators of naltrexone's effects on drinking, urge, and alcohol effects in non-treatment-seeking heavy drinkers in the natural environment. *Alcohol Clin Exp Res*, 32:58-66.

628. Mcgeary, J.E., Monti, P.M., Rohsenow, D.J., Tidey, J., Swift, R., and Miranda, R., Jr. (2006). Genetic moderators of naltrexone's effects on alcohol cue reactivity. *Alcohol Clin Exp Res*, 30:1288-1296.

629. Krystal, J.H., Staley, J., Mason, G., Petrakis, I.L., Kaufman, J., Harris, R.A., Gelernter, J., and Lappalainen, J. (2006). Gamma-aminobutyric acid type A receptors and alcoholism: Intoxication, dependence, vulnerability, and treatment. *Arch Gen Psychiatry*, 63:957-968.

630. Pierucci-Lagha, A., Covault, J., Feinn, R., Nellissery, M., Hernandez-Avila, C., Oncken, C., Morrow, A.L., and Kranzler, H.R. (2005). GABRA2 alleles moderate the subjective effects of alcohol, which are attenuated by finasteride. *Neuropsychopharmacology*, 30:1193-1203.

631. Loh, E.W. and Ball, D. (2000). Role of the GABA(A)beta2, GABA(A)alpha6, GABA(A)alpha1 and GABA(A)gamma2 receptor subunit genes cluster in drug responses and the development of alcohol dependence. *Neurochem Int*, 37:413-423.

632. Monti, P.M., Rohsenow, D.J., Hutchison, K.E., Swift, R.M., Mueller, T.I., Colby, S.M., Brown, R.A., Gulliver, S.B., Gordon, A., and Abrams, D.B. (1999). Naltrexone's effect on cue-elicited craving among alcoholics in treatment. *Alcohol Clin Exp Res*, 23:1386-1394.

633 O'Malley, S.S., Krishnan-Sarin, S., Farren, C., Sinha, R., and Kreek, M.J. (2002). Naltrexone decreases craving and alcohol self-administration in alcohol-dependent subjects and activates the hypothalamo-pituitary-adrenocortical axis. *Psychopharmacology (Berl)*, 160:19-29.

634. Kiefer, F., Jahn, H., Tarnaske, T., Helwig, H., Briken, P., Holzbach, R., Kampf, P., Stracke, R., Baehr, M., Naber, D., and Wiedemann, K. (2003). Comparing and combining naltrexone and acamprosate in relapse prevention of alcoholism: A double-blind, placebo-controlled study. *Arch Gen Psychiatry*, 60:92-99.

CHAPTER 6

Contribution of Animal Models and Preclinical Human Studies to Medication Development for Nicotine Dependence

Athina Markou[1], Christian V. Chiamulera[2] and Robert J. West[3]

[1]Department of Psychiatry, University of California at San Diego, School of Medicine, La Jolla, CA, USA
[2]Section of Pharmacology, Department of Medicine and Public Health, University of Verona, Verona, Italy
[3]Department of Epidemiology and Public Health, University College London, London, UK

Animal and Translational Models for CNS Drug Discovery.
Vol. 3 of 3: Reward Deficit Disorders
Robert McArthur and Franco Borsini (eds), Academic Press, 2008

INTRODUCTION

The Problem of Tobacco Smoking

Tobacco smoking is estimated to lead to the premature death of 5 million people worldwide, and this figure is expected to rise to 10 million by 2020.[1] In developed countries smokers may expect to lose an average of 10 years of healthy life unless they stop.[2] One large prospective study in the United Kingdom found that, after the age of 40 years, every year that smoking cessation is delayed reduces life expectancy by 3 months.[3] Table 6.1 lists the fatal and non-fatal disorders linked to smoking. It is clear from this list that apart from causing premature death, smoking causes substantial chronic morbidity.

Until quite recently, smoking was considered primarily a lifestyle choice, and it was not regarded as appropriate for medical science to get involved in developing ways of helping people to stop. However, this picture has radically altered. It is now recognized that ingesting nicotine, and possibly other psychoactive compounds from tobacco, leads to changes in brain function that make it extremely difficult for even the most strong-willed individuals to stop smoking.[16,17] Developing treatments that can mitigate the effects of, or help with undoing, these functional changes has the potential to save many millions of lives.

Tobacco Addiction and Nicotine Dependence

There is no unique, universally accepted definition of addiction, but a useful working definition that captures its essential features is: "a chronic disposition to experience such powerful motivation to engage in a reward-seeking activity, and/or a weakened disposition to inhibit that activity, that it is maladaptive." In relation to tobacco smoking, probably the most important manifestation of addiction is that serious attempts to stop using tobacco are undermined by powerful motivational forces that conflict with, undermine, and weaken the resolve to remain abstinent.[18] Note that this definition of

Table 6.1 Fatal and serious non-fatal disorders for which tobacco use is a known or probable cause or exacerbating factor. Adapted from West[4]

Smoking	Infertility
Aortic aneurism	Leukemia
Asthma attacks	Low back pain
Cancer of the lung	Low birth weight
Cancer of the larynx	Macular degeneration
Cancers or the oral cavity	Osteoporosis
Cancer of the nasopharynx	Peptic ulcer disease
Cancer of the oropharynx[a]	Peripheral vascular disease
Cancer of the oesophagus	Pneumonia/Sudden infant death syndrome
Cancer of the liver	Spontaneous abortion
Cancer of the cervix	Stillbirth
Cancer of the pancreas	Surgical complications
Cancer of the stomach	Tuberculosis
Cancer of the urinary tract[b]	Type II diabetes
Cataracts	Vascular dementia
Cerebrovascular disease	
Conduct disorder[c]	**Smokeless tobacco use[e]**
COPD[d]	Cancer of the oral cavity
Coronary heart disease	Cancer of the pancreas
Hearing loss	

Sources: All [5,6] *except vascular dementia,*[7] *macular degeneration,*[8] *low back pain,*[9] *tuberculosis,*[10] *diabetes,*[11] *cancer of the pancreas,*[12] *conduct disorder,*[13] *surgical complications,*[14] *and smokeless tobacco.*[15]

[a]*hypopharynx,*

[b]*kidney, ureter, and bladder,*

[c]*in offspring of women who were smoking during pregnancy,*

[d]*chronic obstructive pulmonary disease,*

[e]*these vary greatly in concentrations of carcinogens and therefore risk. Smokeless tobacco is also potentially implicated in heart disease, but the data are conflicting. See www.deathsfromsmoking.net for estimates of numbers for each country and region.*

addiction involves a move away from an earlier one that construed addiction purely in terms of physiological adaptation to a drug leading to unpleasant or life-threatening withdrawal symptoms. The reason for this shift in emphasis is that it is now recognized that the most damaging aspect of the phenomenon is the continuation of maladaptive behaviors, and that the withdrawal symptoms are one of several motivational forces contributing to the continuation of these problem behaviors.

The powerful motivational forces that undermine attempts to control an addictive behavior typically arise from an interaction between: (i) psychological and physiological susceptibility in the individual; (ii) the actions of a psychoactive drug; and (iii) the social and physical environment in which the individual resides. In the case of tobacco addiction, individuals appear to be more susceptible if they are predisposed to depression,[19] have problems with inattention and impulse control,[20] or have increased sensitivity to the effects of nicotine,[21] all of which are influenced by their genotype.[21] Nicotine self-administration leads to profound brain changes that make the act of smoking and the sensations associated with it highly rewarding and make abstinence punishing. These include upregulation of nicotinic acetylcholine receptors (nAChR),[22] abnormal functioning of the mesolimbic dopamine pathway,[23] and changes in the function of multiple neurotransmitter systems.[24] There is probably more than one mechanism involved in tobacco addition, and these mechanisms are important to differing degrees for different smokers.

In the first mechanism, rapid nicotine ingestion from cigarettes leads to changes in neurotransmission, including increases in dopamine release in the nucleus accumbens and glutamate throughout the brain, that with repetition appear to establish increasingly powerful impulses to smoke in the presence of cues previously associated with smoking. One may think of this process as establishing a "habit" mechanism. Evidence for this habit mechanism in humans comes with the observation that addicted smokers often report smoking "automatically" in the presence of cues that have previously been associated with smoking,[25] and they report experiencing heightened "urges" to smoke in the presence of smoking cues.[26]

In the second mechanism, repeated ingestion of nicotine from cigarettes appears to create a kind of "nicotine hunger," a motivational drive state similar in many ways to the hunger for food, that motivates nicotine-seeking and nicotine-taking behavior. Evidence for this mechanism is manifest in the fact that many smokers feel a need to smoke at any time when their brain nicotine concentrations are depleted or lowered below some critical level, such as first thing in the morning,[27] and reinstating their nicotine levels through nicotine patches and other nicotine replacement therapies reduces this need.[28]

A third putative mechanism involves adaptation to nicotine leading to adverse mood and somatic symptoms emerging during abstinence as shown in Table 6.2.[29]

Table 6.2 Acute effects of stopping smoking

Effect	Prevalence	Typical Duration
Irritability or aggression	>25%	<4 weeks
Increased appetite	>25%	>10 weeks
Difficulty concentrating	>25%	<4 weeks
Restlessness	>25%	<4 weeks
Depressed mood	<25%	<4 weeks
Urge to smoke	>25%	>10 weeks
Sleep disturbance	<25%	<4 weeks
Weight gain	>25%	permanent
Cough/sore throat	<25%	<4 weeks
Mouth ulcers	<25%	>2 weeks
Constipation	<25%	>2 weeks

Source: Reprinted from West and Shiffman.[18]

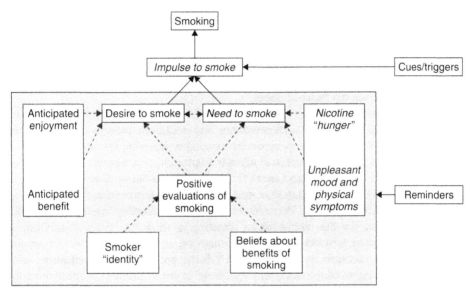

Figure 6.1   Our behaviors at any given moment are controlled by the balance between impulses and inhibitions acting at that moment which are in turn driven by stimuli that serve as triggers and reminders.[33] There are several different pathways generating the impulse to smoke. Smokers differ in how important each of these is, but any one of them could be enough to cause a person to smoke unless there is a good reason not to. Reprinted from.[34] Entries in italics represent pathways influenced by nicotine dependence

It is apparent that all of these symptoms can occur for other reasons as well, and this is probably an important factor in reinstating smoking behavior even after the acute withdrawal period has ended. If smokers learn repeatedly that smoking often reduces "stress," and there are no obvious discriminative stimuli to signal when smoking will or will not have this effect, they are experiencing a variable ratio schedule of negative reinforcement that can establish a behavior pattern that is resistant to extinction. Thus, stressful events are known to be among the most common causes of relapse to smoking weeks, months, or even years after the person stopped.[30]

A fourth putative mechanism involves withdrawal from nicotine leading to reduced ability to exercise response inhibition. Thus, there is some evidence in humans that abstaining smokers suffer from impaired capacity to inhibit responses (in this case saccades orienting toward the appearance of a visual stimulus).[31] Impaired impulse control has been postulated as an important contributory factor to other addictions as well.[32][i]

The data in Figure 6.1 show how the three putative mechanisms generating motivation to smoke fit into the moment-to-moment balance of motivational forces that govern our behavior. In this more complete model of behavior based on the PRIME theory of motivation (Plans, Responses, Impulses, Motives, and Evaluations[33]), it is

---

[i] Please refer to Heidbreder, Impulse and reward deficit disorders: drug discovery and development, or Williams *et al.*, Current concepts in the classification, treatment, and modeling of pathological gambling and other impulse control disorders, in this Volume for further discussion of relationships between impulse control and addiction.

important to note that non-pharmacological factors also play an important role both independently and in interaction with the pharmacological factors. Thus, psychological, social and environmental interventions clearly have a key role to play in shifting the balance of motivational forces away from smoking and toward abstinence. Treatment for nicotine dependence is just one, *albeit* important, factor that feeds into this dynamic motivational balance sheet.

It should be noted here that although there are over 4000 chemicals in tobacco products that could contribute to dependence and addiction, there is little debate that nicotine is one of the major components in tobacco responsible for addiction.[16,17,35,36] Nicotine alone can lead to preferential self-administration of a tobacco product over a control vehicle (i.e., an inert substance).[16,17,37-39] Humans show a greater preference for nicotine over a control substance in studies examining intravenous,[37,40] nasal,[41] and medicinal gum[42] administration. Furthermore, if levels of nicotine are altered, smokers tend to compensate for this alteration by smoking more if the levels of nicotine are reduced or blocked by a nicotinic receptor antagonist, or by smoking less if exogenous or higher levels of nicotine are administered.[17] More specifically, compensatory smoking involves adjusting smoking behaviors by changing the (i) number of puffs on a cigarette, (ii) duration of puffs, (iii) interpuff intervals, or (iv) number of cigarettes smoked. For example, researchers observed this compensatory smoking behavior in smokers who had switched from cigarettes with a high nicotine yield to cigarettes with a low nicotine yield[43] or who had reduced the number of cigarettes smoked.[44,45] The resulting levels of the nicotine metabolite cotinine and other biochemical indicators of tobacco exposure were proportionately less than what would have been expected based on the reduction in nicotine yield of the cigarette or the number of cigarettes smoked.

Accordingly, much research aimed at promoting our understanding of the neurobiology of tobacco dependence and discovering new medications for tobacco addiction focuses on nicotine dependence and animal models of nicotine dependence. Nevertheless, there are also other factors that contribute to the tobacco smoking habit, such as sensory stimuli (e.g., smell and taste of cigarettes), conditioned stimuli (e.g., other people smoking, ashtrays), and discriminatory stimuli (e.g., situations in which the person usually smokes such as in a bar or after a meal; see above). Furthermore, other ingredients in tobacco smoke (e.g., tar, acetaldehyde, tobacco alkaloids, nornicotine, monoamine oxidase inhibitors) other than nicotine either have synergistic effects with nicotine or reinforcing effects of their own.[46-58] Considering though the critical role of nicotine in tobacco addiction and the financial issues involved in drug discovery (see below), at present it appears sufficient for drug discovery efforts to focus on animal models of nicotine dependence with little consideration for the contributing properties of other tobacco ingredients to nicotine dependence.

Criteria for Defining Nicotine Dependence in Humans

The 1988 Surgeon General's report lists the following "criteria for drug dependence," including nicotine dependence[17] p. 7:

Primary Criteria

- Highly controlled or compulsive use
- Psychoactive effects
- Drug-reinforced behavior.

Additional Criteria

- Addictive behavior often involves
  - Stereotypic patterns of use
  - Use despite harmful effects
  - Relapse following abstinence
  - Recurrent drug cravings.
- Dependence-producing drugs often produce
  - Tolerance
  - Physical dependence
  - Pleasant (euphoriant) effects.

These symptoms constitute the descriptors for dependence provided in the *Diagnostic and Statistical Manual of Mental Disorders*, 4th edition (DSM-IV)[59] and the *International Statistical Classification of Diseases*, 10th Revision (ICD-10) (Table 5.1).[60]

ANIMAL MODELS OF NICOTINE DEPENDENCE

What is a Model?

An animal model is an experimental preparation designed to mimic aspects of a human disorder.[61,62] Animal models of nicotine dependence with good construct validity mimic in experimental animals the cardinal features of tobacco addiction summarized in Figure 6.1, Table 6.2, and the diagnostic manuals (see above). However, considering the difficulties inherent in reproducing both the physical act of tobacco smoking and the social factors that are involved in the acquisition, maintenance, and relapse to smoking, certain compromises must be made in the development and validation of suitable animal models. Moreover, priorities must be set regarding what features of tobacco addiction are of the greatest importance to the model, as well as of relevance of these features to the purpose for which the model is designed[61] (see below). Furthermore, it is important to recognize that each animal model is likely to be analogous or homologous to a specific feature of nicotine dependence[62] and that a multitude of animal models are required to mimic the various aspects of the human phenomenon.

An animal model consists of both the inducing condition and the dependent measure or measures.[61] Animal models of drug dependence, including nicotine dependence, are easier to design than animal models of other psychiatric disorders, such as schizophrenia, depression, and anxiety. The reason for this is that the etiology of drug dependence is known to be excessive exposure to the drug, while the etiologies of other mental disorders are mostly unknown or at best hypotheses. Thus, animal models of drug dependence include exposure to the drug as the inducing condition and assess the effects of such drug exposure on measures analogous or homologous to behaviors relevant to addiction in humans.

Models of Acquisition and Maintenance of Nicotine Self-Administration and Dependence

Perhaps the most intuitive approach to nicotine dependence involves the use of the rodent self-administration model,[63] ubiquitous in the preclinical drug addiction literature (for review, see[64][ii]). In this model, nonhuman animal subjects (typically rodents and primarily rats) are trained to perform an operant response that results in the delivery of an intravenous nicotine infusion to the subject. The self-administration model provides a framework in which to examine the acquisition and maintenance of nicotine self-administration, as well as withdrawal and relapse when used in conjunction with the reinstatement model. The use of this protocol in the case of nicotine, however, raises some unique challenges. For logistical reasons, rodent nicotine self-administration studies typically employ an intravenous route of self-administration rather than inhalation as experienced by human tobacco smokers. Fortunately, analysis of the pharmacokinetic profile of intravenous tobacco administration and tobacco smoke inhalation has revealed similar absorption and elimination of nicotine, and formation and elimination of its metabolites.[65]

Social Factors for Smoking and Initiation of Smoking

Perhaps the most apparent differences between human tobacco smoking and rat nicotine self-administration are the conditions under which nicotine use is initiated. While it may be unrealistic to expect animal models to offer good analogy to the human situation with respect to many of the psychosocial factors that have been implicated in smoking, there is good evidence that some characteristics that may predict susceptibility to nicotine self-administration are modeled well in experimental subjects. Individuals that initiate smoking between the ages of 14 and 16 have been reported to be 1.6 times more likely to become dependent compared with those who begin smoking at a later age,[66] an observation that extends to other drugs of abuse.[67] Similarly, rats pre-exposed to nicotine during adolescence (postnatal day 34–43) exhibit increased nicotine intake in a self-administration protocol,[68] as well as enhanced locomotor activity in a novel environment,[69] a phenomenon that is generally predictive of increased susceptibility to drug self-administration.[70]

Impulsivity is a character trait that has been associated with smoking.[71,72] Adult smokers have been shown to exhibit increased "delayed discounting," an index of impulsivity in which subjects choose between a small immediate reward and a large delayed reward, compared with non-smokers.[73][iii] Impulsivity has been suggested to be a predisposing factor in the initiation of tobacco use.[74] However, the finding that ex-smokers do not perform significantly differently than individuals who have never smoked suggests that increased impulsivity may develop with nicotine use rather than predict susceptibility to smoking.[75] Consistent with this observation are data from parent and teacher ratings of impulsivity failing to predict tobacco use in adolescents.[76] Research using the five-choice serial reaction time task, an animal model of impulsivity, has also shown that impulsivity is associated with increased susceptibility to initiate and maintain nicotine use.[77] Specifically, "impulsive action," as indicated by premature

---

[ii] Please refer to Heidbreder, Impulse and reward deficit disorders: drug discovery and development, or Rocha *et al.*, Development of medications for heroin and cocaine addiction and regulatory aspects of abuse liability testing, in this Volume for further discussion and description of self-administration procedures in rodents and nonhuman primates.

[iii] See also Williams *et al.*, Current concepts in the classification, treatment and modeling of pathological gambling and other impulse control disorders, in this Volume.

responding on the five-choice serial reaction time task, was predictive of enhanced nicotine acquisition as well as enhanced motivation for nicotine. Specifically, under self-administration conditions maintained on a fixed-ratio schedule of reinforcement, rats in the upper quartile of impulsivity on this task exhibited greater responding during the first 10 sessions of acquisition and also exhibited higher breakpoints on a progressive-ratio schedule of reinforcement, indicative of enhanced motivation for nicotine.[77] Results from the "delayed reward task," analogous to "delayed discounting" in human participants, indicated a similar finding to that derived from the five-choice serial reaction time task and the progressive-ratio test. However, unlike in the five-choice serial reaction time task, performance on the delayed reward task was not predictive of increased responding during acquisition. Furthermore, unlike the five-choice serial reaction time task, impulsivity as assessed by the delayed reward task did predict increased resistance to extinction and enhanced conditioned-cue elicited reinstatement for lever responding.[77]

Maintenance of Nicotine Self-Administration and Dependence

The intravenous nicotine self-administration procedure allows the assessment of both acquisition of nicotine self-administration and maintenance of self-administration and potentially dependence. Unfortunately, due to practical and financial reasons, typically nicotine self-administration sessions are only 1 h long and are conducted 5 days per week with 2 days per week off nicotine. The issue with this protocol is that in most cases the subjects are not dependent on nicotine in terms of signs of nicotine withdrawal emerging upon cessation of nicotine self-administration. It has been shown that somatic signs of nicotine withdrawal do emerge if the subjects are tested for several weeks 1 h per day, 5 days per week, and dependence develops faster if rats are tested for 1 h daily or for longer periods of time daily.[78] However, when 1 h nicotine self-administration sessions are used, very seldom dependence (in terms of withdrawal signs) is assessed. The interest in most of these studies is whether a test compound decreases the rewarding effects of acute nicotine that are probably analogous to the euphorigenic effects of acute nicotine administration in humans once the initial aversive effects show tolerance. Thus, this protocol of 1 h nicotine self-administration sessions may be most relevant to the discovery of medications or conditions that may block the acute effects of nicotine during early exposure to tobacco smoking, and thus interrupt acquisition of a strong tobacco habit and development of nicotine dependence.

More recently, investigators have used 6–23 h access to nicotine self-administration sessions that more closely mimic the human access conditions than 1 h sessions.[78-81] Indeed it was shown that these conditions lead to nicotine dependence, as defined by the emergence of somatic signs of nicotine withdrawal upon cessation of nicotine self-administration or administration of nAChR antagonists to the subjects.[78,81,82] Moreover, under these conditions, rats self-administer approximately 0.25–1.3 mg/kg/day nicotine base[78,81,82] compared with 3.16 mg/kg/day nicotine base that is usually experimentally administered to rats to induce dependence.[83-85] Nevertheless, this dose range is comparable to nicotine intake in human smokers (0.14–1.14 mg/kg/day). As far as the authors are aware, there have been no systematic studies to investigate the effects of pharmacological treatments on nicotine self-administration under short

(1 h) and long (more than 6 h) daily access conditions. Such studies are critical to determine whether the more laborious long access conditions would provide different experimental outcomes or whether the 1 h access conditions that are easier and more practical to use provide identical results.

Models of Various Aspects of Nicotine Withdrawal

In dependent smokers, abstinence from tobacco smoking leads to a constellation of somatic and affective symptoms that comprise the nicotine withdrawal syndrome. Withdrawal symptoms typically emerge within a few hours after the last cigarette, peak within a few days to 1 week, and return to pre-quit baseline levels after 2 to 4 weeks.[86] The symptoms of nicotine withdrawal in humans are affective disturbances that include irritability/anger, anxiety, and depressed mood; behavioral symptoms that include restlessness, sleep disturbance, and increased appetite; cognitive disturbances such as difficulty concentrating, as well as somatic changes, such as decreased heart rate and mild gastrointestinal disturbances.[42,86] This withdrawal syndrome is believed to be a major factor impairing an individual's ability to remain abstinent from smoking, and thus motivate relapse.[86]

Several rodent models have been developed that serve to investigate the neurobiology of nicotine withdrawal and potentially evaluate medications for treating withdrawal. One of the first and most widely used measures of nicotine withdrawal in rodents is the measurement of the frequency of somatic sings. These signs are reliably seen in rats, but are less reliably observed in mice.[83,87-93] The most prominent signs of the rat withdrawal syndrome include abdominal constrictions (writhing), gasps, ptosis, facial fasciculation, and eye blinks, while miscellaneous other signs, such as escape attempts, foot licks, genital grooming, shakes, scratches, and yawns are rarely observed.[89,94-96] These somatic signs are both centrally and peripherally mediated.[89,91,95,97]

Although the somatic components of nicotine withdrawal are certainly unpleasant, it is hypothesized that avoidance of the affective components of withdrawal plays a more important role in the maintenance of nicotine dependence than the somatic aspects of withdrawal.[24,98] Affective symptoms of nicotine withdrawal in humans include depressed mood, increased irritability, and increased anxiety.[99]

A very reliable and critical measure of the affective and motivational aspects of drug withdrawal is elevations in brain reward thresholds observed after cessation of chronic nicotine administration.[83-85,96,100] iv Elevations in brain reward thresholds are considered to be an operational measure of "diminished interest or pleasure" (i.e., anhedonia[101,102]) in rewarding stimuli that is a symptom of nicotine withdrawal.[59,103] Similar reward threshold elevations are seen during withdrawal from all major drugs of abuse, such as cocaine,[101] ethanol,[104] morphine,[105] amphetamine,[84,85,106-108] and phencyclidine.[109] Interestingly, several dissociations have been observed between the threshold elevations and the somatic signs associated with nicotine withdrawal,[84,85,110] suggesting that the various aspects of withdrawal are mediated by different substrates. Importantly, these anhedonia-like aspects of rat nicotine withdrawal are reversed

---

iv See also Koob, The role of animal models in reward deficit disorders: Views from academia, in this Volume for detailed discussion of brain stimulation reward.

by several antidepressant drug treatments[84,111,112] (Paterson, Semenova and Markou, unpublished observations), as well as experimental glutamatergic compounds.[113] Thus, this animal model of the anhedonia characterizing nicotine withdrawal is likely to have predictive validity in terms of identifying compounds that would reverse the depressed mood that is experienced by some humans who quit smoking (see above).

Other rodent models that may be of relevance to disruption of behavioral performance in humans involve nicotine abstinence-induced disruptions of food-maintained learned behaviors in rats,[114] and decreased overall performance in the five-choice serial reaction time task.[115] Both of these effects are likely to reflect general behavioral suppression during nicotine withdrawal. Most recently, it was shown that light-potentiated startle is increased in nicotine-withdrawing rats,[116] reflecting increased reactivity to stress that may be analogous to increased irritability/anger and anxiety experienced by abstinent smokers. The reported increases in startle reactivity during nicotine withdrawal are not reliably seen across laboratories,[116,117] a variability that is likely attributable to the conditions (high or low stress) under which startle was measured, with increases in startle seen only under high stress conditions as in the light-potentiated startle procedure.[116] Thus, there are several animal, primarily rat, models of the affective (depressed mood and irritability/anxiety) of nicotine withdrawal, as well of the somatic signs of withdrawal. These models can be used to discover compounds that would alleviate nicotine withdrawal symptoms, and thus decrease the motivation to relapse. Unfortunately, mouse models of the nicotine withdrawal syndrome are not as well developed yet, although there are ongoing efforts to establish reliable models of nicotine withdrawal in the mouse.[92,118-120]

Models of Reinstatement of Nicotine-Seeking Behavior with Relevance to Relapse

The primary putative rodent model of relapse to drug-seeking behavior is the reinstatement model (for review, see[121]). In this model, animals that have been previously trained for nicotine self-administration undergo extinction, during which time responding is no longer reinforced, and then tested for subsequent reinstatement of responding, indicative of drug-seeking behavior, elicited by a nicotine infusion,[122] a nicotine-associated conditioned cue,[123,124] or potentially a stressor. Increased nicotine-seeking is considered to be analogous to craving in abstinent human tobacco smokers.

Although this model does have some features relevant to the human relapse situation, the suitability of this procedure as an animal model of relapse has been both criticized[125] and defended[121] in the literature. Some issues with this model include questions relating to the model's validity. The face validity of the model is generally accepted as being good. However, superficial face validity is not necessarily a very desirable feature of a model because it may easily lead to erroneous conclusions.[61] One criticism is that while the stress-induced relapse version of the model successfully models human reports that negative affect predicts drug craving or relapse, the model does not account for reports that positive moods may also predict craving.[126] Stress-induced drug-seeking behavior has been demonstrated with nicotine[127,128] and multiple times with other drugs of abuse (for review, see[121]). Another criticism of this putative "relapse" model is that while the human literature shows that the risk to relapse in human drug addicts decreases with larger periods of abstinence,[129-132] data from the reinstatement model show that the reinstatement effect is, in fact, more robust with increasing periods of abstinence.[133,134]

Nevertheless, this reinstatement model appears to have good construct, and perhaps etiological, validity in the case of nicotine. Data from the human literature show that in abstinent tobacco smokers, relapse was associated with increases in negative affect (related to stress-induced reinstatement in animals) as well as smoking-related cues.[99,135] Furthermore, Shiffman and colleagues[135] have shown that brief exposure to nicotine was associated with relapse to tobacco smoking.

Reinstatement models as putative models of relapse are currently very important as regulatory agencies consider relapse as the most critical measure for showing that a compound has efficacy for smoking cessation (see below). Thus, it is encouraging that this model may have some predictive validity in terms of predicting medications that will be efficacious in preventing relapse. The CB_1 receptor antagonist rimonabant (Acomplia®)/Zimulti®) has been reported to be effective in preventing relapse in abstinent smokers.[136] Similarly, rimonabant has been shown to reduce reinstatement of nicotine-seeking for nicotine-conditioned cues.[137] Interestingly, bupropion (Zyban®/Wellbutrin®), the only non-nicotine-based treatment approved by the regulatory agencies for smoking cessation, increases, rather than decreases, cue-induced reinstatement[138] and increases responding for stimuli associated with nicotine,[139] while it reverses the anhedonia characterizing nicotine withdrawal[111,112] and has inconsistent effects on intravenous nicotine self-administration.[139,140] These findings highlight the fact that, neurobiologically, it appears that the various stages of nicotine dependence, withdrawal, and relapse may respond differently to various types of pharmacological treatments and that different medications may be required to treat the various phases of dependence and withdrawal and facilitate prevention of relapse. As extensively discussed below, this situation creates a market issue, as well as medication compliance issues that need to be dealt with in order to design highly efficacious treatments for nicotine dependence. In terms of drug discovery, we are in the fortunate situation of having animal models in place that would allow such an approach to drug discovery for smoking cessation.

Animal Models of Motivation for Nicotine

Because nicotine addiction is considered a motivational disorder (see above), measures of motivation to self-administer nicotine during the different stages of dependence (e.g., maintenance, short-term abstinence, protracted abstinence when relapse may occur, during exposure to contexts or stimuli associated with nicotine administration) can be directly analogous to human situations where relapse occurs. The progressive-ratio schedule of reinforcement involves successive increases in the number of responses required for the subject to receive one injection of nicotine (for review, see[141 v]). The highest fixed-ratio attained before the subject quits responding for a

---

[v] See also Koob, The role of animal models in reward deficit disorders: Views from academia, Heidbreder, Impulse and reward deficit disorders: Drug discovery and development, or Rocha *et al.*, Development of medications for heroin and cocaine addiction and regulatory aspects of abuse liability testing, in this Volume for detailed discussion of the progressive-ratio procedure.

criterion period of time (usually 1 h) is called the breakpoint. Effects of putative treatments for smoking cessation on breakpoints for nicotine have not been assessed extensively. Studies have shown that a γ-aminobutyric acid (GABA)$_B$ receptor agonist[142] or a metabotropic glutamate mGlu$_5$ receptor antagonist[143] reduced breakpoints for nicotine, while bupropion[140] increased breakpoints for nicotine.

Another animal procedure that reflects the motivational impact of contexts and stimuli previously associated with nicotine is the conditioned place preference procedure (for review, see[141]). In this procedure, animals are confined to one compartment immediately after treatment with nicotine, and confined to another compartment after treatment with vehicle.[vi] After a few pairings, exposure to the apparatus while in a drug-free state results in a preference for the nicotine-paired compartment.[144] When one of the compartments is associated with the aversive effects of nicotine withdrawal, the subject exhibits an aversion for the compartment associated with the aversive aspects of nicotine withdrawal.[95]

HISTORICAL BACKGROUND: WHY DRUG DISCOVERY ON NICOTINE DEPENDENCE WAS DIFFERENT FROM THAT ON OTHER DRUGS OF ABUSE

The process of research and development (R&D) of drug candidates for the treatment of tobacco addiction has a fairly recent history in drug discovery. Only during the last 10–15 years, the pharmaceutical industry has started to develop a specific discovery strategy for nicotine and tobacco dependence. Initially, the R&D strategy for tobacco addiction was similar to that for the discovery of treatments for other drugs of abuse, such as stimulants, alcohol, and opiates. It should be noted though that direct industrial investment in drug addiction research has been very limited compared with research on other neuropsychiatric therapeutic areas, such as anxiety, depression, and schizophrenia.

In the early 1990s, a World Health Organization/United Nations joint conference was held in Vienna on drug craving. Academic experts on drug addiction and major pharmaceutical companies involved in central nervous system research gathered to discuss data and evidence on prevention of drug craving and drug-seeking relapse as the most relevant targets for therapeutic intervention in drug addiction.[145] Data were already available at that time showing the existence of neurobiological mechanisms underlying craving and relapse, and that most of them may be in common with all of the different types of addictive drugs, including nicotine. This conference "formally" stimulated a greater research interest in the pharmaceutical industry for investigations on the molecular mechanisms underlying relapse to drugs of abuse as novel potential targets for pharmacotherapy.

Since that time, the interest of the industry was, and still is today, focused on nicotine as the best-known psychoactive substance responsible for cigarette smoking and

---

[vi] See also Rocha *et al.*, Development of medications for heroin and cocaine addiction and regulatory aspects of abuse liability testing, in this Volume for detailed discussion of the conditioned place-preference procedure.

tobacco addiction for the reasons and evidence outlined above. Nicotine dependence is a "more-than-one-billion patients" market that could guarantee a return-of-investment for research. The major issue was how successful R&D could be on this therapeutic area. Predictive animal models of reinforcing properties of drugs were already available, such as drug self-administration and conditioned place preference (see above), whereas other models were at that time just published or under development, such as protocols for drug priming or cue-induced drug-seeking relapse.[141] According to the drug addiction literature at that time,[146,147] the involvement of dopaminergic mechanisms was well established, whereas neurochemical research was showing the involvement of additional neurotransmitters other than dopamine, such as glutamate, serotonin, and endogenous opioids.[vii] Within a few years, behavioral, cellular, and neurochemical studies contributed to describing drug addiction as a complex brain-behavioral disorder, characterized by acute and prolonged effects leading to gradual neuroadaptation of cellular and molecular mechanisms in common among drugs of abuse.[148]

Although nicotine was known to induce similar behavioral and neurochemical effects as other drugs of abuse, the advancement of nicotine research was slightly delayed compared with other drugs of abuse. Although pioneering studies were published in the 1960s and 1970s, the first reliable protocol of nicotine self-administration in rats was published in 1989.[149] Nicotine self-administration in rats, however, was shown in tobacco industry laboratories long before 1989, but this knowledge was not in the public domain and was only published in the mid-2000s.[150] For several reasons (mostly methodological), successful replication followed only around the mid-1990s, both in industry and academia.[63,151,152] Although nicotine shares reinforcing properties with stimulants, central nervous system depressants, and other drugs of abuse, it is self-administered under somewhat different methodological conditions.[79,153-155] Similarly, nicotine-induced conditioned place preference appeared variable across studies.[156-159] The development of reliable nicotine self-administration protocols in industry was therefore a priority and a necessary step toward the set-up of nicotine-seeking relapse models to test potential drug candidates.

Nevertheless, it soon became apparent that these models would require long training and testing periods to best approximate the human conditions of dependence on nicotine. That is, acquisition of intravenous nicotine self-administration in rats requires several weeks (at least 3 weeks) for the establishment of stable nicotine intake and patterns of self-administration, then a couple of weeks of extinction and/or abstinence is required, and finally reinstatement testing may be initiated together with the assessment of the putative anti-relapse properties of test compounds. Time and technical issues (such as time and effort to maintain intravenous catheters patented) made results from such models "slow like a clinical trial." Further, as discussed above, in most cases rats self-administering nicotine in 1 h sessions, 5 days per week, are not dependent on nicotine.[78] These advantages and disadvantages of the intravenous nicotine self-administration, nicotine-seeking and conditioned place preference procedures (indicated below as "nicotine conditioning models") could not fit all stages of drug discovery.

[vii] For a comprehensive review of this subject, see *Trends in Pharmacological Sciences* special issue, May 1992, 13: 169–219.

PRECLINICAL ANIMAL MODELS OF NICOTINE DEPENDENCE IN DRUG DISCOVERY

Early drug discovery consists of a series of stages ranging from the identification of a drug target, the validation of that target, chemical synthesis program to generate series of lead compounds and the optimization of two or more lead compounds that will be selected for further investment.[160][viii] At the exploratory project stage, multidisciplinary experiments are designed to test hypotheses on novel mechanisms potentially involved in nicotine dependence.[112] These experiments are carried out using either standard pharmacological tools or drugs that already exist in the clinic and are known to interact with the molecular target of interest. Because of the importance of preparing and presenting a cohesive drug discovery program to which considerable resources will be placed, the constraints of time and high throughput of data are relatively small. This relative freedom allows the behavioral pharmacologist to use time- and labor-intensive nicotine dependence models such as those described above to generate the necessary confirmatory data to justify the establishment of the drug discovery project and the allocation of resources for further target validation, assay development, lead identification, and lead characterization or lead discovery.

Once a drug discovery project is established though, there is an essential requirement that pharmacological experiments should provide data quickly in order to direct a chemical synthesis program effectively and generate lead compounds. Medicinal chemists are synthesizing, or providing from chemical libraries, hundreds of compounds for testing; the results of which are then used to modify chemical structures and the generation of more selective and/or potent molecules. Procedures such as intravenous nicotine self-administration, reinstatement of nicotine-seeking or conditioned place preference procedures, clearly cannot cope with the demands of the project at this stage. Other faster throughput procedures are needed at this stage that are based upon molecular or cellular interactions with the target being pursued. These are typically neurochemical and simple behavioral assays.

Neurochemical and Behavioral Assays

Neurochemical assays assess the effects of molecules on the drug target of interest using biochemical or physiological measures such as dopamine release in the nucleus accumbens. Activation of or interaction with these neurochemical systems may also have behavioral correlates that can be exploited. Activation of nicotinic nAChR receptor subtypes by their ligands produces responses to nociceptive stimuli (e.g., hot plate or tail flick tests[161]), seizures,[162-164] hyper- or hypo-locomotion,[165,166] motor alternation (e.g., Y-maze[167]), one-trial learning (e.g., passive avoidance task[168]), memory (e.g., conditioned fear response[169]), or neurovegetative measures (e.g., hypothermia[170]). For

---

[viii] The reader is invited to refer to Heidbreder, Impulse and reward deficit disorders: Drug discovery and development, in this Volume; Lindner *et al.*, Development, optimization and use of preclinical behavioral models to maximize the productivity of drug discovery for Alzheimer's Disease, in Volume 2, *Neurologic Disorders*, or Bartz *et al.*, Preclinical animal models of Autistic Spectrum Disorders (ASD), in Volume 1, *Psychiatric Disorders*, for further discussion of the drug discovery process.

example, antagonism of nicotine-induced seizures may be considered a very simple neurochemical/behavioral assay of α7 nAChR antagonism *in vivo*.[164,171]

These tests may be performed either in the mouse or in the rat. Mice are the species of choice in early screening as its smaller body requires lower amounts of test drug, easier housing (more subjects/cage, smaller cages than for rats), and easier test management (smaller equipment required than that used for testing rats) to obtain large sample sizes per experiment. Since the introduction of mutant mice into the armamentarium of biological research, gene modulation in mice is the primary tool for specific alteration of gene expression of central nervous system targets of interest.[172] Mutant mice, for example, have been generated for different genes expressing receptors relevant to drug dependence,[173] and in particular for genes expressing nAChRs[174-177] (for review, see[178-180]). For example, nicotinic subtype β2 transgenic mice showed a behavioral phenotype to nicotine challenge in several relatively fast behavioral models, such as passive avoidance and hot-plate tests,[174,181] similarly to the effects induced by β2 nAChR antagonists in wild type or control mice.[182] Consequently, behavioral data obtained with lead compounds in the mouse are considered pharmacological analogs of behavioral phenotype data observed in transgenic/knockout mice for the same target. Moreover, in the second half of the 1990s, intravenous nicotine self-administration was developed in the mouse notwithstanding the greater technical difficulties such as keeping catheters patent in such small animals.[183,184] This technical advancement has allowed comparability of self-administration results across species.

As indicated above, nicotine-induced hyperlocomotion is a simple and rapid behavioral assay used extensively for initial *in vivo* screening of potential therapeutics for nicotine dependence.[ix] Acute nicotine injection increases locomotor activity in rats. This effect has been widely characterized across different rat strains, nicotine doses, and routes of administration.[185,186] The correspondence between nicotine-induced hyperlocomotion and increased dopaminergic transmission is common among drugs of abuse.[187-189] Nicotine-induced hyperlocomotion is not a model of nicotine dependence, but is a simple and reliable indicator of nicotine-induced dopamine release and of its psychomotor stimulant properties.[190,191]

A related neurochemical/behavioral assay is the nicotine-induced locomotor sensitization test.[166,192,193] Sensitization to the stimulatory effects of nicotine on locomotion,[165,166] as well as dopamine overflow in the nucleus accumbens and prefrontal cortex,[186,194-197] is induced by repeated daily exposure to nicotine. A test compound may be administered along with nicotine once or twice daily for 4–10 days to assess potential inhibition of sensitization induction. It can also be used to assess potential inhibition of the expression of sensitization following 4–10 days of nicotine administration. In this case the test compound is administered just once with nicotine on a test day following sensitization.

A positive result obtained with this procedure suggests an interaction of the test compound with the neurochemical mechanisms (e.g., dopaminergic, glutamatergic)

[ix] Please refer to McCann *et al.*, Drug discovery and development for reward disorders: Views from Government, in this Volume for further discussion of screening assays in drug discovery.

involved in neuroadaptive changes induced by chronic nicotine exposure.[198,199] The incentive sensitization theory of drug craving suggests that locomotor sensitization and drug-seeking relapse share common neuroadaptive changes.[200] Consequently, inhibition of nicotine-induced motor sensitization by a test compound suggests that it should also be effective in preventing nicotine relapse. However, as we do not know the full extent of the overlap between substrates mediating locomotor sensitization and drug relapse, the use of this assay may lead to erroneous conclusions. Generally, due to the limited exposure to nicotine in sensitization procedures, locomotor sensitization is more likely involved in processes of acquisition of nicotine self-administration rather than reflecting the state of nicotine dependence. Thus, although this behavioral assay is relatively fast and could reflect dopaminergic processes in meso-limbic brain sites, it is not likely to predict the efficacy of compounds to alleviate nicotine withdrawal or prevent relapse in animal models or in human subjects.

Promising lead compounds emerge from *in vitro* and *in vivo* neurochemical models/assays as potential drug candidates, together with a preliminary set of pharmacokinetic and toxicological data during the late lead discovery stage. Usually in drug discovery projects, more than one chemical class is progressed for a target. Consequently more than one potential candidate is advanced to be tested in disease models having higher etiological and construct validity for nicotine dependence. At this stage, resources (and deadlines) may be adjusted accordingly in order to test these molecules, which must now compete with each other and with other compounds from other drug discovery projects for testing and advancement.

Utility of Models of Nicotine Dependence

The face, construct and etiological validity of these animal models have been discussed above and in the literature.[39,201] Intravenous nicotine self-administration, reinstatement of nicotine-seeking as a model of relapse, and conditioned place preference models have been shown to be useful for assessing the behavioral effects of experimental manipulations (e.g., drugs, lesions, gene manipulations, etc.) on different targets: dopaminergic transmission (pathways, receptors, metabolic enzymes, carrier),[202-206] ionotropic and metabotropic glutamate receptors,[207-209] muscarinic receptors and nAChRs,[63,210] GABAergic mechanisms,[143,211,212] serotonergic mechanisms,[213] endogenous opioids,[211] and several other neurochemical mechanisms.[214-216] Compounds interacting with these systems have been tested on these models[96,111,140,143,217-219] (for review, see[112,201]) to assess their potential for the treatment of nicotine dependence. Some of these compounds are now in clinical development (e.g., rimonabant, SB277011), or already in the clinic (e.g., varenicline (Chantix®/Champix®), baclofen (Kemstro®/Lioresal®), bupropion, etc.) for smoking cessation. These results indicate a degree of predictive validity of the models that can be used for further drug discovery initiatives for the treatment of nicotine dependence.

In addition to the traditional pipeline progression of lead discovery and optimization from novel compounds to candidate drugs (bench to bedside), smoking cessation or nicotine dependence drug discovery projects can also draw from clinical observations (bedside to bench). Smoking is a diffuse habit affecting millions of people, both healthy and ill. Therefore, there is a high chance that smokers taking medications may perceive a subjective reduction of smoking satisfaction, and report decreased smoking behavior. These reports have been followed up and, in the case of Zyban® (bupropion[220]), resulted in a registered drug. Having good, reliable and predictive animal models of nicotine dependence promotes the direct testing of diverse chemical entities for smoking cessation and providing a comprehensive preclinical dossier to support proof-of-concept studies in the experimental medicine or more costly controlled clinical trials.[x]

There are many sources of ideas for potential therapeutics for smoking cessation and nicotine dependence. Notwithstanding the genomic revolution, the number of drug targets has been limited concentrating mostly on G-protein-coupled receptors (GPCRs), ion channels, enzymes and hormones.[221] Consequently, it is not surprising that many of the drug targets for nicotine dependence are the same as for other disorders. This felicitous state of affairs not only provides preclinical and clinical investigators in the field with an ample availability of selective and/or potent tools with good drug-like properties, but also provides them with valuable information of ADME (absorption, distribution, metabolism, and excretion), toxicological and clinical data from other disease projects.

Some of these drug targets have already been incorporated into drug discovery projects for smoking cessation and nicotine dependence. Glutamatergic involvement in nicotine dependence, for example, was suggested by the publication on the effects of a glutamate $mGluR_{2/3}$ partial receptor agonist in a rat model of nicotine withdrawal-induced anxiogenesis,[222] which stimulated industrial and academic interest on this family of metabotropic glutamate receptors.[124] A similar interest in alternative drug targets was stimulated by the publication of the potentially selective effect of a dopamine D_3 receptor partial agonist on drug-seeking behavior.[223]

Alternative indications are important considerations for the life-cycle management of established drugs. Interest in the antidepressant nortriptyline (Aventyl®/Pamelor®), which is a noradrenaline/serotonin reuptake inhibitor, and other monoamine reuptake inhibitors for smoking cessation was stimulated by the observation that bupropion also blocks monoamine carriers, including the noradrenaline transporter.[224] This fact offered the opportunity for noradrenaline transporter blockers to be tested in human proof-of-concept studies. Pharmaceutical companies have chemical libraries of thousands of compounds on the shelf that could be tested directly on disease models.

Limitations of Drug Discovery Assays and Models for Nicotine Dependence

Time and Throughput

Models of nicotine dependence, in particular nicotine self-administration and reinstatement of nicotine-seeking, require several weeks of training and testing, separated by wash-out periods, in order to perform dose–response experiments with a within-subjects treatment design. Technical issues, such as the maintenance of patent intravenous catheters (a very difficult procedure in mice) for months, can make a within-subjects design difficult to complete due to loss of viable subjects. An alternative between-subjects design requires larger numbers of subjects to test concurrently. In practical

---

[x] Please refer to McEvoy and Freudenreich, Issues in the design and conductance of clinical trials, in Volume 1, *Psychiatric Disorders*, for a more detailed discussion of proof-of-concept studies.

terms, this means that an investigator needs to either invest in more equipment or carry out several replications to increase the sample size. Notwithstanding the size and number of transgenic or knockout mice that can be generated, experiments in these subjects are usually designed to assess acquisition of nicotine self-administration. It is often quite difficult to establish a stable self-administration baseline and then test these mice for reinstatement. Consequently, this long timeline and low throughput are not easily reconcilable with a high throughput drug discovery perspective where several compounds need to be tested in order to select a chemical class with which to proceed.

Some short-term procedures have been developed[206,208] to assess the effects of dopamine D_3 receptor and glutamate $mGlu_5$ receptor antagonists on reinstatement. A concurrent reinstatement schedule is used in which rats are placed in an operant conditioning box at the beginning of the reinstatement test session 24 h after the last nicotine self-administration session in order to expose them to the same context where nicotine was available, but now without rewarding contingencies (i.e., nicotine delivery) upon responding. After 15 min the rats receive a nicotine priming injection. Burst responding induced by context re-exposure to the operant box where the rat had previously self-administered nicotine is partially extinguished during the initial 15 min component. Consequently, the effect of nicotine priming can be monitored without the interference of background "noise" responding. With this procedure, two doses of test drug can be studied weekly, with at least two daily wash-out self-administration sessions in between.

There are two major protocols used for testing reinstatement,[225] (i) the within-session protocols, where reinstatement of responding is induced after a few hours of extinction from the last drug self-administration session and (ii) the commonly used between-session protocol, where the animals are allowed to extinguish responding for the nicotine-paired lever over several sessions under extinction conditions when nicotine is not available.[122,226]

Within-session protocols, originally developed with cocaine,[227] were not successful in rats trained to self-administer nicotine because it is difficult to extinguish responding for nicotine within 1–3 h (Chiamulera, unpublished data). Between-session protocols allow responding to drop to low levels so that after 10–15 days of extinction, one may observe a robust increase in responding after the presentation of the cue previously associated with nicotine administration or after a nicotine priming injection. By inducing extinction, reinstatement and concomitant drug testing more than once in the same subjects, separated by a few days of testing under saline conditions to re-extinguish the behavior to low levels without returning rats to nicotine self-administration sessions between drug testing, has been shown to work well.[123,124,207] Although this protocol modification significantly improves the time/throughput of testing of potential candidate drugs, they are still perceived in the industry as "small clinical trials" because of the time and effort required.

Genetic and Pharmacological Phenotyping

Behavioral phenotyping of newly generated mutant mice with batteries of tests designed to detect and characterize many altered behaviors[228] has introduced a new level of complexity to behavioral pharmacologists in drug discovery. Not only is there a need to characterize the changes in behaviors induced by genetic manipulation, but also a need to characterize the behavioral effects of compounds designed to interact with novel drug targets such as novel GCPRs.[229,230] The behavioral pharmacologist is called upon to discover function for targets as well as to generate behavioral profiling for an ideal lead compound at a strategic point in the project.

Validity and Extrapolation

Drug discovery projects must always be aware of the possibility of false positive or false negative results being generated from their first-line neurochemical/behavioral assays.[xi] False positive results are generated and expected, and are detected when the compound(s) is subsequently tested in a model of the disorder. False negatives, however, represent a much greater danger as a negative result being generated by the assays leads to compounds to be categorized as inactive and then discarded. False positive/negative rates are usually estimated using clinically active drugs during the set-up stage of a procedure/model, but when such comparators either do not exist, or are few, such an estimate is by definition impossible to make. One factor contributing to false positives may be the limited correspondence of the molecular, cellular and neurochemical basis of front-line neurochemical/behavioral assays to the behavioral changes induced in animals modeling a complex disorder such as nicotine dependence. False negatives, on the other hand, may be generated simply because the compound is exerting its putative anti-dependence effects through a non-dopaminergic mechanism—mechanisms that the assay was not designed to evaluate. Other factors contributing to the generation of false results may be methodological. For example, neurochemical/behavioral assays often include acute or sub-chronic, but not the prolonged nicotine treatments used in the models described above. These assays may be useful, therefore, to study nicotine-induced receptor-specific effects that are useful for compound profiling *in vivo*.[xii]

Species-related factors may also contribute to the generation of false positive/negative effects. Most models of nicotine dependence use rats or nonhuman primates, while, as discussed above, neurochemical/behavioral assays may be performed in either mice or in rats. Pharmacokinetic and toxicity studies are usually performed in mice, whereas only some putative candidate compounds are tested in rats. Species-related pharmacokinetic properties may influence the pharmacodynamic profile of the drug. For example, the active metabolite hydroxy-bupropion that is generated in mice and humans after bupropion administration is not seen in rats.[231] Luckily the smoking cessation properties of bupropion were suggested from results from the clinic and then confirmed in the lab. Bupropion was subsequently tested in rat models

---

[xi] For further discussion regarding the generation of false positive and negative results in early behavioral screening and the consequences thereof for drug discovery, the reader is invited to refer to Large *et al.*, Developing therapeutics for bipolar disorder (BPD): From animal models to the clinic, in Volume 1, *Psychiatric Disorders*.

[xii] The authors allude to another issue brought up in various chapters of the book series, that of *pharmacological Isomorphism*, by which the predictive validity of a model is determined by the effects of clinically active drugs for the disorder that is being modelled. These models may be also limited in their ability to detect activity accurately in novel compounds with mechanisms of action different from the clinically active reference.

of nicotine dependence (for review, see[201]) and is active in some models such as nicotine withdrawal.[96,111] However, bupropion yields inconsistent effects in other models such as nicotine self-administration,[140] and even opposite effects than those predicted in the reinstatement model (e.g., increase, rather than decrease, in cue-induced reinstatement[138]). Interestingly, bupropion showed abuse liability potential in rats (drug discrimination studies) but not in humans.[232,233]

What may account for these divergent results? What is the relevance of hydroxybupropion for smoking cessation, for example, and how closely do these results of bupropion observed in rats resemble those of bupropion in smokers? Perhaps some of the reasons for divergent results may be methodological. Most of the animal research on bupropion was done with an instant release administration involving a systemic injection and not as a slow release formulation as it is given to smokers (but see a recent study where bupropion was administered via minipump[139]). The different formulation may have different pharmacokinetic/pharmacodynamic profiles in rats compared with humans. Therefore, qualitative and quantitative pharmacokinetic differences may cause a different pharmacodynamic profile, including significant differences in effects, such as generalization to stimulant discriminative stimulus and psychomotor stimulation.

Pharmacogenomic and environmental factors are acknowledged as major contributors to diversity to pharmacological response. Nicotine dependence in humans has "emergent" properties that may not be possible to model in experimental animals. Emergence refers to the onset of complexity at a given level of phenomena that may not be predicted by an analysis of the level immediately below.[234] Animal models of a behavioral disorder are used to define a degree of event probability such as drug action on abnormal behavior in humans. What is the probability of predicting a pharmacological effect in non-smoking volunteers or smokers (tobacco addicted subjects) with data obtained with nicotine-addicted experimental animals?[xiii] This is an issue common to all of the behavioral disorders and their models discussed in book series, where the interaction between genotype and environment in humans is highly complex compared with laboratory animals.[235] It is not surprising therefore that individual phenotype variability may mask the pharmacological efficacy of a candidate drug in clinical trials, and argues for a more individual approach to drug development.

HUMAN PRECLINICAL AND CLINICAL STUDIES

As noted above, at this point in time with the limited validation that we have of our animal models and with the small number of medications available for smoking cessation, animal models are likely to provide only a weak indication of to what extent a particular medication will aid cessation of tobacco use. Tests with human volunteers will always be needed. Full-scale clinical trials are expensive and time consuming, and it is not economic to undertake such trials on all treatments that look promising from the animal research. Thus, there is a need to develop measures and experimental paradigms that are economical to implement but will have the maximum predictive value in terms of improved cessation rates. At present, a variety of measures and paradigms are used, each

---

xiii See discussion above of nicotine self-administration model.

with strengths and weaknesses. Furthermore, each one of these measures and paradigms assesses a different aspect of nicotine dependence. This is important for two reasons. First and most importantly, as discussed above, the neurobiology of the different stages of nicotine dependence is likely to be characterized by different neurochemical substrates, and thus different medications would be expected to affect different smoking-related behaviors and motivations differently. Second, animal models mimic a limited – and often only one – aspect of nicotine dependence. That is, a model may be assessing the rewarding effects of nicotine or the motivational impact of stimuli previously associated with nicotine administration in reinitiating the nicotine-seeking behavior, or different affective disturbances associated with nicotine withdrawal. Thus, human preclinical testing of a particular test compound should be based on the known neurobiology of various aspects of nicotine dependence, the neurochemical mechanism of action of the compound, and finally the effects on the relevant animal models that mimic the state of nicotine dependence that the compound is predicted to ameliorate. Unfortunately, it does not appear that such a translational approach has been adapted frequently.

Ad Libitum Smoking Studies

In these studies, the medication or comparator (such as placebo) is given to smokers who are requested not to make any conscious effort to reduce their smoking. This procedure is analogous to assessing the effects of a test compound in intravenous nicotine self-administration in rats. The aim in the human studies is to determine how far the medication reduces the motivation to smoke. The assumption is that if it reduces strength or frequency of the intrinsic motivation to smoke while smoking is not being consciously limited, it will make it easier not to smoke when smokers try to abstain permanently. The outcome measures can be numbers of cigarettes smoked[236] or biochemical measures of smoke intake such as expired-air carbon monoxide or salivary, urinary, or plasma cotinine, or plasma or urinary nicotine. In theory, one could also measure smoking behavior in terms of frequency of puffing and depth of inhalation, but recent research has not found close associations between these parameters and biochemical measures of smoke intake, suggesting that the technology for measuring them needs further development.[237]

When choosing between measures of smoking behavior or smoke intake, it is important to note that different aspects of the behavior may come under the control of different motivational forces. Thus, smokers may continue to smoke the same number of cigarettes as before by virtue of the development of a cue-driven habit but inhale less smoke from each cigarette because the "nicotine hunger" is diminished.[238] Different treatments may affect different smoking parameters over different time courses.

One may also use self-report measures of motivation to smoke such as ratings of time spent with and strength of urges to smoke from the Mood and Physical Symptoms Scale (MPSS).[239] In a laboratory setting, it may also be possible to examine the effect of the medication on the "response cost" that smokers are willing to tolerate to be able to smoke, in terms of money, effort, foregoing other reinforcers, or discomfort. This measure is analogous to the progressive-ratio schedule of reinforcement used in laboratory animals that provides breakpoints as a measure of the motivation to consume the drug (for review, see[141] and discussion above). A limitation of self-report measures in humans is subject to bias and error depending on how far smokers are able to introspect about their motivation to smoke and have expectations based on their perception of the demands of the experimental situation. Thus, it may be that smokers would experience a diminution of motivation to smoke which would fail to be captured because smokers would be basing their ratings on the expectations that they would want to smoke in the experimental situation rather than deeper motivational forces that are less accessible to conscious awareness.

In these *ad libitum* smoking studies, the duration of medication dosing period may vary from 10 s to minutes to weeks. Obviously the short-term, laboratory-based studies would normally be the least expensive, but they might also be least likely to be sensitive to effects of treatments on nicotine dependence because in the short-term smokers' perceptions and motivations may be more under the control of superficial cues and expectations rather than deeper underlying motivations. However, animal measures of motivation to consume nicotine are not likely to be affected by the initial "superficial cues and expectations" that characterize the early abstinence phase in humans. Unfortunately, very few studies in animal models assess the chronic effects of putative medications in animal models. This is not done due to cost and time constraints. However, it is important to assess the effects of chronic administration of test compounds on relevant behaviors because the medication would be given chronically to humans and there is a need to determine whether there may be tolerance to initially observed therapeutic effects. Indeed, chronic administration of an $mGlu_{2/3}$ or a $GABA_B$ receptor agonist resulted in tolerance to the induced decreases in intravenous nicotine self-administration that were observed faster for the $mGlu_{2/3}$ receptor agonist.[123,124,212]

The usefulness of the *ad libitum* smoking paradigm in humans for screening putative cessation aids is not yet clear, but there are some positive signs, including the direct analogy to the intravenous nicotine self-administration procedure in rats, that makes it amenable to translational work. In humans, nicotine replacement therapy has been shown to reduce measures of smoke intake from *ad libitum* smoking,[238,240] and measures of urges to smoke taken during *ad libitum* smoking have been found to be predictive of relapse in a subsequent quit attempt.[241,242]

Studies of Withdrawal Symptoms and "Craving" During Abstinence

In these studies, the medication or comparator is given to smokers who have agreed to undergo a period of abstinence. The aim is to determine how far the medication ameliorates the withdrawal symptoms or "craving." The assumption is that these factors play a role in precipitating relapse. There are numerous self-report withdrawal and craving scales available, all of which have been shown to be sensitive to abstinence and responsive to treatments that aid cessation.[243] Some of the scales use multiple items to measure each withdrawal symptom, while others use a single item. The single-item scales have been found to perform at least as well as the multiple-item scales.[243] It is important in these studies that nearly all subjects entered into the study complete the abstinence period because if smokers drop out it could be because of craving or withdrawal symptoms. If a medication is effective in controlling these factors, there would be more such drop-outs in the placebo group, thus mitigating any observed medication effect.

This approach has considerable intuitive appeal because it appears to target the mediators of relapse. The approach is supported by the fact that medications that aid cessation also mitigate craving and withdrawal symptoms.[244-247] Moreover, recent research has found that a medication, varenicline, that improves cessation rates more than another medication, bupropion, also reduces craving to a greater extent.[248] Finally, there are animal models of aspects of nicotine withdrawal making these procedures amenable to translational work and providing an opportunity for cross-talk and information exchange between the preclinical animal and human domains.

A variant on this theme is to measure the subjective response to a cigarette and/or intensity with which a cigarette is smoked after a period of abstinence. The assumption is that a medication could aid cessation by making the cigarette smoked during a lapse less rewarding and so less likely to lead to further smoking. Evidence in support of this approach comes from a study showing that varenicline, to a significantly greater extent than bupropion, reduced reported satisfaction and reward from a cigarette smoked during a lapse.[248] Using withdrawal symptoms and craving during abstinence as a screening paradigm will be undermined to the extent that smokers' expectations influence their ratings or they are unable to introspect and report their internal states.

Short-Term Abstinence Studies

Probably the most obvious screening study is the classic randomized controlled trial with abstinence as the primary outcome measure, but using a short follow-up period of a few weeks rather than 6 or more months. The assumption is that if the treatment can be shown to be effective in the short-term, the effect is likely to last in the long-term. Thus, while smokers in the medication and comparison groups will relapse as part of a stochastic process in which the motivation to smoke at some point, by chance, exceeds the motivation not to in a proportion of smokers, the rate will be similar. These studies require fewer subjects and a shorter overall duration, and they provide an opportunity to test the medication in conditions as close to a full-scale research clinical trial as possible. The disadvantage is that these trials are still quite costly; and given the administrative costs and fixed costs of setting up clinical trials, regardless of the number of subjects and duration of study, the savings can be quite modest.

It can also be the case that a treatment has an effect while it is being administered, but the treatment effect is reversed when the treatment is discontinued. This could happen, for example, if the treatment maintains the functional abnormalities in the motivational system, not allowing it to recover as long as the treatment is being maintained. It could happen, for example, that a medication would suppress craving and withdrawal symptoms as effectively as smoking a cigarette but only achieve this by maintaining and preserving the brain processes that precisely mediate the smoking state. It could also be the case that the treatment would lead to a transfer of dependence from the cigarette to the treatment and when the treatment is no longer available, the smoker would relapse to cigarettes as the default behavior. Neurochemical studies in experimental animals, as well as studies in disease models, should be able to provide information about the risk of relapse to smoking once the medication is no longer taken. A comprehensive use of the disease models available in which the test compound has positive results should be able to provide answers to such issues before expensive studies are undertaken in humans. Although studies in disease models are expensive and time-consuming, such studies are still less expensive than studies in humans. If we learn to use the most appropriate disease models for the particular mechanism of action under the most appropriate conditions (e.g., chronic treatment with the test compound) and information flows in both directions between human and animal studies over a period of time, we should be able to use information from the disease models to guide designs of human preclinical studies to maximize the chance of positive results.

One approach that can be adopted is to plan a full-scale clinical trial with a "stop-go" decision after analyses of short-term outcome data on a certain number of subjects. There is a cost to be paid in terms of increased sample size for the full trial because of "data peeking," which reduces the power of the analysis, but this could be a price worth paying.[xiv]

Small-Scale Open-Label Studies

These are important in gaining familiarity with the medication and its usage and a sense of its promise in terms of acceptability of side effects, relief from withdrawal symptoms and craving, and even, to some extent, abstinence. In these studies, subjects are put on one or more dosing regimens and observed closely. Given that in particular populations and contexts the relapse rates are quite predictable, observing a substantially lower relapse rate may provide sufficient confidence in treatment efficacy to move ahead with a full-scale clinical trial. An example of this approach was a recent observational study of cytisine (marketed as Tabex® in eastern Europe) in which a 12-month carbon monoxide-verified continuous abstinence rate of approximately 15% was found, which is about three times what would be expected in that sample if they were not receiving an effective treatment.[249]

Commentary

The emphasis in the foregoing discussion has been on efficacy, but of course safety and acceptability are equally important. Clearly, the longer-term field studies will provide a better indication of these parameters than the short-term laboratory studies. It is apparent that there are no simple rules that can be followed when it comes to the design of screening studies, and the choice of design will depend on the budget available and the putative mechanism of action of the treatment, and hopefully information from the animal disease models. Even with the best design, the process has a considerable amount of uncertainty associated with it. A treatment may look very promising in a short-term abstinence study but fail to show effects on long-term abstinence.[250]

[xiv] Please refer to McEvoy and Freudenreich, Issues in the design and conductance of clinical trials, in Volume 1, *Psychiatric Disorders*, for a discussion of adaptive clinical trial designs and the use of Bayesian statistical methods developed to incorporate new data to update the probabilities under investigation.

Arguably, one could improve the predictive validity of screening studies by combining two or more designs or using multiple measures within a given design and only proceeding on the basis of a positive outcome on more than one measure or in more than one paradigm. Arguably, that conservative approach would be rational in the current situation where there already exist quite effective treatments for nicotine dependence and so we should be looking for treatments that are significantly better than those we already have.

One highly important comment is that the choice of paradigm may also be influenced by a desire to provide data from human studies that would lead to improved paradigms in translational research with other species. The goal would be to improve the paradigms for screening treatments at that earlier stage. Unfortunately, such studies do not appear to be conducted, and obviously this is the area where more research is needed to enhance the translational value of preclinical measures in both humans and rats. Such translational value could be further increased by designing in humans laboratory-based objective measures of nicotine reward, motivation to smoke, affective symptoms of withdrawal, and other aspects of nicotine dependence that are amenable to being "translated" to measures used in laboratory animals, such as rodents.

An issue that is beyond the aims of this chapter is the high rates of tobacco smoking among psychiatric populations, particularly schizophrenia.[251,252] This comorbidity is seen for all drugs of abuse and several psychiatric disorders (see other chapters in this volume), but appears to be particularly pervasive for tobacco smoking. This may be the case either because cigarettes are legally sold and/or because of special properties of nicotine and other ingredients in tobacco smoke that may self-medicate aspects of psychiatric disorders that are not well treated by currently available treatments or antagonize undesirable side-effects of psychiatric medications.[253] In terms of drug discovery and development of medications for smoking cessation, psychiatric populations present a particular challenge as they may be more sensitive or have adverse reactions to potential anti-smoking medications. Alternatively, the use of anti-smoking medications may serendipitously improve psychiatric symptoms. It has also been shown that the atypical antipsychotic clozapine decreased tobacco smoking in schizophrenia patients without any encouragement to do so.[254] The above suggest that studies in healthy human volunteer smokers may not necessarily predict the effects of anti-smoking medications in mentally ill patients in terms of either smoking rates or psychiatric symptoms. The pharmacological mechanisms of action of putative anti-smoking medications need to be considered in the context of the hypothesized neurobiology of psychiatric disorders.

SUCCESSES, FAILURES AND SERENDIPITY IN TRANSLATIONAL RESEARCH

The above discussion assumes a "classical" model of drug discovery in which a compound is designed or discovered which has promising properties in terms of basic pharmacology, and preclinical studies are undertaken that also produce promising results. These are followed by human preclinical studies and then human clinical trials. In practice, this path has only been followed successfully with one medication, varenicline. The other popular, non-nicotine medication that has been licensed for smoking cessation, bupropion, was already being used as an antidepressant and was observed clinically to have properties that might aid smoking cessation; in particular, it appeared to make smoking less rewarding.[246] The other medication that clearly aids cessation, nortriptyline, was also licensed as an antidepressant long before it was evaluated as an aid to smoking cessation.[246]

Nicotine replacement therapy has followed a more complex model of development. Early research involved open-label observational studies, *ad libitum* smoking studies, and withdrawal studies undertaken prior to pivotal Phase II and Phase III trials, but even once these had been carried out, further preclinical work continues to examine the mechanism of action.

CHALLENGES, OPPORTUNITIES, AND PERSPECTIVES

Translational Medicine: From Mechanism of Action to Clinical Efficacy

As extensively discussed above, neurobiological evidence so far indicates that there are different neural substrates mediating the various phases and aspects of nicotine dependence. Indeed, tobacco addiction is a multifaceted, complex disease. Basic research, both in animals and humans, has shown that different neurobiological mechanisms may be involved in the brain-behavioral changes underlying tobacco addiction and the various phases of nicotine dependence, and that each one of these mechanisms may be a potential target for pharmacological intervention. Thus, the industry, as well as medical practitioners, has to face the issue of patient compliance in terms of taking the appropriate medication at the appropriate phases of dependence/withdrawal, which may lead to a decrease in medication sales. The smaller the market size, the less the interest of the industry to invest in R&D for a particular therapeutic indication. Nevertheless, with the advance of pharmacogenetics, this is a global issue that the industry has to face (i.e., how to make a profit out of smaller, but better-targeted patient populations; see above). On the other hand, however, if there are several mechanisms of nicotine dependence and different pharmacotherapies are efficacious for the different phases of nicotine dependence, do we need medications with highly specific targets and mechanisms of action, or would a more "dirty" pharmacological profile be more efficacious?

All generated mechanistic hypotheses need to be tested with a rational "traditional" drug design process. As discussed above, there should be extensive cross-validation between human and animal studies. Unfortunately, the industry almost never invests in determining why and how there were failures. There is much to be learned from failures but no time or resources to invest in such endeavors. Fortunately, however, both industry and academia invest time and resources in understanding the mechanism of action of the success stories and the full characterization of the neurochemical and behavioral effects of the success compounds. As indicated above, the nicotine dependence field is in a unique situation compared with most other drug addictions. We have approved compounds for this indication that we can use to understand, validate, and further develop and improve our animal models, as well as preclinical measures in humans.

Product Profile and Return-of-Research-Investment

Based on facts and assumptions summarized above, the pharmaceutical industry may consider marketing of a pharmacological armamentarium that would increase the therapeutic area share (the technical term is "a franchise in the area") in order to have a return-of-investment for R&D. The current scenario shows that it is extremely difficult to develop a "magic bullet" that in the present economic situation could be a "blockbuster" (i.e., a drug with estimated revenues >1 billion USD/year). On the other hand, to invest in the development of a pharmacological armamentarium (e.g., ideally a couple of nicotine replacement products, a bupropion-like drug, a proprietary vaccine slightly different from the original one, a varenicline me-too drug, etc.) even if the market share for each single medication is small, may be a "franchise" in the smoking cessation area to guarantee a mid- to long-term return-of-investment. Remarkably, the smoking cessation therapeutic area may be in commercial synergy with pneumology and cardiovascular markets, so that a novel drug with a limited indication (e.g., acting only on nicotine withdrawal-induced anxiogenesis but well-tolerated and cardiovascularly safe) may be prescribed in cardiovascular smoker patients.

Experimental Medicine Paradigm Shift and Translational Research

As discussed above, human studies with laboratory preclinical models that use objective measures would add more confidence to data obtained from animal disease models. Thus, resources need to be invested in this endeavor. Furthermore, it appears that the trend in the industry is to increase the probability of predictive validity, not necessarily by increasing the validity of a specific model, but by increasing the confidence for the *types* of data generated by a specific model at different stages of drug discovery. Thus, such feedback may increase the accuracy of predictions made from the models, and thus decrease costly unanticipated effects in clinical trials. It is recommended, though, that such feedback is also used to improve the protocols of both the animal models and those of the preclinical human studies so that their predictive value is gradually increased. The bigger the data-sets on a specific compound, hopefully from different animal models, human experimental paradigms, and species and drug formulations, the higher the probability of making good predictions from the preclinical translational work to the efficacy of the medication in the clinic. The issue is, of course, financial. To improve the situation, we need to implement fundamental changes. First, we need to incorporate human preclinical models for different phenomena relating to tobacco addiction into the drug discovery process before traditional clinical trials are undertaken. Several examples are described in the literature and in the sections above that are currently used in academia (e.g., research conducted by Ken Perkins and colleagues, Saul Shiffman and colleagues, Robert West and colleagues). Second, we need to develop experimental medicine models for proof-of-concept studies specific to the mechanistic target of interest, where the assessment of changes in surrogate markers – from imaging, neurovegetative, and biological measures – may give reliable evidence supporting potential efficacy for prevention of relapse to tobacco smoking.

In conclusion, there is not an "ideal story" for drug discovery, from target validation to the clinic, to be told about tobacco addiction and smoking cessation. The selective and specific "magic bullet" is not likely to work for all phases of nicotine dependence. A pharmacological armamentarium is likely to be needed. Based on this scenario which is already clear to the industry, changes are needed in the way we conduct translational preclinical research in both experimental animals and humans. Furthermore, regulatory agencies need to accept therapeutic targets other than just relapse prevention, such as withdrawal relief, craving inhibition, and weight control, as relevant to smoking cessation, because addiction is a multifaceted disorder. Finally, an emphasis on motivational measures for nicotine during the different phases of dependence and withdrawal in preclinical experiments in both experimental animals and humans, and eventually in clinical trials, is needed because nicotine dependence is a motivational disorder (see Introduction to this chapter).

ACKNOWLEDGEMENTS

For reparation of this chapter, AM was supported by NIDA grants R01 DA11946, R01 DA023209, NIMH/NIDA grant U01 MH69062 and Tobacco-Related Disease Research Program grant 15RT-0022 from the State of California. The authors would like to thank Dr. Harinder Aujla for his input on sections of this chapter and Mr. Michael Arends for editorial assistance.

REFERENCES

1. World Health Organisation (2002). *World Health Report 2002: Reducing Risks, Promoting Healthy Life.* World Health Organisation, Geneva.
2. Bronnum-Hansen, H. and Juel, K. (2002). [Health life years lost due to smoking]. *Ugeskr Laeger*, 164:3953–3958.
3. Doll, R., Peto, R., Boreham, J., and Sutherland, I. (2004). Mortality in relation to smoking: 50 years' observations on male British doctors. *BMJ*, 328:1519.
4. West, R. (2006). Tobacco control: present and future. *Br Med Bull*, 77-78:123–136.
5. International Agency for Research on Cancer (2002). *Tobacco Smoke and Involuntary Smoking.* IARC Press, Lyon (France).
6. U.S. Department of Health and Human Services (2004). *The Health Consequences of Smoking: A Report of the Surgeon General*, U.S. Public Health Service, National Center for Chronic Disease Prevention and Health Promotion, Atlanta.
7. Roman, G.C. (2005). Vascular dementia prevention: A risk factor analysis. *Cerebrovasc Dis*, 20(Suppl. 2):91–100.
8. Seddon, J.M., George, S., and Rosner, B. (2006). Cigarette smoking, fish consumption, omega-3 fatty acid intake, and associations with age-related macular degeneration: The US Twin Study of Age-Related Macular Degeneration. *Arch Ophthalmol*, 124:995–1001.
9. Power, C., Frank, J., Hertzman, C., Schierhout, G., and Li, L. (2001). Predictors of low back pain onset in a prospective British study. *Am J Public Health*, 91:1671–1678.
10. Watkins, R.E. and Plant, A.J. (2006). Does smoking explain sex differences in the global tuberculosis epidemic? *Epidemiol Infect*, 134:333–339.
11. Meisinger, C., Doring, A., Thorand, B., and Lowel, H. (2006). Association of cigarette smoking and tar and nicotine intake with development of type 2 diabetes mellitus in men and women from the general population: The MONICA/KORA Augsburg Cohort Study. *Diabetologia*, 49:1770–1776.

12. Boffetta, P., Aagnes, B., Weiderpass, E., and Andersen, A. (2005). Smokeless tobacco use and risk of cancer of the pancreas and other organs. *Int J Cancer*, 114:992–995.

13. Wakschlag, L.S., Pickett, K.E., Kasza, K.E., and Loeber, R. (2006). Is prenatal smoking associated with a developmental pattern of conduct problems in young boys? *J Am Acad Child Adolesc Psychiatry*, 45:461–467.

14. Moller, A.M., Villebro, N., Pedersen, T., and Tonnesen, H. (2002). Effect of preoperative smoking intervention on postoperative complications: A randomised clinical trial. *Lancet*, 359:114–117.

15. Critchley, J.A. and Unal, B. (2003). Health effects associated with smokeless tobacco: A systematic review. *Thorax*, 58:435–443.

16. Royal College of Physicians of London (2000). *Nicotine Addiction in Britain: A Report of the Tobacco Advisory Group of the Royal College of Physicians*. Royal College of Physicians of London, London.

17. U.S. Department of Health and Human Services (1988). *The Health Consequences of Smoking: Nicotine Addiction - A Report of the Surgeon General* [DHHS Publication No. (CDC) 88-8406]. Rockville, MD: US Department of Health and Human Services, Public Health Service, Centers for Disease Control, Center for Health Promotion and Education, Office on Smoking and Health.

18. West, R. and Shiffman, S. (2007). *Smoking Cessation*. Health Press, Abingdon.

19. Karp, I., O'Loughlin, J., Hanley, J., Tyndale, R.F., and Paradis, G. (2006). Risk factors for tobacco dependence in adolescent smokers. *Tob Contr*, 15:199–204.

20. Rodriguez, D., Tercyak, K.P., and Audrain-McGovern, J. (2007). Effects of inattention and hyperactivity/impulsivity symptoms on development of nicotine dependence from mid adolescence to young adulthood. *J Pediatr Psychol* (in press).

21. Pomerleau, O.F. (1995). Individual differences in sensitivity to nicotine: Implications for genetic research on nicotine dependence. *Behav Genet*, 25:161–177.

22. Buisson, B. and Bertrand, D. (2002). Nicotine addiction: The possible role of functional upregulation. *Trends Pharmacol Sci*, 23:130–136.

23. Di Chiara, G. (1998). A motivational learning hypothesis of the role of mesolimbic dopamine in compulsive drug use. *J Psychopharmacol*, 12:54–67.

24. Kenny, P.J. and Markou, A. (2001). Neurobiology of the nicotine withdrawal syndrome. *Pharmacol Biochem Behav*, 70:531–549.

25. West, R.J. and Russell, M.A. (1985). Pre-abstinence smoke intake and smoking motivation as predictors of severity of cigarette withdrawal symptoms. *Psychopharmacology*, 87:334–336.

26. Waters, A.J., Shiffman, S., Sayette, M.A., Paty, J.A., Gwaltney, C.J., and Balabanis, M.H. (2004). Cue-provoked craving and nicotine replacement therapy in smoking cessation. *J Consult Clin Psychol*, 72:1136–1143.

27. Haberstick, B.C., Timberlake, D., Ehringer, M.A., Lessem, J.M., Hopfer, C.J. et al. (2007). Genes, time to first cigarette and nicotine dependence in a general population sample of young adults. *Addiction*, 102:655–665.

28. Shiffman, S., Elash, C.A., Paton, S.M., Gwaltney, C.J., Paty, J.A. et al. (2000). Comparative efficacy of 24-hour and 16-hour transdermal nicotine patches for relief of morning craving. *Addiction*, 95:1185–1195.

29. Hughes, J.R. (2007). Effects of abstinence from tobacco: valid symptoms and time course. *Nicotine Tob Res*, 9:315–327.

30. Shiffman, S. (2005). Dynamic influences on smoking relapse process. *J Pers*, 73:1715–1748.

31. Dawkins, L., Powell, J.H., West, R., Powell, J., and Pickering, A. (2007). A double-blind placebo-controlled experimental study of nicotine: II. Effects on response inhibition and executive functioning. *Psychopharmacology*, 190:457–467.

32. Lubman, D.I., Yucel, M., and Pantelis, C. (2004). Addiction, a condition of compulsive behaviour? Neuroimaging and neuropsychological evidence of inhibitory dysregulation. *Addiction*, 99:1491–1502.

33. West, R. and Hardy, A. (2006b). *Theory of Addiction*. Blackwell, Oxford.

34. West, R. (2006a). COPD Reviews: New approaches to smoking cessation [Online]. *Prous Science, Timely Topics in Medicine: Respiratory Diseases* [cited December 11 2007], available from: http://www.ttmed.com/respiratory/reviews.cfm?ID_Dis=8andID_Cou=23andID_TA=80

35. Balfour, D.J. (2004). The neurobiology of tobacco dependence: A preclinical perspective on the role of the dopamine projections to the nucleus accumbens. *Nicotine Tob Res*, 6:899–912.

36. Stolerman, I.P. and Jarvis, M.J. (1995). The scientific case that nicotine is addictive. *Psychopharmacology*, 117:2–10.

37. Henningfield, J. E. and Goldberg, S.R. (1983). Nicotine as a reinforcer in human subjects and laboratory animals. *Pharmacol Biochem Behav*, 19:989–992.

38. Swedberg, M.D., Henningfield, J.E., and Goldberg, S.R. (1990). Nicotine dependency: animal studies. In Wonnacott, S., Russell, M.A.H., and Stolerman, I.P. (eds.), *Nicotine Psychopharmacology: Molecular, Cellular, and Behavioural Aspects*. Oxford University Press, Oxford, pp. 38–76.

39. Rose, J.E. and Corrigall, W.A. (1997). Nicotine self-administration in animals and humans: Similarities and differences. *Psychopharmacology*, 130:28–40.

40. Harvey, D.M., Yasar, S., Heishman, S.J., Panlilio, L.V., Henningfield, J.E., and Goldberg, S.R. (2004). Nicotine serves as an effective reinforcer of intravenous drug-taking behavior in human cigarette smokers. *Psychopharmacology*, 175:134–142.

41. Perkins, K.A., Grobe, J.E., Caggiula, A., Wilson, A.S., and Stiller, R.L. (1997). Acute reinforcing effects of low-dose nicotine nasal spray in humans. *Pharmacol Biochem Behav*, 56:235–241.

42. Hughes, J.R., Gust, S.W., Keenan, R.M., and Fenwick, J.W. (1990). Effect of dose on nicotine's reinforcing, withdrawal-suppression and self-reported effects. *J Pharmacol Exp Ther*, 252:1175–1183.

43. Scherer, G. (1999). Smoking behaviour and compensation: A review of the literature. *Psychopharmacology*, 145:1–20.

44. Fagerstrom, K.O. and Hughes, J.R. (2002). Nicotine concentrations with concurrent use of cigarettes and nicotine replacement: A review. *Nicotine Tob Res*, 4(Suppl. 2):S73–S79.

45. Hecht, S.S., Murphy, S.E., Carmella, S.G., Zimmerman, C.L., Losey, L., Kramarczuk, I. *et al.* (2004). Effects of reduced cigarette smoking on the uptake of a tobacco-specific lung carcinogen. *J Natl Cancer Inst*, 96:107–115.

46. Butschky, M., Bailey, D., Henningfield, J.E., and Pickworth, W.B. (1995). Smoking without nicotine delivery decreases withdrawal in 12-hour abstinent smokers. *Pharmacol Biochem Behav*, 50:91–96.

47. Rose, J.E., Behm, F.M., Westman, E.C., Mathew, R.J., London, E.D., Hawk, T.C. *et al.* (2003). PET studies of the influences of nicotine on neural systems in cigarette smokers. *Am J Psychiatry*, 160:323–333.

48. Shahan, T.A., Bickel, W.K., Madden, G.J., and Badger, G.J. (1999). Comparing the reinforcing efficacy of nicotine containing and de-nicotinized cigarettes: a behavioral economic analysis. *Psychopharmacology*, 147:210–216.

49. Crooks, P.A. and Dwoskin, L.P. (1997). Contribution of CNS nicotine metabolites to the neuropharmacological effects of nicotine and tobacco smoking. *Biochem Pharmacol*, 54:743–753.

50. Jacob, P., Yu, L., Shulgin, A.T., and Benowitz, N.L. (1999). Minor tobacco alkaloids as biomarkers for tobacco use: Comparison of users of cigarettes, smokeless tobacco, cigars, and pipes. *Am J Public Health*, 89:731–736.

51. Bardo, M.T., Bevins, R.A., Klebaur, J.E., Crooks, P.A., and Dwoskin, L.P. (1997). (−)-Nornicotine partially substitutes for (+)-amphetamine in a drug discrimination paradigm in rats. *Pharmacol Biochem Behav*, 58:1083–1087.

52. Bardo, M.T., Green, T.A., Crooks, P.A., and Dwoskin, L.P. (1999). Nornicotine is self-administered intravenously by rats. *Psychopharmacology*, 146:290–296.

53. Caggiula, A.R., Donny, E.C., White, A.R., Chaudhri, N., Booth, S., Gharib, M.A. *et al.* (2001). Cue dependency of nicotine self-administration and smoking. *Pharmacol Biochem Behav*, 70:515–530.

54. Caggiula, A.R., Donny, E.C., White, A.R., Chaudhri, N., Booth, S., Gharib, M.A. *et al.* (2001). Environmental stimuli promote the acquisition of nicotine self-administration in rats. *Psychopharmacology*, 163:230–237.

55. Perkins, K.A., Gerlach, D., Vender, J., Grobe, J., Meeker, J., and Hutchison, S. (2001). Sex differences in the subjective and reinforcing effects of visual and olfactory cigarette smoke stimuli. *Nicotine Tob Res*, 3:141–150.

56. Belluzzi, J.D., Wang, R., and Leslie, F.M. (2005). Acetaldehyde enhances acquisition of nicotine self-administration in adolescent rats. *Neuropsychopharmacology*, 30:705–712.

57. Cao, J., Belluzzi, J.D., Loughlin, S.E., Keyler, D.E., Pentel, P.R., and Leslie, F.M. (2007). Acetaldehyde, a major constituent of tobacco smoke, enhances behavioral, endocrine, and neuronal responses to nicotine in adolescent and adult rats. *Neuropsychopharmacology*, 32:2025–2035.

58. Guillem, K., Vouillac, C., Koob, G.F., Cador, M., Stinus, L. (2008). Monoamine oxidase inhibition dramatically prolongs the duration of nicotine withdrawal-induced place aversion. *Biol Psychiatry*, 63:158–163.

59. American Psychiatric Association. (1994). *Diagnostic and Statistical Manual of Mental Disorders*, fourth ed. American Psychiatric Press, Washington DC.

60. World Health Organization (1992). *International Statistical Classification of Diseases and Related Health Problems*, 10th revision, World Health Organization, Geneva.

61. Geyer, M.A. and Markou, A. (1995). Animal models of psychiatric disorders. In Bloom, F.E. and Kupfer, D.J. (eds.), *Psychopharmacology: The Fourth Generation of Progress*. Raven Press, New York (NY), pp. 787–798.

62. Geyer, M.A. and Markou, A. (2002). The role of preclinical models in the development of psychotropic drugs. In Davis, K.L., Charney, D., Coyle, J.T., and Nemeroff, C. (eds.), *Neuropsychopharmacology: The Fifth Generation of Progress*. Lippincott Williams and Wilkins, Philadelphia, pp. 445–455.

63. Watkins, S.S., Epping-Jordan, M.P., Koob, G.F., and Markou, A. (1999). Blockade of nicotine self-administration with nicotinic antagonists in rats. *Pharmacol Biochem Behav*, 62:743–751.

64. Lerman, C., LeSage, M.G., Perkins, K.A., O'Malley, S.S., Siegel, S.J., Benowitz, N.L. *et al.* (2007). Translational research in medication development for nicotine dependence. *Nat Rev Drug Discov*, 6:746–762.

65. Rotenberg, K.S., Miller, R.P., and Adir, J. (1980). Pharmacokinetics of nicotine in rats after single-cigarette smoke inhalation. *J Pharm Sci*, 69:1087–1090.

66. Breslau, N., Fenn, N., and Peterson, E.L. (1993). Early smoking initiation and nicotine dependence in a cohort of young adults. *Drug Alcohol Depend*, 33:129–137.

67. Anthony, J.C. and Petronis, K.R. (1995). Early-onset drug use and risk of later drug problems. *Drug Alcohol Depend*, 40:9–15.

68. Adriani, W., Spijker, S., Deroche-Gamonet, V., Laviola, G., Le Moal, M., Smit, A.B. *et al.* (2003). Evidence for enhanced neurobehavioral vulnerability to nicotine during periadolescence in rats. *J Neurosci*, 23:4712–4716.

69. Adriani, W., Deroche-Gamonet, V., Le Moal, M., Laviola, G., and Piazza, P.V. (2006). Preexposure during or following adolescence differently affects nicotine-rewarding properties in adult rats. *Psychopharmacology*, 184:382–390.

70. Suto, N., Austin, J.D., and Vezina, P. (2001). Locomotor response to novelty predicts a rat's propensity to self-administer nicotine. *Psychopharmacology*, 158:175–180.

71. Doran, N., Spring, B., and McChargue, D. (2007). Effect of impulsivity on craving and behavioral reactivity to smoking cues. *Psychopharmacology*, 194:279–288.

72. Mitchell, S.H. (2004). Measuring impulsivity and modeling its association with cigarette smoking. *Behav Cogn Neurosci Rev*, 3:261-275.

73. Reynolds, B., Patak, M., and Shroff, P. (2007). Adolescent smokers rate delayed rewards as less certain than adolescent nonsmokers. *Drug Alcohol Depend*, 90:301-303.

74. Lipkus, I.M., Barefoot, J.C., Feaganes, J., Williams, R.B., and Siegler, I.C. (1994). A short MMPI scale to identify people likely to begin smoking. *J Pers Assess*, 62:213-222.

75. Bickel, W.K., Odum, A.L., and Madden, G.J. (1999). Impulsivity and cigarette smoking: Delay discounting in current, never, and ex-smokers. *Psychopharmacology*, 146:447-454.

76. Burke, J.D., Loeber, R., White, H.R., Stouthamer-Loeber, M., and Pardini, D.A. (2007). Inattention as a key predictor of tobacco use in adolescence. *J Abnorm Psychol*, 116:249-259.

77. Diergaarde, L., Pattij, T., Poortvliet, I., Hogenboom, F., de Vries, W., Schoffelmeer, A.N. *et al.* (2008). Impulsive choice and impulsive action predict vulnerability to distinct stages of nicotine seeking in rats. *Biol Psychiatry*, 63:301-308.

78. Paterson, N.E. and Markou, A. (2004). Prolonged nicotine dependence associated with extended access to nicotine self-administration in rats. *Psychopharmacology*, 173:64-72.

79. Valentine, J.D., Hokanson, J.S., Matta, S.G., and Sharp, B.M. (1997). Self-administration in rats allowed unlimited access to nicotine. *Psychopharmacology*, 133:300-304.

80. LeSage, M.G., Keyler, D.E., Shoeman, D., Raphael, D., Collins, G., and Pentel, P.R. (2002). Continuous nicotine infusion reduces nicotine self-administration in rats with 23-h/day access to nicotine. *Pharmacol Biochem Behav*, 72:279-289.

81. O'Dell, L.E., Chen, S.A., Smith, R.T., Specio, S.E., Balster, R.L., Paterson, N.E. *et al.* (2007). Extended access to nicotine self-administration leads to dependence: Circadian measures, withdrawal measures, and extinction behavior in rats. *J Pharmacol Exp Ther*, 320: 180-193.

82. Kenny, P.J. and Markou, A. (2006). Nicotine self-administration acutely activates brain reward systems and induces a long-lasting increase in reward sensitivity. *Neuropsychopharmacology*, 31:1203-1211.

83. Epping-Jordan, M.P., Watkins, S.S., Koob, G.F., and Markou, A. (1998). Dramatic decreases in brain reward function during nicotine withdrawal. *Nature*, 393:76-79.

84. Harrison, A.A., Liem, Y.T., and Markou, A. (2001). Fluoxetine combined with a serotonin-1A receptor antagonist reversed reward deficits observed during nicotine and amphetamine withdrawal in rats. *Neuropsychopharmacology*, 25:55-71.

85. Semenova, S. and Markou, A. (2003). Clozapine treatment attenuated somatic and affective signs of nicotine and amphetamine withdrawal in subsets of rats exhibiting hyposensitivity to the initial effects of clozapine. *Biol Psychiatry*, 54:1249-1264.

86. Shiffman, S., West, R., and Gilbert, D. (2004). Recommendation for the assessment of tobacco craving and withdrawal in smoking cessation trials. *Nicotine Tob Res*, 6:599-614.

87. Malin, D.H., Lake, J.R., Newlin-Maultsby, P., Roberts, L.K., Lanier, J.G., Carter, V.A. *et al.* (1992). Rodent model of nicotine abstinence syndrome. *Pharmacol Biochem Behav*, 43:779-784.

88. Malin, D.H., Lake, J.R., Carter, V.A., Cunningham, J.S., Hebert, K.M., Conrad, D.L. *et al.* (1994). The nicotinic antagonist mecamylamine precipitates nicotine abstinence syndrome in the rat. *Psychopharmacology*, 115:180-184.

89. Hildebrand, B.E., Nomikos, G.G., Bondjers, C., Nisell, M., and Svensson, T.H. (1997). Behavioral manifestations of the nicotine abstinence syndrome in the rat: Peripheral versus central mechanisms. *Psychopharmacology*, 129:348-356.

90. Isola, R., Vogelsberg, V., Wemlinger, T.A., Neff, N.H., and Hadjiconstantinou, M. (1999). Nicotine abstinence in the mouse. *Brain Res*, 850:189-196.

91. Carboni, E., Bortone, L., Giua, C., and Di Chiara, G. (2000). Dissociation of physical abstinence signs from changes in extracellular dopamine in the nucleus accumbens and in the prefrontal cortex of nicotine dependent rats. *Drug Alcohol Depend*, 58:93-102.

92. Semenova, S., Bespalov, A., and Markou, A. (2003). Decreased prepulse inhibition during nicotine withdrawal in DBA/2 J is reversed by nicotine self-administration. *Eur J Pharmacol*, 472:99-110.

93. Jonkman, S., Henry, B., Semenova, S., and Markou, A. (2005). Mild anxiogenic effects of nicotine withdrawal in mice. *Eur J Pharmacol*, 516:40-45.

94. Malin, D.H., Lake, J.R., Upchurch, T.P., Shenoi, M., Rajan, N., and Schweinle, W.E. (1998). Nicotine abstinence syndrome precipitated by the competitive nicotinic antagonist dihydro-ß-erythroidine. *Pharmacol Biochem Behav*, 60:609-613.

95. Watkins, S.S., Stinus, L., Koob, G.F., and Markou, A. (2000). Reward and somatic changes during precipitated nicotine withdrawal in rats: Centrally and peripherally mediated effects. *J Pharmacol Exp Ther*, 292:1053-1064.

96. Cryan, J.F., Bruijnzeel, A.W., Skjei, K.L., and Markou, A. (2003). Bupropion enhances brain reward function and reverses the affective and somatic aspects of nicotine withdrawal in the rat. *Psychopharmacology*, 168:347-358.

97. Malin, D.H., Lake, J.R., Schopen, C.K., Kirk, J.W., Sailer, E.E., Lawless, B.A. *et al.* (1997). Nicotine abstinence syndrome precipitated by central but not peripheral hexamethonium. *Pharmacol Biochem Behav*, 58:695-699.

98. Hughes, J.R. (1992). Tobacco withdrawal in self-quitters. *J Consult Clin Psychol*, 60:689-697.

99. Shiffman, S. and Waters, A.J. (2004). Negative affect and smoking lapses: a prospective analysis. *J Consult Clin Psychol*, 72:192-201.

100. Skjei, K.L. and Markou, A. (2003). Effects of repeated withdrawal episodes, nicotine dose, and duration of nicotine exposure on the severity and duration of nicotine withdrawal in rats. *Psychopharmacology*, 168:280-292.

101. Markou, A. and Koob, G.F. (1991). Postcocaine anhedonia: An animal model of cocaine withdrawal. *Neuropsychopharmacology*, 4:17-26.

102. Markou, A. and Koob, G.F. (1992). Construct validity of a self-stimulation threshold paradigm: Effects of reward and performance manipulations. *Physiol Behav*, 51:111-119.

103. Covey, L.S., Glassman, A.H., and Stetner, F. (1997). Major depression following smoking cessation. *Am J Psychiatry*, 154:263-265.

104. Schulteis, G., Markou, A., Cole, M., and Koob, G.F. (1995). Decreased brain reward produced by ethanol withdrawal. *Proc Natl Acad Sci USA*, 92:5880-5884.

105. Schulteis, G., Markou, A., Gold, L.H., Stinus, L., and Koob, G.F. (1994). Relative sensitivity to naloxone of multiple indices of opiate withdrawal: A quantitative dose-response analysis. *J Pharmacol Exp Ther*, 271:1391-1398.

106. Lin, D., Koob, G.F., and Markou, A. (1999). Differential effects of withdrawal from chronic amphetamine or fluoxetine administration on brain stimulation reward in the rat: Interactions between the two drugs. *Psychopharmacology*, 145:283-294.

107. Lin, D., Koob, G.F., and Markou, A. (2000). Time-dependent alterations in ICSS thresholds associated with repeated amphetamine administrations. *Pharmacol Biochem Behav*, 65:407-417.

108. Paterson, N.E., Myers, C., and Markou, A. (2000). Effects of repeated withdrawal from continuous amphetamine administration on brain reward function in rats. *Psychopharmacology*, 152:440-446.

109. Spielewoy, C. and Markou, A. (2003). Withdrawal from chronic phencyclidine treatment induces long-lasting depression in brain reward function. *Neuropsychopharmacology*, 28:1106-1116.

110. Watkins, S.S., Koob, G.F., and Markou, A. (2000). Neural mechanisms underlying nicotine addiction: Acute positive reinforcement and withdrawal. *Nicotine Tob Res*, 2:19-37.

111. Paterson, N.E., Balfour, D.J., and Markou, A. (2007). Chronic bupropion attenuated the anhedonic component of nicotine withdrawal in rats via inhibition of dopamine reuptake in the nucleus accumbens shell. *Eur J Neurosci*, 25:3099-3108.

112. Cryan, J.F., Gasparini, F., van Heeke, G., and Markou, A. (2003). Non-nicotinic neuropharmacological strategies for nicotine dependence: Beyond bupropion. *Drug Discov Today*, 8:1025–1034.

113. Kenny, P.J., Gasparini, F., and Markou, A. (2003). Group II metabotropic and α-amino-3-hydroxy-5-methyl-4-isoxazole propionate (AMPA)/kainate glutamate receptors regulate the deficit in brain reward function associated with nicotine withdrawal in rats. *J Pharmacol Exp Ther*, 306:1068–1076.

114. Carroll, M.E., Lac, S.T., Asencio, M., and Keenan, R.M. (1989). Nicotine dependence in rats. *Life Sci*, 45:1381–1388.

115. Semenova, S., Stolerman, I.P., and Markou, A. (2007). Chronic nicotine administration improves attention while nicotine withdrawal induces performance deficits in the 5-choice serial reaction time task in rats. *Pharmacol Biochem Behav*, 87:360–368.

116. Jonkman, S., Risbrough, V.B., Geyer, M.A., and Markou, A. (2007). Spontaneous nicotine withdrawal potentiates the effects of stress in rats. *Neuropsychopharmacology*. (in press).

117. Helton, D.R., Modlin, D.L., Tizzano, J.P., and Rasmussen, K. (1993). Nicotine withdrawal: a behavioral assessment using schedule controlled responding, locomotor activity, and sensorimotor reactivity. *Psychopharmacology*, 113:205–210.

118. Salas, R., Pieri, F., and De Biasi, M. (2004). Decreased signs of nicotine withdrawal in mice null for the β4 nicotinic acetylcholine receptor subunit. *J Neurosci*, 24:10035–10039.

119. Jonkman, S. and Markou, A. (2006). Blockade of nicotinic acetylcholine or dopamine D1-like receptors in the central nucleus of the amygdala or the bed nucleus of the stria terminalis does not precipitate nicotine withdrawal in nicotine-dependent rats. *Neurosci Lett*, 400:140–145.

120. Damaj, M.I., Kao, W., and Martin, B.R. (2003). Characterization of spontaneous and precipitated nicotine withdrawal in the mouse. *J Pharmacol Exp Ther*, 307:526–534.

121. Epstein, D.H., Preston, K.L., Stewart, J., and Shaham, Y. (2006). Toward a model of drug relapse: An assessment of the validity of the reinstatement procedure. *Psychopharmacology*, 189:1–16.

122. Chiamulera, C., Borgo, C., Falchetto, S., Valerio, E., and Tessari, M. (1996). Nicotine reinstatement of nicotine self-administration after long-term extinction. *Psychopharmacology*, 127:102–107.

123. Paterson, N.E., Froestl, W., and Markou, A. (2005). Repeated administration of the GABA$_B$ receptor agonist CGP44532 decreased nicotine self-administration, and acute administration decreased cue-reinstatement of nicotine-seeking in rats. *Neuropsychopharmacology*, 30:119–128.

124. Liechti, M.E., Lhuillier, L., Kaupmann, K., and Markou, A. (2007). Metabotropic glutamate 2/3 receptors in the ventral tegmental area and the nucleus accumbens shell are involved in behaviors relating to nicotine dependence. *J Neurosci*, 27:9077–9085.

125. Katz, J.L. and Higgins, S.T. (2003). The validity of the reinstatement model of craving and relapse to drug use. *Psychopharmacology*, 168:21–30.

126. McKay, J.R., Rutherford, M.J., Alterman, A.I., Cacciola, J.S., and Kaplan, M.R. (1995). An examination of the cocaine relapse process. *Drug Alcohol Depend*, 38:35–43.

127. Buczek, Y., Le, A.D., Wang, A., Stewart, J., and Shaham, Y. (1999). Stress reinstates nicotine seeking but not sucrose solution seeking in rats. *Psychopharmacology*, 144:183–188.

128. Zislis, G., Desai, T.V., Prado, M., Shah, H.P., and Bruijnzeel, A.W. (2007). Effects of the CRF receptor antagonist D-Phe CRF$_{(12-41)}$ and the α2-adrenergic receptor agonist clonidine on stress-induced reinstatement of nicotine-seeking behavior in rats. *Neuropharmacology*, 53:958–966.

129. Gossop, M., Green, L., Phillips, G., and Bradley, B. (1990). Factors predicting outcome among opiate addicts after treatment. *Br J Clin Psychol*, 29:209–216.

130. Gilpin, E.A., Pierce, J.P., and Farkas, A.J. (1997). Duration of smoking abstinence and success in quitting. *J Natl Cancer Inst*, 89:572–576.

131. Higgins, S.T., Badger, G.J., and Budney, A.J. (2000). Initial abstinence and success in achieving longer term cocaine abstinence. *Exp Clin Psychopharmacol*, 8:377–386.

132. McKay, J.R. and Weiss, R.V. (2001). A review of temporal effects and outcome predictors in substance abuse treatment studies with long-term follow-ups: Preliminary results and methodological issues. *Eval Rev*, 25:113–161.

133. Grimm, J.W., Hope, B.T., Wise, R.A., and Shaham, Y. (2001). Neuroadaptation: Incubation of cocaine craving after withdrawal. *Nature*, 412:141–142.

134. Tran-Nguyen, L.T., Fuchs, R.A., Coffey, G.P., Baker, D.A., O'Dell, L.E., and Neisewander, J.L. (1998). Time-dependent changes in cocaine-seeking behavior and extracellular dopamine levels in the amygdala during cocaine withdrawal. *Neuropsychopharmacology*, 19: 48–59.

135. Fagerstrom, K. and Balfour, D.J. (2006). Neuropharmacology and potential efficacy of new treatments for tobacco dependence. *Expert Opin Investig Drugs*, 15:107–116.

136. Shiffman, S., Ferguson, S.G., and Gwaltney, C.J. (2006). Immediate hedonic response to smoking lapses: Relationship to smoking relapse, and effects of nicotine replacement therapy. *Psychopharmacology*, 184:608–618.

137. De Vries, T.J., de Vries, W., Janssen, M.C., and Schoffelmeer, A.N. (2005). Suppression of conditioned nicotine and sucrose seeking by the cannabinoid-1 receptor antagonist SR141716A. *Behav Brain Res*, 161:164–168.

138. Liu, X., Caggiula, A.R., Palmatier, M.I., Donny, E.C., and Sved, A.F. (2008). Cue-induced reinstatement of nicotine-seeking behavior in rats: effect of bupropion, persistence over repeated tests, and its dependence on training dose. *Psychopharmacology*, 196:365–375.

139. Paterson, N.E., Balfour, D.J., and Markou, A. (2007). Chronic bupropion partially attenuated the anhedonic component of nicotine withdrawal in rats via inhibition of dopamine reuptake in the nucleus accumbens shell. *Nicotine Tob Res*, 25:3099–3108.

140. Bruijnzeel, A.W. and Markou, A. (2003). Characterization of the effects of bupropion on the reinforcing properties of nicotine and food in rats. *Synapse*, 50:20–28.

141. Markou, A., Weiss, F., Gold, L.H., Caine, S.B., Schulteis, G., and Koob, G.F. (1993). Animal models of drug craving. *Psychopharmacology*, 112:163–182.

142. Paterson, N.E. and Markou, A. (2005). The metabotropic glutamate receptor 5 antagonist MPEP decreased break points for nicotine, cocaine and food in rats. *Psychopharmacology*, 179:255–261.

143. Paterson, N.E., Semenova, S., Gasparini, F., and Markou, A. (2003). The mGluR5 antagonist MPEP decreased nicotine self-administration in rats and mice. *Psychopharmacology*, 167:257–264.

144. Mombereau, C., Lhuillier, L., Kaupmann, K., and Cryan, J.F. (2007). GABA$_B$ receptor-positive modulation-induced blockade of the rewarding properties of nicotine is associated with a reduction in nucleus accumbens ΔFosB accumulation. *J Pharmacol Exp Ther*, 321:172–177.

145. United Nations International Drug Control Programme and World Health Organization (1992). *Informal Expert Group Meeting on the Craving Mechanism* [report no. V92-54439T]. United Nations International Drug Control Programme and World Health Organization, Geneva.

146. Wise, R.A. (1988). The neurobiology of craving: implications for the understanding and treatment of addiction. *J Abnorm Psychol*, 97:118–132.

147. Koob, G.F. (1992). Drugs of abuse: anatomy, pharmacology and function of reward pathways. *Trends Pharmacol Sci*, 13:177–184.

148. Hyman, S.E. (1993). Molecular and cell biology of addiction. *Curr Opin Neurol Neurosurg*, 6:609–613.

149. Corrigall, W.A. and Coen, K.M. (1989). Nicotine maintains robust self-administration in rats on a limited-access schedule. *Psychopharmacology*, 99:473–478.
150. DeNoble, V.J. and Mele, P.C. (2006). Intravenous nicotine self-administration in rats: Effects of mecamylamine, hexamethonium and naloxone. *Psychopharmacology*, 184:266–272.
151. Tessari, M., Valerio, E., Chiamulera, C., and Beardsley, P.M. (1995). Nicotine reinforcement in rats with histories of cocaine self-administration. *Psychopharmacology*, 121:282–283.
152. Donny, E.C., Caggiula, A.R., Knopf, S., and Brown, C. (1995). Nicotine self-administration in rats. *Psychopharmacology*, 122:390–394.
153. Shoaib, M., Schindler, C.W., and Goldberg, S.R. (1997). Nicotine self-administration in rats: Strain and nicotine pre-exposure effects on acquisition. *Psychopharmacology*, 129:35–43.
154. Donny, E.C., Caggiula, A.R., Mielke, M.M., Jacobs, K.S., Rose, C., and Sved, A.F. (1998). Acquisition of nicotine self-administration in rats: The effects of dose, feeding schedule, and drug contingency. *Psychopharmacology*, 136:83–90.
155. Corrigall, W.A. (1999). Nicotine self-administration in animals as a dependence model. *Nicotine Tob Res*, 1:11–20.
156. Clarke, P.B. and Fibiger, H.C. (1987). Apparent absence of nicotine-induced conditioned place preference in rats. *Psychopharmacology*, 92:84–88.
157. Fudala, P.J., Teoh, K.W., and Iwamoto, E.T. (1985). Pharmacologic characterization of nicotine-induced conditioned place preference. *Pharmacol Biochem Behav*, 22:237–241.
158. Fudala, P.J. and Iwamoto, E.T. (1986). Further studies on nicotine-induced conditioned place preference in the rat. *Pharmacol Biochem Behav*, 25:1041–1049.
159. Fudala, P.J. and Iwamoto, E.T. (1987). Conditioned aversion after delay place conditioning with nicotine. *Psychopharmacology*, 92:376–381.
160. Gershell, L.J. and Atkins, J.H. (2003). A brief history of novel drug discovery technologies. *Nat Rev Drug Discov*, 2:321–327.
161. Damaj, M.I. and Martin, B.R. (1993). Is the dopaminergic system involved in the central effects of nicotine in mice? *Psychopharmacology*, 111:106–108.
162. Dixit, K.S., Dhasmana, K.M., Saxena, R.C., and Kohli, R.P. (1971). Antagonism of intracerebrally induced nicotinic convulsions in mice: A method for measuring the central antinicotinic activity of CNS acting agents. *Psychopharmacologia*, 19:67–72.
163. Miner, L.L. and Collins, A.C. (1988). The effect of chronic nicotine treatment on nicotine-induced seizures. *Psychopharmacology*, 95:52–55.
164. Miner, L.L., Marks, M.J., and Collins, A.C. (1985). Relationship between nicotine-induced seizures and hippocampal nicotinic receptors. *Life Sci*, 37:75–83.
165. Clarke, P.B. and Kumar, R. (1983). The effects of nicotine on locomotor activity in non-tolerant and tolerant rats. *Br J Pharmacol*, 78:329–337.
166. Ksir, C., Hakan, R., Hall, D.P. Jr. and Kellar, K.J. (1985). Exposure to nicotine enhances the behavioral stimulant effect of nicotine and increases binding of [^{3}H]acetylcholine to nicotinic receptors. *Neuropharmacology*, 24:527–531.
167. Hatchell, P.C. and Collins, A.C. (1980). The influence of genotype and sex on behavioral sensitivity to nicotine in mice. *Psychopharmacology*, 71:45–49.
168. Nordberg, A. and Bergh, C. (1985). Effect of nicotine on passive avoidance behaviour and motoric activity in mice. *Acta Pharmacol Toxicol (Copenh)*, 56:337–341.
169. Gould, T.J. and Wehner, J.M. (1999). Nicotine enhancement of contextual fear conditioning. *Behav Brain Res*, 102:31–39.
170. Mansner, R., Alhava, E., and Klinge, E. (1974). Nicotine hypothermia and brain nicotine and catecholamine levels in the mouse. *Med Biol*, 52:390–398.
171. Damaj, M.I., Glassco, W., Dukat, M., and Martin, B.R. (1999). Pharmacological characterization of nicotine-induced seizures in mice. *J Pharmacol Exp Ther*, 291:1284–1291.
172. Aguzzi, A., Brandner, S., Sure, U., Rüedi, D., and Isenmann, S. (1994). Transgenic and knockout mice: models of neurological disease. *Brain Pathol*, 4:3–20.

173. Pich, E.M. and Epping-Jordan, M.P. (1998). Transgenic mice in drug dependence research. *Ann Med*, 30:390-396.

174. Picciotto, M.R., Zoli, M., Rimondini, R., Lena, C., Marubio, L.M., Pich, E.M. *et al.* (1998). Acetylcholine receptors containing the β2 subunit are involved in the reinforcing properties of nicotine. *Nature*, 391:173-177.

175. Marubio, L.M. and Paylor, R. (2004). Impaired passive avoidance learning in mice lacking central neuronal nicotinic acetylcholine receptors. *Neuroscience*, 129:575-582.

176. Tapper, A.R., McKinney, S.L., Nashmi, R., Schwarz, J., Deshpande, P., Labarca, C. *et al.* (2004). Nicotine activation of α4* receptors: sufficient for reward, tolerance, and sensitization. *Science*, 306:1029-1032.

177. Molles, B.E., Maskos, U., Pons, S., Besson, M., Guiard, P., Guilloux, J.P. *et al.* (2006). Targeted in vivo expression of nicotinic acetylcholine receptors in mouse brain using lentiviral expression vectors. *J Mol Neurosci*, 30:105-106.

178. Marubio, L.M. and Changeux, J. (2000). Nicotinic acetylcholine receptor knockout mice as animal models for studying receptor function. *Eur J Pharmacol*, 393:113-121.

179. Champtiaux, N. and Changeux, J.P. (2004). Knockout and knockin mice to investigate the role of nicotinic receptors in the central nervous system. *Prog Brain Res*, 145:235-251.

180. Lester, H.A., Fonck, C., Tapper, A.R., McKinney, S., Damaj, M.I., Balogh, S. *et al.* (2003). Hypersensitive knockin mouse strains identify receptors and pathways for nicotine action. *Curr Opin Drug Discov Devel*, 6:633-639.

181. Marubio, L.M., del Mar Arroyo-Jimenez, M., Cordero-Erausquin, M., Léna, C., Le Novère, N., de Kerchove d'Exaerde, A. *et al.* (1999). Reduced antinociception in mice lacking neuronal nicotinic receptor subunits. *Nature*, 398:805-810.

182. Damaj, M.I., Fei-Yin, M., Dukat, M., Glassco, W., Glennon, R.A., and Martin, B.R. (1998). Antinociceptive responses to nicotinic acetylcholine receptor ligands after systemic and intrathecal administration in mice. *J Pharmacol Exp Ther*, 284:1058-1065.

183. Rasmussen, T. and Swedberg, M.D. (1998). Reinforcing effects of nicotinic compounds: Intravenous self-administration in drug-naive mice. *Pharmacol Biochem Behav*, 60:567-573.

184. Stolerman, I.P., Naylor, C., Elmer, G.I., and Goldberg, S.R. (1999). Discrimination and self-administration of nicotine by inbred strains of mice. *Psychopharmacology*, 141:297-306.

185. Museo, E. and Wise, R.A. (1990). Locomotion induced by ventral tegmental microinjections of a nicotinic agonist. *Pharmacol Biochem Behav*, 35:735-737.

186. Benwell, M.E. and Balfour, D.J. (1992). The effects of acute and repeated nicotine treatment on nucleus accumbens dopamine and locomotor activity. *Br J Pharmacol*, 105:849-856.

187. Clarke, P.B., Fu, D.S., Jakubovic, A., and Fibiger, H.C. (1988). Evidence that mesolimbic dopaminergic activation underlies the locomotor stimulant action of nicotine in rats. *J Pharmacol Exp Ther*, 246:701-708.

188. Fung, Y.K. (1990). The importance of nucleus accumbens in nicotine-induced locomotor activity. *J Pharm Pharmacol*, 42:595-596.

189. Imperato, A., Mulas, A., and Di Chiara, G. (1986). Nicotine preferentially stimulates dopamine release in the limbic system of freely moving rats. *Eur J Pharmacol*, 132:337-338.

190. Marshall, D.L., Redfern, P.H., and Wonnacott, S. (1997). Presynaptic nicotinic modulation of dopamine release in the three ascending pathways studied by in vivo microdialysis: Comparison of naive and chronic nicotine-treated rats. *J Neurochem*, 68:1511-1519.

191. Yu, Z.J. and Wecker, L. (1994). Chronic nicotine administration differentially affects neurotransmitter release from rat striatal slices. *J Neurochem*, 63:186-194.

192. Ksir, C., Hakan, R.L., and Kellar, K.J. (1987). Chronic nicotine and locomotor activity: Influences of exposure dose and test dose. *Psychopharmacology*, 92:25-29.

193. Reavill, C. and Stolerman, I.P. (1990). Locomotor activity in rats after administration of nicotinic agonists intracerebrally. *Br J Pharmacol*, 99:273-278.

194. Nisell, M., Nomikos, G.G., and Svensson, T.H. (1994). Systemic nicotine-induced dopamine release in the rat nucleus accumbens is regulated by nicotinic receptors in the ventral tegmental area. *Synapse*, 16:36–44.

195. Nisell, M., Nomikos, G.G., Hertel, P., Panagis, G., and Svensson, T.H. (1996). Condition-independent sensitization of locomotor stimulation and mesocortical dopamine release following chronic nicotine treatment in the rat. *Synapse*, 22:369–381.

196. Pontieri, F.E., Tanda, G., Orzi, F., and Di Chiara, G. (1996). Effects of nicotine on the nucleus accumbens and similarity to those of addictive drugs. *Nature*, 382:255–257.

197. Balfour, D.J., Benwell, M.E., Birrell, C.E., Kelly, R.J., and Al-Aloul, M. (1998). Sensitization of the mesoaccumbens dopamine response to nicotine. *Pharmacol Biochem Behav*, 59:1021–1030.

198. Shoaib, M. and Stolerman, I.P. (1992). MK801 attenuates behavioural adaptation to chronic nicotine administration in rats. *Br J Pharmacol*, 105:514–515.

199. Chergui, K., Charléty, P.J., Akaoka, H., Saunier, C.F., Brunet, J.L., Buda, M. *et al.* (1993). Tonic activation of NMDA receptors causes spontaneous burst discharge of rat midbrain dopamine neurons in vivo. *Eur J Neurosci*, 5:137–144.

200. Robinson, T.E. and Berridge, K.C. (1993). The neural basis of drug craving: an incentive-sensitization theory of addiction. *Brain Res Brain Res Rev*, 18:247–291.

201. Le Foll, B. and Goldberg, S.R. (2006). Nicotine as a typical drug of abuse in experimental animals and humans. *Psychopharmacology*, 184:367–381.

202. Corrigall, W.A. and Coen, K.M. (1991). Selective dopamine antagonists reduce nicotine self-administration. *Psychopharmacology*, 104:171–176.

203. Corrigall, W.A. and Coen, K.M. (1991). Cocaine self-administration is increased by both D1 and D2 dopamine antagonists. *Pharmacol Biochem Behav*, 39:799–802.

204. Corrigall, W.A., Franklin, K.B., Coen, K.M., and Clarke, P.B. (1992). The mesolimbic dopaminergic system is implicated in the reinforcing effects of nicotine. *Psychopharmacology*, 107:285–289.

205. Corrigall, W.A., Coen, K.M., and Adamson, K.L. (1994). Self-administered nicotine activates the mesolimbic dopamine system through the ventral tegmental area. *Brain Res*, 653:278–284.

206. Andreoli, M., Tessari, M., Pilla, M., Valerio, E., Hagan, J.J., and Heidbreder, C.A. (2003). Selective antagonism at dopamine D_3 receptors prevents nicotine-triggered relapse to nicotine-seeking behavior. *Neuropsychopharmacology*, 28:1272–1280.

207. Bespalov, A.Y., Dravolina, O.A., Sukhanov, I., Zakharova, E., Blokhina, E., Zvartau, E. *et al.* (2005). Metabotropic glutamate receptor (mGluR5) antagonist MPEP attenuated cue- and schedule-induced reinstatement of nicotine self-administration behavior in rats. *Neuropharmacology*, 49(Suppl. 1):167–178.

208. Tessari, M., Pilla, M., Andreoli, M., Hutcheson, D.M., and Heidbreder, C.A. (2004). Antagonism at metabotropic glutamate 5 receptors inhibits nicotine- and cocaine-taking behaviours and prevents nicotine-triggered relapse to nicotine-seeking. *Eur J Pharmacol*, 499:121–133.

209. Dravolina, O.A., Zakharova, E.S., Shekunova, E.V., Zvartau, E.E., Danysz, W., and Bespalov, A.Y. (2007). mGlu1 receptor blockade attenuates cue- and nicotine-induced reinstatement of extinguished nicotine self-administration behavior in rats. *Neuropharmacology*, 52:263–269.

210. Liu, X., Caggiula, A.R., Yee, S.K., Nobuta, H., Sved, A.F., Pechnick, R.N. *et al.* (2007). Mecamylamine attenuates cue-induced reinstatement of nicotine-seeking behavior in rats. *Neuropsychopharmacology*, 32:710–718.

211. Corrigall, W.A., Coen, K.M., Adamson, K.L., Chow, B.L., and Zhang, J. (2000). Response of nicotine self-administration in the rat to manipulations of mu-opioid and γ-aminobutyric acid receptors in the ventral tegmental area. *Psychopharmacology*, 149:107–114.

212. Paterson, N.E., Froestl, W., and Markou, A. (2004). The $GABA_B$ receptor agonists baclofen and CGP44532 decreased nicotine self-administration in the rat. *Psychopharmacology*, 172:179–186.

213. Corrigall, W.A. and Coen, K.M. (1994). Nicotine self-administration and locomotor activity are not modified by the 5-HT$_3$ antagonists ICS 205-930 and MDL 72222. *Pharmacol Biochem Behav*, 49:67-71.

214. Rauhut, A.S., Mullins, S.N., Dwoskin, L.P., and Bardo, M.T. (2002). Reboxetine: attenuation of intravenous nicotine self-administration in rats. *J Pharmacol Exp Ther*, 303:664-672.

215. Fu, Y., Matta, S.G., Brower, V.G., and Sharp, B.M. (2001). Norepinephrine secretion in the hypothalamic paraventricular nucleus of rats during unlimited access to self-administered nicotine: An in vivo microdialysis study. *J Neurosci*, 21:8979-8989.

216. Guillem, K., Vouillac, C., Azar, M.R., Parsons, L.H., Koob, G.F., Cador, M. *et al.* (2005). Monoamine oxidase inhibition dramatically increases the motivation to self-administer nicotine in rats. *J Neurosci*, 25:8593-8600.

217. Rauhut, A.S., Neugebauer, N., Dwoskin, L.P., and Bardo, M.T. (2003). Effect of bupropion on nicotine self-administration in rats. *Psychopharmacology*, 169:1-9.

218. Shoaib, M., Sidhpura, N., and Shafait, S. (2003). Investigating the actions of bupropion on dependence-related effects of nicotine in rats. *Psychopharmacology*, 165:405-412.

219. Cohen, C., Perrault, G., Voltz, C., Steinberg, R., and Soubrié, P. (2002). SR141716, a central cannabinoid (CB$_1$) receptor antagonist, blocks the motivational and dopamine-releasing effects of nicotine in rats. *Behav Pharmacol*, 13:451-463.

220. Ferry, L.H. (1999). Non-nicotine pharmacotherapy for smoking cessation. *Prim Care*, 26:653-669.

221. Bartfai, T. and Lees, G.V. (2006). *Drug Discovery: From bedside to Wall Street*. Elsevier Academic Press, New York.

222. Helton, D.R., Tizzano, J.P., Monn, J.A., Schoepp, D.D., and Kallman, M.J. (1997). LY354740: A metabotropic glutamate receptor agonist which ameliorates symptoms of nicotine withdrawal in rats. *Neuropharmacology*, 36:1511-1516.

223. Pilla, M., Perachon, S., Sautel, F., Garrido, F., Mann, A., Wermuth, C.G. *et al.* (1999). Selective inhibition of cocaine-seeking behaviour by a partial dopamine D$_3$ receptor agonist. *Nature*, 400:371-375.

224. Balfour, D.J. (2001). The pharmacology underlying pharmacotherapy for tobacco dependence: A focus on bupropion. *Int J Clin Pract*, 55:53-57.

225. Shaham, Y., Shalev, U., Lu, L., De Wit, H., and Stewart, J. (2003). The reinstatement model of drug relapse: History, methodology and major findings. *Psychopharmacology*, 168:3-20.

226. Shaham, Y., Adamson, L.K., Grocki, S., and Corrigall, W.A. (1997). Reinstatement and spontaneous recovery of nicotine seeking in rats. *Psychopharmacology*, 130:396-403. [erratum: 133:106].

227. de Wit, H. and Stewart, J. (1983). Drug reinstatement of heroin-reinforced responding in the rat. *Psychopharmacology*, 79:29-31.

228. Vitaterna, M.H., Pinto, L.H., and Takahashi, J.S. (2006). Large-scale mutagenesis and phenotypic screens for the nervous system and behavior in mice. *Trends Neurosci*, 29:233-240.

229. Tecott, L.H. and Nestler, E.J. (2004). Neurobehavioral assessment in the information age. *Nat Neurosci*, 7:462-466.

230. Crabbe, J.C. and Morris, R.G. (2004). Festina lente: Late-night thoughts on high-throughput screening of mouse behavior. *Nat Neurosci*, 7:1175-1179.

231. Suckow, R.F., Smith, T.M., Perumal, A.S., and Cooper, T.B. (1986). Pharmacokinetics of bupropion and metabolites in plasma and brain of rats, mice, and guinea pigs. *Drug Metab Dispos*, 14:692-697.

232. Lamb, R.J. and Griffiths, R.R. (1990). Self-administration in baboons and the discriminative stimulus effects in rats of bupropion, nomifensine, diclofensine and imipramine. *Psychopharmacology*, 102:183-190.

233. Wilens, T.E., Spencer, T.J., Biederman, J., Girard, K., Doyle, R., Prince, J. *et al.* (2001). A controlled clinical trial of bupropion for attention deficit hyperactivity disorder in adults. *Am J Psychiatry*, 158:282–288.

234. Morowitz, H. (1995). The emergence of complexity. *Complexity*, 1:4.

235. Swan, G.E. (1999). Implications of genetic epidemiology for the prevention of tobacco use. *Nicotine Tob Res*, 1(Suppl. 1):S49–S56.

236. Houtsmuller, E.J., Thornton, J.A., and Stitzer, M.L. (2002). Effects of selegiline (L-deprenyl) during smoking and short-term abstinence. *Psychopharmacology*, 163:213–220.

237. Strasser, A.A., Ashare, R.L., Kozlowski, L.T., and Pickworth, W.B. (2005). The effect of filter vent blocking and smoking topography on carbon monoxide levels in smokers. *Pharmacol Biochem Behav*, 82:320–329.

238. Foulds, J., Stapleton, J., Feyerabend, C., Vesey, C., Jarvis, M., and Russell, M.A. (1992). Effect of transdermal nicotine patches on cigarette smoking: A double blind crossover study. *Psychopharmacology*, 106:421–427.

239. West, R. and Hajek, P. (2004). Evaluation of the mood and physical symptoms scale (MPSS) to assess cigarette withdrawal. *Psychopharmacology*, 177:195–199.

240. Benowitz, N.L., Zevin, S., and Jacob, P. (1998). Suppression of nicotine intake during ad libitum cigarette smoking by high-dose transdermal nicotine. *J Pharmacol Exp Ther*, 287:958–962.

241. Killen, J.D. and Fortmann, S.P. (1997). Craving is associated with smoking relapse: Findings from three prospective studies. *Exp Clin Psychopharmacol*, 5:137–142.

242. West, R.J., Hajek, P., and Belcher, M. (1989). Severity of withdrawal symptoms as a predictor of outcome of an attempt to quit smoking. *Psychol Med*, 19:981–985.

243. West, R., Ussher, M., Evans, M., and Rashid, M. (2006). Assessing DSM-IV nicotine withdrawal symptoms: A comparison and evaluation of five different scales. *Psychopharmacology*, 184:619–627.

244. Cahill, K., Stead, L.F., and Lancaster, T. (2007). Nicotine receptor partial agonists for smoking cessation. *Cochrane Database Syst Rev*, 1:CD006103.

245. Gourlay, S.G., Stead, L.F., and Benowitz, N.L. (2004). Clonidine for smoking cessation. *Cochrane Database Syst Rev*, 3:CD000058.

246. Hughes, J., Stead, L.F., and Lancaster, T. (2007). Antidepressants for smoking cessation. *Cochrane Database Syst Rev*, 1:CD000031.

247. Silagy, C., Lancaster, T., Stead, L., Mant, D., and Fowler, G. (2004). Nicotine replacement therapy for smoking cessation. *Cochrane Database Syst Rev*, 3:CD000146.

248. West, R., Baker, C., Capellerri, J., Bushmakin, A. (2008). Effect of varenicline and bupropion SR on craving, nicotine withdrawal symptoms, and rewarding effects of smoking during a quit attempt. *Psychopharmacology*, 197:371–377.

249. Zatonski, W., Cedzynska, M., Tutka, P., and West, R. (2006). An uncontrolled trial of cytisine (Tabex) for smoking cessation. *Tob Control*, 15:481–484.

250. West, R., Hajek, P., and McNeill, A. (1991). Effect of buspirone on cigarette withdrawal symptoms and short-term abstinence rates in a smokers clinic. *Psychopharmacology*, 104:91–96.

251. Unrod, M., Cook, T., Myers, M.G., and Brown, S.A. (2004). Smoking cessation efforts among substance abusers with and without psychiatric comorbidity. *Addict Behav*, 29:1009–1013.

252. West, R. and Jarvis, M. (2005). Tobacco smoking and mental disorder. *Ital J Psychiatry Behav Sci*, 15:10–17.

253. Markou, A., Kosten, T.R., and Koob, G.F. (1998). Neurobiological similarities in depression and drug dependence: A self-medication hypothesis. *Neuropsychopharmacology*, 18:135–174.

254. George, T.P., Sernyak, M.J., Ziedonis, D.M., and Woods, S.W. (1995). Effects of clozapine on smoking in chronic schizophrenic outpatients. *J Clin Psychiatry*, 56:344–346.

Development of Medications for Heroin and Cocaine Addiction and Regulatory Aspects of Abuse Liability Testing

Beatriz A. Rocha[1], Jack Bergman[2], Sandra D. Comer[3], Margaret Haney[3] and Roger D. Spealman[4]

[1]Merck Research Laboratories, Rahway, NJ, USA
[2]Harvard Medical School, ADARC-McLean Hospital, Belmont, MA, USA
[3]College of Physicians and Surgeons of Columbia University, Department of Psychiatry, New York, NY, USA
[4]Harvard Medical School, New England Primate Research Center, Southborough, MA, USA

Animal and Translational Models for CNS Drug Discovery,
Vol. 3 of 3: Reward Deficit Disorders
Robert McArthur and Franco Borsini (eds), Academic Press, 2008

INTRODUCTION

Drug addiction, and in particular cocaine and heroin addiction, represents a terrible social burden and public health issue due to its high incidence among the youth, poor treatment outcomes, and high rate of relapse. According to the 2006 National Survey on Drug Dependence, sponsored by the Substance Abuse and Mental Health Services

Administration (SAMHSA) of the US Department of Health and Human Services,[1] 2.4 million persons were estimated to be current cocaine users, and 338 000 current heroin users. While the rate of cocaine use has remained stable between 2002 and 2006, the prevailing rate of current heroin use has increased from 0.06% to 0.14% of the US population.

Estimates of first use within the last year suggest that among the population aged 12 years or older, 977 000 persons used cocaine for the first time, and 91 000 persons initiated heroin use. Even though the incidence of cocaine initiation was approximately 10 times that of heroin, the average age at first use among recent cocaine and heroin first users was approximately 20 years for both drugs. In regard to treatment, while an estimated 7.8 million persons aged 12 years or older (3.2% of the total population in this age group) needed treatment for an illicit drug use problem in 2006, only 1.6 million actually received treatment in the past year. Of those, 928 000 were treated for cocaine abuse, and 466 000 were treated for heroin abuse. These data highlight the fact that the vast majority of substance abusers who need treatment do not receive it. Medication approaches to treating addiction may vary at different points in the treatment process. If the treatment process includes an initial detoxification phase, as in the case of drugs like heroin that produce physical dependence and a related abstinence syndrome, pharmacotherapeutic management often can be symptomatic. For example, alpha-2 adrenergic drugs, such as clonidine, can be useful for relieving the immediate physical distress associated with opioid withdrawal, including watery eyes and nose, diarrhea, and irritability.[2] Such drugs also have been proposed to have utility in alleviating symptoms associated with withdrawal from other types of drug dependence, for example, in alcoholism.

When maintenance rather than withdrawal is the primary concern, an effective therapeutic dosing regimen of the treatment drug must be established in the individual. Sometimes, this can be controversial. For example, if an opioid receptor antagonist is chosen for treatment of heroin addiction, the dosing regimen is more or less rationally based on the goal of blocking the effects of doses of heroin to which the individual is likely to be exposed. On the other hand, if an opioid such as methadone is chosen for treatment of heroin addiction, the dosage may be dictated not only by the goal of reducing heroin-seeking behavior, but also by legal restrictions on the dosage that can be administered in the particular treatment program.

There is consensus among the medical/scientific community and regulators that successful management of drug addiction depends not only on the pharmacological efficacy of a given medication, but also on a composite of patient and treatment-related factors. These include psychosocial therapy, psychiatric comorbidities, patient motivation, and the general environment in which treatment occurs. In particular, psychosocial counseling, which is intended to assist the patient in developing control over drug seeking and prevent relapse into abuse, is currently considered to be an essential component of treatment. Emphasizing this point, the US and European labels of the products currently approved for the treatment of opioid dependence, Subutex® (buprenorphine) and Suboxone® (fixed-dose combination of buprenorphine and naloxone (Narcan®)), include the statement that "… all efficacy data were generated in conjunction with psychosocial counseling as part of a comprehensive addiction treatment program," and "that there have been no clinical studies conducted to assess the efficacy of buprenorphine as the only component of treatment." At present, there is no clearly prescribed course of either pharmacotherapeutic or psychosocial treatment for different types of drug addiction. Nevertheless, several studies stress the point that the longer that patients remain in treatment, the better the treatment outcome (for review see[3]).

The limited number of medications approved for the treatment of heroin dependence and the lack of medications approved for the treatment of cocaine dependence suggest the many obstacles that must be overcome in developing drug treatments for these indications. Beyond the inherent difficulty of translating preclinical data into clinical efficacy (as detailed throughout the present chapter series), the development of medications for heroin and cocaine dependence provides some unique challenges, among them attenuating the clinical features of both acute and chronic drug intoxication that may involve – to varying extents in different individuals – drug-induced tolerance and physical dependence. In addition, the multifaceted presentation of drug addiction – somatic and/or effectual signs and symptoms that may be short-lived or long-lived – makes it especially difficult to address with a single medication. Moreover, the reluctance of the pharmaceutical industry to prioritize development of compounds for the indications of heroin and cocaine dependence adds "marketplace complexity" to the problem[i] (cf.,[4]).

Notwithstanding the formidable challenge of developing medications for heroin and cocaine dependence, the pharmaceutical industry retains a vested interest in assessing abuse liability issues in drug development. In response to regulatory requirements the pharmaceutical industry has relied on preclinical assays with accepted face validity, such as drug self-administration and drug discrimination, for the initial evaluation of the abuse liability of new chemical entities. From an industry perspective, such assays are indispensable for assessing the abuse potential of new compounds – independently of their application in the search for medications to combat cocaine or heroin abuse.

Assessing abuse potential is a process that extends beyond the introduction of a new compound into clinical use. Thus, abuse potential assessment encompasses multidisciplinary approaches from animal pharmacology and human laboratory tests for compounds under development, and continues with post-marketing surveillance of approved compounds to measure abuse liability. For compounds under development, the main goal is to determine – based on abuse potential assessments – the level of regulatory control that may be needed to assure the best risk/benefit ratio between medical and non-medical use. The assessment of the abuse potential of compounds under development provides data to regulatory agencies and guides labeling, and is an integral aspect of drug abuse prevention.[5] For approved compounds that nonetheless are subject to abuse, the goals change and become those of identifying the scope of the problem as it emerges through post-marketing surveillance techniques (e.g., spontaneous reports of adverse events, surveillance of safety databases), and of reviewing existing levels of regulatory control as necessary. Such post-marketing surveillance has

---

[i] Please refer to McCann *et al.*, Drug discovery and development for reward disorders: Views from government, in this volume for further discussion on industry-governmental perceptions of development of compounds for the treatment of heroin and cocaine dependence.

little, if any, impact on preclinical abuse liability evaluation. However, post-marketing surveillance has become an integral component of the regulatory approval process, offering both institutional accountability and some assurance that errors in preclinical abuse liability assessment can be corrected in a timely manner.

While the concept of abuse liability applies to any compound that enters the central nervous system (CNS) (i.e., "… the likelihood that a drug with CNS or anabolic effects will sustain patterns of non-medical self-administration that result in disruptive or undesirable consequences"),[6] much of the historical framework for classification of drugs with abuse liability has depended on the field's development. Thus, early classification of psychoactive compounds always included an evaluation of similarities with "stimulants" and "depressants." Advances in pharmacology subsequently led to a more detailed classification of compounds with abuse liability based on drug type, for example, amphetamine-like, morphine-like or pentobarbital-like. However, increasing diversity in the pharmacology of psychoactive substances has made it clear that such traditional designations are inadequate for regulatory purposes. For example, the club drug MDMA, which has both amphetamine-like and D-lysergic acid diethylamide (LSD)-like properties, did not easily fit into a single designation and, perhaps as a consequence, was not closely regulated until its abuse liability was already widely reported. The proper classification of novel psychoactive drugs thus remains an important challenge for the pharmaceutical industry and for regulatory agencies that review new compounds. Ideally, such classification could be translated into a risk/benefit profile for each new compound, which then could be disclosed to practitioners and patients. The practical need for consistency in determining the balance between legitimate medical uses versus misuse is addressed by regulatory guidelines that have been developed at national and international levels.

The present chapter focuses initially on the clinical aspects of heroin and cocaine dependence and on the types of preclinical assays that can be used to study the abuse-related effects of these drugs. Subsequently, the information that has been obtained with these behavioral methodologies is discussed in two contexts of drug discovery: (1) the assessment of medications to treat heroin and cocaine addiction and (2) the assessment of the abuse liability of compounds in development for indications other than drug addiction. Finally, a brief overview of the current regulatory environment is given to illustrate the increasing importance of assessing abuse liability during drug development, and consequently the industry's obligations in limiting drug abuse liability.

Opioid and Cocaine Addiction

Even though, according to the *Diagnostic and Statistical Manual of Mental Disorders* of the American Psychiatric Association, 4th edition, text revision (DSM-IV-TR),[7] the definition of drug addiction goes beyond the pharmacological properties of a particular class of drugs, the specific clinical features of a particular drug, in part, define drug-specific therapeutic goals. Thus, understanding of the clinical features and diagnosis of heroin and cocaine addiction is critical for appreciating how animal and human laboratory models play a role in the assessment of medications for these conditions.

According to the DSM-IV-TR, substance abuse is characterized by "... a maladaptive pattern of substance use leading to clinically significant impairment or distress, as manifested by recurrent substance use resulting in a failure to fulfill major role obligations, recurrent substance use in situations in which it is physically hazardous, recurrent substance-related legal problems, and/or continued substance use despite persistent or recurrent social or interpersonal problems." Drug dependence is characterized by a more severe constellation of symptoms including repeated drug self-administration that is associated with tolerance and withdrawal (see above), and compulsive drug taking behavior (substance taken in larger amounts or over longer periods of time than intended, persistent desire or failed efforts to cut down use, excessive amount of time spent trying to obtain the drugs, giving up important social, work, or recreational activities to use opioids, or continued use despite knowledge of problems associated with use).

Heroin

Acute intoxication and toxicity

When given acutely, opioid agonists can induce a feeling of intense euphoria. Sedation also occurs during acute intoxication, and one characteristic effect of opioids is "nodding," in which the intoxicated individual seems to "nod off" into a dreamy state. Although they may appear to be asleep, these individuals are usually responsive to verbal or tactile stimuli. Paradoxically, some users also report increased stimulation after they receive opioids. They become much more talkative and have bursts of energy. This latter effect generally occurs at moderate doses and soon after drug administration. In inexperienced users, the effects of opioids are sometimes reported to be unpleasant, predominantly due to nausea and vomiting. In addition to changes in mood, opioids also produce itchiness, constipation, urinary retention, and decreases in pupil diameter (miosis). Respiratory depression is also a hallmark of opioid intoxication, and high doses may cause death as a result of respiratory failure.

Tolerance and dependence

Repeated administration of opioids can result in tolerance, which is characterized by a decrease in effect with repeated administration of a fixed dose of drug or a need for increasing doses to achieve the same level of effect observed with the initial dose. Typically, heroin abuse is characterized by a regular pattern of daily use consisting of a 4–6 h inter-dose interval. Although tolerance develops to most of the effects of opioids, it does not do so uniformly. For example, little tolerance may develop to opioid-induced miosis or constipation. In addition to tolerance, physical dependence may develop after repeated opioid administration. Physical dependence is revealed by a characteristic withdrawal syndrome when either an opioid antagonist is administered or the opioid agonist is abruptly discontinued. Symptoms of opioid withdrawal include: sweating, runny nose, sneezing, watery eyes, nausea, vomiting, fatigue, yawning, stomach pain, diarrhea, piloerection ("gooseflesh"), feelings of changes in temperature (hot/cold flashes), decreased appetite, increased pupil size, muscle pain, restlessness, insomnia, irritability, dysphoria, and anxiety (e.g.,[8]). Mild increases in blood pressure, pulse, and body temperature are also associated with opioid withdrawal.

The phrase "kicking the habit" was derived from the involuntary leg movements or muscle spasms that often occur during withdrawal from opioids. "Going cold turkey" is another colloquialism that grew out of observations of the pronounced piloerection observed during opioid withdrawal.

The severity of withdrawal depends on several factors including the specific drug that had been used, the dose that had been self-administered, the interval between doses, and the duration of drug use. Generally, withdrawal from longer-acting drugs, such as methadone, is milder in severity than withdrawal from shorter-acting drugs, such as morphine or heroin. And, as expected, withdrawal is milder when the individual is maintained on a smaller dose compared to a larger dose. The duration of withdrawal also varies among different opioid agonists. Typically, the duration of withdrawal is several weeks to months in methadone-maintained individuals, compared to several days to weeks in fentanyl- or morphine-maintained individuals (see[78] for a comprehensive discussion of opioid withdrawal).

Cocaine

Acute intoxication and patterns of use

Cocaine intoxication has been described as a "rush," including feelings of elation, clarity, well-being, energy, and confidence. Symptoms such as paranoia, intensified sounds and light sensitivity may also occur, as well as increased heart rate, blood pressure, and pupil dilation. The pattern in which cocaine is abused is in bouts or "binge-crash cycles," where doses are repeatedly administered with short inter-dose intervals (5–30 min). Once cocaine use is initiated, the probability is high that more cocaine will be used shortly thereafter. After termination of a bout of cocaine use, there may be a period described as a "crash," characterized by anxiety, irritability, and depression.[9]

Tolerance and dependence

Acute tolerance to many of cocaine's effects develops during a single episode of cocaine use. For example, the first dose produces larger cardiovascular and subjective effects than do subsequent doses, despite sustained elevations in cocaine plasma levels.[10] Within a binge, cocaine users will often use large amounts of cocaine in an attempt to achieve the effect obtained by the first dose. In fact, the development of acute tolerance may contribute to the binge pattern of cocaine abuse.

Whether long-term tolerance to cocaine's reinforcing and subjective effects develops is less clear. Although continuous, chronic cocaine exposure under experimental conditions can produce tolerance to cocaine's reinforcing or subjective effects,[11,12] the development of tolerance in cocaine users, who binge on cocaine rather than maintain constant plasma cocaine levels, has not been well demonstrated. In one study cocaine- and opioid-dependent volunteers were less sensitive to the subjective and cardiovascular effects of a single administration of intravenous (iv) cocaine (0.4 mg/kg) than were occasional cocaine users.[13] However, the cocaine-dependent group was also opioid dependent, and only one dose of cocaine was tested, making comparisons between groups difficult. In other studies by Ward and colleagues,[14,15] there was little evidence of a decrease in cocaine's subjective effects when smoked and iv cocaine were compared within the same individual across repeated bouts of cocaine administration.

Thus, although tolerance is a criterion for cocaine dependence in the DSM-IV, tolerance to the reinforcing effects of cocaine has not been shown to occur under conditions in which the drug is self-administered.

Although abstinence following chronic cocaine use is not associated with physical withdrawal symptoms comparable to those of opioids, both clinical observations and controlled laboratory studies have shown that abstinence following the repeated use of cocaine is associated with increased depression, fatigue, anxiety, and irritability.[16,17] Thus, cocaine withdrawal is primarily characterized by disorders in mood, which typically resolve within several weeks.

Relapse

Clinical outcome for the treatment of both opioid and cocaine dependence is characterized by high rates of relapse, defined as the resumption of drug use after a period of abstinence. In the case of cocaine dependence, for example, most patients relapse within 1 year of initiating abstinence. In fact, relapse is a defining feature of drug addiction, and in part reveals the strength of internal and external cues linked to previous drug use. Although the precise factors influencing the high rate of relapse are poorly understood, clinical and laboratory studies have demonstrated that, in addition to withdrawal symptoms, exposure to the abused drug,[18-20] stress,[21,22] and drug-associated cues[23-25] can serve as triggers to induce relapse. Genetic polymorphisms in dopamine receptor subtypes also may be associated with the effects of cues and cocaine exposure on ratings of craving (e.g.,[26]). Of course, craving itself does not invariably predict drug use or relapse, and there are still unanswered fundamental questions about genetic and environmental factors that influence relapse and about the implications of these factors for the development of medications.

PHARMACOTHERAPY FOR DRUG ADDICTION AND ABUSE POTENTIAL OF NEW COMPOUNDS

Animal Laboratory Models

Preclinical testing of potential pharmacotherapies for drug addiction in laboratory animals is an integral component of most medication development efforts. Viewed from the perspective of model development, preclinical testing contributes most effectively to drug discovery when salient features of the targeted pathology can be modeled directly in animals. For example, dopamine depletion in Parkinson's disease can be modeled meaningfully in rats and monkeys by exposure to neurotoxins such as 6-OHDA or MPTP, and these animal models have been used to predict the clinical utility of various dopamine- as well as non-dopamine-based medications.[27-29] As a result, this approach has become an essential component of the evaluation of candidate anti-Parkinson drugs.[ii] The medication discovery process may become more

---

[ii] Please refer to Merchant *et al.*, Animal models of Parkinson's disease to aid drug discovery and development, in Volume 2, Neurologic Disorders for further discussion of animal models of Parkinson's disease.

complicated when the etiology of a disease cannot be identified definitively or when multiple etiologies exist (e.g., in schizophrenia). In such cases, the clinical utility of a candidate medication often is forecast from preclinical assays that only partially mirror functional aspects of the target pathology. For example, preclinical procedures using prepulse inhibition of the startle reflex in animals are thought to mimic aspects of the sensorimotor or attentional deficits in schizophrenic patients.[iii] [30,31] Although such functional parallels in laboratory animals and humans greatly facilitate the development of robust preclinical models, the discovery of novel antipsychotic drugs in preclinical laboratory testing still relies heavily on models whose therapeutic predictive validity is based largely upon pharmacological isomorphism (i.e., the effects of a candidate drug are comparable to successful pharmacotherapies such as haloperidol or clozapine (cf.,[32]). Predictably, this approach often results in the development of novel compounds that offer only modest advantages over existing prototypes.

The discovery process becomes even more complex when neither clinically effective drugs nor well-validated laboratory models are available to guide preclinical testing. To some extent, this situation applies to the discovery of pharmacotherapies for drug addiction. Moreover, the identification of pharmacotherapies that target all forms of drug addiction is likely to be an unrealistic goal, as drug addiction is heavily dependent on the pharmacological properties of the particular class of drugs in question. Treatment agents for one type of drug addiction (e.g., methadone for opioid addiction) may have little relevance for treating other forms of addiction (e.g., stimulant addiction). In addition, although a number of preclinical assays have been proposed for the screening of candidate medications to treat drug addiction, their predictive validity for identifying successful treatments may be unknown, especially in cases where no broadly effective treatment currently exists (e.g., cocaine addiction). These difficulties notwithstanding, it is possible to identify some therapeutic goals for the management of drug addiction that can be addressed experimentally in preclinical testing. One goal might be to decrease the level of drug consumption during periods in which the drug is more or less readily available for self-administration. Another goal might be to decrease the level of drug-seeking behavior during withdrawal from drug use. Yet a third goal might be to forestall resumption of drug seeking following a period of drug abstinence. Pharmacotherapies are likely to differ widely in effectiveness across these potential applications and, as already mentioned, across different classes of abused drug. Thus, an important challenge for preclinical researchers has been to develop models that are sufficiently malleable to address therapeutic goals for multiple classes of addictive drugs.[iv]

Most preclinical testing programs have addressed this challenge with modifications of methodologies that are currently used for testing the abuse potential and other abuse-related effects of new drugs. This has been a useful approach because

---

[iii] Please refer to Jones *et al.*, Developing new drugs for schizophrenia: From animals to the clinic, in Volume 1, Psychiatric Disorders for further discussion of prepulse inhibition and other animal models of schizophrenia.

[iv] Please refer to Little *et al.*, Pharmacotherapy of alcohol dependence: Improving translation from the bench to the clinic, in this volume for further discussion on criteria and therapeutic goals for alcohol abuse and dependence.

such procedures provide well-established experimental frameworks for determining how candidate medications might modify the addictive impact of abused drugs. For the present purpose, four of these laboratory procedures will be described briefly and, as a convenience, are divided into "direct" and "indirect" assay procedures. Direct assay procedures invariably involve some form of drug self-administration, which can provide essential information for gauging the reinforcing effects of drug injections. Indirect assay procedures are often used to infer aspects of drug's subjective or rewarding effects that may be related to abuse liability. The most widely used of these latter assays include intracranial self-stimulation (ICSS), place conditioning, and drug discrimination procedures.[v][33]

In regards to the assessment of the abuse potential of new compounds in development, use of both direct and indirect assays is accepted as complimentary by regulatory agencies, such as the Food and Drug Administration (FDA) in the United States, the European Medicines Agency (EMEA) in the European Union (EU), and the Pharmaceutical and Medical Devices Agency (PMDA) in Japan. Historically, these agencies have based their initial evaluation of abuse liability of a new chemical entity on three core datasets: (1) reinforcing effects, as assessed by drug self-administration, (2) subjective or interoceptive effects, as assessed by drug discrimination, and (3) potential to induce physical dependence, as determined by occurrence of a withdrawal syndrome after discontinuation of repeated administration. Animal behavioral data from these three core sets are in general used to provide go/no go decisions into further human laboratory-based assessment, and are described in labeling whenever relevant. The recent adoption of a new guideline in the EU[34] has introduced new challenges for global development of new compounds, as discussed under the section "Assessment of abuse liability of compounds: the regulatory environment". In the United States, even though the FDA has provided guidance to the industry on this process since 1990 through several draft documents, a new draft version is expected to be released soon. Several elements that have been maintained throughout the different versions of the FDA guidance are in agreement with the EU guideline, and are likely to be present in the new draft, such as the following: no single procedure can provide a complete evaluation of the abuse potential of a compound, testing should be flexible, and regulatory assessment and decision-making should be based on the full array of data available. Both preclinical and clinical assessment procedures have been acknowledged to provide relevant information, but human data, as with all medical concerns, are still expected to carry the greatest weight in the regulatory review and decision-making process.

Indirect Assays for Medications Development

Intracranial self-stimulation

The discovery that delivery of a specified intensity and duration of electric current to a targeted brain region can be used to maintain operant behavior in experimental

---

[v] Please refer also to Heidbreder, Impulse and reward deficit disorders: Drug discovery and development, or Koob, The role of animal models in reward deficit disorders: Views of academia, in this volume for detailed descriptions of these procedures.

subjects led to the rapid development of ICSS procedures to study the neurobiology of brain reward processes. From an early point in their development ICSS procedures also were proposed for the evaluation of pharmacotherapies for psychiatric disorders.[35] In one widely used application, the threshold intensity at which a given duration and frequency of electrical stimulation to a targeted brain region maintains behavior is established, and how the threshold value is altered by drug treatment is taken as a measure of the drug's effect on the brain reward system. A lowering of ICSS threshold, an effect often observed after administration of opioids, stimulants, and some other drugs of abuse, often is taken as an indication of convergent rewarding actions of drug administration and brain stimulation.[36] Conversely, increased ICSS threshold values – for example, following treatment with κ-opioid receptor agonists – are considered to reflect adverse effects of a drug on endogenous reward systems.[37] Using these procedures, then, an effective pharmacotherapy for drug addiction might be one that is relatively innocuous when administered alone but that nonetheless restores drug-lowered ICSS thresholds back to baseline values. Several lines of investigation have provided instances of such effects. For example, doses of the 5-HT$_3$ receptor antagonist granisetron (Kytril®) that did not systematically alter ICSS threshold when tested alone attenuated the threshold-lowering effects of morphine.[38] Similarly, dopamine D$_3$ receptor-preferring antagonists such as SB-277011-A and NGB 2904 appear to reverse the effects of cocaine on ICSS threshold values, consistent with their ability to attenuate the effects of cocaine in other types of procedures.[39,40] Other reports also have pointed to glutamatergic or GABAergic ligands, for example, the metabotropic glutamate mGlu5 receptor antagonist MPEP or the GABA$_B$ receptor antagonist baclofen (Kemstro® and Lioresal®), as promising candidate medications on the basis of their ability to attenuate cocaine's threshold-lowering effects in ICSS procedures.[41,42] Thus, ICSS procedures may have utility for evaluating candidate medications for stimulant and opioid addiction.

Limitations in the application of ICSS procedures for this purpose most often lie in reliance on threshold measures that evaluate the effects of a single dose of a drug as the stimulus intensity is varied. This necessarily reduces the information that can be obtained regarding the effects of abused drugs and how those effects may be modified by a candidate medication. For example, selected doses of morphine have been reported to decrease threshold values for ICSS. Yet, these same doses of morphine also may reduce response rates maintained by a higher intensity of stimulation.[43] Thus, it can be difficult to evaluate a candidate medication's therapeutic potential simply on the basis of how it alters the ICSS threshold. Whether or not a more comprehensive approach, such as one based on full intensity–response and dose–response analysis, can improve the utility of ICSS procedures for evaluating of candidate medications remains to be determined.

Place conditioning

Place conditioning procedures offer a second type of indirect assay that has become increasingly used in preclinical research. In place conditioning procedures, different (physically distinctive) compartments of an experimental chamber are paired with different drug treatments. During conditioning, the subject is confined to one compartment of the chamber after drug injections and, in alternate sessions, to a second compartment after vehicle injections. Subsequently, the influence of the pairings can be studied by allowing the subject access to both compartments during a test session in which no injections are given. Place preference can be inferred when the subject spends more time in the drug-paired compartment, whereas place aversion can be inferred when the subject spends a greater percentage of time in the vehicle-paired compartment. The magnitude of preference in typical place conditioning experiments is relatively small and less influenced by changes in dose compared to the effects observed in drug self-administration procedures (see below). Nevertheless, place conditioning procedures are technically simple to run and, consequently, have been used extensively in studies with rodents. The preponderance of data from place conditioning studies indicates that conditioning with drugs of abuse typically leads to the expression of place preference.[44,45] Presumably, an effective candidate medication would prevent the development and/or inhibit the expression of place preference depending on the particular treatment regimen.

The apparent simplicity of this type of preclinical assay notwithstanding, the utility of place conditioning procedures for medications development is controversial. Some drugs that attenuate the development and/or expression of drug-induced place conditioning likely interfere with conditioning processes in a behaviorally non-selective manner or may have aversive side effects. For example, adding naloxone to morphine can block the ability of morphine to induce a conditioned place preference,[46] and this can be understood in terms of naloxone's antagonism of morphine's actions at the μ-opioid receptor. Additionally, however, morphine conditioned place preference can be attenuated by a wide variety of other drugs including κ-opioid receptor agonists, dopamine receptor antagonists, $GABA_B$ receptor agonists, $5-HT_{1A}$ receptor agonists, nicotinic receptor antagonists, cannabinoid CB1 receptor antagonists, anticholinergics, glutamatergic ligands, and other classes of drugs. While it is likely that each of these behaviorally active drugs interferes with some aspect of the conditioning processes, it is less likely that such interference, of itself, is a reliable predictor of therapeutic potential. Thus, data from place conditioning experiments should continue to be viewed cautiously in the evaluation of candidate medications for drug addiction.

Drug discrimination

Drug discrimination procedures provide a third type of indirect assay of abuse potential and, historically, have been useful for categorizing drugs on the basis of similarities in their interoceptive effects (possibly analogous to subjective effects in humans). In these procedures, subjects typically are trained to respond differentially on two manipulanda (e.g., press either of two levers) depending on the pre-session treatment. Under training conditions, responding on one lever is reinforced (e.g., with food pellets) when the pre-session treatment is an injection of the training drug, whereas responding on the second lever is reinforced when the pre-session treatment is an injection of vehicle. Under testing conditions, responding on either (or, in some variants, neither) lever is reinforced, and the effect of the treatment can be expressed as a percentage of responses that are directed to the drug-associated lever. Early studies with morphine in both rats and monkeys showed that the discriminative-stimulus effects of morphine are dose-related, stereospecific, generalizable to other morphine-like opioids (but not to drugs from most other pharmacological classes), and can be surmountably antagonized by prior administration of μ-opioid receptor antagonists such as naloxone (see[47] for review). Similarly comprehensive studies have made comparable observations regarding the discriminative-stimulus effects of other abused drugs, including stimulants such as cocaine and methamphetamine.[48-51] For example, the effects of cocaine are dose related, can be mimicked fully by other indirectly acting dopamine receptor agonists such as methamphetamine or methylphenidate and partially by directly acting dopamine receptor agonists, can be surmountably antagonized by dopamine D_1 or D_2 receptor blockers, and can be modulated by other monoaminergic ligands.[52-54]

In view of the successful application of drug discrimination procedures for characterizing the effects of opioids and stimulants, it is reasonable to ask whether the attenuation or reduction of such effects might serve as a useful preclinical indicator of therapeutically beneficial actions. Some data from opioid discrimination research are encouraging in this regard. For example, the morphine-like and morphine-antagonist effects of methadone (Dolophine®) and naltrexone (Revia®), respectively, in drug discrimination studies clearly imply utility for modulating the subjective effects of abused opioids. Similarly, the partial agonist actions of buprenorphine have been well characterized in drug discrimination studies and underlie its recent introduction as a medication for the management of heroin abuse and addiction.[55] On the other hand, data from cocaine discrimination research are less encouraging. Clinical evaluations of dopamine receptor blockers that surmountably antagonize the discriminative-stimulus effects of cocaine have been disappointing to date, and several drugs that attenuate the discriminative-stimulus effects of cocaine (e.g., κ-opioid receptor agonists) are not effective therapeutics for cocaine addiction.[56,57] Furthermore, although the utility of methadone for the management of opioid addiction is well appreciated, drugs that substitute for cocaine in drug discrimination procedures are only beginning to receive attention as candidate medications.[58] Overall, then, drug discrimination procedures appear to be of some value in identifying candidate medications, especially when the pharmacological strategy that underlies drug discovery is clearly defined. It is worth considering two aspects of these types of procedures that are particularly appealing. First, it is likely that the subjective effects of an abused drug (which by their nature defy rigorous definition) contribute to its discriminative-stimulus effects. This likelihood suggests that drug discrimination procedures may be of particular value for helping to identify candidate medications that are intended to blunt the subjective effects of an abused drug. Second, drug discrimination procedures have been successfully applied across a variety of species, including humans. Direct comparisons between the discriminative-stimulus effects of abused opioids and stimulants in laboratory animals and humans have improved our understanding of the behavioral pharmacology of these drug classes and may be useful for guiding early development of candidate medications to treat opioid or stimulant addiction.

Indirect Assays for Assessment of Abuse Potential

Among the indirect assays described above, drug discrimination is the only one that, historically, has been considered by worldwide regulatory agencies as a core assay in the assessment of abuse potential of new compounds. Consequently, a vast drug discrimination database exists and has been pivotal in providing the basis for labeling of new compounds. The most common challenge for industry in the use of drug discrimination procedures has been the choice of appropriate positive comparators, particularly for compounds that belong to new chemical classes. In such cases, evaluating the substance's own discriminative-stimulus effects and, as well, its generalization to known drugs of abuse might provide an acceptable strategy.

A common assumption in assessing new compounds has been that generalization in drug discrimination procedures to substances that cause dependence itself is indicative of dependence potential, but the recent EU guideline challenge this assumption.[34] The guideline accepts that drug discrimination may be used to assess the similarity of the substance's interoceptive cue with prototypic substances known to cause dependence, but it specifically acknowledges that substitution for substances that cause dependence itself is not necessarily indicative of dependence potential. At the present time, it is uncertain how the EMEA will implement this guidance.

In regard to ICSS and place conditioning, these assays have not been routinely used in the evaluation of abuse liability, and thus limited data are available in this regard. However, probably due to an increasing number of compounds under development with new mechanisms of action for which prototypic active substances do not exist, the regulatory bodies have lately been more open to new approaches. The EU guideline acknowledges that a flexible approach can be considered when new classes of substances are tested, and specifically mentions that conditioned place preference should be acceptable for evaluating the rewarding properties of active substances in such cases.

Suggestion of alternative assays by a regulatory agency provides significant incentive for the industry to start using such assays, and thus to expand the database on the evaluation of the assay's predictive validity. If one considers that this first step has been embraced by the EMEA in the new EU guideline,[34] one can expect the use of place conditioning in the assessment of abuse liability to expand. However, since the EU guideline has been only recently adopted and companies have had limited interactions with the agency so far on the topic, it is not clear how data from place conditioning will influence the overall evaluation of abuse liability, or how they will be accepted in the regulatory filings of New Marketing Applications. Likewise, it is unclear whether the EU position will influence other regulatory agencies throughout the world, and in particular the FDA in the United States.

Direct Assays for Medications Development
Drug self-administration

Although indirect behavioral evaluations can clearly provide useful data on abuse-related properties of a drug, the most powerful method for assessing abuse potential involves direct measurement of a drug's reinforcing effects using drug self-administration procedures. Models involving drug self-administration in rodents, nonhuman primates, and humans have been used extensively to investigate the effects of candidate medications for the treatment of drug dependence. These self-administration procedures can provide meaningful behavioral data on the efficacy of potential treatment medications in a relatively small number of subjects under controlled conditions. It has been hypothesized that medications that selectively decrease self-administration of drugs in laboratory animals would be similarly effective in decreasing drug use in humans.

Drug self-administration techniques can trace their origins to the seminal work on morphine dependence in chimpanzees by Spragg,[59] and their proliferation to the development of reliable automated methods for iv self-administration of drugs in rats and monkeys in the 1960s (e.g.,[60-64]). The widespread acceptance of animal drug self-administration procedures as research tools derives not only from the face validity of the technique, but even more importantly, from the finding that animals (especially nonhuman primates) will reliably self-administer most drugs that are abused by humans (e.g.,[65,66]). The predictive validity of drug self-administration techniques in animals for identifying drugs with high abuse liability in people[67-69] has made the procedure virtually indispensable for the screening of investigational drugs for abuse potential, and has promoted the widely held view that these techniques are also useful for identifying effective pharmacotherapies to treat drug addiction.

Drugs can be self-administered by various routes of administration, but in the case of stimulants and opioids, the majority of studies have been conducted using the intravenous route. In most studies of iv drug self-administration in animals, a catheter is surgically introduced into a major vein (typically, a jugular or femoral vein) and secured to permit delivery of drug solutions over the course of the study. The externalized portion of the catheter is protected by a jacket, harness, head mount, or similar device, and is connected to an infusion pump to provide consistent and accurate injections of drug solutions. Drug injections are delivered to the subject contingent on performance of a specified operant response such as pressing a lever.

The contingencies that determine when and how many responses are required to produce an injection are determined by the schedule of reinforcement, which is simply a set of rules that govern the sequential and temporal relations between responses and reinforcers. For example, a reinforcer (e.g., injection of cocaine or heroin) may be scheduled to follow a specified number or responses (ratio schedules) or to follow a response after a specified period of time has elapsed (interval schedules). The majority of studies involving stimulant and opioid self-administration have employed simple fixed-ratio (FR) schedules, with a substantial minority of studies using progressive-ratio (PR) schedules (see reviews by[70,71]), second-order schedules (see reviews by[72,73]) and to a lesser extent fixed-interval schedules (see reviews by[74,75]). Each type of schedule engenders its own characteristic temporal pattern and rate of responding – performances that are remarkably reproducible across species, type of operant response, and type of reinforcer (e.g., food, drug, etc.). These characteristic schedule-controlled performances provide a meaningful way to compare behavior maintained by drugs and other types of reinforcers, but may constrain direct comparisons across experiments using different schedules of drug self-administration. In general, the reinforcing effects of drugs with high abuse liability are readily evident across a wide range of schedules; however, effects of a potential medication to treat drug addiction may differ under different scheduling procedures. For example, a candidate medication with antagonist properties might increase drug self-administration under a single-response FR schedule but decrease self-administration under a schedule with greater response demands or intermittency (e.g., PR or second-order schedules).

As mentioned previously, the most desirable assays are those that can be used to model specific aspects of the pathology or clinical condition under consideration. Over the past several years, researchers have increasingly modified drug self-administration procedures to bring them more in line with (i.e., model) the conditions that are encountered by drug-dependent individuals. These modifications include provisions for "binge" patterns of cocaine self-administration or more widely spaced patterns of heroin self-administration and pairing environmental stimuli such as lights or sounds that are explicitly associated with drug injection or drug availability to mimic environmental cues associated with drug procurement, preparation, and consumption. In addition, self-administration studies with laboratory animals may include alternative sources of reinforcement, such as food, in order to mirror the non-drug choices available to drug-dependent individuals.

Evaluated over a sufficiently wide range of doses, drug self-administration data in animal studies are frequently characterized by an inverted U-shaped function relating the dose of drug available for each injection and an appropriate measure of behavior, such as response rate or number of self-administered injections. The characteristic inverted U-shaped dose–response curves for cocaine, heroin, and other drugs typically reflect an interaction between the reinforcing effects of the drug, which are often considered to be monotonically related to dose, and its other direct effects on operant performance, which generally become more pronounced over successive injections of high doses. Consequently, the ascending portion of the inverted U-shaped curve frequently is considered to provide the most unambiguous information regarding a drug's reinforcing effects, and it is this portion of the dose–response curve that is typically studied in the human laboratory, where safety concerns override the testing of potentially dangerous doses. Some drug abusers do, however, self-administer very high doses, and full dose–response curve evaluations in animal studies can provide pertinent information about the effectiveness of a candidate medication against low or high doses of an abused drug.

Biphasic dose–response curves do not always lend themselves well to simple analysis because the different effects of high and low doses of a drug may be reflected in the ascending and descending limbs of the function. Some researchers circumvent the problem by limiting drug intake to minimize the emergence of other, complicating drug effects. For example, Griffiths *et al.*[76] limited drug intake to only a single injection per day, imposing a sufficiently long interval between injections to permit drug washout before the next day's self-administration session. This type of approach addresses the problems produced by accumulation of the self-administered drug and can result in a dose–response curve that incorporates a much wider range of doses in the ascending limb. However, this approach substantially lengthens the time needed to conduct comprehensive studies and may not be the most practical way to evaluate candidate medications. Other investigators have addressed the problem of analyzing biphasic dose–response curves by focusing on dependent variables that are related more monotonically to dose, such as the proportion of responses allocated to the drug self-administration lever in choice procedures involving drug and non-drug alternatives or "break point" in the case of PR schedules.[77-80] In addition to facilitating analysis of dose–response relationships, dependent variables such as these can provide meaningful estimates of the reinforcing strength of a drug either in the face of alternative reinforcers or increased response cost. The utility of these approaches for evaluating candidate medications for opioid and stimulant addiction is based on the reasonable assumption that an effective medication strategy would be one that reduces the relative reinforcing strength of an abused drug, thereby allowing other options to gain more effective control of behavior.

In principle, the basic design of drug self-administration studies in laboratory animals is similar to designs used in human laboratory studies. In a typical animal study, once stable drug self-administration is established, test sessions are conducted by administering a candidate pharmacotherapy or its vehicle as a pretreatment before the self-administration session and measuring the effect of the medication relative to vehicle. As discussed earlier, studies of this type are most informative when they evaluate an appropriate range of doses of both the candidate medication and the self-administered drug to determine how the shape and position of the self-administration dose–response curve are altered as a result of drug pretreatment. Depending on the outcome, a candidate medication also may be evaluated for its own ability to maintain self-administration. Collectively, such studies can identify drugs that alter opioid or stimulant self-administration in a manner consistent with potential clinical utility, as well as provide information concerning the drug's own potential for abuse or, viewed from another perspective, the drug's acceptability by patients and likelihood for compliance with taking the medication.

In practice, drug self-administration studies focusing on medication development in animals are often paired with corresponding studies involving non-drug reinforcers, such as food, to determine the specificity with which a candidate medication affects drug-reinforced behavior. Additional studies using drug discrimination procedures and/or other quantitative behavioral assessments also are frequently conducted in parallel with drug self-administration studies to provide relevant information about how the candidate medication may alter the subjective effects of the self-administered drug and about potential side effects. Such supplemental studies provide data that can be compared with information from rating scales, questionnaires, drug discrimination experiments, and other observations that are often used to supplement drug self-administration data in human laboratory studies.

Assessment of Abuse Liability in Animals

A similar rationale is employed for abuse liability studies of new compounds. While drug self-administration is considered as the first line assay by worldwide regulatory agencies as mentioned above, drug discrimination is also carried in parallel to provide relevant information about the compound's interoceptive effects in addition to its reinforcing effects. Based on the extensive use of drug self-administration and the availability of historical data, the regulatory agencies have been relatively flexible on their acceptance of study designs. However, overall expectancy is that testing will assess a broad range of doses, including maximally tolerated doses, and that studies will include a positive control. The recently approved EU guideline[34] further indicates that a flexible approach to study design (i.e., use of timeouts, type of schedule of reinforcement, etc.) is acceptable.

Human Laboratory Models

Indirect Assays for Medications Development

Subjective-effects questionnaires

Early studies of the subjective effects of opioids were conducted in patients during assessments of side effects, such as nausea and itchiness. For the most part, these data were collected in a non-systematic fashion.[81-86] Early abuse liability studies did assess euphoria, a subjective effect that is thought to contribute to the abuse of a drug, but the observers rather than the participants recorded this response.[87] One of the first questionnaires developed to allow the participants to record their own subjective responses used a 54-item 7-point bipolar rating scale with adjectives (e.g., "Sad–Happy") at either end of the scale.[88] The 54 items were grouped into three main classifications: mood (including ratings of euphoria and dysphoria), mentation, and sedation.

In 1961, the Single Dose Questionnaire (SDQ) was developed by Fraser, Isbell and colleagues at the Addiction Research Center in Lexington, Kentucky.[89] This was the first questionnaire that was designed specifically to measure the abuse liability of drugs. It was used in experienced drug abusers who were asked to (1) identify if they felt an effect, (2) identify the drug from a list of seven abused drugs (including "other"), (3) report if they experienced any of the 12 symptoms listed, and (4) rate their degree of liking for what had been administered to them on a scale of 0 (not at all) to 4 (an awful lot). Simultaneously, observers were asked to rate any observable "signs" associated with those symptoms. The SDQ continues to be widely used in opioid abuse liability studies.

In addition to the SDQ, a 500-item true–false questionnaire, called the Addiction Research Center Inventory (ARCI), was developed in Lexington to examine the effects of drugs from a variety of pharmacological classes.[90] Subsequently, a shorter 49-item version of the ARCI was developed[91] and continues to be widely used in abuse liability testing. In the short form of the ARCI, six subscales are used to characterize the effects of drug(s). These include the LSD scale, a measure of dysphoria and somatic and sensory disturbances, the Morphine-Benzedrine Group (MBG), a measure of euphoria, the Pentobarbital-Chlorpromazine-Alcohol Group (PCAG), a measure of sedation (sometimes called apathetic sedation), and two scales, the Amphetamine and Benzedrine Group scales, measures of stimulant effects.

Finally, a number of laboratories use visual analog scales (VAS), in which participants rate the effects of a drug on a continuous scale. Typically, a statement, such as "I like the drug" or "I feel high," is presented and participants are instructed to make a mark on a 100-mm line that is anchored on one end with "Not at all" and on the other end with "Extremely."

Drug discrimination procedure

An additional procedure that has been used in humans to characterize subjective effects is one that was originally developed in laboratory animals (see above). In this procedure, participants are first given "Drug A" and "Drug B" during "sample" sessions and are instructed to pay attention to the effects produced by the drugs. During subsequent "training" sessions, Drug A or Drug B is administered, although they are not labeled as such, and participants are instructed to indicate whether they received Drug A or Drug B. At the end of each training session, participants are informed of which drug was administered. Typically, correct responses are reinforced with money, and incorrect responses are not reinforced. Participants are considered to have learned the discrimination when they have attained a specified level of accuracy during the training phase. After the criterion level of correct responding is reached, novel drugs are administered during test sessions in order to determine whether they share discriminative-stimulus effects with the training drug. All responses are reinforced during the test phase, and it is only after the test session is completed that the participant is informed that it was a test session. Variations of this procedure include the use of two different doses of the same drug (e.g., Drug A is a low dose and Drug B is a higher dose), and the use of two or even three different drugs (e.g., Drug A, Drug B, and Drug C), one of which may be placebo.

Using these types of drug discrimination procedures, a great deal of useful information has been obtained on the similarities and differences among various mixed-action opioids (e.g.,[92-95]) and for assessing the utility of various candidate medications for treating opioid withdrawal symptoms.[96-98] In one such study, the calcium channel blocker isradipine (Dynacirc®) and the *N*-methyl-d-aspartate (NMDA) receptor antagonist dextromethorphan were studied in opioid-dependent participants, who discriminated the effects of placebo and small doses of naloxone, which precipitated mild withdrawal symptoms.[98] Under these experimental conditions, isradipine attenuated naloxone-appropriate responding and some of the naloxone-precipitated withdrawal symptoms. Dextromethorphan moderately decreased naloxone-appropriate responding but increased naloxone-induced subjective responses, consistent with other studies reporting equivocal effects of NMDA antagonists on opioid withdrawal symptoms in humans (e.g.,[99-101]).

Direct Assays for Medications Development
Drug self-administration

Single- and forced-choice procedures One of the simplest types of drug self-administration paradigm is a single-choice procedure in which participants are given a sample dose of drug and then asked whether or not they would like to take the dose again. In this type of study, the behavior is a verbal response. If participants choose the active dose on more occasions than they choose placebo, then the active dose is considered to be reinforcing. A major limitation of the single-choice procedure is that a high placebo response rate is often seen.[102] A more common approach is to use a forced-choice procedure, whereby participants are given a sample of Drug A and then, during a separate session, a sample of Drug B. Drug A could be placebo and Drug B could be a dose of active drug in this paradigm. Participants are instructed to pay attention to the effects produced by Drugs A and B and different stimuli, such as the color of the pill, are used to facilitate the association between the sample dose and the effects it produces. During subsequent choice sessions, participants are instructed to indicate verbally whether they would like to ingest Drug A or Drug B. Typically, a minimum of five choice tests are provided, and the number or percentage of choices of Drug A and Drug B is determined. If the active dose is chosen significantly more than placebo, then the drug is considered to be a reinforcer. An adaptation of the drug versus drug forced-choice procedure is a drug versus money forced-choice procedure. This type of study is similar to the drug versus drug procedure, except that participants are asked to choose between a dose of drug and a fixed amount of money. The advantages of a drug versus money choice procedure are that fewer sample sessions are required and participants have the opportunity to choose a reinforcer other than drug. This procedure also corresponds conceptually with the drug versus non-drug choice procedure used in the animal laboratory.

Multiple-choice procedure Although the above procedures are the most commonly used methods for examining the reinforcing effects of drugs in humans, one potential problem is that they can be time-consuming. In response to this limitation, a multiple-choice procedure (MCP) was developed as a first attempt to provide a quick and efficient estimate of the reinforcing effects of a drug.[103] As in the case of the other procedures, participants generally are first given doses of the test drug by the experimenter. They subsequently are asked to make a series of choices on a questionnaire between either two doses of drug (Drug A versus Drug B, Drug A versus Drug C, Drug B versus Drug C, etc.) or between drug and money (Drug A versus $0.50, Drug A versus $0.75, Drug A versus $1, etc.). After the questionnaire is completed, all of the choices are compiled, and one randomly selected choice is given to the participant. Using this procedure, Griffiths and colleagues[103] demonstrated that pentobarbital was chosen over money and that higher doses of pentobarbital were chosen over lower doses or placebo. The MCP appears to be an efficient alternative to the more traditional self-administration procedures but has some disadvantages as well. Although drug and money are actually delivered during this procedure (which is similar to traditional self-administration procedures), the disadvantages are that only one of many choices are reinforced, and there is often a substantial delay between the time that the choices are made and the time that the reinforcer is actually delivered. The fact that there is a low probability that a choice will be reinforced, and there is a delay between choice and the delivery of drug or money diminishes the strength of the MCP as a measure of reinforcement.

Operant procedures The second major type of self-administration procedure involves the use of operant responses. Specifically, participants are instructed to make responses on a manipulandum (computer mouse, joy stick, etc.) in order to receive drug. As described above for studies conducted in laboratory animals, operant schedules such as the FR and PR schedules are commonly used in self-administration studies with human research volunteers. The main advantage of the PR schedule over the FR schedule is that it is better suited to examine the relative reinforcing effectiveness or value of a range of different drugs and doses. Theoretically, drugs that maintain higher break point values are also drugs that have greater abuse liability.[104,79] Typically, a "timeout" period, during which drug is unavailable, follows each drug delivery and a maximum number of drug deliveries that can be self-administered is imposed for safety reasons.

In recent years, a behavioral economic approach to analyzing the data from FR and PR studies of the reinforcing effects of drugs in humans has received a fair amount of attention (e.g.,[105]). In this analytical approach, drug use is viewed from an economic perspective, which takes into consideration the cost of the drug, income, open versus closed economies, and the presence of competing or complementary reinforcers, among other variables (see[106] for a review of behavioral economics). The primary advantage of a behavioral economic approach, compared to the traditional measures (break point, rate of responding, and drug preference), is that it may provide a more realistic characterization of the reinforcing effects of drugs. Traditional perspectives on relative reinforcing efficacy have assumed that it is a homogenous phenomenon, whereby break point, peak response rate, and drug preference generally are greater for higher doses than for lower doses.[107] However, empirical evidence suggests that relative reinforcing efficacy may be a heterogeneous phenomenon that could be better explained from the perspective of economic theory. Ongoing research to compare behavioral economic approaches with more traditional measures of drug reinforcement continue to shed light on the comparability of findings and practicality of these different approaches.

Assessment of Abuse Liability in Humans

If any of the data generated in animals are suggestive of abuse liability, and/or adverse events of relevance (e.g., euphoria) are identified in any preclinical or clinical trial, human laboratory-based assessments are likely to be required. Human or clinical studies of abuse liability were first compiled by Fischman and Mello,[107] and have served to guide drug development and regulation for traditional classes of drugs, such as stimulants, opioids, or depressants. However, new pharmacological classes of substances that do not fit into these categories have challenged the generality of the models. The rationale and procedures for conducting such studies were recently discussed by Griffiths and colleagues.[108] Briefly, an acute dose–effect trial, as described above, compares the profile of acute effects of a range of doses of the new compound to those of placebo and a range of doses of the active comparator (i.e., compound with known abuse liability). Ideally, such trials are conducted in subjects with extensive histories of polydrug abuse (including abuse of drugs from the same pharmacological class of the new compound), and in a closed residential research unit that provides constant clinical monitoring of the subjects. Selection of the appropriate subject population, an appropriate positive control, and a range of doses are of major importance in these trials. Testing is expected to assess a broad range of doses, including maximally tolerated doses or doses at least several multiples of the therapeutic doses. Studies are anticipated to include both a positive (i.e., a drug of known significant abuse liability) and a negative control (e.g., placebo), and results are expected to document that subjects detected the positive control drug as one having significant abuse potential. Currently, many researchers consider such classic acute dose–effect comparison study in volunteers with histories of drug abuse as the "gold standard" for the initial testing of abuse liability of a novel compound.

Because final regulatory guidance statements on clinical assessment of abuse liability are limited (guideline from the FDA is currently under review), pharmaceutical companies are constantly challenged by decisions on whether or not clinical studies should be pursued, in particular for substances with new mechanisms of action that do not have a CNS target. General consensus in the field does not exist, and public information is limited due to the fact that the majority of these decisions are made in case-by-case dialogs between the sponsor and the regulatory agency. Many public forums have provided discussions on the topic, and the general impression remains that, particularly for brain penetrant compounds with novel mechanisms of action, the regulatory agencies will not fully discard the need for clinical studies evaluating abuse liability until comprehensive preclinical and clinical datasets are available, and the human safety profile of the compound is defined. The perception is that the new guidelines will maintain the emphasis on determining if the new drug is likely to be abused by current drug abusers, and the recommendation for conducting human abuse liability studies in experienced drug abuser volunteers.

ASSESSMENT OF ABUSE LIABILITY OF COMPOUNDS: THE REGULATORY ENVIRONMENT

From a regulatory perspective, abuse liability testing of compounds depends on the stage of development of the compound. For compounds under global development, worldwide regulatory guidelines define requirements for approval and labeling information on the risk/benefit ratio between medical use and possibility of misuse. For compounds that have been approved and have reached the marketplace, monitoring of adverse events through post-marketing surveillance triggers, at the national level, labeling safety updates related to the risk of abuse liability. In the international arena, the World Health Organization (WHO), through its Expert Committee on Drug Dependence, is responsible for reviewing the medical and scientific characteristics of marketed psychoactive substances, and for making recommendations to the Commission on Narcotic Drugs of the United Nations on the appropriate level of international control.[114]

Most relevant for the present review is the understanding of the current regulatory requirements for assessing abuse liability of compounds under development, and the international control of compounds currently approved for the treatment of opioid dependence. The description of the actual regulatory environment in the United States and the EU is presented to illustrate some of the actual challenges sponsors face for assessing abuse liability during drug development, and the WHO recommendation for scheduling of buprenorphine illustrates some of the intricacies of the international process.

Regulatory Environment in the United States and the EU

In the United States, the FDA has set the stage for scheduling and controlling substances since the passage of the Controlled Substances Act (CSA) in 1970. In 2003, the Controlled Substance Staff (CSS) was created within the FDA to oversee the evaluation of abuse liability, drug dependence, risk management and drug scheduling of new compounds. During the New Drug Application (NDA) review, the CSS assesses preclinical and clinical data to determine whether the compound under review requires further abuse liability studies, scheduling under the CSA, and/or a risk management program (RMP) directed toward reducing abuse, overdose, or diversion. As part of the NDA, every new compound, regardless of the indication, is required to undergo a risk–benefit analysis, which among others includes the risk evaluation for misuse, physical dependence, and addiction.[109] If a compound is determined to have potential for abuse, the FDA/CSS and the Drug Enforcement Agency (DEA), under the CSA of 1970, evaluate the balance between patient access and protection of the public, and schedule the compound accordingly. That is, they assign the compound to one of five schedules of controlled substances (from the most restrictive Schedule I for substances that have a high abuse liability and no approved medical use, to the least restrictive Schedule V for substances that have no recognized abuse liability and an approved medical use).

In the United States, scheduling is often perceived by physicians as an unfavorable attribute, and therefore abuse liability of a compound may be a point of market differentiation. In contrast, in the EU, a scheduling process is not implemented, but labeling language on risk of abuse and dependence is included in Section 4.4 of the Summary of Product Characteristics, under Special Warnings and Precautions for Use.

The abuse liability profile is required to be part of the NDA in the United States and the Marketing Application in the EU, and should include a description of the drug's properties, the full preclinical and clinical dataset. It should also include an integrated summary of safety and efficacy, and the sponsor's proposal to schedule the compound (in the United States) or to include specific warning language in the label (in the EU).

Besides the drug classes known to be subject to regulation as controlled substances or that have warning labeling based upon a full preclinical and clinical assessment (i.e., opioids, depressants, stimulants, hallucinogens, and cannabinoids), substances with affinity at any of the receptor systems associated with drug abuse (e.g., dopamine, $GABA_A$/benzodiazepine, NMDA, serotonin), anabolic steroids, and substances that belong to novel pharmacological classes are of concern for the FDA/CSS and the EU/EMEA. The EU guideline specifies that for substances for which no class-specific standards are available for reference and for which dependence potential has yet to be determined, sponsors should follow a two-tiered approach. First, the substance's pharmacological profile (receptor binding and *in vitro* activity) should be evaluated, and subsequently behavioral studies should be conducted to evaluate the substance's reinforcing properties and liability for physical dependence. These considerations apply for all substances and corresponding metabolites that enter the CNS, independent of their primary indication.[34]

In all cases, the expected preclinical dataset for all substances that are of concern is expected to include behavioral pharmacology studies designed to evaluate reinforcing (drug self-administration) and/or subjective (drug discrimination) effects, as well as physical dependence (withdrawal symptoms). Clinical pharmacology studies are usually carried out in polydrug abusers, and would similarly assess subjective and reinforcing effects in addition to toxicity/performance impairment, tolerance, and physical dependence. Both preclinical behavioral pharmacology and clinical pharmacology studies are expected to use a pharmacologically similar substance as an active comparator in all appropriate domains, which may be very challenging for compounds that have new mechanisms of action. It is clear that, depending on the target indication, findings from abuse liability studies can be of high impact on go/no go decisions in the course of drug development. The decisions that sponsors constantly face are not so much whether to evaluate the abuse liability of a new compound but, rather, how early and at what level of intensity to do so since there is no validated or accepted early indicator of abuse potential.

Two additional recommendations, which run somewhat counter to common practice, were introduced in the EU guideline. One recommendation was the clear preference for using rodents instead of nonhuman primates in behavioral pharmacology assays. The guideline clearly states that where rodents or nonhuman primates are available, rodents are preferred, provided their appropriateness is demonstrated and the model is validated. The guideline further states that there is no need for routine confirmatory studies in nonhuman primates. These recommendations are, to say the very least, controversial, and cogent arguments against primary reliance on rodent testing have received a great deal of international attention (e.g.,[110]).

The other recommendation was to conduct behavioral pharmacology assays under GLP conditions to the greatest extent possible, in compliance with ICH S7A.[111] This requirement is based on the fact that the regulators and EMEA reviewers, in contrast to the experts in the field, consider the testing of abuse liability under the scope of safety and toxicology instead of behavioral pharmacology, which reflects the renewed focus of regulatory agencies worldwide on safety issues. In the United States, a new drug safety bill, recently signed into law, grants the FDA more power over the safety of prescription drugs. This is expected to shift the FDA's attention away from experimental drugs pending approval toward those already on the market, giving the agency more power to act when worrisome problems arise.

Overall, the current environment provides only limited guidance on how regulatory agencies are evaluating abuse liability of new compounds and making labeling recommendations. This is not surprising, as the EU guideline has only recently been adopted and the FDA guidance in the United States is under development. The increasing focus of the regulatory agencies on safety aspects of drug development suggests that decisions on the approval of new compounds will be heavily supported by very detailed risk management plans, post-marketing commitments, and larger safety databases worldwide.[vi]

World Health Organization

The WHO scheduling criteria for psychoactive substances are dictated by the 1961 Convention on Narcotic drugs[112] and the 1971 Convention on Psychotropic drugs,[113] and are based on similarity rules.[114] In accordance with the 1961 Convention, the WHO Expert Committee determines whether the substance under review has morphine-, cocaine-, or cannabis-like effects or can be converted into a scheduled substance having such effects. On the other hand, under the 1971 Convention, besides the similarity rule, the scheduling criteria for psychotropic substances have an additional requirement for "... evidence that the substance is being or is likely to be abused so as to constitute a significant public health and social problem warranting the placing of the substance under international control."

Major hurdles arise from the lack of specific guidance on how to define the similarity between substances, and the ambiguity of the criteria for choosing between the 1961 or the 1971 Conventions, particularly when the drug under review exhibits

---

[vi] For further discussion of regulatory guidance for development of therapeutics with abuse liability potential for disorders such as ADHD in a pediatric population, please refer to Tannock *et al.*, Towards a biological understanding of ADHD and the discovery of novel therapeutic approaches, in Volume 1, Psychiatric Disorders.

some similarity with both narcotic and psychotropic drugs. Most potent analgesics are controlled as narcotics under the 1961 Convention, but few others are controlled under the 1971 Convention. While cocaine is controlled under the 1961 Convention, amphetamines are controlled under the 1971 Convention. Further complexity arises because of the different ordering of the Schedules in the two Conventions. In the 1961 Convention, Schedule IV is the most restrictive, whereas it is the least restrictive in the 1971 Convention.

The complexity of the WHO scheduling and recommendation system can be illustrated by the case of buprenorphine, which is currently approved in about 60 countries as an analgesic, and in about 30 countries for substitution treatment for opioid dependence. Buprenorphine was critically reviewed at the twenty-fifth meeting of the Committee,[115] which recommended its placement in Schedule III of the 1971 Convention. However, in 1995, due to evidence of significant abuse and illicit trafficking, the International Narcotics Control Board (INCB) requested a revision of the control system. On the basis of the similarity criterion, buprenorphine should be reclassified under the 1961 Convention since it has greater similarity to morphine than to the nonspecific opioid agonist lefetamine (Santenol®), which was the only analgesic controlled as a psychotropic substance at the time the 1971 Convention was adopted.

However, the lack of guidance on transferring a substance from one Convention to the other, as well as a provision of simultaneous deletion from one Convention and addition to the other, did not provide the procedural legal basis for rescheduling. In addition, there would be consequences in implementation of the conventions in national laws and an unintended effect of restricting access to buprenorphine for use in opioid agonist therapy was possible in certain countries. Based on the unique pharmacological actions of buprenorphine, and its expanded role in the treatment of opioid dependence, the Committee did not recommend any change in the present scheduling of the substance in Schedule III of the 1971 Convention.[116]

APPROVED AND PROPOSED MEDICATIONS TO TREAT OPIOID AND COCAINE ADDICTION

Heroin (Opioid) Addiction

The use of heroin has continued to increase since the mid-1980s, and approximately one-quarter of all individuals who try heroin eventually meet the DSM-IV-TR criteria for dependence (e.g.,[117]). Although there are currently more approved medications to treat heroin addiction than to treat addiction involving other major drugs of abuse, none of these medications are effective in the absence of additional supportive measures. Medications to treat heroin addiction, therefore, should be viewed as pharmacological adjuncts that increase the effectiveness of counseling, drug management training, and other forms of psychosocial interventions. Apart from a few drugs, such as clonidine, which can help alleviate withdrawal symptoms, most medications for heroin addiction come from the opioid family of drugs and are intended for long-term use. This section provides a brief overview of animal and human laboratory studies of three such drugs: the opioid agonist methadone, the partial agonist buprenorphine, and the antagonist naltrexone.

Methadone

The greatest success in the development of medications against drug addiction thus far has come with the use of methadone to treat heroin dependence. Methadone was originally introduced in the 1960s as a replacement therapy in heroin-addicted individuals,[118] and methadone maintenance remains the most widely used pharmacotherapy for the management of opioid addiction in the United States. It is instructive to consider how closely the clinical efficacy of methadone is mirrored in laboratory studies of heroin self-administration.

Animal laboratory studies

Studies since the mid-1970s have investigated the effects of methadone on self-administration of heroin and other abused opioids in laboratory animals.[76,119-121,79] Using a discrete-trial choice procedure in which baboons could select either an injection of heroin or delivery of food, Griffiths and colleagues[76] found that continuous iv infusion of methadone for 10 days or longer resulted in a consistent decrease in self-administered heroin and an increase in the number of food deliveries, suggesting that chronic methadone maintenance selectively decreased the reinforcing effects of heroin. Harrigan and Downs[120] also reported that continuous iv infusion of methadone results in a dose-related decrease in heroin self-administration by rhesus monkeys. However, when the infusion dose of methadone was sufficiently high to reduce self-administration over a reasonably broad range of heroin doses, the subjects appeared "debilitated and depressed," suggesting a generalized suppression of behavior rather than a selective effect on heroin reinforcement. Related behavioral side effects of methadone also were observed in a study by Mello et al.[121] in which self-administration behavior by rhesus monkeys was evaluated during alternating daily sessions of either opioid (heroin or hydromorphone) or food reinforcement. In that study, daily treatment with gradually increasing doses of methadone over a 4-month period did not consistently reduce heroin self-administration even at doses that disrupted the subjects' ability to perform the food-reinforced operant response. The findings of Harrigan and Downs[120] and Mello et al.[121] are largely consistent with those reported in dogs allowed to self-administer morphine 24h/day.[119] In that study, continuous iv infusion of methadone resulted in a temporary reduction of morphine self-administration that was accompanied by marked sedation.

Heroin addiction in humans is thought to be maintained both by the positive reinforcing effects of the self-administered drug and by amelioration of the aversive effects induced by opioid withdrawal. Based on these considerations, Negus[79] investigated the effects of 5-day blocks of continuously infused methadone on heroin self-administration by rhesus monkeys using a choice procedure involving concurrent options of heroin and food reinforcement. Subjects initially were studied under a condition of limited heroin access, which did not induce appreciable physical dependence. Under these conditions, methadone had little or no effect on either heroin choice or total heroin intake. The subjects were then made physically dependent on heroin by adding supplemental periods of daily heroin self-administration, which increased the daily heroin intake 7- to 8-fold. In these heroin-dependent subjects, termination of supplemental heroin increased heroin self-administration during choice periods and induced overt signs of opioid withdrawal. Methadone (0.56 mg/kg/h) prevented both the withdrawal-associated increase in heroin choice and the emergence of withdrawal signs, findings that appear to model key effects of methadone in heroin-dependent people accurately.

The different profiles of effects of methadone maintenance reported in the above studies illustrate the influence of methodological factors (e.g., choice versus non-choice procedures) and differences in the degree to which subjects were physically dependent on opioids (presumably less severe under conditions of limited heroin access) in laboratory self-administration studies. It is perhaps relevant to note that, as in laboratory studies, methadone maintenance is not invariably effective in its clinical application. Depending on dosage, treatment conditions, and outcome measures, success rates in well-controlled clinical studies may range from 20% to 70% (e.g.,[122]). Thus, the challenge of developing laboratory models to predict successful medications is formidable. Nevertheless, the positive results with methadone reported by Griffiths *et al.*[76] and Negus[79] suggest that one successful approach might be to identify treatments that promote the replacement of drug self-administration by other active behavioral options, as in concurrently programmed schedules of drug and non-drug reinforcement.[123]

Human laboratory studies

One of the earliest studies to demonstrate the effectiveness of methadone in reducing opioid self-administration by human research volunteers was conducted by Jones and Prada,[119] who demonstrated that hydromorphone self-administration and ratings of drug liking were reduced in participants who were maintained on doses up to 100 mg/day methadone. A more recent study, using a procedure in which participants could choose to self-administer drug or receive money, showed that methadone reduced heroin choice in a dose-related manner.[124] These data collected in the human laboratory correspond well with the results obtained using similar choice procedures in animals and predict the clinical experience with methadone. In both the clinical situation and the laboratory, doses in the 50 mg/day range are less effective in reducing heroin use than are higher doses in the 100 mg/day range (e.g.,[125,126] for review). In addition, similar to clinical experience, the laboratory data show that the full therapeutic benefit of methadone is achieved after a 3- 4-week period of maintenance.

Buprenorphine

Buprenorphine is a potent, long-acting partial agonist at the μ-opioid receptor, where it exhibits a profile of mixed agonist and antagonist properties. Its antagonist-like properties attenuate the effects of heroin and other μ-receptor agonists, while its agonist-like properties likely contribute to its improved compliance compared to conventional μ-receptor antagonists (see[127]).

Animal laboratory studies

As with conventional μ-antagonists, buprenorphine (iv) has been found to decrease self-administration of heroin and related μ-agonists in laboratory animals.[121,128–131,80] Comparing across studies, the results are consistent with the conclusion that buprenorphine induces a relatively selective antagonism of the reinforcing effects of μ-agonists. In a recent study by Negus,[79] which involved choice between concurrently available heroin and food, buprenorphine induced a dose-dependent, rightward shift in the dose–response curve for heroin choice in non-dependent rhesus monkeys – an effect virtually identical to that of naloxone. After the subjects were rendered physically dependent on and then withdrawn from heroin, buprenorphine blunted overt signs of withdrawal and withdrawal-induced increases in heroin choice, effects similar to but considerably less pronounced than those of methadone. Aceto[132] also reported that buprenorphine induced a partial suppression of withdrawal signs in morphine-dependent rhesus monkeys.

As expected of a partial agonist, buprenorphine (iv) serves as a reinforcer in non-human primates with a history of opioid self-administration.[133-138] Direct comparisons of buprenorphine self-administration with self-administration of other μ-receptor agonists suggest that buprenorphine's reinforcing effectiveness is low.[136,134] Collectively, these findings suggest that although buprenorphine functions as a reinforcer, it does not induce abuse to a degree comparable to that of heroin. In fact, the modest reinforcing effects of buprenorphine may be an important factor in the improved compliance reported for buprenorphine compared to full antagonists such as naltrexone.

Human laboratory studies

Buprenorphine has been studied extensively in human laboratory settings for its ability to reduce self-administration of heroin and other opioids. When participants were maintained on subcutaneously administered buprenorphine, heroin self-administration was reduced by 69–98%, compared to when these same participants were maintained on placebo.[139,140] Subsequent research, using sublingual formulations of buprenorphine, also demonstrated that it was effective in reducing opioid self-administration. In a study using an alternating inpatient and outpatient design, Greenwald and colleagues[141] showed that those individuals who abstained from heroin use during outpatient phases were also the ones who were successful in abstaining from hydromorphone use during inpatient laboratory sessions. In addition, the abstainers showed reduced positive subjective effects after hydromorphone administration. Similarly, studies using the marketed preparations of sublingual buprenorphine alone[142,143] or the buprenorphine/naloxone combination,[144] showed dose-related reductions in the reinforcing effects of heroin and hydromorphone. The reductions in opioid self-administration by the marketed preparations of sublingual buprenorphine were more modest than those obtained with injected buprenorphine, which may have been due to the lower bioavailability of sublingual buprenorphine. In addition, the experimental design may have played a role in the effect obtained in the more recent studies. That is, sublingual buprenorphine was administered in the evening, 19 h prior to heroin self-administration sessions, which likely resulted in the more limited effectiveness of buprenorphine in the studies by Comer and colleagues.[142,144]

Naltrexone

The relatively early development of safe and effective μ-opioid antagonists, such as naloxone and naltrexone,[145] provided important tools for investigating receptor-mediated mechanisms that were involved in the behavioral actions of heroin and other opioid agonists, which includes their reinforcing effects. Moreover, these developments prompted speculation that such antagonists might be a broadly effective treatment for heroin addiction and, as a result, laboratory studies in animals and man to evaluate this possibility.

Animal laboratory studies

Early interest in the use of opioid antagonists as medications for heroin addiction was encouraged by the consistent finding that naloxone and naltrexone effectively reduced opioid self-administration in rodents[146-148] and nonhuman primates.[76,149] Follow-up studies involving comprehensive dose–response analyses, the availability of alternative reinforcers, and chronic treatment regimens have continued to offer compelling evidence that naloxone and naltrexone induce a selective antagonism of heroin's reinforcing effects.[150,151,79] Doses of the antagonists typically used in these studies produce few adverse effects in non-dependent subjects, but are sufficient to precipitate withdrawal signs in animals rendered physically dependent to heroin.

Although naltrexone, like buprenorphine, clearly induces pharmacological blockade of heroin's reinforcing effects, it does not maintain self-administration behavior in laboratory animals, consistent with its μ-antagonist effects. To the extent that the presence of reinforcing effects in animal studies provides information relevant to the key issue of compliance, one would predict greater clinical compliance with buprenorphine than with naltrexone. In fact, as discussed below, naltrexone has not been a particularly effective pharmacotherapy in practice because most patients simply do not comply with prescribed treatment regimens.

Human laboratory studies

Early laboratory studies showed that oral naltrexone virtually eliminated the reinforcing effects of heroin.[152,153,134] More recently, an injectable naltrexone formulation also was found to be quite effective in reducing both the subjective and reinforcing effects of heroin in laboratory studies, with a duration of action of up to 1 month.[154,155] Despite the unequivocal pharmacological effectiveness of oral naltrexone in antagonizing the effects of heroin, previous clinical experience with naltrexone has been less than satisfactory, primarily due to medication noncompliance. The recent development of sustained-release naltrexone, which eliminates the need for repeated daily patient compliance, may offer more hope for the effectiveness of this medication for treating opioid dependence. A 2-month clinical trial of this formulation of depot naltrexone was encouraging in this regard, with treatment retention rates rivaling those seen with agonist replacement therapies.[156] Overall, the greatest success in the development of medications against drug addiction thus far has occurred in the development of opioid pharmacotherapies for opioid addiction.

Cocaine Addiction

There are currently no effective medications to treat cocaine addiction. Although numerous drugs have been proposed as potential pharmacotherapies (sometimes with very little supporting evidence), few have been evaluated systematically in animal and human laboratory studies. This section provides a brief overview of animal and human laboratory studies of four such drugs: ecopipam, which has proven to be ineffective in clinical trials, baclofen, aripiprazole (Abilify®) and modafinil (Provigil®), which are currently under active investigation.

Ecopipam and Related Dopamine D₁ Receptor Antagonists

Given that cocaine's reinforcing effects in animal studies are closely linked to its inhibition of dopamine re-uptake,[157-159] one medication development strategy for reducing cocaine's effects has been to target antagonists of the dopamine receptor system. Two main families of dopamine receptor subtypes (D_1 and D_2) have been identified.[160,161] Although both D_1 and D_2 receptors contribute to cocaine's reinforcing effects,[162,163] only D_1 receptor antagonists have been evaluated from preclinical animal and human studies through Phase II clinical trials.

Animal laboratory studies

Acute pretreatment with D_1 receptor antagonists appears to block cocaine's reinforcing effects in rats[164-166] and monkeys[167,168] in a partially surmountable manner. Some studies report that acute pretreatment with the D_1 receptor antagonists, SCH 23390[169-171] or SCH 39166 (ecopipam)[172,173] only suppressed cocaine self-administration at doses that induced other forms of behavioral disruption. Other studies, however, have reported that D_1 antagonists can act selectively by altering cocaine self-administration at doses that do not result in motor incapacitation or impaired operant behavior in rats[164,174] or monkeys.[175] In general, however, such selectivity is observed over a relatively narrow dose range (see review by[176]).

Chronic administration of dopamine antagonists over days or weeks may result in a diminution or reversal of the effects seen following acute administration in rodents and monkeys.[177-180] Kleven and Woolverton,[175] for example, found that although continuous infusion with SCH 23390 initially decreased cocaine self-administration by rhesus monkeys, the effects diminished over a period of several days, and self-administration eventually returned to control levels. Moreover, following termination of treatment with SCH 23390, self-administration of cocaine eventually increased above control levels in the majority of monkeys. Similarly, termination of chronic D_1 antagonist administration in rodents has been reported to exacerbate behavioral effects of cocaine as well as other indirect or direct dopamine agonists.[181-183] These behavioral changes appear to be associated with an increased density of D_1 receptors,[184-186] and enhanced D_1 receptor sensitivity.[187] Collectively, data from laboratory animal studies suggest that maintenance on a D_1 antagonist, rather than producing a sustained reduction in cocaine's effects, appears to be transiently effective and may lead to increased cocaine self-administration after chronic treatment. As discussed below, these findings are consistent with the results of clinical trials in which long-term treatment with ecopipam failed to decrease cocaine self-administration.

Human laboratory studies

Acute oral (po) pretreatment with ecopipam (10-100 mg), dose dependently decreased the effects of cocaine (30 mg iv) on ratings of "high" and "good drug effect," and decreased the reported desire for cocaine in cocaine-dependent research volunteers.[188] By contrast, maintenance on ecopipam (10-100 mg po) for 5-7 days prior to smoked or iv cocaine neither decreased cocaine's subjective effects (0-50 mg/70 kg,[189]) nor reduced cocaine self-administration (12–50 mg[190]). In fact, ecopipam actually increased self-administration of a low cocaine dose (12 mg) and increase ratings of "high," and "good drug effect," as well as the perceived quality of higher cocaine doses (25, 50 mg[190]).

Consistent with the results from human laboratory studies, a controlled, multi-site clinical trial testing ecopipam maintenance in cocaine-dependent treatment seekers was terminated due to a lack of efficacy (see[191]). Thus, the effects of ecopipam maintenance in the clinic are consistent with those in both animal and human cocaine self-administration studies that incorporated chronic D_1 antagonist treatment regimens.

Baclofen

In addition to reducing the dopaminergic actions of cocaine by directly blocking dopamine receptors, other approaches for the development of potential medications have included indirect modulation of cocaine's dopaminergic actions. One group of drugs that has been well characterized in laboratory studies includes the direct and indirect γ-aminobutyric acid (GABA) agonists.[192] GABA is thought to modulate the effects of cocaine and other stimulant drugs by tonically inhibiting the firing of dopamine neurons in the ventral tegmental area (VTA).[193] Consistent with this view, administration of the $GABA_B$ receptor agonist baclofen into the VTA decreased cocaine-induced elevations in extracellular dopamine in the nucleus accumbens.[194] Such findings have encouraged the systematic evaluation of baclofen over the past several years in both animal and human laboratory studies.

Animal laboratory studies

Consistent with its neurochemical actions (described above), acute systemic administration of baclofen has been shown to decrease cocaine self-administration in rats. These latter effects are particularly evident at low doses of cocaine;[195-197] higher doses of cocaine appear to surmount the effects of baclofen.[198] Selective $GABA_B$ receptor antagonists have been reported to reverse the effects of baclofen, demonstrating its pharmacological specificity.[199] Some studies in rats suggest that the effects of baclofen are behaviorally specific in that baclofen decreased self-administration of cocaine at doses that did not alter food-reinforced behavior.[195,200,197] However, other studies in rats[201] and nonhuman primates[202] suggest that acute baclofen administration decreases cocaine self-administration only at doses that produce sedation.

Although acutely administered baclofen can unquestionably decrease cocaine self-administration in an apparently pharmacological (but perhaps not behaviorally) selective manner, there are as yet no systematic animal studies of long-term treatment with baclofen. In light of recent findings that chronic baclofen administration can reduce cocaine use both in the human laboratory and in the clinic (see below), studies involving chronic administration of baclofen in animal self-administration studies appear warranted and may provide needed information regarding the concordance between preclinical and clinical effects of this type of candidate medication.

Human laboratory studies

In human laboratory studies, acute baclofen pretreatment (up to 30 mg po) did not alter the reinforcing or subjective effects of intranasal cocaine (4 or 45 mg) in occasional cocaine users.[203] However, maintenance on baclofen altered smoked cocaine self-administration compared to placebo in cocaine-dependent volunteers.[204] Specifically, baclofen administration (60 mg/day) for at least 3 days prior to cocaine self-administration sessions decreased self-administration of a low dose of cocaine (12 mg), but did not alter the self-administration of higher cocaine doses. Maintenance on baclofen also did not alter cocaine's subjective effects (i.e., ratings of "high" or "stimulated") in human volunteers. Regarding the issue of baclofen's specificity, there was no evidence that these doses of baclofen were sedating or that they nonspecifically decreased cocaine self-administration in humans.

In a randomized clinical pilot study,[205] cocaine-dependent patients maintained on baclofen (60 mg/day) used significantly less cocaine than those maintained on placebo. Baclofen, as well as other GABA receptor agonists, appears most effective in decreasing cocaine use after several weeks of maintenance. Baclofen-treated patients were most likely to provide urine samples free from cocaine metabolites after 3–8 weeks of study participation.[205] Similarly, another GABA agonist, tiagabine, was effective in modifying cocaine use in a clinical sample only after at least 3 weeks of continuous administration.[206] These encouraging clinical outcomes are generally consistent with the effects of acute baclofen in animals and the effects of relatively short-term chronic baclofen in human laboratory studies. The clinical effects of baclofen are generally more robust than might have been predicted from the preclinical data, perhaps reflecting differences between the effects of acute or short-term treatment in the laboratory and sustained treatment in clinical studies. It may be that the laboratory data would have been more compelling if baclofen had been administered for several weeks prior to the re-evaluation of cocaine's effects. Taken together, findings with baclofen and ecopipam illustrate the important role of chronic administration studies in the preclinical evaluation of candidate therapies for cocaine abuse and addiction.

Aripiprazole

Various dopamine D_1 and D_2 receptor partial agonists have been proposed as candidate pharmacotherapies for cocaine addiction based on their ability to modulate iv cocaine self-administration in rodents and/or nonhuman primates (see review by[176]). However, none of these drugs has yet shown clinical efficacy in human cocaine abusers. Despite such findings, aripiprazole, an atypical D_2 partial agonist with low incidence of extrapyramidal side effects, has been proposed for clinical trials for the treatment of cocaine dependence. It is instructive to consider the extent to which existing data in animal and human laboratory studies support this proposal.

Animal laboratory studies

Surprisingly, there appears to be only one published report that specifically examined the effects of aripiprazole in animals (rats) trained to self-administer cocaine.[207] Although intraperitoneal (ip) aripiprazole (0.5–2.5 mg/kg) attenuated the reinstatement of extinguished cocaine seeking in that study, it did not significantly reduce active self-administration of cocaine. A trend toward increased intake of cocaine (arguably, an undesirable clinical end point) at higher doses of aripiprazole is consistent with findings using conventional D_2 partial agonists, such as terguride, in rats (e.g.,[208]). Although the limited data from animal studies should be viewed cautiously, they do not suggest that aripiprazole would be an effective medication to treat cocaine addiction in people.

Human laboratory studies

The effect of the partial dopamine D_2 receptor agonist, aripiprazole, on cocaine self-administration and subjective effects is currently being tested in cocaine-dependent volunteers by Haney and colleagues (personal information). Aripiprazole has a long half-life, so participants were maintained on both placebo and aripiprazole (15 mg/day) capsules for 17 days prior to the onset of cocaine self-administration sessions. Smoked cocaine dose–response curves (0–50 mg) were determined using a PR schedule of reinforcement. Preliminary data suggest that aripiprazole maintenance significantly increased cocaine (12 and 25 mg) self-administration compared to placebo maintenance. Under placebo cocaine conditions, aripiprazole also increased ratings of "anxious" and "stimulated." Furthermore, in combination with cocaine, aripiprazole increased cocaine and alcohol craving, as well as peak ratings of "good drug effect" and "stimulated." Thus, maintenance on aripiprazole appears to increase rather than decrease the self-administration and subjective effects of cocaine. It does not appear that aripiprazole increased cocaine self-administration by antagonizing the reinforcing effects of cocaine, because there were also enhanced ratings of cocaine "high" under aripiprazole maintenance compared to placebo.

There are no published clinical trials testing aripiprazole to treat cocaine dependence. However, a recent clinical trial for amphetamine dependence was terminated early because the aripiprazole group showed higher rates of amphetamine use than those receiving placebo.[209] Based on these initial clinical results and the consistently negative findings with aripiprazole in animal and human laboratory studies, it does not appear that aripiprazole would be broadly beneficial for the treatment of cocaine dependence.

Modafinil

The FDA-approved wake-promoting agent modafinil is an example of an "agonist" approach to the treatment of cocaine dependence, which is in some ways conceptually similar to the use of methadone maintenance therapy to treat heroin dependence. In this regard, certain of modafinil's pharmacological effects overlap with those of cocaine. Modafinil occupies both dopamine and norepinephrine transporter sites,[210] and clinically relevant doses of modafinil increase extracellular dopamine levels.[211] Furthermore, modafinil both enhances glutamate release and inhibits GABA release,[212] as well as firing of midbrain dopamine neurons.[213] Modafinil appears to have a cocaine-like profile of behavioral effects in laboratory animals,[214] but there are currently no published reports concerning the effects of modafinil treatment on cocaine self-administration or other abuse-related effects in rodents or nonhuman primates. Thus, the discussion of its effects in laboratory studies is necessarily confined to studies in human subjects.

Human laboratory studies

In human laboratory studies, modafinil (up to 400 mg po) produced stimulant-like subjective effects,[215] and was self-administered more than placebo under certain conditions,[216] consistent with its effects in animals. However, even at high doses (600–800 mg) modafinil did not produce cocaine- or amphetamine-like discriminative-stimulus or subjective effects,[217-219] and post-marketing surveillance indicates that modafinil misuse is low.[220] The first human laboratory studies that tested modafinil combined with cocaine (iv) demonstrated that the medication was safe and that it decreased cocaine's intoxicating and cardiovascular effects.[221-223] A recent human study of smoked cocaine self-administration similarly demonstrated that modafinil maintenance (200–400 mg/day) decreased high dose cocaine self-administration (25 and 50 mg), as well as cocaine's intoxicating (e.g., high, good drug effect) and cardio-vascular effects.[224] These laboratory findings correspond well with a pilot clinical trial, which showed that modafinil (400 mg/day) significantly reduced the number of cocaine-positive urines.[225] A more recent multi-site clinical trial confirmed modafinil's efficacy to promote abstinence in cocaine-dependent patients, provided that they were not co-dependent on alcohol and did not have a history of alcohol dependence (Frank Vocci, personal communication, 2007).

Concordance Between Animal and Human Laboratory Studies and Clinical Outcome

An essential question is whether medications that decrease the subjective, discrimi-native-stimulus, and/or reinforcing effects of cocaine decrease drug use in the clinic. Further, if a medication is efficacious in the clinic, will patients be compliant in taking the medication? As discussed above, for example, naltrexone capsules fully block both the reinforcing and subjective effects of heroin, but oral naltrexone is not an effective pharmacotherapy because most patients stop taking the capsules.

Compounds showing promise in preclinical research typically have been tested in open-label trials in which both physicians and patients know the medication to be administered. Positive open-label studies are then followed up by double-blind, placebo-controlled trials. These clinical trials are the standard by which the efficacy of a new med-ication is assessed, yet they take several years to complete, require hundreds of patients, and are costly and potentially risky (see[226]). A vast number of medications have been proposed and tested for the treatment of cocaine dependence in either open or double-blind clinical trials. These drugs include antidepressants, anticonvulsants, antipsychotics, and mood stabilizers, among others (see[227]). None of these proposed medications have as yet proven to be broadly effective and none have so far been FDA approved for this indi-cation. Thus, better indicators are needed to select medications for cocaine dependence in large clinical trials. Prior to exposing a large number of treatment seekers to an inves-tigational medication, there needs to be both a strong scientific rational for the effective-ness of the medication for the treatment of drug dependence. In addition, there should be strong evidence that the co-administration of the medication and the abused drug is safe and selectively modifies the target behavior (i.e., drug taking).

With regard to opioids, good concordance appears to exist among results of studies conducted in laboratory animals, human research volunteers, and clinical settings. In addition, data from animal and human laboratory studies of cocaine self-administration are consistent with clinical outcome, particularly when medications are given repeatedly prior to self-administration sessions (see[228]). Acute ecopipam administration blocked cocaine's reinforcing effects in rats and monkeys, and decreased cocaine's positive subjective effects in human subjects. Yet chronic ecopipam administration had little effect on or actually increased cocaine self-administration in animal and human laboratory studies. Predictably, ecopipam maintenance did not improve clinical outcome in individuals seeking treatment for their cocaine use in clinical trials. Similarly, aripiprazole maintenance increased cocaine self-administration in human laboratory studies and appeared to worsen outcome for amphetamine treatment (no clinical data on cocaine are available to date).

The clinical data with baclofen and modafinil, although still largely preliminary, appear to validate the human laboratory model of cocaine self-administration for medications development by demonstrating that both baclofen and modafinil maintenance can decrease cocaine self-administration in the laboratory, and decrease cocaine use in the clinic. The clinical findings with baclofen also appear to be predicted by acute administration studies in animals. However, no animal studies have yet examined the effects of repeated treatment with baclofen or modafinil prior to daily opportunities to self-administer cocaine. Such studies may be particularly valuable for validating the preclinical self-administration model as a tool for drug discovery with respect to cocaine pharmacotherapy.

Improving the Predictive Value of Existing Models

Self-administration Versus Other Models for Medications Development

The self-administration model clearly predicts medications that effectively treat opioid dependence, and appears to be the best current behavioral paradigm for preclinical testing of medications to treat cocaine dependence. For opioids, clinically effective medications decrease both the subjective and reinforcing effects of opioids, but the same is not necessarily true for cocaine. Several human laboratory studies have assessed the ability of medications to dampen the discriminative-stimulus and subjective effects of cocaine, hypothesizing that blocking these effects will predict utility in the clinic (see[229] for review;[230]). However, cocaine's subjective effects appear to be more sensitive to modulation by medications than its reinforcing effects, resulting in a high rate of false positives when only subjective effects are measured. The same appears to be largely true in animal studies involving direct comparisons of drug discrimination and drug self-administration results.

Cocaine use is extraordinarily difficult to disrupt both in the human laboratory and in the clinic. Medications rarely decrease cocaine self-administration, even when they substantially decrease ratings of cocaine "high" or "good drug effect" (see[231-234]). Measures of craving are also not a proxy for self-administration, as craving is far easier to disrupt than drug taking. For example, maintenance on the antidepressant, desipramine, resulted in a 40% decrease in ratings of cocaine craving yet had no effect on the amount of cocaine self-administered in laboratory animals,[179] in human research volunteers[231] or in the clinic (e.g.,[235]). This dissociation between craving and actual drug use emphasizes the importance of measuring cocaine self-administration when investigating potential pharmacotherapies for drug dependence. In contrast to the large number of false positives obtained in studies that do not assess self-administration, no medication has been found to significantly decrease the self-administration of high doses of cocaine in the human laboratory until recently.[224] Drug self-administration studies in laboratory animals have been of significant value for identifying candidate medications that fail to modify drug self-administration or that display undesirable properties such as high abuse potential.[176] Human and animal self-administration studies appear to predict medication failure or success in the clinic with much better accuracy than studies that rely solely on drug discrimination data, self-reported subjective effects, or measures of craving (see[228]).

Alternative Reinforcers

The majority of drug self-administration studies in laboratory animals have not provided explicit alternative reinforcers as part of the experimental design, whereas human drug users may be exposed to a wide range of alternative reinforcers. Most individuals who use drugs are faced with at least minimally reinforcing alternatives to drug use, and animal self-administration procedures involving drug and non-drug alternatives can provide a useful paradigm for evaluating the impact of such alternatives. A consistent finding in such studies is that increasing the relative availability of alternative reinforcers results in a reduction of drug choice.[236-238] Similarly, increasing the value of an alternative reinforcer in human studies decreases drug choice,[239,240] although the value of the alternative reinforcer may have less impact once drug self-administration is initiated.[20] These observations suggest that medications that decrease the reinforcing effects of an abused drug might be expected to increase the relative reinforcing strength of non-drug alternatives, resulting in a shift in the allocation of behavior from drug taking to non-drug alternatives. Shifts of this type have been observed for heroin self-administration in studies with opioid antagonists and partial agonists[79] and encourage the continued use of choice procedures for evaluating potential pharmacotherapies for drug addiction.

The immediacy of alternative reinforcement also appears to play a key role in the degree to which a candidate medication shifts behavior away from drug self-administration. For example, rhesus monkeys responding to self-administer either cocaine or an alternative reinforcer chose less cocaine when the alternative reinforcer was delivered immediately than when its delivery was delayed.[241] It is noteworthy that in most human laboratory procedures the alternative reinforcer is not available until completion of the test session, whereas drug delivery is typically immediate. Additional research into the effects of delays in reinforcement on drug versus non-drug choice procedures in both animal and human laboratory situations is likely to provide a more comprehensive picture of how potential medications may influence the relationship between temporal parameters of drug and non-drug reinforcer delivery and the allocation of behavior to those choices.

Medication Regimen

Medications targeting drug dependence require long-term treatment regimens to be maximally effective. Yet evaluations of potential pharmacotherapies for cocaine dependence in animal drug self-administration studies most often use acute medication pretreatment paradigms. One factor that may be especially relevant for improving the predictive validity of laboratory self-administration models is the duration of medication administration (see[57,228,242,243]). Based on findings from human laboratory studies, treatment durations of just several days can reveal changes in the effects of a candidate medication that might not be anticipated on the basis of its acute effects. Some medications only decrease cocaine self-administration when given acutely, while others may require repeated administration before selectively attenuating drug self-administration. As discussed above, ecopipam was effective when administered acutely but not chronically, whereas available data suggest that the effectiveness of baclofen increases over the course of repeated treatment. Recent studies with D-amphetamine provide further evidence of the strikingly different results that can be obtained with acute and chronic drug treatment. In these studies, acute D-amphetamine administration suppressed both food intake and cocaine self-administration in nonhuman primates. With repeated D-amphetamine administration, however, tolerance developed to the suppression of food intake but not the suppression of cocaine self-administration.[244] Collectively, the available data support the idea that animal and human laboratory models may gain improved predictive power by adopting protocols that include appropriate periods of medication maintenance.

Improving the Predictive Value of Existing Models in the Assessment of Abuse Liability

The predictive value of animal-to-human models used in the assessment of abuse liability should be evaluated according to the stage of the compound's development, which is in sharp contrast with any other therapeutic indication, and represents different challenges for sponsors. Since assessment of abuse liability is relatively regulated during development (i.e., EMEA guideline on the preclinical assessment of abuse liability[34]), the questions sponsors face frequently are not about the predictive value of one or other test *per se*, but rather the decision of when exactly during development some tests should be initiated in order to best fulfill regulatory requirements.

As discussed above, from the regulatory perspective, the general consideration that abuse liability testing is required for safety assessment has been increasingly demanding of clinical studies earlier in development. In particular for compounds with new pharmacological mechanisms of action, the current tendency is to minimize the number of animal assays and to move into clinical studies in polydrug abusers as soon as possible. The predictive value of the ensemble of preclinical and clinical tests is ultimately measured at the time of the compound's approval, when a final scheduling recommendation (or lack thereof) is made.

After approval, when the compound is available to the public, assay predictability would be reflected in the compound's demonstrated imbalance between use and misuse. Even though scheduling *per se* can impact such balance, prospective studies that evaluate scheduling influence on the non-medical use of a compound have not been systematically completed. Sponsors and regulators are working toward more consistency in the implementation of post-marketing plans for tracking abuse, which will generate data that would facilitate such studies to be conducted.

In summary, throughout the drug development process and post-marketing surveillance, the pharmaceutical industry plays a major role in generating preclinical and clinical data, which ultimately represents a long-term contribution toward confirmation of the predictive validity and translation of the assays described in this chapter.

REFERENCES

1. SAMHSA (2007). Substance Abuse and Mental Health Services Administration 2006 National Survey on Drug Use and Health: National Findings. Rockville, MD: Substance Abuse and Mental Health Services Administration, Office of Applied Studies.
2. Gold, M.S., Pottash, A.C., Sweeney, D.R., and Kleber, H.D. (1980). Opiate withdrawal using clonidine. A safe, effective, and rapid non-opiate treatment. *JAMA*, 243:343–346.
3. Drug Abuse Treatment Toolkit. Contemporary Drug Abuse Treatment – A Review of the Evidence Base. United Nations International Drug Control Program, New York, 2002.
4. Vocci, F., Chiang, C.N., Cummings, L., and Hawks, R. (1995). Overview: Medications development for the treatment of drug abuse. *NIDA Res Monogr*, 149:4–15.
5. Schuster, C.R. and Henningfield, J. (2003). Conference on abuse liability assessment of CNS drugs. *Drug Alcohol Depend*, 70(3 Suppl):S1–S4.
6. Draft Guidelines for Research Involving the Abuse Liability Assessment of New Drugs, Food and Drug Administration (FDA) Center for Drug Evaluation and Research (CDER); Division of Anesthetic, Critical Care and Addiction Drug Products, December 1998.
7. DSM-IV-TR; American Psychiatric Association (2000): *Diagnostic and Statistical Manual of Mental Disorders*, 4th edition, Text Revision. American Psychiatric Association, Washington, DC.
8. Kolb, L. and Himmelsbach, C.K. (1938). Clinical studies of drug addiction. III: A critical review of the withdrawal treatments with method of evaluating abstinence syndromes. *Am J Psychiatry*, 94:759–799.
9. Siegel, R.K. (1982). Cocaine smoking. *J Psychoactive Drugs*, 14(4):271–359.
10. Foltin, R.W. and Fischman, M.W. (1991). Assessment of abuse liability of stimulant drugs in humans: A methodological survey. *Drug Alcohol Depend*, 28:3–48.
11. Emmett-Oglesby, M.W., Peltier, R.L., Depoorter, R.Y., Pickering, C.L., Hooper, M.L., Gong, Y.H., and Lane, J.D. (1993). Tolerance to self-administration of cocaine in rats: Time course and dose-response determination using a multi-dose method. *Drug Alcohol Depend*, 32(3):247–256.
12. Walsh, S.L., Haberny, K.A., and Bigelow, G.E. (2000). Modulation of intravenous cocaine effects of chronic oral cocaine in humans. *Psychopharmacology (Berl)*, 150(4):361–373.
13. Mendelson, J.H., Sholar, M., Mello, N.K., Teoh, S.K., and Sholar, J.W. (1998). Cocaine tolerance: Behavioral, cardiovascular, and neuroendocrine function in men. *Neuropsychopharmacology*, 18(4):263–271.
14. Ward, A.S., Haney, M., Fischman, M.W., and Foltin, R.W. (1997a). Binge cocaine self-administration by humans: Smoked cocaine. *Behav Pharmacol*, 8(8):736–744.
15. Ward, A.S., Haney, M., Fischman, M.W., and Foltin, R.W. (1997b). Binge cocaine self-administration in humans: Intravenous cocaine. *Psychopharmacology (Berl)*, 132(4):375–381.
16. Weddington, W.W., Brown, B.S., Haertzen, C.A., Cone, E.J., Dax, E.M., Herning, R.I., and Michaelson, B.S. (1990). Changes in mood, craving, and sleep during short-term abstinence reported by male cocaine addicts. A controlled, residential study. *Arch Gen Psychiatry*, 47(9):861–868.
17. Foltin, R.W. and Fischman, M.W. (1997). Residual effects of repeated cocaine smoking in humans. *Drug Alcohol Depend*, 47(2):117–124.
18. Jaffe, J.H., Cascella, N.G., Kumor, K.M., and Sherer, M.A. (1989). Cocaine-induced cocaine craving. *Psychopharmacology (Berl)*, 97(1):59–64.
19. Walsh, S.L., Geter-Douglas, B., Strain, E.C., and Bigelow, G.E. (2001). Enadoline and butorphanol: Evaluation of kappa-agonists on cocaine pharmacodynamics and cocaine self-administration in humans. *J Pharmacol Exp Ther*, 299(1):147–158.

20. Donny, E.C., Bigelow, G.E., and Walsh, S.L. (2003). Choosing to take cocaine in the human laboratory: Effects of cocaine dose, inter-choice interval, and magnitude of alternative reinforcement. *Drug Alcohol Depend*, 69:289–301.

21. Sinha, R., Krishnan-Sarin, S., Farren, C., and O'Malley, S. (1999). Naturalistic follow-up of drinking behavior following participation in an alcohol administration study. *J Subst Abuse Treat*, 17(1–2):159–162.

22. Harris, D.S., Reus, V.I., Wolkowitz, O.M., Mendelson, J.E., and Jones, R.T. (2005). Repeated psychological stress testing in stimulant-dependent patients. *Prog Neuropsychopharmacol Biol Psychiatry*, 29(5):669–677.

23. Ehrman, R.N., Robbins, S.J., Childress, A.R., and O'Brien, C.P. (1992). Conditioned responses to cocaine-related stimuli in cocaine abuse patients. *Psychopharmacology (Berl)*, 107(4): 523–529.

24. Childress, A.R., Mozley, P.D., McElgin, W., Fitzgerald, J., Reivich, M., and O'Brien, C.P. (1999). Limbic activation during cue-induced cocaine craving. *Am J Psychiatry*, 156(1):11–18.

25. Carter, B.L. and Tiffany, S.T. (1999). Cue-reactivity and the future of addiction research. *Addiction*, 94(3):349–351.

26. Hutchison, K.E., Lachance, H., Niaura, R., Bryan, A., and Smoken, A. (2002). The DRD4 VNTR polymorphism influences reactivity to smoking cues. *J Abnorm Psychol*, 111:134–143.

27. Arai, N., Isaji, M., Miyata, H., Fukuyama, J., Mizuta, E., and Kuno, S. (1995). Differential effects of three dopamine receptor agonists in MPTP-treated monkeys. *J Neural Trasm Park Dis Dement Sect*, 10(1):55–62.

28. Grondin, R., Goulet, M., Morissette, M., Bedard, P.J., and Di Paolo, T. (1999). Dopamine D1 receptor mRNA and receptor levels in the striatum of MPTP monkeys chronically treated with SKF-82958. *Eur J Pharmacol*, 378(3):259–263.

29. Iravani, M.M., Haddon, C.O., Cooper, J.M., Jenner, P., and Schapira, A.H. (2006). Pramipexole protects against MPTP toxicity in non-human primates. *J Neurochem*, 96:1315–1321.

30. Braff, D.L. and Geyer, M.A. (1991). Sensorimotor gating and schizophrenia. Human and animal model studies. *Arch Gen Psychiatry*, 48(4):379–380.

31. Geyer, M.A., Krebs-Thomson, K., Braff, D.L., and Swerdlow, N.R. (2001). Pharmacological studies of prepulse inhibition models of sensorimotor gating deficits in schizophrenia: A decade in review. *Psychopharmacology (Berl)*, 156(2–3):117–154.

32. Ellenbroek, B., van den Kroonenberg, P.T., and Cools, A.R. (1988). The effects of an early stressful life event on sensorimotor gating in adult rats. *Schizophr Res*, 30(3):251–260.

33. Koob, G.F. (1998). Drug abuse and alcoholism. *Overview Adv Pharmacol*, 42:969–977.

34. CHMP Guideline on the Non-clinical Investigation of the Dependence Potential of Medicinal Products, EMEA/CHMP/SWP94227/2004, March 2006.

35. Olds, J., Killam, K.F., and Bach-Y-Rita, P. (1956). Self-stimulation of the brain used as a screening method for tranquilizing drugs. *Science*, 124(3215):265–266.

36. Kornetsky, C. and Esposito, R.U. (1979). Euphorigenic drugs: Effects on the reward pathways of the brain. *Fed Proc*, 38(11):2473–2476.

37. Todtenkopf, M.S., Marcus, J.F., Portoghese, P.S., and Carlezon, W.A., Jr. (2004). Effects of kappa-opioid receptor ligands on intracranial self-stimulation in rats. *Psychopharmacology (Berl)*, 172(4):463–470.

38. Rompre, P.P., Injoyan, R., and Hagan, J.J. (1995). Effects of granisetron, a 5-HT3 receptor antagonist, on morphine-induced potentiation of brain stimulation reward. *Eur J Pharmacol*, 287(3):263–269.

39. Vorel, S.R., Ashby, C.R., Jr., Paul, M., Liu, X., Hayes, R., Hagen, J.J., Middlemiss, D.N., Stemp, G., and Gardner, E.L. (2002). Dopamine D3 receptor antagonism inhibits cocaine-seeking and cocaine-enhanced brain reward in rats. *J Neurosci*, 22(21):9595–9603.

40. Xi, Z.X. and Gardner, E.L. (2007). Pharmacological actions of NGB 2904, a selective dopamine D3 receptor antagonist, in animal models of drug addiction. *CNS Drug Rev*, 13(2):240-259.

41. Kenny, P.J., Boutrel, B., Gasparini, F., Koob, G.F., and Markou, A. (2005). Metabotropic glutamate 5 receptor blockade may attenuate cocaine self-administration by decreasing brain reward function in rats. *Psychopharmacology (Berl)*, 179(1):247-254.

42. Slattery, D.A., Markou, A., Froestl, W., and Cryan, J.F. (2005). The GABAB receptor-positive modulator GS39783 and the GABAB receptor agonist baclofen attenuate the reward-facilitating effects of cocaine: Intracranial self-stimulation studies in the rat. *Neuropsychopharmacology*, 30(11):2065-2072.

43. Carlezon, A., Jr. and Wise, R.A. (1993). Morphine-induced potentiation of brain stimulation reward is enhanced by MK-801. *Brain Res*, 620(2):339-342.

44. Tzschentke, T.M. (2007). Measuring reward with the conditioned place preference (CPP) paradigm: Update of the last decade. *Addict Biol*, 12(3-4):227-462.

45. Bardo, M.T. and Bevins, R.A. (2000). Conditioned place preference: What does it add to our preclinical understanding of drug reward? *Psychopharmacology (Berl)*, 153(1):31-43.

46. Zarrindast, M.R., Faraji, N., Rostami, P., Sahraei, H., and Ghoshouni, H. (2003). Cross-tolerance between morphine- and nicotine-induced conditioned place preference in mice. *Pharmacol Biochem Behav*, 74(2):363-369.

47. Holtzman, S.G. (1983). Discriminative stimulus properties of opioid agonists and antagonists. In Cooper, S.J. (ed.), *Theory in Psychopharmacology*, Vol. 2. Academic Press, London, pp. 1-46.

48. Kleven, M.S., Anthony, E.W., and Woolverton, W.L. (1990). Pharmacological characterization of the discriminative stimulus effects of cocaine in rhesus monkeys. *J Pharmacol Exp Ther*, 254(1):312-317.

49. Spealman, R.D., Bergman, J., Madras, B.K., and Melia, K.F. (1991). Discriminative stimulus effects of cocaine in squirrel monkeys: Involvement of dopamine receptor subtypes. *J Pharmacol Exp Ther*, 258(3):945-953.

50. Bergman, J. and Tidey, J.W. (1998). Drug discrimination in methamphetamine-trained monkeys: Agonist and antagonist effects of dopaminergic drugs. *J Pharmacol Exp Ther*, 285(3):1163-1174.

51. Munzar, P. and Goldbery, S.R. (2000). Dopaminergic involvement in the discriminative-stimulus effects of methamphetamine in rats. *Psychopharmacology (Berl)*, 148(2):209-216.

52. Cunningham, K.A. and Callahan, P.M. (1991). Monoamine reuptake inhibitors enhance the discriminative state induced by cocaine in the rat. *Psychopharmacology (Berl)*, 104(2):177-180.

53. Spealman, R.D. (1993). Modification of behavioral effects of cocaine by selective serotonin and dopamine uptake inhibitors in squirrel monkeys. *Psychopharmacology (Berl)*, 112(1):93-99.

54. Spealman, R.D. (1995). Discriminative stimulus effects of cocaine in squirrel monkeys: Lack of antagonism by the dopamine D2 partial agonists terguride, SDZ 208-911, and SCZ 208-912. *Pharmacol Biochem Behav*, 51(4):661-665.

55. France, C.P. and Woods, J.H. (1985). Effects of morphine, naltrexone, and dextrorphan in untreated and morphine-treated pigeons. *Psychopharmacology (Berl)*, 85(3):377-382.

56. Spealman, R.D. and Bergman, J. (1992). Modulation of the discriminative stimulus effects of cocaine by mu and kappa opioids. *J Pharmacol Exp Ther*, 261(2):607-615.

57. Negus, S.S., Mello, N.K., and Fivel, P.A. (2000). Effects of GABA agonists and GABA-A receptor modulators on cocaine discrimination in rhesus monkeys. *Psychopharmacology*, 152:398-407.

58. Grabowski, J., Rhoades, H., Stotts, A., Cowan, K., Kopecky, C., Dougherty, A., Moeller, F.G., Hassan, S., and Schmitz, J. (2004). Agonist-like or antagonist-like treatment for cocaine dependence methadone for heroin dependence: Two double-blind randomized clinical trials. *Neuropsychopharmacology*, 29(5):969-981.

59. Spragg, S.D. (1940). Morphine addiction in chimpanzees. *Comparative Psych Monographs*, 15:1–132.
60. Weeks, J.R. (1962). Experimental morphine addiction: Method for automatic intravenous injection in unrestrained rats. *Science*, 138:143–144.
61. Thompson, T. and Schuster, C.R. (1964). Morphine self-administration, food reinforced, and avoidance behaviors in rhesus monkeys. *Psychopharmacology*, 5:87–94.
62. Pickens, R. (1968). Self-administration of stimulants by rats. *Int J Addict*, 3:215–221.
63. Woods, J.H. and Schuster, C.R. (1968). Reinforcement properties of morphine, cocaine and SPA as a function of unit dose. *Int J Addict*, 3:231–237.
64. Deneau, G.A., Yanagita, T., and Seevers, M.H. (1969). Self-administration of psychoactive substances by the monkey. *Psychopharmacologia*, 61:30–48.
65. Schuster, C.R. and Johanson, C.E. (1974). The use of animal models for the study of drug abuse. In Gibbins, R.J., Israel, Y., Kalant, H., Popham, R., Schmidt, W., and Smart, R.G. (eds.), *Research Advances in Alcohol and Drug Problems*, Vol. 1. Wiley, New York, pp. 1–31.
66. Johanson, C.E. and Balster, R.L. (1978). A summary of the results of a drug self-administration study using substitution procedures in rhesus monkeys. *Bulletin on Narcotics*, 30:43–54.
67. Brady, J.V., Griffiths, R.R., Hienz, R.D., Ator, N.A., Lukas, S.E., and Lamb, R.J. (1987). Assessing drugs for abuse liability and dependence potential in laboratory primates. In Bozarth, M.A. (ed.), *Methods of Assessing the Reinforcing Properties of Abused Drugs*. Springer Verlag, New York, pp. 45–86.
68. Balster, R.L. (1991). Drug abuse potential evaluation in animals. *Br J Addict*, 86:1549–1558.
69. Lile, J.A. and Nader, M.A. (2003). The abuse liability and therapeutic potential of drugs evaluated for cocaine addiction as predicted by animal models. *Current Neuropharmacology*, 1:21–46.
70. Richardson, N.R. and Roberts, D.C. (1996). Progressive ratio schedules in drug self-administration studies in rats: A method to evaluate reinforcing efficacy. *J Neurosci Methods*, 66:1–11.
71. Stafford, D., LeSage, M.G., and Glowa, J.R. (1998). Progressive-ratio schedules of drug delivery in the analysis of drug self-administration: A review. *Psychopharmacology*, 139:169–184.
72. Everitt, B.J. and Robbins, T.W. (2000). Second-order schedules of drug reinforcement in rats and monkeys: Measurement of reinforcing efficacy and drug-seeking behaviour. *Psychopharmacology (Berl)*, 153(1):17–30.
73. Schindler, C.W., Panlilio, L.V., and Goldberg, S.R. (2002). Second-order schedules of drug self-administration in animals. *Psychopharmacology (Berl)*, 163(3–4):327–344.
74. Spealman, R.D. and Goldberg, S.R. (1978). Drug self-administration by laboratory animals: Control by schedules of reinforcement. *Annu Rev Pharmacol Toxicol*, 8:313–339.
75. Corrigall, W.A. and Coen, K.M. (1989). Fixed-interval schedules for drug self-administration in the rat. *Psychopharmacology*, 99:136–139.
76. Griffiths, R.R., Wurster, R.M., and Brady, J.V. (1976). Discrete-trial choice procedure: Effects of naloxone and methadone on choice between food and heroin. *Pharmacol Revi*, 27:357–365.
77. Woolverton, W.L. and Balster, R.L. (1981). Effects of antipsychotic compounds in rhesus monkeys given a choice between cocaine and food. *Drug Alcohol Depend*, 8:69–78.
78. Collins, E. and Kleber, H.D. (2004). Opioids: Detoxification. In: Galanter, M. and Kleber, H.D., (eds.), *Textbook of Substance Abuse Treatment*, 3rd ed., The American Psychiatric Publishing, Inc, Washington, DC, pp. 265–290.
79. Negus, S.S. (2006). Choice between heroin and food in nondependent and heroin-dependent rhesus monkeys: Effects of naloxone, buprenorphine, and methadone. *J Pharmacol Exp Ther*, 317:711–723.
80. Bergman, J. and Paronis, C.A. (2006). Measuring the reinforcing strength of abused drugs. *Mol Interv*, 6(5):273–283.

81. Macht, D.I., Herman, N.B., and Levy, C.S. (1916). A quantitative study of the analgesia produced by opium alkaloids, individually and in combination with each other, in normal man. *J Pharmacol Exp Ther*, 8:1–37.

82. Seevers, M.H. and Pfeiffer, C.C. (1936). A study of the analgesia, subjective depression, and euphoria produced by morphine, heroin, dilaudid and codeine in the normal human subject. *Journal of Pharmacology and Experimental Therapeutics*, 56:166–187.

83. Brown, R.R. (1940). The order of certain psycho-physiological events following intravenous injections of morphine. *J Gen Psychol*, 22:321–340.

84. Kuhn, R.A. and Bromliley, R.B. (1951). Human pain thresholds determined by the radiant heat technique, the effect upon them of acetylsalicyclic acid, morphine sulfate and sodium phenobarbital. *J Pharmacol Exp Ther*, 101:47–55.

85. Lee, R.E. and Pfeiffer, C.C. (1951). Influence of analgesics, Dromoran, Nisentil and morphine, on pain threshold in man. *Journal of Applied Physiology*, 4:193–198.

86. Kornetsky, C., Humphries, O., and Evarts, E.V. (1957). Comparison of psychological effects of certain centrally acting drugs in man. *AMA Arch Neurological Psychiatry*, 77(3): 318–324.

87. Isbell, H. (1948). Methods and results of studying experimental human addiction to the newer synthetic analgesics. *Ann NY Acad Sci*, 51:108–122.

88. Lasagna, L., von Felsinger, J.M., and Beecher, H.K. (1955). Drug-induced mood changes in man. 1: Observations on healthy subjects, chronically ill patients, and "postaddicts". *JAMA*, 157:1006–1020.

89. Fraser, H.F., van Horn, G.D., Martin, W.R., Wolbach, A.B., and Isbell, H. (1961). Methods for evaluating addiction liability. (A) "Attitude" of opiate addicts toward opiate-like drugs, (B) a short-term "direct" addiction test. *J Pharmacol Exp Ther*, 133:371–387.

90. Haertzen, C.A., Hill, H.E., and Belleville, R.E. (1963). Development of the Addiction Research Center Inventory (ARCI): Selection of items that are sensitive to the effects of various drugs. *Psychopharmacologia*, 4:155–166.

91. Martin, W.R., Sloan, J.W., Sapira, J.D., and Jasinski, D.R. (1971). Physiologic, subjective and behavioral effects of amphetamine, methamphetamine, ephedrine, phenmetrazine, and methylphenidate in man. *Clin Pharmacol Ther*, 12:245–258.

92. Bickel, W.K., Bigelow, G.E., Preston, K.L., and Liebson, I.A. (1989). Opioid drug discrimination in humans: Stability, specificity and relation to self-reported drug effect. *J Pharmacol Exp Ther*, 251:1053–1063.

93. Preston, K.L., Bigelow, G.E., Bickel, W.K., and Liebson, I.A. (1989). Drug discrimination in human postaddicts: Agonist-antagonist opioids. *J Pharmacol Exp Ther*, 250:184–196.

94. Preston, K.L. and Bigelow, G.E. (2000). Effects of agonist–antagonist opioids in humans trained in a hydromorphone/not hydromorphone discrimination. *J Pharmacol Exp Ther*, 295:114–124.

95. Oliveto, A., Sevarino, K., McCance-Katz, E., and Feingold, A. (2002). Butorphanol and nalbuphine in opioid-dependent humans under a naloxone discrimination procedure. *Pharmacol Biochem Behav*, 71:85–96.

96. Oliveto, A., Benios, T., Gonsai, K., Feingold, A., Poling, J., and Kosten, T.R. (2003). D-cycloserine–naloxone interactions in opioid-dependent humans under a novel-response naloxone discrimination procedure. *Exp Clin Psychopharmacol*, 11(3):237–246.

97. Oliveto, A., Sevarino, K., McCance-Katz, E., Benois, T., Poling, J., and Feingold, A. (2003). Clonidine and yohimbine in opioid-dependent humans responding under a naloxone novel-response discrimination procedure. *Behav Pharmacol*, 14(2):97–109.

98. Oliveto, A., Poling, J., Kosten, T.R., and Gonsai, K. (2004). Isradipine and dextromethorphan in methadone-maintained humans under a naloxone discrimination procedure. *Eur J Pharmacol*, 491(2–3):157–168.

99. Bisaga, A., Gianelli, P., and Popik, P. (1997). Opiate withdrawal with dextromethorphan. *Am J Psychiatry*, 154(4):584.

100. Bisaga, A., Comer, S.D., Ward, A.S., Popik, P., Kleber, H.D., and Fischman, M.W. (2001). The NMDA antagonist memantine attenuates the expression of opioid physical dependence in humans. *Psychopharmacology (Berl)*, 157(1):1-10.

101. Rosen, M.I., McMahon, T.J., Woods, S.W., Pearsall, H.R., and Kosten, T.R. (1996). A pilot study of dextromethorphan in naloxone-precipitated opiate withdrawal. *Eur J Pharmacol*, 307(3):251-257.

102. Roehrs, T., Pedrosi, B., Rosenthal, L., Zorick, F., and Roth, T. (1997). Hypnotic self-administration: Forced choice versus single-choice. *Psychopharmacology (Berl)*, 133(2):121-126.

103. Griffiths, R.R., Troisi, J.R., Silverman, K., and Mumford, G.K. (1993). Multiple-choice procedure: An efficient approach for investigating drug reinforcement in humans. *Behav Pharmacol*, 4:3-13.

104. Katz, J.L. (1990). Models of relative reinforcing efficacy of drugs and their predictive utility. *Behav Pharmacol*, 1(4):283-301.

105. Giordano, L.A., Bickel, W.K., Shahan, T.A., and Badger, G.J. (2001). Behavioral economics of human drug self-administration: Progressive ratio versus random sequences of response requirements. *Behav Pharmacol*, 12(5):343-347.

106. Bickel, W.K., Marsch, L.A., and Carroll, M.E. (2000). Deconstructing relative reinforcing efficacy and situation the measure of pharmacological reinforcement with behavioral economics: A theoretical proposal. *Psychopharmacology*, 153:44-56.

107. Fischman, M.W. and Mello, N.K. (1989), Testing for abuse liability of drugs in humans. National Institute on Drug Abuse Research Monograph No. 92. Rockville, MD, National Institute on Drug Abuse.

108. Griffiths, R.R., Bigelow, G.E., and Ator, N.A. (2003). Principles of initial experimental drug abuse liability assessment in humans. *Drug Alcohol Depend*, 70(3 Suppl):S41-S54.

109. Code of Federal Regulations, Title 21, Part 314.50.

110. Conference on Preclinical Abuse Liability Testing sponsored by the College on Problems of Drug Dependence, Annapolis, USA, 2006; Negus, S.S, Fantegrossi, W.E. Overview of conference on preclinical abuse liability testing: Current methods and future challenges. *Drug Alcohol Depend* (in press); The Use of Non-human Primates in Research: A Working Group Report sponsored by The Academy of Medical Sciences, Medical Research Council, The Royal Society, and the Wellcome Trust, UK, 2006; Weerts, E.M., Fantegrossi, W.E., and Goodwin, A.K. (2007). The value of nonhuman primates in drug abuse research. *Exp Clin Psychopharmacol*, 15:309-327.

111. International Conference on Harmonization of Technical Requirements for Registration of Pharmaceuticals for Human Use – ICH Harmonized Tripartite Guideline: Safety Pharmacology Studies for Human Pharmaceuticals S7A.

112. Single Convention on Narcotic Drugs, 1961. United Nations, 80-43678.

113. Convention on Psychotropic Substances, 1971. United Nations, V. 93-88220.

114. Guidelines for the WHO review of dependence-producing psychoactive substances for international control (reprinted from document EB 105/2000/ REC/1, Annex 9).

115. WHO Technical Report Series, No. 775.

116. WHO Technical Report Series, No. 942.

117. Kleber, H.D. (2003). Pharmacologic treatments for heroin and cocaine dependence. *Am J Addict*, 12(Suppl 2):S5-S18.

118. Dole, V.P. and Nyswander, M. (1965). A medical treatment for diacetylmorphine (heroin) addiction. A clinical trial with methadone hydrochloride. *JAMA*, 193:646-650.

119. Jones, B.E. and Prada, J.A. (1977). Effects of methadone and morphine maintenance on drug-seeking behavior in the dog. *Psychopharmacology (Berl)*, 54(2):109-112.

120. Harrigan, S.E. and Downs, D.A. (1981). Pharmacological evaluation of narcotic antagonist delivery systems in rhesus monkeys. *NIDA Res Monogr*, 28:77-92.

121. Mello, N.K., Bree, M.P., and Mendelson, J.H. (1983). Comparison of buprenorphine and methadone effects on opiate self-administration in primates. *J Pharmacol Exp Ther*, 225:378-386.

122. Johnson, R.E., Chutuape, M.A., Strain, E.C., Walsh, S.L., Stitzer, M.L., and Bigelow, G.E. (2000). A comparison of levomethadyl acetate, buprenorphine, and methadone for opioid dependence. *New Engl J Med*, 343(18):1290-1297.

123. Bergman, J. and Paronis, C.A. (2006). Measuring the reinforcing strength of abused drugs. *Mol Interv*, 6(5):273-283.

124. Donny, E.C., Brasser, S.M., Bigelow, G.E., Stitzer, M.L., and Walsh, S.L. (2005). Methadone doses of 100 mg or greater are more effective than lower doses at suppressing heroin self-administration in opioid-dependent volunteers. *Addiction*, 100:1496-1509.

125. Strain, E.C., Bigelow, G.E., Liebson, I.A., and Stitzer, M.L. (1999). Moderate- vs. high-dose methadone in the treatment of opioid dependence: A randomized trial. *JAMA*, 281:1000-1005.

126. Faggiano, F., Vigna-Taglianti, F., Versino, E., and Lemma, P. (2003). Methadone maintenance at different dosages for opioid dependence. *Cochrane Database System Review*, 3:CD002208.

127. Walsh, S.L. and Eissenberg, T. (2003). The clinical pharmacology of buprenorphine: Extrapolating from the laboratory to the clinic. *Drug Alcohol Depend*, 70:S13-S27.

128. Winger, G., Skjoldager, P., and Woods, J.H. (1992). Effects of buprenorphine and other opioid agonists and antagonists on alfentanil- and cocaine-reinforced responding in rhesus monkeys. *J Pharmacol Exp Ther*, 261:311-317.

129. Winger, G. and Woods, J.H. (1996). Effects of buprenorphine on behaviour maintained by heroin and alfentanil in rhesus monkeys. *Behav Pharmacol*, 7(2):155-159.

130. Mello, N.K. and Negus, S.S. (1998). The effects of buprenorphine on self-administration of cocaine and heroin "speedball" combinations and heroin alone by rhesus monkeys. *J Pharmacol Exp Ther*, 285:444-456.

131. Chen, S.A., O'Dell, L.E., Hoefer, M.E., Greenwell, T.N., Zorrilla, E.P., and Koob, G.F. (2006). Unlimited access to heroin self-administration: Independent motivational markers of opiate dependence. *Neuropsychopharmacology*, 31:2692-2707.

132. Aceto, M.D. (1984). Characterization of prototypical opioid antagonists, agonist-antagonists, and agonists in the morphine-dependent rhesus monkey. *Neuropeptides*, 5:15-18.

133. Mello, N.K., Mendelson, J.H., Kuehnle, J.C., and Sellers, M.S. (1981). Operant analysis of human heroin self-administration and the effects of naltrexone. *J Pharmacol Exp Ther*, 216:45-54.

134. Mello, N.K., Lukas, S.E., Bree, M.P., and Mendelson, J.H. (1988). Progressive ratio performance maintained by buprenorphine, heroin and methadone in Macaque monkeys. *Drug Alcohol Depend*, 21:81-97.

135. Young, A.M., Stephens, K.R., Hein, D.W., and Woods, J.H. (1984). Reinforcing and discriminative stimulus properties of mixed agonist-antagonist opioids. *J Phamacol Exp Ther*, 229:118-126.

136. Mello, N.K. and Mendelson, J.H. (1985). Behavioral pharmacology of buprenorphine. *Drug Alcohol Depend*, 14:283-303.

137. Lukas, S.E., Brady, J.V., and Griffiths, R.R. (1986). Comparison of opioid self-injection and disruption of schedule-controlled performance in the baboon. *J Pharmacol Exp Ther*, 238:924-931.

138. Winger, G. and Woods, J.H. (2001). The effects of chronic morphine on behavior reinforced by several opioids or by cocaine in rhesus monkeys. *Drug Alcohol Depend*, 62:181–189.

139. Mello, N.K. and Mendelson, J.H. (1980). Buprenorphine suppresses heroin use by heroin addicts. *Science*, 207:657–659.

140. Mello, N.K., Mendelson, J.H., and Kuehnle, J.C. (1982). Buprenorphine effects on human heroin self-administration: An operant analysis. *J Pharmacol Exp Ther*, 223:30–39.

141. Greenwald, M.K., Johanson, C.E., and Schuster, C.R. (1999). Opioid reinforcement in heroin-dependent volunteers during outpatient buprenorphine maintenance. *Drug Alcohol Depend*, 56(3):191–203.

142. Comer, S.D., Collins, E.D., and Fischman, M.W. (2001). Buprenorphine sublingual tablets: Effects on IV heroin self-administration by humans. *Psychopharmacology*, 154:28–37.

143. Greenwald, M.K., Schuh, K.J., Hopper, J.A., Schuster, C.R., and Johanson, C.E. (2002). Effects of buprenorphine sublingual tablet maintenance on opioid drug-seeking behavior by humans. *Psychopharmacology (Berl)*, 160(4):344–352.

144. Comer, S.D., Sullivan, M.A., and Walker, E.A. (2005). Comparisons of intravenous buprenorphine and methadone self-administration by recently detoxified heroin-dependent individuals. *J Pharmacol Exp Ther*, 315:1320–1330.

145. Blumberg, H. and Dayton, H.B. (1974). Naloxone, naltrexone, and related noroxymorphones. *Adv Biochem Psychopharmacol*, 8:33–43.

146. Weeks, J.R. and Collins, R.J. (1976). Changes in morphine self-administration in rats induced by prostaglandin E1 and naloxone. *Prostaglandins*, 12:11–19.

147. Ettenberg, A., Pettit, H.O., Bloom, F.E., and Koob, G.F. (1982). Heroin and cocaine intravenous self-administration in rats: Mediation by separate neural systems. *Psychopharmacology*, 78:204–209.

148. Koob, G.F., Pettit, H.O., Ettenberg, A., and Bloom, F.E. (1984). Effects of opiate antagonists and their quaternary derivatives on heroin self-administration in the rat. *J Pharmacol Exp Ther*, 229:481–486.

149. Harrigan, S.E. and Downs, D.A. (1978). Continuous intravenous naltrexone effects on morphine self-administration in rhesus monkeys. *J Pharmacol Exp Ther*, 204:481–486.

150. Bertalmio, A.J. and Woods, J.H. (1989). Reinforcing effect of alfentanil is mediated by mu opioid receptors: Apparent pA2 analysis. *J Pharmacol Exp Ther*, 251:455–460.

151. Rowlett, J.K., Wilcox, K.M., and Woolverton, W.L. (1998). Self-administration of cocaine-heroin combinations by rhesus monkeys: Antagonism by naltrexone. *J Pharmacol Exp Ther*, 286:61–69.

152. Meyer, R.E., McNamee, H.B., Mirin, S.M., and Altman, J.L. (1976). Analysis and modification of opiate reinforcement. *Int J Addict*, 11:467–484.

153. Altman, J.L., Meyer, R.E., Mirin, S.M., McNamee, H.B., and McDougle, I. (1976). Opiate antagonists and the modification of heroin self-administration behavior in man: An experimental study. *Int J Add*, 11:485–499.

154. Comer, S.D. and Collins, E.D. (2002). Self-administration of intravenous buprenorphine and the buprenorphine/naloxone combination by recently detoxified heroin abusers. *J Pharmacol Exp Ther*, 303:695–703.

155. Sullivan, M.A., Vosburg, S.K., and Comer, S.D. (2006). Depot naltrexone: Antagonism of the reinforcing, subjective, and physiological effects of heroin. *Psychopharmacology (Berl)*, 189(1):37–46.

156. Comer, S.D., Sullivan, M.A., Yu, E., Rothernberg, J.L., Kleber, H.D., Kampman, K., Dackis, C., and O'Brien, C.P. (2007). Injectable, sustained-release naltrexone for the treatment of opioid dependence: A randomized, placebo-controlled trial. *Arch Gen Psychiatry*, 64(7):855.

157. Ritz, M.C., Lamb, R.J., Goldberg, S.R., and Kuhar, M.J. (1987). Cocaine receptors on dopamine are related to self-administration of cocaine. *Science*, 237:1219–1223.

158. Bergman, J., Madras, B.K., Johnson, S.E., and Spealman, R.D. (1989). Effects of cocaine and related drugs in nonhuman primates. III: Self-administration by squirrel monkeys. *J Pharmacol Exp Ther*, 251:150-155.

159. Madras, B.K., Fahey, M.A., Bergman, J., Canfield, D.R., and Spealman, R.D. (1989). Effects of cocaine and related drugs in nonhuman primates. I: [3H] cocaine binding sites in caudate-putamen. *J Pharmacol Exp Ther*, 251:131-141.

160. Jackson, D.M. and Weslind-Danielsson, A. (1994). Dopamine receptors: Molecular biology, biochemistry and behavioural aspects. *Pharmacol Ther Practice*, 64:291-370.

161. Kebabian, J.W. and Calne, D.B. (1979). Multiple receptors for dopamine. *Nature*, 277:93-96.

162. Spealman, R.D., Bergman, J., Madras, B.K., Kamien, J.B., and Melia, K.F. (1992). Role of D1 and D2 dopamine receptors in the behavioral effects of cocaine. *Neurochem Int*, 20:147S-152S.

163. Woolverton, W.L. and Johnson, K.M. (1992). Neurobiology of cocaine abuse. *Trends Pharmacol Sci*, 13:193-200.

164. Caine, S.B. and Koob, G.F. (1994). Effects of dopamine D-1 and D-2 antagonists on cocaine self-administration under different schedules of reinforcement in the rat. *J Pharmacol Exp Ther*, 270:209-218.

165. Depoortere, R.Y., Li, D.H., Lane, J.D., and Emmett-Oglesby, M.W. (1993). Parameters of self-administration of cocaine in rats under a progressive-ratio schedule. *Pharmacol Biochem Behav*, 45:539-548.

166. Hubner, C.B. and Moreton, J.E. (1991). Effects of selective D1 and D2 dopamine antagonists on cocaine self-administration in the rat. *Psychopharmacology*, 105:151-156.

167. Bergman, J., Kamien, J.B., and Spealman, R.D. (1990). Antagonism of cocaine self-administration by selective dopamine D(1) and D(2) antagonists. *Behav Pharmacol*, 1:355-363.

168. Campbell, U.C., Lac, S.T., and Carroll, M.E. (1999). Effects of baclofen on maintenance and reinstatement of intravenous cocaine self-administration in rats. *Psychopharmacology*, 143:209-214.

169. Howell, L.L. and Byrd, L.D. (1991). Characterization of the effects of cocaine and GBR 12909, a dopamine uptake inhibitor, on behavior in the squirrel monkey. *J Pharmacol Exp Ther*, 258:178-185.

170. Woolverton, W.L. (1986). Effects of D1 and D2 dopamine antagonists on the self-administration of cocaine and piribedil by rhesus monkeys. *Pharmacol Biochem Behav*, 24:531-535.

171. Woolverton, W.L. and Virus, R.M. (1989). The effects of a D1 and a D2 dopamine antagonist on behavior maintained by cocaine or food. *Pharmacol Biochem Behav*, 32:691-697.

172. Winger, G. (1994). Dopamine antagonist effects on behavior maintained by cocaine and alfentanil in rhesus monkeys. *Behav Pharmacol*, 5:141-152.

173. Platt, D.M., Rowlett, J.K., and Spealman, R.D. (2001). Modulation of cocaine and food self-administration by low- and high-efficacy D1 agonists in squirrel monkeys. *Psychopharmacology*, 157:208-216.

174. McGregor, A. and Roberts, D.C.S. (1993). Dopaminergic antagonism within the nucleus accumbens or the amygdala produces differential effects on intravenous cocaine self-administration under fixed and progressive ratio schedules of reinforcement. *Brain Res*, 624:245-252.

175. Kleven, M.S. and Woolverton, W.L. (1990). Pharmacological characterization of the discriminative stimulus effects of cocaine in rhesus monkeys. *J Pharmacol Exp Ther*, 254:312-317.

176. Platt, D.M., Rowlett, J.K., and Spealman, R.D. (2002). Behavioral effects of cocaine and dopaminergic strategies for preclinical medication development. *Psychopharmacology*, 163:265-282.

177. Emmett-Oglesby, M. and Mathis, D. (1988). Chronic administration of SCH 23390 produces sensitization to the discriminative stimulus properties of cocaine. In Harris, L.S. (ed.), *Problems of Drug Dependence. Proceedings of the 50th Annual Scientific Meeting,* The College on Problems and Drug Dependence, Rockville, MD, United States Department of Health and Human Services, p. 367.

178. Kleven, M.S. and Woolverton, W.L. (1990). Effects of continuous infusions of SCH 23390 on cocaine- or food-maintained behavior in rhesus monkeys. *Behav Pharmacol,* 1:365–373.

179. Kosten, T.A., DeCaprio, J.L., and Nester, E.J. (1996). Long-term haloperidol administration enhances and short-term administration attenuates the behavioral effects of cocaine in a place conditioning procedure. *Psychopharmacology,* 128:304–312.

180. Kosten, T.A. (1997). Enhanced neurobehavioral effects of cocaine with chronic neuroleptic exposure in rats. *Schizophr Bull,* 23:203–213.

181. Barone, P., Tucci, I., Parashos, S.A., and Chase, T.N. (1988). Supersensitivity to a D-1 dopamine receptor agonist and subsensitivity to a D-2 receptor agonist following chronic D-1 receptor blockade. *Eur J Pharmacol,* 49:225–232.

182. Braun, A.R., Laruelle, M., and Mouradian, M.M. (1997). Interactions between D1 and D2 dopamine receptor family agonists and antagonists: The effects of chronic exposure on behavior and receptor binding in rats and their clinical implications. *J Neural Transm,* 104:341–362.

183. Vaccheri, A., Dall'Olio, R., Gandolfi, O., Roncada, P., and Montanaro, N. (1987). Enhanced stereotyped response to apomorphine after chronic D-1 blockade with SCH 23390. *Psychopharmacology,* 91:394–396.

184. Creese, I. and Chen, A. (1985). Selective D-1 dopamine receptor increase following chronic treatment with SCH 23390. *Eur J Pharmacol,* 109:127–128.

185. Gui-Hua, C., Perry, B.D., and Woolverton, W.L. (1992). Effects of chronic SCH 23390 or acute EEDQ on the discriminative stimulus effects of SKF 38393. *Pharmacol Biochem Behav,* 41:321–327.

186. Hess, E.J., Albers, L.J., Le, H., and Creese, I. (1986). Effects of chronic SCH23390 treatment on the biochemical and behavioral properties of D1 and D2 dopamine receptors: Potentiated behavioral responses to a D2 dopamine agonist after selective D1 dopamine receptor upregulation. *J Pharmacol Exp Ther,* 238:846–854.

187. White, F.J., Joshi, A., Koeltzow, T.E., and Hu, X.T. (1998). Dopamine receptor antagonists fail to prevent induction of cocaine sensitization. *Neuropsychopharmacology,* 18(1):26–40.

188. Romach, M.K., Glue, P., Kampman, K., Kaplan, H.L., Somer, G.R., Poole, S., Clarke, L., Coffin, V., Cornish, J., O'Brien, C.P., and Sellers, E.M. (1999). Attenuation of the euphoric effects of cocaine by the dopamine D1/D5 antagonist ecopipam (SCH 39166). *Arch Gen Psychiatry,* 56(12):1101–1106.

189. Nann-Vernotica, E., Donny, E.C., Bigelow, G.E., and Walsh, S.L. (2001). Repeated administration of the D1/5 antagonist ecopipam fails to attenuate the subjective effects of cocaine. *Psychopharmacology (Berl),* 155(4):338–347.

190. Haney, M., Ward, A.S., Foltin, R.W., and Fischman, M.W. (2001). Effects of ecopipam, a selective D1 antagonist, on smoke cocaina self-administration by humans. *Psychopharmacology,* 155:330–337.

191. Grabowski, J., Rhoades, H., Silverman, P., Schmitz, J.M., Stotts, A., Creson, D., and Bailey, R. (2000). Risperidone for the treatment of cocaine dependence: Randomized, double-blind trial. *J Clin Psychopharmacol,* 20(3):305–310.

192. Cousins, M.S., Roberts, D.C.S., and de Wit, H. (2002). GABA B receptor agonists for the treatment of drug addicition: A review of recent findings. *Drug Alcohol Depend,* 65:209–220.

193. Korotkova, T.M., Ponomarenko, A.A., Brown, R.E., and Haas, H.L. (2004). Functional diversity of ventral midbrain dopamine and GABAergic neurons. *Mol Neurobiol,* 29(3):243–259.

194. Fadda, P., Scherma, M., Fresu, A., Collu, M., and Fratta, W. (2003). Baclofen antagonizes nicotine-, cocaine-, and morphine-induced dopamine release in the nucleus accumbens of rat. *Synapse*, 50:1-6.

195. Roberts, D.C., Andrews, M.M., and Vickers, G.J. (1996). Baclofen attenuates the reinforcing effects of cocaine in rats. *Neuropsychopharmacology*, 15:417-423.

196. Roberts, D.C.S. and Andrews, M.M. (1997). Baclofen suppression of cocaine self-administration: demonstration using a discrete trials procedure. *Psychopharmacology*, 131:271-277.

197. Brebner, K., Phelan, R., and Roberts, D.C.S. (2000). Effect of baclofen on cocaine self-administration in rats reinforced under fixed-ration and progressive-ratio schedules. *Psychopharmacology*, 148:314-321.

198. Roberts, D.C.S. and Brebner, K. (2000). GABA modulation of cocaine self-administration. *Ann NY Acad Sci*, 909:145-158.

199. Brebner, K., Phelan, R., and Roberts, D.C.S. (2000). Intra-VTA baclofen attenuates cocaine self-administration on a progressive ration schedule of reinforcement. *Pharmacol Biochem Behav*, 66:857-862.

200. Shoaib, M., Swanner, L.S., Beyer, C.E., Goldberg, S.R., and Schindler, C.W. (1998). The BABAB agonist baclofen modifies cocaine self-administration in rats. *Behav Pharmacol*, 9:195-206.

201. Barrett, A.C., Negus, S., Mello, N.K., and Caine, S.B. (2005). Effect of GABA agonists and GABA-A receptor modulators on cocaine-and food-maintained responding and cocaine discrimination in rats. *J Pharmacol Exp Ther*, 315:858-878.

202. Weerts, E.M., Froestl, W., and Griffiths, R.R. (2005). Effects of GABAergic modulators on food and cocaine self-administration in baboons. *Drug Alcohol Depend*, 80:369-376.

203. Lile, J.A., Stoops, W.W., Allen, T.S., Glaser, P.E.A., Hays, L.R., and Rush, C.R. (2004). Baclofen does not alter the reinforcing, subject-rated or cardiovascular effects of intranasal cocaine in humans. *Psychopharmacology*, 171:441-449.

204. Haney, M., Hart, C.L., and Foltin, R.W. (2006). Effects of baclofen on cocaine self-administration: Opioid- and nonopioid-dependent volunteers. *Neuropsychopharmacology*, 31(8):1814-1821.

205. Shoptaw, S., Yang, X., Rotheram-Fuller, E.J., Hsieh, Y.C., Kintaudi, P.C., Charuvastra, V.C., and Ling, W. (2003). Randomized placebo-controlled trial of baclofen for cocaine dependence: Preliminary effects for individuals with chronic patterns of cocaine use. *J Clin Psychiatry*, 64(12):1440-1448.

206. Gonzalez, G., Sevarino, K., Sofuoglu, M., Poling, J., Oliveto, A., Gonsai, K., George, T.P., and Kosten, T.R. (2003). Tiagabine increases cocaine-free urines in cocaine-dependent methadone-related patients: Results of a randomized pilot study. *Addiction*, 98(11):1625-1632.

207. Feltenstein, M.W., Altar, C.A., and See, R.E. (2007). Aripiprazole blocks reinstatement of cocaine seeking in an animal model of relapse. *Biol Psychiatry*, 61:582-590.

208. Pulvirenti, L., Balducci, C., Piercy, M., and Koob, G.F. (1998). Characterization of the effects of the partial dopamine agonist terguride on cocaine self-administration in the rat. *J Pharmacol Exp Ther*, 286(3):1231-1238.

209. Tiihonen, J., Kuoppasalmi, K., Fohr, J., Tuomola, P., Kuikanmaki, O., Vorma, H., Sokero, P., Haukka, J., and Meririnne, E. (2007). A comparison of aripiprazole, methylphenidate, and placebo for amphetamine dependence. *Am J Psychiatry*, 164(1):160-162.

210. Mignot, E., Nishino, S., Guilleminault, C., and Dement, W.C. (1994). Modafinil binds to the dopamine uptake carrier site with low affinity. *Sleep*, 17(5):436-437.

211. Madras, B.K., Xie, Z., Lin, Z., Jassen, A., Panas, H., Lynch, L., Johnson, R., Livni, E., Spencer, T.J., Bonab, A.A., Miller, G.M., and Fischman, A.J. (2006). Modafinil occupies dopamine and norepinephrine transporters in vivo and modulated the transporters and trace amine activity in vitro. *J Pharmacol Exp Ther*, 319(2):561-569.

212. Ferraro, L., Antonelli, T., Tanganelli, S., O'Conner, W.T., Perez de la Mora, M., Mendez-Franco, J., Rambert, F.A., and Fuxe, K. (1999). The vigilance promoting drug modafinil increases extracellular glutamate levels in the medial preoptic area and the posterior hypothalamus of the conscious rat: Prevention by local GABAA receptor blockade. *Neuropsychopharmacology*, 20(4):346–356.

213. Korotkova, T.M., Klyuch, B.P., Ponomarenko, A.A., Lin, J.S., Haas, H.L., and Sergeeva, O.A. (2007). Modafinil inhibits rat midbrain dopaminergic neurons through D2-like receptors. *Neuropharmacology*, 52(2):626–633.

214. Gold, L.H. and Balster, R.L. (1996). Evaluation of the cocaine-like discriminative stimulus effects and reinforcing effects of modafinil. *Psychopharmacology (Berl)*, 126(4): 286–292.

215. Hart, C.L., Haney, M., Vosburg, S.K., Comer, S.D., Gunderson, E., and Foltin, R.W. (2006). Modafinil attenuates disruptions in cognitive performance during simulated night-shift work. *Neuropsychopharmacology*, 31(7):1526–1536.

216. Stoops, W.W., Lile, J.A., Fillmore, M.T., Glaser, P.E., and Rush, C.R. (2005). Reinforcing effects of modafinil: Influence of dose and behavioral demands following drug administration. *Psychopharmacology (Berl)*, 182(1):186–193.

217. Jasinski, D.R. (2000). An evaluation of the abuse potential of modafinil using methylphenidate as a reference. *J Psychopharmacol*, 14(1):53–60.

218. Rush, C.R., Kelly, T.H., Hays, L.R., and Wooten, A.F. (2002). Discriminative-stimulus effects of modafinil in cocaine-treated humans. *Drug Alcohol Depend*, 67(3):311–322.

219. Rush, C.R., Kelly, T.H., Hays, L.R., Baker, R.W., and Wooten, A.F. (2002). Acute behavioral and physiological effects of modafinil in drug abusers. *Behav Pharmacol*, 13(2):105–115.

220. Myrick, H., Malcolm, R., Taylor, B., and LaRowe, S. (2004). Modafinil: Preclinical, clinical, and post-marketing surveillance – a review of abuse liability issues. *Ann Clin Psychiatry*, 16(2):101–109.

221. Dackis, C.A., Lynch, K.G., Yu., E., Samara, F.F., Kampman, K.M., Cornish, J.W., Rowan, A., Poole, S., White, L., and O'Brien, C.P. (2003). Modafinil and cocaina: A double-blind, placebo-controlled drug interaction study. *Drug Alcohol Depend*, 70(1):29–37.

222. Donovan, J.L., DeVane, C.L., Malcolm, R.J., Mojsiak, J., Chiang, C.N., Elkashef, A., and Taylor, R.M. (2005). Modafinil influences the pharmacokinetics of intravenous cocaina in healthy cocaina-dependent volunteers. *Clin Pharmacokinet*, 44(7):753–765.

223. Malcolm, R., Swayngim, K., Donovan, J.L., DeVane, C.L., Elkshef, A., Chiang, N., Khan, R., Mojsiak, J., Myrick, D.L., Hedden, S., Cochran, K., and Wooson, R.F. (2006). Modafinil and cocaine interactions. *Am J Drug Alcohol Abuse*, 32(4):577–587.

224. Hart, C.L., Haney, M., Vosburg, S.K., Rubin, E., and Foltin, R.W. Human. (2008). Cocaine self-administration is decreased by modafinil. *Neuropsychopharmcology*, 33(4):761–768.

225. Dackis, C.A., Kampman, K.M., Lynch, K.G., Pettinati, H.M., and O'Brien, C.P. (2005). A double-blind, placebo-controlled trial of modafinil for cocaine dependence. *Neuropsychopharmacology*, 30(12):2298.

226. Fischman, M.W. and Foltin, R.W. (1998). Cocaine self-administration research: Implications for rational pharmacotherapy. In Higgins, S.T. and Katz, J.L. (eds.), *Cocaine Abuse Research: Pharmacology, Behavior and Clinical Applications*. Academic Press, San Diego, CA, pp. 181–207.

227. Vocci, F.J. and Elkashef, A. (2005). Pharmacotherapy and other treatments for cocaine abuse and dependence. *Curr Opin Psychiatry*, 18(3):265–270.

228. Haney, M., Spealman, R. (2008). Controversies in translational research: Drug self-administration. *Psychopharmacology* (Berl), 2008 Feb 19. [Epub ahead of print] PMID: 18283437 [Pub Med].

229. Johanson, C.E., Lundahl, L.H., and Schubiner, H. (2007). Effects of oral cocaine on intravenous cocaine discrimination in humans. *Exp Clin Psychopharmacol*, 15(3): 219–227.

230. Comer, S.D., Ashworth, J.B., Foltin, R.W., Johanson, C.E., Zacny, J.P., and Walsh, S.L. (2008). The role of human drug self-administration procedures in the development of medications. *Drug and Alcohol Dependence*, 96:1–15.

231. Fischman, M.W., Foltin, R.W., Nestadt, G., and Pearlson, G.D. (1990). Effects of desipramine maintenance on cocaine self-administration by humans. *J Pharmacol Exp Ther*, 253(2):760–770.

232. Haney, M., Foltin, R.W., and Fischman, M.W. (1998). Effects of pergolide on intravenous cocaine self-administration in men and women. *Psychopharmacology (Berl)*, 137(1):15–24.

233. Haney, M., Collins, E.D., Ward, A.S., Foltin, R.W., and Fischman, M.W. (1999). Effect of a selective dopamine D1 agonist (ABT-431) on smoked cocaine self-administration in humans. *Psychopharmacology (Berl)*, 143(1):102–110.

234. Hart, C.L., Ward, A.S., Collins, E.D., Haney, M., and Foltin, R.W. (2004). Gabapentin maintenance decreases smoked cocaine-related subjective effects, but not self-administration by humans. *Drug Alcohol Depend*, 73(3):279–287.

235. Campbell, J., Nickel, E.J., Penick, E.C., Wallace, D., Gabrielli, W.F., Rowe, C., Liskow, B., Powell, B.J., and Thomas, H.M. (2003). Comparison of desipramine or carbamazepine to placebo for crack cocaine-dependent patients. *Am J Addict*, 12(2):122–136.

236. Nader, M.A. and Woolverton, W.L. (1992). Effects of increasing the magnitude of an alternative reinforcer on drug choice in a discrete-trials choice procedure. *Psychopharmacology*, 105:169–174.

237. Woolverton, W.L., English, J.A., and Weed, M.R. (1997). Choice between cocaine and food in a discrete-trials procedure in monkeys: A unit price analysis. *Psychopharmacology*, 133:269–274.

238. Campbell, U.C. and Carroll, M.E. (2000). Reduction of drug self-administration by an alternative non-drug reinforcer in rhesus monkeys: Magnitude and temporal effects. *Psychopharmacology*, 147:418–425.

239. Higgins, S.T., Bickel, W.K., and Hughes, J.R. (1994). Influence of an alternative reinforcer on human cocaine self-administration. *Life Sci*, 55:179–187.

240. Hatsukami, D.K., Pentel, P.R., Glass, J., Nelson, R., Brauer, L.H., Crosby, R., and Hanson, K. (1994). Methodological issues in the administration of multiple doses of smoked cocaine-base in humans. *Pharmacol Biochem Behav*, 47:531–540.

241. Woolverton, W.L. and Anderson, K.G. (2006). Effects of delay to reinforcement on the choice between cocaine and food in rhesus monkeys. *Psychopharmacology*, 186:99–106.

242. Mello, M.K. and Negus, S.S. (1996). Preclinical evaluation of pharmocotherapies for treatment of cocaine and opioid abuse using drug self-administration procedures. *Neuropsychopharmacology*, 14:375–424.

243. McCance-Katz, E., Kosten, T.A., and Kosten, T.R. (2001). Going from the bedside back to the bench with ecopipam: A new strategy for cocaine pharmacotherapy development. *Psychopharmacology*, 155:327–329.

244. Negus, S.S. and Mello, N.K. (2003). Effects of chronic d-amphetamine treatment on cocaine and food-maintained responding under a second-order schedule in rhesus monkeys. *Drug Alcohol Depend*, 70:39–52.

CHAPTER 8

Anti-obesity Drugs: From Animal Models to Clinical Efficacy

Colin T. Dourish[1], John P.H. Wilding[2] and Jason C.G. Halford[3]

[1]P1vital, Department of Psychiatry, University of Oxford,
Warneford Hospital, Oxford, UK
[2]School of Clinical Sciences, Clinical Sciences Centre,
University Hospital Aintree, Liverpool, UK
[3]Human Ingestive Laboratory, School of Psychology,
University of Liverpool, Liverpool, UK

Animal and Translational Models for CNS Drug Discovery,
Vol. 3 of 3: Reward Deficit Disorders
Robert McArthur and Franco Borsini (eds), Academic Press, 2008

INTRODUCTION

The current obesity epidemic and its severe consequences for health provide a strong rationale for developing safe and effective drugs to help manage obesity. Existing drugs have modest efficacy, and may cause clinically significant adverse effects that limit their widespread acceptance. Recent developments in the science of body weight regulation have provided a wealth of potential targets for drug development, and many candidates are now reaching the clinical development stages, however it remains to be seen which of these will be successful. There have also been many recent developments in the assessment of altered energy balance, adiposity, and its pathophysiological consequences that give researchers a wealth of potential markers of treatment success, although it is likely that only hard clinical endpoints will eventually convince regulators and physicians that obesity treatment is worthwhile.

Despite the lack of a specific obese eating style, obesity and weight gain are associated with heightened preference for dietary fat, weakened satiety mechanisms, high disinhibition and hunger, and in some instances, binge eating. These may constitute distinct obesity behavioral phenotypes. By using models of dietary-induced obesity (DIO), food choice and satiety, and the detailed assessment of drug effects on the expression of human appetite, early indications of the efficacy of appetite-suppressing anti-obesity drugs can be gauged. How we use these models to best indicate clinical success remains to be determined.

CLINICAL ASPECTS OF OBESITY AND CARDIOMETABOLIC DISEASE

Background – Do We Need Drugs for Obesity?

There is no doubt that the world is facing a major epidemic of obesity, largely as a result of rapidly changing lifestyles, with increased availability of energy dense, palatable foods, and declining levels of physical activity being widely recognized as the key culprits.[1] The major health consequences of this obesity epidemic are now recognized to be significant. A very wide range of obesity-related co-morbidities are becoming more prevalent in the population, and this will have serious consequences for the health of these individuals, and for the health care economies that will have to fund the costs of this increased burden of disease. The greatest contributor to this is undoubtedly type 2 diabetes,[2] closely followed by the related problems of dyslipidemia and hypertension which lead to greater risk of macrovascular diseases, such as myocardial infarction, stroke, and peripheral vascular disease. Together with obesity, these make up the components of the metabolic syndrome which is now thought to affect about 25% of the population in the United States and many other developed nations.[3] Other consequences of obesity are also important – these include increased risk of several gastrointestinal and genitourinary cancers, obstructive sleep apnea, osteoarthritis, back pain, infertility, congenital abnormalities and mood disorders, especially depression.

Given this significant disease burden, it is not surprising that attempts are being made both to address the causes of the epidemic at a population level through public health initiatives,[4] and to provide guidance to clinicians and health care professionals who are seeking effective ways to intervene to help individual patients lose weight. At present the most effective approaches are surgical, but these are generally reserved for the most severely obese patients and still carry significant risk in terms of morbidity and to some extent mortality. Various dietary approaches, with and without physical activity programs have been shown to have limited success, with even the most comprehensive programs only producing a mean weight loss of about 4 kg after 2 years of intervention.[5,6] This has led to a genuine clinical need for pharmacological treatments that at present sit between lifestyle approaches and surgery in terms of their clinical effectiveness.

Current Drugs for Obesity and Their Limitations

Orlistat (Xenical®/Alli®) is an intestinal lipase inhibitor. It is not systemically absorbed, but acts within the gut lumen to block the digestion of approximately 30% of ingested dietary fat; for a person on a typical Western diet containing 40% energy from fat, this will result in an energy loss of about 200 kcal per day. Orlistat is licensed for the treatment of obesity in patients with a body mass index, BMI $> 30 \text{kg/m}^2$ (or $>28 \text{kg/m}^2$ if co-morbidities, such as diabetes, hypertension, or dyslipidemia are present). The usual dose is 120 mg three times daily with meals. It should only be used for longer than 3 months if a 5% weight loss has been achieved. Possible side-effects include loose, fatty stools, oily spotting, and occasionally fecal incontinence. Small falls in circulating concentrations of fat-soluble vitamins have also been observed.[7] Clinical trials have lasted for up to 4 years, and have consistently shown greater weight loss with orlistat than placebo (mean placebo–subtracted weight loss = 3.5 kg after 2 years of treatment).[8,9] Weight loss is slightly less in those patients with type 2 diabetes.[10] In general a broad range of cardiovascular risk factors are improved with orlistat treatment, including LDL cholesterol, blood pressure, and blood glucose. In patients with impaired glucose tolerance, orlistat has been shown to reduce the risk of progression to overt diabetes by 46% in a subgroup of patients in a larger, 4-year study.[8] A low dose preparation (60 mg three times daily) of orlistat with about 80% of the efficacy of the standard preparation has recently been licensed for over-the-counter sale in the United States and Europe.

Sibutramine (Reductil®/Meridia®) is a centrally acting serotonin and noradrenaline reuptake inhibitor. It acts to increase satiety, and may also have a modest effect to reduce the fall in metabolic rate seen after weight loss. It is licensed for up to 12 months use for patients with a BMI $> 30 \text{kg/m}^2$ (or $>27 \text{kg/m}^2$ if co-morbidities are present). The starting dose is 10 mg daily. This can be increased to 15 mg after 1 month if 2 kg weight loss has not been achieved; the drug should be discontinued at 2 months if 2 kg weight loss has not been reached. It should only be used longer than 3 months if a 5% weight loss has been achieved. Side-effects include constipation, dry mouth, tachycardia, and sometimes a rise in blood pressure. Blood pressure and pulse rate should be monitored before starting therapy with sibutramine; the drug should not be used if blood pressure exceeds 145/90 mmHg, and discontinued if the systolic or diastolic blood pressure rises by more than 10 mmHg or if the pulse rises by more than 10 beats per minute on two consecutive visits. Monitoring of blood pressure and pulse should be every 2 weeks for the first 3 months of treatment, monthly for the second 3 months, and 3 monthly thereafter. It is also important to note that there are a number of contraindications for sibutramine, including congestive cardiac failure and concomitant use with antidepressants. The average placebo-subtracted weight loss in clinical trials (where all patients received lifestyle advice) was 5.3 kg after 12 months, compared to 0.5 kg for placebo.[11] Slightly less weight loss is seen in patients with pre-existing diabetes.[12] Whilst sibutramine can improve lipid profiles, specifically increasing HDL cholesterol and lowering triglycerides it may, in some patients cause a rise in blood pressure, and often a small (1–3 beats per minute) increase in heart rate.[13] This effect on blood pressure has been the cause of some concern, but in general only affects a small percentage of patients.[14]

Rimonabant (Acomplia®) is an antagonist at the cannabinoid-1 (CB1) receptor and has recently been licensed in the European Union for the treatment of obesity in patients with a BMI $> 30 \text{kg/m}^2$ (or $>27 \text{kg/m}^2$ if co-morbidities are present). The dose is 20 mg daily. The most common side-effects ($>5\%$ patients in trials) were nausea, dizziness, diarrhea, anxiety, upper respiratory tract infection, nasopharyngitis, influenza, pain in extremities and arthralgia. The side-effects that have received most attention relate to alterations in mood and an increased risk of anxiety and depressive disorders, including suicidality. This led to a US Food and Drug Administration (FDA) advisory committee voting against approval of rimonabant in the United States until longer term safety and efficacy data are available, despite good evidence of clinical effectiveness in clinical trials, with a mean placebo-subtracted weight loss of 5.4 kg, evidence of efficacy in type 2 diabetes and dyslipidemia, and improvements across the full range of cardiovascular risk factors, including lipids, blood pressure, and diabetes control, some of which may be due to peripheral metabolic effects that may be independent of weight loss.[15-18]

Other Drugs – Withdrawn or No Longer Recommended for Routine Clinical Use

There are many drugs that were originally approved in the 1950s and 1960s for the treatment of obesity, at a time when much less stringent regulations were in place than at present for the licensing of such agents. These included centrally acting amines such as dexamphetamine and metamphetamine (now withdrawn), phentermine and diethylpropion (which remain available in some countries, but are not recommended for routine use). These agents, whilst effective for short-term weight loss, have significant cardiovascular central nervous system (CNS) and side-effects, including tachycardia, palpitations, sleep disturbance, anxiety, and a potential for dependence, and in general, their risk–benefit ratio is considered unfavorable. The second generation of drugs was based on the serotonin system, and included mazindol, fenfluramine, and dexfenfluramine.[19] The latter two were withdrawn in 1997, following reports of cardiac valve abnormalities, particularly when the drugs were used in combination with phentermine.[20] Thyroxine and diuretics have been used in some clinical settings, but have no place in the management of obesity.

The Need for New Drugs

Given the increasing burden of disease associated with obesity, and steadily increasing evidence on the benefits of weight loss, there is a clear medical need for effective treatment. However the limitations of existing drugs, as described above in terms of limited efficacy and adverse effects, makes development of new alternatives an important scientific goal. Achieving this goal requires a concerted effort to fully understand the biology of human energy balance, a complex biological system that is sensitive to psychological and societal influences, as well as biological feedback mechanisms. The regulatory hurdles are also now set very high, such that any potential new drug must pass stringent tests of both efficacy and safety before it can be made available to patients. The remainder of this chapter will describe current understanding of energy balance regulation, how changes can be detected and monitored, the role of psychobiological factors, and their assessment, and finally, a description of current targets and the regulatory steps required before a new drug for obesity can be approved.

BIOLOGY AND GENETICS OF ENERGY REGULATION AND IMPLICATIONS FOR OBESITY

Energy balance is very tightly regulated over the adult lifetime of an organism. This is a particularly remarkable feat given the variations in energy intake and expenditure that may occur on a day to day basis in most adult humans. For example, for a person with energy requirements of 2000 kcal per day, a 10 kg weight gain over 40 years would equate to an error of only 0.0024% in overall energy balance. It has been essential from an evolutionary perspective for humans to maintain sufficient energy stores to cover immediate and medium term future metabolic requirements, to deal with times of shortage, and increased requirements because of changes in temperature, physical activity, and the increased nutritional demands (on the female) of carrying a fetus to term and providing for the child's early nutritional needs. Short-term energy stores exist in the form of muscle and liver glycogen, and medium and long-term energy stores as lipid, stored in adipose tissue.

A complex regulatory system has evolved to preserve energy balance. The key components of this system are the energy stores, the nutrient sensing, neural and hormonal pathways that signal whether or not there is sufficient energy to meet immediate and future needs, and the CNS components that integrate these signals and determine energy intake and help regulate metabolism and energy expenditure (Figure 8.1). Two important points about these systems are that firstly, whilst there are strong evolutionary pressures to increase food intake and conserve energy, those that might regulate weight to a healthy level and protect against chronic disease are much weaker, as many of these conditions develop beyond reproductive ages. Secondly, the capacity of adipose tissue to expand and store energy is very large, such that a very high positive energy balance is possible.

Short-Term (Episodic) Signals

The short-term, episodic signals regulating energy intake are generally those that signal either the onset or the termination of a meal.

Meal Onset

There is some evidence that small decrements in blood glucose may trigger meal onset, although this hypothesis has not been widely accepted.[21] Most recent interest has focused on the hormone ghrelin. Ghrelin was initially identified as the ligand for

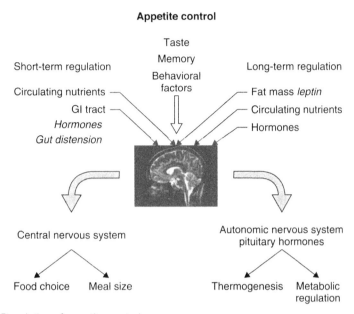

Figure 8.1   Regulation of appetite control

an orphan growth hormone secretagogue receptor (GHS-1), but the observation that it was predominantly expressed in the stomach, and that administration produced a robust increase in food intake in animals and humans, led to its identification as an important regulator of food intake.[22-24] Circulating ghrelin concentrations rise in anticipation of meals, even in the absence of external time cues, suggesting it may be a meal initiation signal.[25] Ghrelin concentrations are reduced in obesity, and restored by weight loss, suggesting it is unlikely to be a major etiological factor in most obesity,[26] although the observation of increased ghrelin concentrations in Prader-Willi syndrome, one of the commonest monogenic obesity syndromes is intriguing.[27]

Meal Termination

Many gut-derived hormones have been implicated in the process of post-meal satiation; the first of these to be identified was cholecystokinin (CCK), which has been shown to reduce food intake in a number of studies in rodents and humans.[28,29] This effect occurs via both central and peripherally located receptors. Agonist drugs at the CCK-A receptor have been developed,[30] but none have proceeded beyond Phase II clinical studies. More recently identified physiological peripheral satiety signals include peptide YY 3-36 (PYY), and two preproglucagon-derived peptides, glucagon-like peptide 1 (7-36) amide (GLP-1) and oxyntomodulin, both of which act via the GLP-1 receptor.[31-33] One long-acting peptide agonist at the GLP-1 receptor has already entered clinical use as a treatment for diabetes mellitus (Exenatide®), and this agent both stimulates insulin secretion and results in modest weight loss.[34] It remains to be seen whether small molecule agonists or long-acting peptide analogs of PYY or oxyntomodulin will prove to be sufficiently potent for clinical use.

Thermogenesis

Neural signals from the CNS to the periphery that regulate thermogenesis can also be considered short-term signals that are reacting to a change or threatened change in the organism's energy balance. The most important signal here is the sympathetic outflow to brown adipose tissue. This is activated after meals and in the cold (to generate heat) and its output is reduced under conditions of energy restriction. This system is especially important in smaller organisms, such as rodents, which have a large surface area to body mass ratio, but they do probably only have a limited role in adult humans; it has proven difficult to identify agents that are sufficiently selective to be useful drugs in humans,[35] and the limited capacity of adults to increase thermogenesis makes this a relatively unattractive option, although many existing drugs that decrease energy intake have a modest effect to increase thermogenesis, as these two biological processes appear to be closely coupled.[36]

Long-Term (Tonic) Signals

The most studied long-term signal is undoubtedly leptin. Leptin is the gene product of the leptin gene, a defect in which causes obesity in the ob/ob mouse.[37] Leptin is produced almost exclusively in adipose tissue, and acts via the blood stream at sites in the hypothalamus to reduce energy intake. Soon after the discovery of leptin, its receptor was also cloned; mutations in the leptin receptor were found to be responsible for obesity in the db/db mouse and the Zucker fatty rat.[38,39] In rodents and humans without mutations in the leptin receptor, circulating leptin concentrations are proportional to fat mass, so leptin deficiency is unlikely to account for most human obesity, and clinical trials in obese humans have shown leptin to be ineffective at reducing body weight.[40] Leptin is however of undoubted biological importance in body weight regulation in humans – Farooqi and colleagues have described two children with severe early onset obesity associated with a homozygous mutation in the leptin gene resulting in an absence of functional leptin, and a dramatic near normalization of body weight with recombinant leptin therapy.[41,42] These children also have abnormalities in gonadotrophic and immune function, consistent with earlier observations made in leptin-deficient rodents. Heterozygotes have also been found, and tend to lower leptin and have a BMI higher than those with two normal copies of the leptin gene. Rare cases of leptin receptor mutations have also been described; unfortunately these are not amenable to treatment.[43]

Insulin can also be considered a tonic signal as it is taken up into the cerebrospinal fluid (CSF) through an active transport mechanism and acts on central appetite regulating pathways to reduce food intake, interacting with leptin signaling at several levels.[44] In baboons, CSF insulin appears to mirror long-term changes in energy balance, further supporting its potential role.[45] It has been more difficult to study effects of insulin in humans, but some insulin analogs in clinical use for diabetes treatment appear to cause less weight gain than conventional insulin, possibly because of increased CNS access.

CNS Integrating Pathways

Most of the pathways described above converge on the hypothalamus, which integrates these signals to determine eating behavior and to regulate peripheral energy expenditure. However these pathways can be overridden by higher centers and the social and psychological context is of great importance when trying to understand energy balance regulation fully. The hypothalamus has several distinct clusters of neurons that are interconnected and link with brainstem afferent neurons that receive peripheral input from the vagus and direct contact with the blood, it also contains specialized areas that are considered "outside" the blood brain barrier and can therefore directly sense peripheral concentrations of nutrients. It is also in close proximity and has direct and indirect links with the CSF. Bilateral links are also present with higher centers, which may be able to modulate the final effects of these complex signals. The hypothalamus alone contains over 50 neurotransmitters and other molecules that may affect energy intake and/or energy expenditure, and although many of the major pathways are now well characterized, current understanding of the role of the hypothalamus in energy regulation is incomplete.

Major hypothalamic pathways that are regulated by leptin include neurons in the arcuate nucleus that synthesize neuropeptide Y (NPY) or Agouti-related peptide (AGRP), both of which increase food intake and decrease energy expenditure and are suppressed by leptin, and neurons that synthesize proopiomelanocortin (POMC) and cocaine and amphetamine-related transcript (CART), which decrease food intake and increase energy expenditure and are activated by leptin.[46] These pathways have mainly been identified by the study of rodents, particularly those that develop obesity such as the Agouti mouse, which overexpresses Agouti, an antagonist at the melanocortin 4 (MC4) receptor. Interestingly, the commonest monogenic form of early onset obesity occurs as a result of mutations in the MC4 receptor, which is activated by POMC the products α and β melanocyte stimulating hormone (MSH), and other defects in this pathway, including prohormone convertase 1. Defects in POMC itself have also been described, again showing close parallels between animal models and human disease.[47,48] Many of the other peripheral signals also converge on these key set of pathways; for example PYY is thought to mainly act by activation of the Y2 receptor which decreases NPY secretion from the arcuate nucleus, and ghrelin may also act via neurons in the arcuate nucleus, where there is a high density of the GHS receptor.[32,49]

The monoamine neurotransmitters, noradrenaline, serotonin, and dopamine are all implicated in appetite regulation, and have been found to interact with both leptin dependent and leptin independent pathways in the brainstem and hypothalamus.[50] Many existing and potential drugs act by modulation of these systems, but the potential for "off target" effects has proved problematic, with only phentermine, diethylpropion, and sibutramine remaining available for clinical use, and with concerns about abuse potential and cardiovascular adverse effects and a remarkable paucity of long-term safety and efficacy data limiting the use of phentermine and diethylpropion to short-term use only.[i] New developments, focusing on the serotonin 5-HT_{1A}, 5-HT_{2C}, and

---

[i] Please refer to Rocha *et al.*, Development of medications for heroin and cocaine addiction and regulatory aspects of abuse liability testing, in this Volume for further discussion of regulatory aspects of abuse potential of drugs.

5-HT$_6$ receptors that are specifically involved in appetite control may prove to be of value in the future (see below).[51-54]

The endocannabinoid system is relatively recently characterized, but has already resulted in the development of the CB1 receptor blocking drug, rimonabant, that has now been licensed in Europe and some other countries. The endocannabinoid receptors CB1 and CB2 were cloned as receptors for the active ingredient of cannabis, tetrahydrocannabinol, which is well known to increase food intake. CB1 receptors are distributed widely in the CNS and periphery, whereas CB2 receptors are only found peripherally, and are thought to be concerned with immune and nociceptive functions.[55] Endogenous ligands for the CB1 receptor are synthesized on demand from arachidonic acid and include anandamide and 2-archidonylglycerol (2-AG). They increase food intake, and their concentrations are increased in rodent models of dietary obesity and in leptin receptor mutations, suggesting a physiological role.[56,57] Unfortunately the CB1 receptor has other roles, including effects on mood and anxiety, which is the basis for some of adverse effects seen in rimonabant clinical trials. A number of other CB1 receptor antagonists are in development (see below), but it remains to be seen if these will have significant benefits over rimonabant in terms of efficacy and adverse effect profiles.

Whilst much of the basic scientific work on the regulation of energy balance has been conducted in rodents, studies of rare obesity syndromes in humans have shown the importance of similar pathways, and many of the regulatory pathways identified so far in rodents appear to have a similar functional role in humans. Nevertheless it is essential to consider potential species differences, and conduct work in humans as well as rodents, particularly when validating potential drug targets in obesity. This can be particularly problematic when considering CNS targets, as it is often the case that the only way a target can be validated is by means of a clinical trial. It is therefore important to consider whether biomarkers or other surrogates can be used to help study the regulation of body weight, and help validate potential drug targets.

BIOMARKERS OF ENERGY REGULATION AND OBESITY

Anthropometric Measures

BMI (calculated as weight in kg/height^2 in meters) has been considered the standard measure of adiposity for clinical use and for epidemiological studies for many years. A considerable amount of data based on BMI have been used to show the increased morbidity and mortality associated with obesity and the WHO current definitions of obesity, still rely predominantly on BMI.[1] However use of BMI as the sole measure of fatness can classify some physically fit, muscular individuals as obese because it does not directly measure body fat or fat distribution. Other simple anthropometric measures that can be used to complement BMI, and may be better markers of some (particularly diabetes and cardiovascular) disease risk include measures of waist circumference and waist–hip ratio.[58] However these measures probably add little to disease risk estimation in people with a BMI significantly above 35 kg/m^2. Other more detailed measures of body composition, derived from skin-fold thickness measurements, can be accurate when used by experienced researchers, but are certainly too time consuming for routine clinical use, and have largely been superseded by non-invasive measurement of body fat, using modern imaging technology.

Measures of Adiposity

The gold standard measures of body fatness are underwater weighing and double-labeled water techniques; the former requires specialist equipment and is an awkward technique for both subjects and researchers. Double-labeled water is expensive, and availability is limited by access to isotope.[59] The newer technique of air-displacement plethysmography ("Bod-Pod") overcomes some of these limitations, and has been reasonably well validated against gold-standard techniques.[60] Dual beam X-ray absorbitometry (DEXA) is widely available, and uses a low radiation dose (less than 10% of a standard chest X-ray). Using a three compartment model it provides an accurate estimate of lean body mass, fat mass and body water and an estimate of fat distribution. Bioimpedence techniques estimate body fatness using prediction equations from measures of body water and lean body mass derived from electrical conductivity of the body. The prediction equations and techniques are improving all the time.[61] Bioimpedence methods are popular because of low cost of the equipment and ease of use, and may be of particular use in large, long-term longitudinal studies, and in clinical practice to measure changes within individuals, but limitations mean they are unlikely to be accepted as a gold standard for clinical research studies.

Most of these measures do not take into account fat distribution, which is important when assessing risk of diabetes and cardiovascular disease, so any study which wishes to assess changes in risk, should measure both change in total body fat, and use a measure of fat distribution. Measures of fat distribution should take into account both subcutaneous, and "visceral" fat – the latter may include fat that is deposited within non-adipose cells within tissues, such as muscle and liver. Subcutaneuous and visceral fat area can be measured using CT or MRI scanning. Most protocols measure visceral and subcutaneous fat area from a single or multiple slices. MRI is preferred for research because it does not involve radiation exposure. Techniques that estimate visceral fat volume are also available, but remain at the experimental research stage at present.[62] All of these techniques are of limited value in the most extremely obese individuals, who are often too heavy or are of too great girth to be scanned safely, even with modern scanners. Direct measurement of intracellular fat in tissues such as muscle and liver using MRI is of considerable research interest, but its use is currently limited to a few very specialist centers worldwide.[63]

Leptin correlates approximately with total fat mass, and may be considered a crude biomarker reflecting this, but gender differences and wide variation between individuals for a given fat mass limits its usefulness in this regard. The recently identified hormone omentin has recently been suggested to inversely reflect visceral fat mass, but this requires further validation.[64,65]

Measures of Energy Expenditure

The three main components of energy expenditure are the basal resting metabolic rate (RMR), the thermic effect of food, and the energy requirements of voluntary physical activity. Accurate measurement of energy expenditure is an important biomarker, and can be used both in the study of the etiology of obesity and in the evaluation of therapy. Calorimetric methods rely on measurement of oxygen consumption and carbon dioxide production, with a correction for protein metabolism based on measurement of urinary nitrogen excretion. Whole body calorimetry in a sealed room is considered the gold standard, but requires considerable technical support and is expensive to set up and maintain; it can however be used to study subjects over a period of time, and can therefore include the effects of meals and episodes of physical activity. Indirect calorimetry using ventilated hood systems is a popular alternative, which correlates well with whole body measurements of resting energy expenditure and is useful for making short-term measurements of RMR and the thermic effect of food. Physical activity can be measured to a reasonable degree of accuracy using modern triaxial accelerometers, but these can only provide estimates of the energy expenditure.[66]

The gold standard for free living individuals requires the use of doubly labeled water ($^2H_2^{18}O$), which can be used to estimate energy expenditure over periods of 3–10 days with a reasonable degree of accuracy ($\pm 4\%$).[67]

Risk Factors

When evaluating the clinical effectiveness of anti-obesity drugs there is no doubt that measurement of conventional cardiovascular and metabolic disease risk factors is essential. At a minimum, evaluations should include measurements of blood pressure, fasting measurements of total, HDL and (calculated) LDL cholesterol, triglycerides, and blood glucose.

Insulin Resistance and Systemic Inflammation

Other biomarkers of CV and metabolic risk may also be measured, although their usefulness in routine clinical practice is not yet proven. Improvements in insulin resistance are expected with effective weight loss and it is helpful to demonstrate that this improves, especially as insulin resistance may precede and usually co-exists with type 2 diabetes. Several methods exist – the simplest involve measurement of fasting insulin and glucose, such as the HOMA (homeostasis model assessment) and QUICKI, but these are probably less accurate than the hyperinsulinemic euglycemic clamp or frequently sampled intravenous glucose tolerance tests.[68,69] In general, for large clinical trials a simple fasting test will probably suffice, reserving the more sophisticated methods if it is necessary to answer a specific question; however innovative study designs are needed to convincingly demonstrate "weight loss independent" effects on insulin resistance.

Other "non-classical" risk factors such as uric acid, fibrinogen, plasminogen activator inhibitor 1 (PAI-1), C-reactive protein (CRP), tumor necrosis factor α, interleukins, adiponectin, and liver function tests (particularly alanine transferase as a measure of fatty liver) are also likely to improve with weight loss and could also be considered helpful surrogates of metabolic improvement, although recent experiences with thiazolidinediones (e.g., rosiglitazone) for diabetes, may temper enthusiasm for reliance on such surrogate outcomes.

THE PSYCHOBIOLOGY OF APPETITE EXPRESSION

Despite the complexity of the biology underpinning energy regulation, many individuals experience great difficulty controlling their own body weight. The system appears not to be able to control excess energy intake. In theory, only a small reduction in daily food intake, over time, could yield significant weight loss. However, in practice this small behavioral change is difficult to achieve for the majority of those seeking to lose weight. This suggests a specific need for pharmacological assistance to gain better control of appetite and not just produce weight loss. Underpinning the structure of feeding behavior in animals and eating behavior in humans are the numerous systems which inform the brain of the body's energy status. These signals can be classed as either episodic or tonic.[70,71]

Episodic Signals in the Regulation of Appetite Expression

The differentiation between episodic and tonic regulatory signals has previously been described. However, it is worth considering these signals, critical to appetite regulation, from the psychological perspective. Episodic feedback is generated by signals arriving from sensory systems (from oral cavity to gut) and the detection of changes in metabolic parameters caused by ingestion. From the anticipation of intake caused by the perception of food, through the oro-sensory experience of consumption, the distension of the gut as food is swallowed, the chemo-detection of nutrients in the gut, and finally the post-ingestive effect of circulating products of digestion, these episodic signals provide a rich array of information on recent energy intake. This transient input allows the body to estimate near instantaneously what has been eaten and thereby adjust eating behavior appropriately. Deficiencies or abnormalities in episodic systems can promote over-consumption and may sustain abnormal eating behavior.[70,71]

Tonic Signals in the Regulation of Appetite Expression

Tonic signals are derived from the conversion and storage of energy, and are responsive to deficits or excesses of energy stored in adipose tissue. In this respect they differ from episodic signals, representing a less immediate but more accurate representation of the bodies energy needs. Despite their tonic nature they show circadian periodicity (i.e., some meal by meal flux). Moreover, absence of tonic inhibitory signals can cause dramatic effects on the expression of eating behavior.[72] Thus, tonic signals are equally important in controlling the expression of eating behavior.[71]

From the behavioral perspective, both episodic and tonic systems contribute to the expression of eating behavior, determining when and how much is eaten. Fluctuations in both produce strong feelings of either hunger or satiety. Moreover, a deficiency in either can cause marked aberrations in the expression of eating; thus they have a common output. Where the episodic and tonic systems differ is in the nature of the input and the duration of their effects.

PSYCHOLOGICAL AND BEHAVIORAL ASPECTS OF OBESITY

The constant metabolic need for energy requires co-ordinated behavioral responses appropriate for the food environment. Given fluctuating environmental demands and the idiosyncrasies of human nature, the fact that individuals can often maintain a fairly constant weight over prolonged periods of time suggests considerable regulatory behavior control. As leptin deficiency shows us, an intimate link between the status of energy stores and the expression of eating behavior must exist.

Genetics Versus Environment?

The rapid rise in levels of obesity has focused interest on factors such as diet, lifestyle, and the obesogenic environment. Nonetheless, a genetic component to obesity is difficult to deny.[73,74] For genetically identical individuals the same phenotype can persist in quite different environments. However, some environments do not permit the expression of a genetic propensity to gain weight.[75] Human obesity has a strong polygenic component that is expressed in both vulnerable populations and in individuals in a permissive environment.[76]

Over the last 14 years hundreds of gene polymorphisms have been linked to human obesity.[77] Whilst our understanding of genetics and the physical phenotype has improved, behavioral phenotypes have been somewhat overlooked. Genes underpinning appetite expression, that is, hunger and satiety, may be critical in explaining individual susceptibility to obesity.[78] Whether obesity can be considered less of a metabolic and more of a neurobehavioral disease[78] is currently uncertain, but behavioral traits linked to obesity constitute phenotypes worthy of examination.

The focus on metabolic rather than behavioral markers for obesity has hindered research into behavioral phenotypes, but progress has been made. For instance, Blundell *et al.*[80] have identified subpopulations of high fat consumers, either disposed or resistant to diet-induced weight gain. What differs between the two groups is not habitual diet *per se* but the expression of their eating behavior. Specifically the obese high fat consumers have a weaker satiety response to fatty meals, a preference for fat that is not diminished by satiety, a stronger hedonic attraction to palatable foods and to eating and higher trait hunger and disinhibition scores. These would all promote a positive energy balance and adiposity.

Obese Eating Style?

From the beginnings of the study of obesity, researchers have sought to define characteristic differences in eating behavior between obese and normal weight individuals accurately. Early laboratory-based observational studies tried to define a specific "obese eating style".[79] These early studies showed that the obese had a faster eating rate (g/min), chewed each mouthful less and consumed more mouthfuls per minute.[79] However, due to the small sample sizes, differing methodologies employed, and the demand characteristics of these studies, no real agreement on what characterized "obese eating style" could be reached.

There were theoretical as well as methodological problems with these early studies. Stunkard and Kaplan[79] suggested that assumptions of homogeneity (i.e., that all obese, irrespective of weight-status, age, and gender would react in the same way) and robustness (that these eating styles would persist despite variations in eating circumstances) underpinning these studies were problematic. Despite this, Stunkard and

Kaplan also noted that some traits were commonly observed in these studies. Firstly, obese individuals often failed to demonstrate the decrease in eating rate (deceleration) that normally characterizes the end of a meal. Secondly, obese individuals, when offered a range of foods, tended to choose a greater number of differing food items during their meal, an effect that was dependent on the palatability of the food items offered.

Even with the decline in the notion of an "obese eating style," the idea that traits, such as eating rate and responsiveness to highly palatable food, constitute risk factors for over-consumption remain.[71,80] Once the mechanisms underpinning these traits, or behavioral phenotypes, have been identified, therapeutic approaches may be possible.

Underperformance of Appetite Control and Deficits in the Specific Satiety Mechanisms

From the earliest studies of eating rate and meal size, data have suggested that ingested food may impact less on the eating behavior of obese individuals than their lean counterparts. For instance, the normal physiological development of satiety, a deceleration in cumulative intake during a meal, is not always seen in the obese.[81-83] Thus it is possible that the obese may not experience the same satiating power of food. Certainly, weight gaining individuals have been shown to eat 1650 kJ a day more than similar sized weight stable individuals, an intake that fails to produce any proportional decrease in post-meal hunger or increase in post-meal fullness.[84] Similarly, in identical twins discordant for weight status, the obese twin reported eating significantly more food at each meal,[85] a difference not corrected for through appetite regulation. These data suggest a comparative deficiency or "underperformance" in the processes of within-meal satiation in these individuals.

A number of factors related to the episodic control of appetite could contribute to this comparatively poor satiety response. Firstly, the obese tend to consume more energy dense foods usually high in fat and/or refined sugar.[86,87] These are foods which produce a poorer satiety response both because of their high energy density,[88,89] and the specific deficits in the satiety response to fatty meals which characterizes obese high fat consumers.[80] Secondly, there is evidence that some obese individuals have large gastric capacities, particularly the morbidly obese and binge eaters,[90,91] allowing the consumption of greater volumes before fullness is reached. Finally, lean/obese differences in the gastrointestinal peptide response to nutrients within the gut may constitute another physiological mechanism underpinning weakened satiation in the obese.[71,92,93] Thus food type, gastric capacity, and physiological response to ingestion could all combine to promote continued over-consumption.

Differences in Palatability/Hedonic Experience and Over-Responsiveness to High Fat Foods

Obese/lean differences in satiety may explain some disparity in eating behavior; however, can obese individuals over consume purely because of an inherent weakness in satiety? Eating behavior is a sensory and hedonic experience; and the pleasure of consumption can overwhelm processes of satiation. Lowe and Levine[94] have argued that the food environment chronically activates the hedonic appetite system, producing a motivation to "over-eat." Certainly, palatable food stimulates hunger[95-97] and also appears to delay fullness.[96] Indeed, Yeomans *et al.*[98] have commented that palatability exerts the strongest stimulatory effect on eating when sated.

Given the relationship between the consumption of dietary fat and obesity, it is not surprising to find experimental evidence that the obese display heightened preference for high fat foods. Food preferences in the obese and weight gainers have all been characterized by a preference for fat.[84,99,100] Similarly, pleasantness ratings for high fat foods have been shown to correlate with percent body fat.[101] These observations suggest that differences in hedonic experiences of certain foods may underpin obesity-promoting food choices. The hedonic capacity to enjoy high fat foods, together with a weakened response to them, constitute key behavioral risk factors, which interact to promote and sustain obesity.[80,102]

Trait Hunger, Disinhibition, and Binge Eating

It is clear that appetite regulation goes beyond the physiological processes underpinning hunger, satiation, and satiety. In fact "uncontrolled eating" (comprising high trait hunger and disinhibition) may characterize the eating behavior of many obese individuals to a greater or lesser extent.[103,104] The notion that the obese have a disposition to respond more to environmental stimuli to eat than to their physiological need is not a new concept. Numerous studies have shown that the obese demonstrate higher levels of trait hunger, that is, experiences such as feelings of constant/excessive hunger and over-responsiveness to the sensory characteristics of food (e.g.,[105,106]). However, constant and provokable hunger is not the only trait associated with obesity.

Disinhibition, excessive and poorly controlled eating behavior, is also associated with adiposity.[105,107-110] Disinhibition is associated with weight gain,[109] weakened satiety response to fat,[81] and with dietary relapse[111] suggesting it is a key obesity-promoting behavioral characteristic. However, do hunger and disinhibition constitute obesity behavior phenotypes? Both trait hunger and disinhibition appear to have a genetic component. Provencher *et al.*[112] estimated generalized heritability for hunger and disinhibition to be 28% and 18% respectively. Moreover, allele variation has been shown to link behavior traits such as hunger and disinhibition with obesity and liability to gain body fat.[113]

An inability to control one's weight satisfactorily lies not only at the route of obesity, but also of many eating disorders. Aberrant and disordered eating behavior, usually linked to anorexia and bulimia nervosa, is also associated with obesity. Binge eating at sub-clinical levels is very common in the obese and full blown clinical binge eating disorder (BED) is far more prevalent in the obese than other weight status populations. Moreover, the occurrence of both behavior and disorder increases in the obese population with the degree of adiposity,.[114,115-119] Binge eating is a significant clinical issue. Obese individuals with BED do appear to consume more calories than weight matched non-BED counterparts.[120]

The behavior traits most linked with binge eating in the obese are disinhibition and hunger.[116,118,121] However, whilst disinhibition and hunger serve to undermine efforts to control intake and prevent weight loss, the aberrant behavior of binge eating also serves to perpetuate the disorder itself. Binge eating undermines the physiological processes, such as gastric capacity and gut hormone release, which underpin normal appetite expression.[90,91,122] Normal appetite regulation and eating behavior may be partially restored with satiety-enhancing drugs.[82,123-126]

BEHAVIORAL INDICES FOR ASSESSING THE ACTION AND EFFICACY OF ANTI-OBESITY DRUGS

Even though obesity is ultimately a matter of energy balance, that is, caloric intake exceeding energy expenditure, susceptibility is clearly linked to food choices and eating behavior. Factors such as weakened satiety, over-responsiveness to palatability, high trait hunger, high disinhibition, and binge eating are etiological and symptomological factors that require addressing during drug development.

Susceptibility to Obesity and Hyperphagia

Animal models of obesity provide important indices for the assessment of potential therapeutic effects such as changes in adiposity and body fat distribution, and in key endocrine and metabolic factors.[127,128] Behavioral indices such as hyperphagia are necessary for assessing the potential efficacy of appetite-suppressing, anti-obesity drugs. Genetic models of obesity produce very reliable weight gain and aberrant eating behavior against which the efficacy of various drugs can be tested. Similarly, drug-induced weight gain also produces robust changes in feeding behavior and body weight. However, whilst mechanistically interesting, these models are not entirely valid representations of the human condition.[233]

Rodents, like humans, show a tendency to gain weight when exposed to a highly palatable energy dense diet and during this period they demonstrate marked hyperphagia.[129] Methods of producing Diet induced obesity (DIO) vary from adding sugar and fat to standard laboratory chow or providing separate sources of fat or sugar in addition to laboratory chow, to providing a selection of a variety of high fat and/or sugar foods in a cafeteria paradigm. The latter approaches also allow the assessment of drug effects on food choice. Some rodent strains appear more susceptible to weight gain than others, and even within groups of animals from a single strain, some animals appear to respond to the DIO exposure, whilst others do not.[130,131] This has the potential to complicate research, but is analogous to the human condition; therefore the tendency of many investigators to solely select DIO responders for studies may not be appropriate. Critically to the treatment of the human condition, the reversal of DIO in the rodents reduces morbidity and mortality.[128,132]

Currently, anti-obesity drugs are licensed to treat obesity and not to prevent it. Therefore, it is critical to consider whether the DIO model tests the drugs ability to prevent rather than treat obesity. Nonetheless, previously and currently licensed appetite-suppressing anti-obesity drugs have all been shown to reverse DIO and associated hyperphagia.[82] It is important to note that in the DIO model, the effects of most drugs on daily caloric intake weaken as the study progresses, even if drug effects on body weight remain. It is difficult to say whether a loss of hyperphagia in rodents is mirrored in humans, or how it would relate to clinical efficacy, but species differences in hypophagia response may cause problems when trying to judge the clinical potential of a drug. As so few drugs have successfully reached the clinic, it is still difficult to ascertain what type of DIO model, and which measures, best predict ultimate clinical success.[233]

Structure of Feeding Behavior

The nature of drug-induced reductions in food intake are critical in determining if a potential appetite-suppressing anti-obesity drug can progress into clinical trials. Drugs can reduce food intake in rodents and humans in a variety of ways, through the induction of nausea or malaise, or through CNS related effects such as hyperactivity or sedation. Such effects, even if secondary to drug action on mechanisms of satiety or food preference, prevent the compound being of any clinical value. The concept of using rodent behavior to determine the nature of drug-induced hypophagia is well established. For over 40 years researchers have studied drug effects on feeding and other behavior in rodents, to determine if a drug reduces food intake in a manner consistent with a selective effect on the mechanisms underpinning appetite expression.

The validity of behavior as an indicator of toxic, pathologic, or non-physiological events appears well established.[133,134] Monitoring animal behavior, including feeding and non-feeding activities, provides a powerful bio-behavioral assay of drug action on appetite. It also avoids some problems of validity as it uses the animal's natural behavior rather than attempting to model it on the human condition. Early studies examined drug effects on food intake, frequency of eating bouts, inter-bout intervals, eating rate and meal patterning. Changes in these micro-structural indices were successfully used to discriminate between drugs enhancing satiety and those inducing hypophagia by other means.[135] One of the most detailed behavioral assays of drug action on appetite expression is the behavioral satiety sequence (BSS).[28,136,175] The BSS examines the microstructure of rodent behavior, and the sequence consists of a stochastic progression of behavior from an initial phase of eating, through peaks of active and grooming behavior, to an eventual phase of predominately resting behavior. The BSS appears robustly related to the processes of satiation (meal termination) and the development of satiety (post-ingestive inhibition of eating) as its expression relies on the presence of a caloric load within the gut, and the stimulation of satiety factors that this gastric load produces.[28,137] It should be noted that both current and previously licensed appetite-suppressing anti-obesity drugs, sibutramine and D-fenfluramine and a wide range of other putative satiety compounds and peptides have both been shown to preserve and enhance the BSS.[138,139,140-142,175,205] The BSS is not only a useful screening tool, its use in conjunction with selective agonists and agonists, has helped establish key novel anti-obesity drug targets[138,175-177,181-182,205] (see below).

Assessing the Effects of Drugs on Human Feeding Behavior

A variety of approaches to measuring the effects of anti-obesity drugs on human food intake can be employed. Self-report measures such as food diaries, short-term recall, and food frequency questionnaires are most suited to large population samples and not early clinical trials. Laboratory-based observation studies provide more precision and reliability at the expense of "naturalness".[143] As with rodents, human eating behavior can be considered in terms of macrostructure, i.e., the effects of a drug on total daily caloric intake, caloric intake at each meal, macronutrient intake, meal duration, and the number of eating events occurring during the day. Such studies have been used for nearly 30 years to characterize the effects of drugs on human appetite.[144,145] The inclusion of large *ad libitum* buffet style meals, and/or the availability of snacks throughout the day, also allows researchers to fully assess drug effects on food choice.

In human studies, researchers are able to assess the effects of drugs on subjective experiences of appetite to confirm any satiety-enhancing effect. Self-report scales have been used to determine the nature of a drug's effect on food intake in some of the earliest human studies.[143] These rating scales initially focused on feelings of hunger and fullness, but now encapsulate other aspects of satiety such as desire to eat, prospective consumption (how much could you eat?), and satisfaction. The measures come in a variety of forms but the most widely accepted format is the visual analog scale (VAS). The VAS format presents participants with a 100 mm horizontal line anchored at each end with extremes of the sensation (e.g., "not hungry at all" and "extremely hungry") on which to mark their feeling.[143] It is interesting to note that ratings of appetite sensations are not only predictors of energy intake but also of body weight loss.[146]

The effect on eating behavior and appetite of drugs such as fenfluramine, D-fenfluramine, and sibutramine are well characterized in lean healthy volunteers and in the obese.[147-149] As these three drugs also produce clinically significant weight loss it would appear that behavioral studies may predict clinical efficacy. In reality, behavioral studies can clearly demonstrate proof of concept but not all drugs successfully tested at this stage will ultimately possess the potential to be effective weight loss-inducing agents. For instance the selective serotonin reuptake inhibitor (SSRI) fluoxetine (Prozac®) produces clear effects on appetite and food intake in both lean and obese participants[150,151] along with weight loss for up to 6 months of treatment, however; after that time, weight regain occurs (although it should be noted that fluoxetine went on to be used successfully for the treatment of binge eating in both bulimics and the obese).[82] The case of fluoxetine demonstrates that behavioral studies may reliably predict weight loss over 12 or 24 weeks if not over 12 months. Therefore, as indices of performance in the next phase of clinical testing the results of these studies are still critical. Moreover, failure to determine drug effects on human eating behavior prior to Phase II weight loss trials can lead to costly failure.[152]

Microstructure of Human Eating Behavior: Eating Rate, Cumulative Intake Curves and Deceleration

The effects of drugs on parameters within a meal have also been studied for nearly 30 years. Rogers and Blundell[144] clearly demonstrated that whilst amphetamine and fenfluramine both reduced food intake, the increase in eating rate produced by amphetamine represents the activating effects of the drug, whilst the reduction in eating behavior produced by fenfluramine represents enhanced satiety. To study eating rate, many researchers visually coded eating behavior[144] whilst others developed automated means of assessing intake, such as the universal eating monitor (UEM).[153,154]

UEM curves are influenced by gender and food deprivation, as well as the composition and palatability of test meals, demonstrating that they are valid representations of the changes in appetite that occur within meals, and critically that they can be used to discriminate between factors which influence the expression of appetite.[154,155] The UEM also appears to possess good test-retest reliability.[156]

Cumulative intake curves may not only be useful for examining the effects of drugs and physiological states on appetite. The obese differ from the lean in the respect that they often fail to display the decelerating cumulative intake curves associated with the normal biological development of satiety.[81,157] Failure to demonstrate a reduction in eating rate during a meal has been observed in obese children,[158,159] in syndromes characterized by severe obesity such as Prader-Willi,[160] in the morbidly obese,[161] and in obese highly restrained adults,[155] but not always in obese adults *per se*.[155,162] A recent study has shown that the appetite-suppressing, anti-obesity drug sibutramine significantly slows eating rate and changes within- meal ratings of appetite consistent with satiety in the obese.[83] This was accompanied by a non-significant increase in frequency cumulative intake curves in the drug condition.[83] These data indicate that characteristic differences in the microstructure of human eating behavior may be useful in determining a drug's effects on appetite. In fact, sibutramine-induced reductions in intake were associated with clear reductions in within-meal ratings of hunger and increases in those of fullness during the meal. This is the first data directly confirming that sibutramine enhances within-meal satiation. Combining continuous measurement of intake with changes in appetite provides a powerful assay of drug action. Potentially, the use of differing within-meal appetite ratings could differentiate between drugs acting on the motivation to initiate eating episodes through mechanism of liking or wanting and from those acting on satiety, a difference traditional macro-structural approaches are unlikely to detect. Certainly, opioid receptor antagonists produce distinct effects on within-meal appetite ratings.[163]

MOLECULAR TARGETS FOR ANTI-OBESITY DRUGS

As discussed in detail above, molecular targets for obesity are many and varied, ranging from modifications of current therapies, such as monoamine reuptake and lipase inhibitors, to novel neurotransmitter and neuropeptide receptors. Due to past failures and drug withdrawals (see above) the pharmaceutical industry faces an increasingly uphill task in convincing the regulatory authorities of the efficacy and, in particular, the safety of new drugs to treat obesity. In this section we consider the most interesting new molecular targets for obesity, emerging strategies that can be used by pharmaceutical companies to discover and develop compounds that act on these targets and the challenging regulatory requirements for their approval as drug therapies.

Lipase Inhibitors

Attempts have been made to develop novel lipase inhibitors that reduce body weight but have a lower propensity to cause gastrointestinal side-effects than orlistat (see above). The most advanced such compound in development is cetilistat which Alizyme and Takeda are preparing for Phase III clinical trials. In a recently published report of a Phase II clinical trial,[164] cetilistat produced a significant weight loss and was well tolerated in 442 obese patients in a 12-week study. Alizyme has claimed that cetilistat and orlistat have similar efficacy but cetilistat is better tolerated and causes a lower incidence of gastrointestinal side-effects than orlistat.[165] The superior tolerability and side-effect profile exhibited by cetilistat has been attributed by Alizyme to differences between the molecular structures of orlistat and cetilistat. Nevertheless, as both compounds act on the same molecular mechanism to cause weight loss and the gastrointestinal side-effects are thought to be mechanism based it is difficult to understand why there should be significant differences in the side-effect profile of the two compounds but no difference in efficacy. Therefore, the outcome of the planned Phase III clinical trials with cetilistat is awaited with interest.

Serotonin/Noradrenaline Reuptake Inhibitors

Sibutramine

A number of companies have attempted to develop mixed reuptake inhibitors that retain the weight loss efficacy of sibutramine (see above) but have a reduced propensity to cause cardiovascular side-effects.

Tesofensine

Tesofensine is a dopamine, serotonin, and noradrenaline (triple) reuptake inhibitor originally developed by NeuroSearch for the treatment of Alzheimer's disease and Parkinson's disease. Development of the compound for these neurological indications was unsuccessful but significant weight loss was reported during the clinical trials in Parkinson's disease.[166] Hence, tesofensine is now being developed by NeuroSearch for the treatment of obesity and type 2 diabetes. In September 2007 NeuroSearch reported the outcome of a Phase IIb study with tesofensine for the treatment of obesity. Data from the study in 203 patients showed that 24-weeks' treatment with tesofensine resulted in a dose-dependent weight loss of 6.5–12%. Tesofensine was reported to have a good safety profile and was well tolerated although an increased number of adverse events (e.g., increased heart rate and blood pressure) were observed in the highest dose groups of 0.5 mg and 1.0 mg. NeuroSearch[167] stated that no clinically relevant cardiovascular adverse events or changes in either blood pressure or pulse were seen, according to FDA criteria. Nevertheless, in studies in Parkinson's disease decreased body weight and elevated heart rate were described as common in the 1.0 mg dosage group. Further, a sustained increase in supine systolic blood pressure was recorded in 5.7% of subjects in the combined NS 2330 groups and in no placebo subjects and a sustained increase in supine diastolic blood pressure was recorded in 2.6% of the combined NS 2330 groups and in no placebo subjects. In addition, 56.2% of NS 2330 subjects had a maximum increase in heart rate of at least 10 beats per minute as evaluated on ECGs compared to 18.8% of placebo subjects.[166] The results of pivotal Phase III clinical trials with tesofensine for obesity are awaited to determine whether the compound will have a superior cardiovascular risk-benefit profile compared to sibutramine.

PSN 602

Prosidion is developing compounds that combine 5-HT_{1A} receptor agonist properties with monoamine reuptake inhibition. It is proposed that the increases in heart rate and blood pressure associated with sibutramine may be prevented by adding a 5-HT_{1A} receptor agonist action to monoamine reuptake inhibition. The lead compound from this program PSN 602 is undergoing preclinical testing and was expected to enter clinical studies in 2007. However, although 5-HT_{1A} receptor agonists have been reported to decrease blood pressure in animals under certain experimental conditions[168] there is little or no evidence that 5-HT_{1A} receptor agonists decrease blood pressure in humans.[169-171] Prosidion compounds have been claimed to have demonstrated anti-obesity efficacy and superior cardiovascular safety to sibutramine in rats.[172,173] However, given the species differences in the effects of 5-HT_{1A} receptors agonists on cardiovascular responses described above, the question of whether a combination 5-HT_{1A} receptor agonist/monoamine reuptake inhibitor will be a safe and effective obesity therapy can only be resolved by clinical studies.

Selective Serotonin Receptor Ligands

Non-selective serotonin drugs such as sibutramine and d-fenfluramine have been shown to be effective weight loss agents but their therapeutic utility is limited by cardiovascular side-effects, which in the case of d-fenfluramine led to the worldwide withdrawal of the drug from the market (see above). The serotonergic actions of sibutramine and d-fenfluramine are mediated by their ability to increase extracellular 5-HT levels and thereby act on multiple 5-HT receptors, of which 14 subtypes have been identified.[174] It has been proposed that if the 5-HT receptor subtype at which 5-HT acts to decrease eating can be identified and if this proved to be a different 5-HT receptor subtype to that responsible for the actions of 5-HT on cardiovascular function it may be possible to develop 5-HT drugs that cause weight loss and have an improved cardiovascular side-effect profile.[175-177] During the past 10 years 5-HT_{2C} receptors and 5-HT_6 receptors have been identified as the most attractive 5-HT receptor targets for obesity as both subtypes appear to be present in high densities in areas of the brain that control eating but are present in low densities or are absent in the peripheral cardiovascular system.[174-179]

5-HT_{2C} Receptor Agonists

There is definitive evidence that 5-HT_{2C} receptors play a crucial role in mediating the inhibitory action of d-fenfluramine on eating. Thus, the selective 5-HT_{2C} receptor antagonist SB-242084 attenuates the hypophagic effect of both d-fenfluramine and its active metabolite d-norfenfluramine.[180] Furthermore, fenfluramine-induced reductions in feeding are also attenuated in 5-HT_{2C} receptor knockout mice.[181] 1-(m-chlorophenyl) piperazine (mCPP) is a non-selective 5-HT_{2C} receptor agonist that produces behaviorally selective reductions in food intake in rodents;[138,182] and humans.[184] The effects of mCPP on feeding in rodents are attenuated by the 5-HT_{2C} receptor antagonist SB-242084.[182] Chronic administration of mCPP either subcutaneously or orally significantly attenuates body weight gain in rodents and food intake and body weight gain increase toward control values in animals withdrawn from drug treatment.[54,185] Similarly, when administered chronically for 14 days the compound decreases body weight in obese patients.[184] Furthermore, in agreement with rodent data, the non-selective 5-HT$_{2C}$ receptor antagonist, ritanserin prevents the hypophagic effects of d-fenfluramine on food intake in human volunteers.[183]

Such data provide a compelling rationale for the potential utility of selective 5-HT$_{2C}$ receptor agonists as anti-obesity agents and therefore a number of pharmaceutical companies have initiated research programs to develop selective 5-HT$_{2C}$ receptor agonists for the treatment of obesity.

The first selective 5-HT$_{2C}$ receptor agonists were developed in a collaboration between Cerebrus (later Vernalis) and Roche and one of the first compounds to be reported with selectivity, especially against 5-HT$_{2A}$ receptors, was the indoline molecule VER-3323.[186,187,188] VER-3323 reduced food intake when administered orally, and the hypophagic effect was dose-dependently blocked by pre-treatment with selective 5-HT$_{2C}$ receptor antagonists.[187] This collaboration also yielded the first selective compound, VR 1065, to progress to Phase I clinical trials.[189] However, this compound was subsequently withdrawn from Phase 1 trials, when the pharmacokinetic data in humans identified high levels of a metabolite of the drug.[189] Another drug discovery collaboration reported at this time was that of Biovitrum and GlaxoSmithKline. During the early stages of the program Biovitrum completed Phase II trials with the 5-HT-agonist BVT.993. It was found that BVT.933 significantly reduced body weight in patients without causing any serious side-effects. However, the compound was considered to have insufficient selectivity for the 5-HT$_{2C}$ receptor and in 2003 development of BVT.933 was terminated and the collaborative program focused on finding more selective compounds.[190] Recently, this collaboration appears to have been abandoned and GSK has returned all rights to the project to Biovitrum, which, subsequently, has decided not to develop the compounds further for obesity.[191]

The most advanced 5-HT$_{2C}$ receptor agonist in development is lorcaserin, which is being developed by Arena Pharmaceuticals and began Phase III clinical trials for obesity in North America in 2006.[192] Results from a Phase IIb trial of lorcaserin demonstrated that patients who received the drug experienced significantly greater weight loss than patients on placebo.[193] Lorcaserin was reported to be well tolerated at therapeutic doses and echocardiographic assessment of over 800 patients that participated in the Phase II trials indicated no apparent adverse effects of the drug on cardiovascular measures (heart valves or pulmonary artery pressures).[193] Lorcaserin is reported to have 100-fold selectivity for the 5 HT$_{2C}$ receptor relative to the 5-HT$_{2B}$ receptor, but is only 15-fold selective for the 5 HT$_{2C}$ receptor relative to the 5-HT$_{2A}$ receptor.[193] As insufficient selectivity for the 5-HT$_{2C}$ receptor has led to the withdrawal of previous 5-HT$_{2C}$ receptor agonists from clinical development (e.g., BVT.933, see above) it remains to be seen whether the selectivity, efficacy, and safety profile of lorcaserin will be sufficient for the compound to achieve regulatory approval and market penetration.

5-HT$_6$ Receptor Antagonists

The 5-HT$_6$ receptor is a promising new CNS target for obesity[177] and a number of pharmaceutical companies are developing selective 5-HT$_6$ receptor ligands as potential anti-obesity agents. Interestingly, both selective 5-HT$_6$ receptor agonists and antagonists are being developed for obesity by different companies (see below).

Initially, the discovery of selective 5-HT$_6$ receptor ligands[194] enabled the role of 5-HT$_6$ receptors in ingestive behavior to be characterized. The first selective 5-HT$_6$ receptor antagonist to be disclosed was Ro 04-6790.[195] This compound has relatively low potency at human 5-HT$_6$ receptors and poor brain penetration.[194] Nevertheless, Ro 04-6790 has proved useful in probing 5-HT$_6$ receptor function. Thus, Ro 04-6790 reduced food intake in rats and this effect was resistant to PCPA-induced brain 5-HT depletion indicating that the 5-HT$_6$ receptors involved have a post-synaptic location.[196] Chronic administration of Ro 04-6790 for 2 weeks significantly reduced body weight gain in rats and similarly, 5-HT$_6$ mRNA antisense oligonucleotides have been reported to decrease body weight gain in rats.[197-199] Subsequently, Perez-Garcia and Meneses[200] have reported that two other 5-HT$_6$ receptor antagonists, SB-357134 and SB-399885, also decreased food intake in rats.

Somewhat surprisingly, given the effects of 5-HT$_6$ receptor blockade described above, Esteve is developing selective 5-HT$_6$ receptor agonists for obesity and have described the effects of E -6837 which *in vitro* exhibits partial 5-HT$_6$ receptor agonist properties in rat cell lines and full 5-HT$_6$ receptor agonist properties in human cell lines.[179,201] Recently, 4-week treatment with E-6837 was reported to decrease food intake and body weight in rats with DIO. E-6837-induced weight loss was mediated by a decrease in fat mass, with a concomitant reduction in plasma leptin.[179]

Biovitrum is developing a 5HT$_6$ receptor antagonist for obesity and initially disclosed pre-clinical studies with BVT.5182.[202] BVT.5182 was reported to decrease food intake and body weight when administered orally to mice with DIO and analysis of meal patterns appeared to indicate that the compound reduced food intake via the enhancement of satiety.[202] Interestingly, Caldirola[202] also reported that the effect of a 5-HT$_6$ receptor antagonist on food intake in mice was potentiated by treatment with a 5-HT$_{2C}$ receptor agonist.

Biovitrum disclosed the selection of a 5-HT$_6$ receptor antagonist clinical development candidate, BVT.74316, in June 2005. BVT.74316 has been reported to decrease food intake and reduce body weight and body fat mass in both short and long-term studies in animals.[203] The compound began Phase I clinical trials in healthy volunteers in August 2006.[203]

EPIX Pharmaceuticals is also developing a 5HT$_6$ receptor antagonist for obesity and recently reported results from a Phase Ib clinical trial of their development candidate PRX-07034.[204] In this double-blind, placebo-controlled trial of 21 obese participants PRX-07034, at a dose of 600 mg twice-daily for 28 days, lost a mean of 0.45 kg while participants on placebo gained 1.37 kg. Treatment with PRX-07034 was associated with a significant reduction in plasma leptin levels. After 42 days, the 28-day dosing period and a 14-day follow-up period, participants on PRX-07034 lost a mean of 0.26 kg compared to a mean gain of 1.25 kg in the placebo group. However, there was evidence of QT prolongation in the ECG of participants who received PRX-07034 and therefore further development of the drug is expected to be conducted using the R-enantiomer because EPIX believes, based on preclinical data, that the S-enantiomer is predominantly responsible for the QT prolongation.[204]

Cannabinoid Receptor Ligands

Rimonabant is a CB1 receptor antagonist that has recently been licensed in Europe for the treatment of obesity (see above). The potential psychiatric side-effects of rimonabant (alterations in mood and an increased risk of anxiety and depressive disorders, including suicidality) have recently received much attention and led to an FDA advisory committee voting against approval of the drug in the United States until longer term safety and efficacy data are available. A number of companies are developing CB1 receptor antagonists for obesity and their principal objective is retain the weight loss efficacy of rimonabant but have a reduced propensity to cause psychiatric side-effects. The most advanced CB1 receptor antagonists in development are taranabant (Merck) and CP-945,598 (Pfizer) both of which are undergoing Phase III clinical trials with NDA applications anticipated in 2008–2009. In addition, the CB1 receptor antagonists AVE 1625 (Sanofi-Aventis) and SLV 319 (BMS/Solvay) are both in Phase II clinical trials.

As the psychiatric side-effects of CB1 receptor antagonists appear to be mechanism based it remains to be seen whether the objective of retaining weight loss efficacy with a reduced risk of psychiatric side-effects can be achieved. Clearly, however, it would be prudent to assess the effects of novel CB1 receptor antagonists on both eating behavior and mood at an early stage of clinical development to minimize the risk of failure at the stage of regulatory approval (see discussion below on experimental medicine approaches to this problem).

Neuropeptide Receptor Ligands

A large number of neuropeptide receptors have been explored as potential targets for novel drugs to treat obesity. To date, however, clinical success has been limited and a neuropeptide ligand has yet to be approved to treat obesity. For example there was considerable interest in the development of CCK agonists to treat obesity based on evidence that CCK acted as a satiety factor.[205] Thus, GlaxoSmithKline were developing the non-peptide CCK-A agonist GW-181771 for obesity but development was discontinued due to lack of efficacy in Phase II clinical trials.[235]

Nevertheless, the neuropeptide approach appears to hold considerable promise and several neuropeptide ligands that are currently in clinical development are considered below.

Amylin, Glucagon, and GLP-1

Pramlintide (pramlintide acetate, Symlin®, Amylin Pharmaceuticals) and exenatide (Byetta®, Amylin Pharmaceuticals, and Lilly) are injectable drugs that were approved in 2005 for the treatment of diabetes. Pramlintide is approved for patients with Type 1 and Type 2 diabetes and is a synthetic analog of human amylin (a pancreatic β-cell hormone) that acts in conjunction with insulin to delay gastric emptying and inhibit the release of glucagon. Exenatide is a GLP-1 mimetic that has multiple mechanisms for lowering glucose levels, including the enhancement of insulin secretion and is indicated for use in patients with Type 2 diabetes. Clinical trials have shown that both drugs reduce fasting plasma glucose levels and body weight.[206] Novo Nordisk have a GLP-1 analog, liraglutide that has completed Phase II trials in diabetes, and is also reported to result in weight loss – it remains to be seen whether they will pursue and anti-obesity indication for this agent. Oxyntomodulin, another product of the proglucagon gene, most likely acts via the GLP-1 receptor, and has been found to reduce food intake in acute studies; further Phase II trials are underway.[33]

Amylin has pramlintide in clinical development for the treatment of obesity and in 2004 reported results from a Phase II study in obese subjects evaluating the safety and tolerability of the drug. In the study, obese subjects were able to tolerate higher doses of pramlintide than those previously studied in diabetes trials, and achieved clinically and statistically significant weight loss. In 2006, Amylin reported data from a Phase II study demonstrating that patients completing 52 weeks of pramlintide therapy experienced a 7–8% mean body weight reduction (depending upon dose) compared to a 1% reduction in patients receiving placebo.

A clinical study of pramlintide in combination with phentermine and sibutramine is currently underway. This Phase IIb study is designed to replicate preclinical data showing that pramlintide added to an oral obesity agent produced additive weight loss in animals and results of this study are expected in 2007.[231]

NPY (Y2, Y4, and Y5)

NPY is a heterogeneously distributed neuropeptide that elicits its physiological effects by an action on six different receptor subtypes (Y1–Y6). NPY stimulates food intake, inhibits energy expenditure, and increases body weight by activating Y1 and Y5 receptors in the hypothalamus.[207] Based on these observations, several companies have attempted to develop neuropeptide Y2, Y4, and Y5 receptor ligands as potential anti-obesity agents.

Shionogi & Co is developing S-2367, a selective Y5 receptor antagonist, for obesity. In pre-clinical studies S-2367 increased energy consumption, suppressed visceral fat accumulation, and improved blood glucose and serum lipid levels. Shionogi has successfully completed Phase IIa proof-of-concept studies with S-2367 and Phase IIb studies are currently underway in the United States (1500 patients in two studies). Patient enrollment was initiated in March 2007 and is expected to be completed by year end. An interim analysis of the data was scheduled for March 2008.[238]

7TM Pharma is developing obinepitide, a dual Y2–Y4 agonist and TM30339 a selective Y4 agonist for obesity. Recently 7TM Pharma disclosed positive results from a Phase I/II clinical study with obinepitide.[208] The study was a double-blind, placebo-controlled dose-range finding study in obese patients to evaluate the effects of obinepitide on food intake. At present 7TM Pharma is examining the effects of obinepitide on weight loss in a 28 day Phase II study in obese patients with results expected in the first quarter of 2008.

7TM Pharma has also recently initiated a Phase I/II clinical trial with TM30339, a selective Y4 agonist for obesity.[209]

Melanin-Concentrating Hormone

Melanin-concentrating hormone (MCH) has been implicated in the control of feeding behavior and energy homeostasis. MCH stimulates food intake in rodents and deletion of MCH or the MCH-1 receptor gene decreases body weight whereas over-production of MCH increases weight gain.[210] Blockade of MCH-1 receptors in rodents abolishes MCH-induced feeding and promotes hypophagia and weight loss.[211-214]

Neurogen is developing MCH-1 receptor antagonists for the treatment of obesity. In May 2007, Neurogen reported the results of initial Phase I studies in 71 male and female participants with an MCH-1 receptor antagonist development candidate NGD-4715. NGD-4715 was reported to be safe and well tolerated at a broad range of doses.[237] Neurogen is now planning a multiple ascending dose study in healthy volunteers and then plan to proceed into a Phase II proof-of-concept study in obese patients.

Peptide YY

Peptide YY (PYY) 3-36 is one of the two main endogenous forms of PYY, a hormone released in endocrine cells in the human small intestine after a meal.[215] Infusion of PYY 3-36 has been reported to reduce hunger and food intake in obese volunteers.[215] As PYY is a protein, initial clinical development has focused on PYY delivery by injection (Amylin) and nasal spray (Nastech).

Amylin is developing injectable AC162352, a synthetic analog of human PYY 3-36 for obesity. Subcutaneous injection of AC162352 decreased appetite and increased satiety in obese individuals.[216]

Nastech Pharmaceuticals is developing a PYY 3-36 nasal spray for obesity and is currently conducting a dose ranging study to evaluate its effects on appetite and food intake in obese participants. The objective of the study is to identify appropriate doses for evaluation in a long-term Phase II efficacy and safety trial.[217]

SCREENING STRATEGIES FROM MOLECULAR TARGET TO INITIAL CLINICAL TRIAL

There are a number of recognized steps used by most pharmaceutical companies to establish, validate, and conduct a drug discovery program for obesity. Many programs will begin with the identification and validation of a novel molecular target from the human genome; others will be initiated as "me too" projects when a competitor company provides initial validation of a novel target. Important considerations in the initiation of a drug discovery program are the availability of, or the potential to discover, a medicinal chemistry starting point from which to derive lead compounds (ideally that are small drug-like molecules) that act selectively on the molecular target. Once a chemical lead has been identified an *in vitro* and *in vivo* biological screening cascade will be established and validated. The screening cascade will be used to derive structure–activity relationships and to guide the medicinal chemistry required to identify biologically active lead compounds for clinical development. An example of a screening strategy that could be used for discovery of selective 5-HT$_{2C}$ receptor agonists is illustrated in Figure 8.2.

Initial screening of compounds submitted from the medicinal chemistry program would be carried out in high throughput radioligand binding and functional (measurement of changes in intracellular calcium caused by receptor activation using a fluorescence imaging plate reader, FLIPR) assays. Active compounds that achieved initial

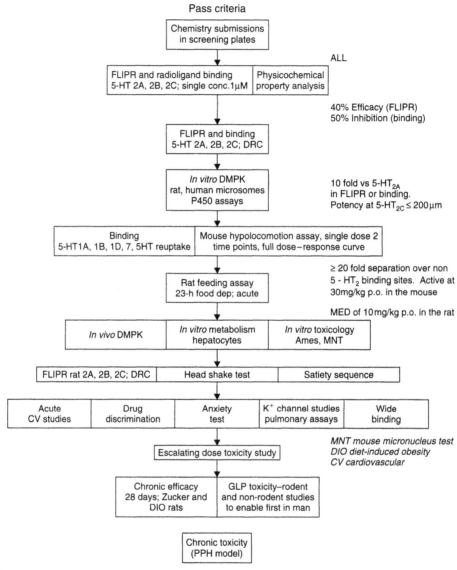

Figure 8.2 Example of a screening cascade for a 5-HT$_{2C}$ receptor agonist anti-obesity drug discovery program

cut-off criteria would move on to dose–response (DRC) assays in radioligand binding and *in vitro* and *in vivo* assessment of drug metabolism and pharmacokinetic (DMPK) properties using liver microsomes.

Successful compounds move on to selectivity screening in radioligand binding assays and to initial *in vivo* efficacy assays using rodent motor responses.

Active compounds are next assessed for efficacy *in vivo* in simple rodent feeding assays and, if successful, their *in vivo* DMPK and *in vitro* toxicology properties are established. Compounds that pass at this stage are examined in a range of increasingly complex rodent behavioral assays to assess efficacy (after both acute and chronic dosing) and side-effects. Success or failure in these assays will determine whether a compound is selected as a development candidate and submitted for the GLP toxicology studies required to enable Phase I clinical trials in man.[ii]

Screening strategies that have incorporated early *in vitro* and *in vivo* DMPK assays have been successful in many companies in reducing the number of clinical development failures due to poor bioavailability and pharmacokinetic properties.[218] Currently, the principal cause of failure in clinical development is efficacy and this is a particular problem for CNS targets where animal models have poor predictive validity.[218] Hence there is an urgent need to identify improved methods to reduce attrition in clinical development and experimental (or translational) medicine is a promising approach to this problem (see below).

PROOF OF CONCEPT FOR NOVEL ANTI-OBESITY DRUGS: THE ROLE OF EXPERIMENTAL MEDICINE STUDIES TO DETERMINE DRUG EFFICACY AND SIDE-EFFECTS

Obesity is currently one of the most active areas of research and development in the pharmaceutical industry, and many novel neurotransmitter and neuropeptide targets are being explored (see above). Approval of a new anti-obesity drug by regulatory authorities such as the FDA and the European Medicines Agency (EMEA) requires efficacy to be demonstrated in clinical trials over a 2-year treatment period. Inevitably, many late phase drug candidates fail and each failure can cost companies hundreds of millions of dollars. As outlined above, the principal cause of failure in clinical development is efficacy and this is a particular problem for CNS targets where animal models have poor predictive validity.[218] Therefore, there is an urgent need for improved experimental medicine methods for early and accurate identification of the potential efficacy of anti-obesity drug candidates. Similarly, it is clear from the progress of recent regulatory applications such as that of rimonabant (see above) that a comprehensive CNS side-effects package will be equally, if not more, important as efficacy measures in an NDA submission. Therefore, incorporation of a CNS side-effect battery in early experimental medicine efficacy studies will be valuable as it may be possible to define a therapeutic window where there are significant drug effects on food intake but minimal CNS adverse effects.

Measurement of Anti-Obesity Drug Efficacy in Experimental Medicine Studies: The UEM Approach

The effect of anti-obesity drugs on the pattern and structure (microstructure) of human eating behavior appears likely to be critical to their efficacy (see above). Most

---

[ii] Please refer to McEvoy and Freudenreich, Issues in the design and conductance of clinical trials, in Volume 1, Psychiatric disorders for further discussion of development strategies following the selection of a drug candidate from preclinical drug discovery programs.

if not all potential anti-obesity drugs, however, have undergone Phase III clinical trials without detailed examination of their effect on human eating patterns in experimental medicine studies. The effects of a drug on the microstructure of eating behavior (cumulative intake, eating rate, appetite, and satiety) can provide an important measure of potential efficacy, in addition to shedding light on mechanism of action and possible side-effects (see above and below).

To measure the microstructure of eating, automated methods of assessing intake have been established that generate cumulative intake curves, from which within-meal changes in eating rate can be identified. The UEM, first reported by Kissileff and colleagues was designed to continually measure food intake through the use of hidden scales, placed underneath a participants' plate.[153] The scales could be linked to a computer to allow continuous recording of food intake and to elicit subjective ratings of appetite and satiety from participants at regular intervals during a meal.[219,220]

Despite the sensitivity of UEM measures, the microstructure of eating behavior has not been routinely used as an experimental medicine assay for potential anti-obesity drugs. Recently, we completed a study to demonstrate that UEM-derived measures of meal intake can provide a valuable measure of the potential efficacy of anti-obesity drugs.[83,234] Sibutramine was used as a reference standard to establish the utility of within-meal analyses of eating behavior. Previous studies had used UEM equipment to measure food intake in volunteers after sibutramine treatment,[162] but had not used the UEM to investigate the effect of the drug on the microstructure of human eating behavior. The study was an outpatient, randomized, double-blind, placebo-controlled crossover trial in 30 obese female participants and examined the effects of two doses of sibutramine (10 mg and 15 mg/day for 7 days) on the microstructure of eating behavior. The results provided the most complete characterization of the anti-obesity effects of sibutramine to date and validated new UEM measures against which novel anti-obesity drugs can be compared in subsequent studies. Sibutramine at 10 mg and 15 mg reduced food intake and eating rate. In addition, 10 mg sibutramine reduced hunger later in the meal whereas 15 mg sibutramine increased fullness early in the meal both actions being consistent with enhanced within-meal satiation. The results provide novel evidence that decreased consumption of a test meal induced by sibutramine is primarily due to reduced eating rate, enhancing the deceleration in cumulative food intake within a meal associated with the development of satiety.[83,234]

The effects of sibutramine on eating behavior in the UEM provide a template with which to compare novel, appetite-reducing, anti-obesity compounds. Drug-induced changes in food intake and appetite appear to provide a reliable indication of potential weight-reducing efficacy. Therefore, the potential for late stage failure could be reduced by incorporation of a UEM efficacy model in Phase I experimental medicine studies.

Measurement of Anti-Obesity Drug Efficacy and CNS Side-Effects in Experimental Medicine Studies: The Emotional Test Battery Approach

There is considerable evidence that drugs used to treat anxiety and/or depression may have effects on the food intake and body weight of patients.[221] Conversely, it is possible that anti-obesity drugs may adversely affect brain reward pathways causing mood changes such as anhedonia and potentially anxiety and/or depression (e.g., CB1 receptor antagonists, see above). The prevalence of depression and anxiety is high among obese persons seeking treatment and in a recent UK study of 253 patients attending specialist obesity services, 48% had elevated scores for depression and 56% had elevated scores for anxiety on the hospital anxiety and depression (HAD) scale.[222] Therefore, it is valuable to ascertain at an early stage of the drug development process whether a candidate compound/mechanism could cause mood changes and/or exacerbate symptoms of depression or anxiety. For example, it has been reported that CB1 receptor knockout mice demonstrate a depressive state and that CB1 receptor blockade could induce depression.[223] Although the eligibility criteria reported in the RIO Phase III clinical trials with rimonabant do not mention a specific HAD score at which patients were excluded, the participants in these trials had a mean subscore for depression of approximately three, suggesting that efforts were made to enroll individuals with minimal or no depressive symptoms.[15-17] Consequently, it is uncertain whether the psychiatric safety data of rimonabant presented in the reports of the RIO trials can be generalized to general clinical practice in which a high percentage of obese patients are likely to have a HAD scale subscore for depression that is significantly higher than three.[224] This evidence has been interpreted as suggesting that there is not sufficient evidence from the RIO trials to determine whether rimonabant is safe for use in obese individuals with even mild depression.[224] This conclusion was reinforced by the decision of an FDA Advisory Panel to recommend that the FDA should not approve the drug in the United States, despite the fact that the drug is already on sale in Europe.

Thus in addition to the detection of early biomarkers for obesity it would be valuable if the early clinical development models were sensitive to the potential CNS side effects of anti-obesity compounds. It is also possible that rimonabant (or other compounds in development to treat obesity) could interact with the underlying psychopathology of obese patients who often have reduced self-esteem and other features of depression (see above). Therefore, incorporation of a CNS side-effect battery in UEM studies could be valuable in dose ranging for novel compounds as it may be possible to define a therapeutic window where there are significant effects on food intake but minimal neuropsychiatric adverse effects.

One potential approach would be to incorporate an emotional test battery such as that developed by Harmer and colleagues[225-227,236] in UEM efficacy studies. Negative biases in thinking, memory, and interpretation of events are believed to be important in maintaining the symptoms of depression and anxiety. Indeed, the main aim of cognitive therapy is to remediate these negative biases in the treatment of these disorders. It has recently been found that antidepressants have similar modulating actions on the neural processing of negative emotional information in healthy volunteers and depressed patients, which may be important in their therapeutic actions. Acute oral or intravenous administration of the SSRI, citalopram (Celexa®), increases the processing of anxiety-related stimuli in healthy volunteers.[226,232] This mechanism could underlie the known tendency of SSRIs to increase anxiety in patients early in their treatment. Similarly, acute administration of the 5-HT precursor, l-tryptophan in healthy volunteers also increases the recognition of fearful facial expressions[228] while tryptophan depletion, a dietary manipulation that lowers brain 5-HT synthesis, has the opposite effect.[229] In contrast, chronic administration of citalopram and the selective noradrenaline reuptake inhibitor, reboxetine (Edronax®/Vestra®,[230] decreases the salience of negative emotional stimuli present in facial expressions.[225-227] These results suggest that it is possible to measure bi-directional changes in emotional behavior (that is relevant to anxiety and depression) in healthy volunteers.

Therefore, experimental medicine studies that incorporate a UEM method together with an emotional test battery may prove valuable to determine the efficacy and CNS side-effect profile of anti-obesity drug candidates in Phase I clinical development.

SUMMARY

Obesity is now recognized as a significant predisposing factor to many important chronic diseases, and as a result there is increasing acceptance that medical therapy with drugs to help reduce and maintain body weight may sometimes be appropriate. Currently available treatments have modest efficacy, but concerns over adverse effects and lack of long-term data with regard to clinically relevant outcomes have limited their routine use. Recent and past experience with drugs in this therapeutic area has resulted in regulatory requirements which set a high standard for safety of such drugs. The high prevalence of obesity and perception of significant unmet need has led to increased scientific understanding of the biology of body weight regulation, particularly how this relates to eating behavior. Whilst potential anti-obesity drugs can target many components of energy regulation, the expression of eating behavior in the obese is critical to understanding the etiology of the condition. Drugs which target heightened preference for dietary fat, weakened satiety mechanisms, high disinhibition, and hunger, may be of particular therapeutic benefit. Assessment of the behavioral action of drugs at the preclinical and early clinical stages may be of particular benefit. Many new therapeutic targets have emerged as a result of this new biological understanding, and are proceeding rapidly to clinical development. To avoid previous pitfalls, obesity drug development programs must therefore be especially rigorous in combining the study of efficacy with understanding of potential adverse effects at each stage of development. As with other chronic conditions, it may also be necessary to look at the effects of combinations of treatments, especially as the best long-term outcomes have now been shown for bariatric surgery, which is more efficacious than any currently available drug therapy. Even for those drugs that prove efficacious and safe in the short-term, it will be necessary to conduct long-term studies to determine the effects on outcomes that matter to patients.

REFERENCES

1. World Health Organization (1998). Obesity: preventing and managing the global epidemic. Report of a WHO consultation on obesity, Geneva, 3–5 June, 1997. Geneva, World Health Organization.
2. Chan, J.M., Stampfer, M.J., Ribb, E.B., Willett, W.C., and Colditz, G.A. (1994). Obesity, fat distribution and weight gain as risk factors for clinical diabetes in man. *Diabetes Care*, 17:961–969.

3. Ford, E.S., Giles, W.H., and Dietz, W.H. (2002). Prevalence of the metabolic syndrome among US adults: Findings from the third National Health and Nutrition Survey. *JAMA*, 287:356–359.

4. National Institute for Health and Clinical Excellence (NICE) (2006). Obesity: Guidance on the prevention, identification, assessment and management of overweight and obesity in adults and children. http://www.nice.org.uk/guidance/index.jsp?action=byID&r=true&o=11000

5. The Treatment of Mild Hypertension Research Group (1991). The treatment of mild hypertension study. A randomized, placebo-controlled trial of a nutritional-hygienic regimen along with various drug monotherapies. Arch Intern Med, 151: 1413–1423.

6. Diabetes Prevention Program Research Group (2002) Reduction in the incidence of type 2 diabetes with lifestyle intervention or metformin. New Engl J Med, 346: 393–403.

7. Sjöstrom, L., Rissanen, A., Andersen, T., Boldrin, M., Golay, A., Koppeschaar, H.P., and Krempf, M. (1998). Weight loss and prevention of weight regain in obese patients: a 2-year, European, randomised trial of orlistat. *Lancet*, 352:167–172.

8. Torgerson, J.S., Hauptman, J., Boldrin, M.N., and Sjostrom, L. (2004). Xenical in the prevention of diabetes in obese subjects (XENDOS) study. *Diabetes Care*, 27:155–161.

9. Avenell, A., Brown, T.J., Mcgee, M.A., Campbell, M.K., Grant, A.M., Broom, J., Jung, R.T., and Smith, W.C.S. (2004). What are the long-term benefits of weight reducing diets in adults? A systematic review of randomized controlled trials. *J Hum Nutr Diet*, 17:317–335.

10. Norris, S.L., Zhang, X., Avenell, A., Gregg, E., Schmid, C.H., and Lau, J. (2005). Pharmacotherapy for weight loss in adults with type 2 diabetes mellitus. *Cochrane Database of Systematic Reviews*, 2.

11. Fujioka, K., Wickens, M., and Chong, E. (2001). Sibutramine vs placebo: Meta-analysis of weight changes in obese patients with and without type 2 diabetes. *Obes Res*, 9:153S.

12. McNulty, S.J., Ur, E., and Williams, G.A. (2003). randomized trial of sibutramine in the management of obese type 2 diabetic patients treated with metformin. *Diabetes Care*, 26:125–131.

13. James, W.P.T., Astrup, A., Finer, N., Hilsted, J., Kopelman, P., Rossner, S., Saris, W.H.M., and Van Gaal, L.F. (2000). Effect of sibutramine on weight maintenance after weight loss: A randomised trial. *Lancet*, 356:2119–2125.

14. Hazenberg, B.P., Wickens, M., and Chong, E. (2002). Meta-analysis of mean changes in resting blood pressure in obese hypertensive patients treated with sibutramine. *Am J Clin Nutr*, 75:89.

15. Van Gaal, L.F., Rissanen, A.M., Scheen, A.J., Ziegler, O., and Rossner, S. (2005). Effects of the cannabinoid-1 receptor blocker rimonabant on weight reduction and cardiovascular risk factors in overweight patients: 1-year experience from the RIO-Europe study. *Lancet*, 365:1389–1397.

16. Pi-Sunyer, X., Despres, J.P., Scheen, A., and Van Gaal, L. (2006a). Improvement of metabolic parameters with rimonabant beyond the effect attributable to weight loss alone: Pooled 1-year data from the RIO (rimonabant in obesity and related metabolic disorders) program. *J Am Coll Cardiol*, 47:362A.

17. Pi-Sunyer, F.X., Aronne, L.J., Heshmati, H.M., Devin, J., Rosenstock, J. *et al.* (2006b). For the RIO-North America Study Group. Effect of rimonabant, a cannabinoid-1 receptor blocker, on weight and cardiometabolic risk factors in overweight or obese patients: RIO-North America: A randomized controlled trial. *JAMA*, 295:761–775.

18. Scheen, A.J., Finer, N., Hollander, P., Jensen, M.D., and Van Gaal, L.F. (2006). Efficacy and tolerability of rimonabant in overweight or obese patients with type 2 diabetes: a randomised controlled study. *Lancet*, 368:1650–1660.

19. Colman, E. (2005). Anorectics on trial: A half century of federal regulation of prescription appetite suppressants. *Ann Intern Med*, 143:380–385.

20. Connolly, H.M., Crary, J.L., McGoon, M.D., Hensrud, D.D., Edwards, B.S., Edwards, W.D., and Schaff, H.V. (1997). Valvular heart disease associated with fenfluramine-phentermine. *New Engl J Med*, 37:581–588.

21. Campfield, L.A. and Smith, F.J. (1990). Transient declines in blood glucose signal meal initiation. *Int J Obes*, 14 (Suppl 3):15-31.

22. Tschop, M., Smiley, D.L., and Heiman, M.L. (2000). Ghrelin induces adiposity in rodents. *Nature*, 407:908-913.

23. Wren, A.M., Small, C.J., Ward, H.L., Murphy, K.G., Dakin, C.L., Taheri, S., Kennedy, A.R., Roberts, G.H., Morgan, D.G., Ghatei, M.A., and Bloom, S.R. (2000). The novel hypothalamic peptide ghrelin stimulates food intake and growth hormone secretion. *Endocrinology*, 141:4325-4328.

24. Asakawa, A., Inui, A., Kaga, T., Yuzuriha, H., Nagata, T., Ueno, N., Makino, S., Fujimiya, M., Niijima, A., Fujino, M.A., and Kasuga, M. (2001). Ghrelin is an appetite-stimulatory signal from stomach with structural resemblance to motilin. *Gastroenterology*, 120:337-345.

25. Cummings, D.E., Purnell, J.Q., Frayo, R.S., Schmidova, K., Wisse, B.E., and Weigle, D.S. (2001). A preprandial rise in plasma ghrelin levels suggests a role in meal initiation in humans. *Diabetes*, 50:1714-1719.

26. English, P.J., Ghatei, M.A., Malik, I.A., Bloom, S.R., and Wilding, J.P.H. (2002). Food fails to suppress ghrelin levels in obese humans. *J Clin Endocrinol Metabol*, 87:2984-2987.

27. Cummings, D.E., Clement, K., Purnell, J.Q., Vaisse, C., Foster, K.E., Frayo, R.S., Schwartz, M.W., Basdevant, A., and Weigle, D.S. (2002). Elevated plasma ghrelin levels in Prader-Willi syndrome. *Nat Med*, 8:643-644.

28. Dourish, C.T., Rycroft, W., and Iversen, S.D. (1989). Postponement of satiety by blockade of cholecystokinin (CCK-B) receptors. *Science*, 245:1509-1511.

29. Ballinger, A., Mcloughlin, L., Medbak, S., and Clark, M. (1995). Cholecystokinin is a satiety hormone in humans at physiological postprandial plasma-concentrations. *Clin Sci*, 89:375-381.

30. Bignon, E., Alonso, R., Arnone, M., Boigegrain, R., Brodin, R., Gueudet, C., Heaulme, M., Keane, P., Landi, M., Molimard, J.C., Olliero, D., Seban, M., Simiand, E., Soubrie, J.P., Pascal, M., Maffrand, J.P., and Le Fur, G. (1999). SR146131: A new potent, orally active, and selective nonpeptide cholecystokinin subtype 1 receptor agonist. II. In vivo pharmacological characterization. *J Pharmacol Exp Ther*, 289:752-761.

31. Turton, M.D., Oshea, D., Gunn, I., Beak, S.A., Edwards, C.M.B., Meeran, K., Cho, S.J., Taylor, G.M., Heath, M.M., Lambert, P.D., Wilding, J.P.H., Smith, D.M., Ghatei, M.A., Herbert, J., and Bloom, S.R. (1996). A role for glucagon-like peptide-1 in the central regulation of feeding. *Nature*, 379:69-72.

32. Batterham, R.L., Cowley, M.A., Small, C.J., Herzog, H., Cohen, M.A., Dakin, C.L., Wren, A.M., Brynes, A.E., Low, M.J., Ghatei, M.A., Cone, R.D., and Bloom, S.R. (2002). Gut hormone PYY3-36 physiologically inhibits food intake. *Nature*, 418:650-654.

33. Wynne, K., Park, A.J., Small, C.J., Patterson, M., Ellis, S.M., Murphy, K.G., Wren, A.M., Frost, G.S., Meeran, K., Ghatei, M.A., and Bloom, S.R. (2005). Subcutaneous oxyntomodulin reduces body weight in overweight and obese subjects – A double-blind, randomized, controlled trial. *Diabetes*, 54:2390-2395.

34. Riddle, M.C., Henry, R.R., Poon, T.H., Zhang, B., Mac, S.M., Holcombe, J.H., Kim, D.D., and Maggs, D.G. (2006). Exenatide elicits sustained glycaemic control and progressive reduction of body weight in patients with type 2 diabetes inadequately controlled by sulphonylureas with or without metformin. *Diabetes Metabol Res Rev*, 22:483-491.

35. Arch, J.R.S. (2002). Beta(3)-adrenoceptor agonists: Potential, pitfalls and progress. *Eur J Pharmacol*, 440:99-107.

36. Bray, G.A. (2000). Reciprocal relation of food intake and sympathetic activity: Experimental observations and clinical implications. *Int J Obes*, 24:S8-S17.

37. Zhang, Y., Proenca, R., Maffei, M., Barone, M., Leopold, L., and Friedman, J.M. (1994). Positional cloning of the mouse obese gene and its human homologue. *Nature*, 372:425-432.

38. Tartaglia, A., Dembski, M., Weng, X., Deng, N., Culpepper, J., Devos, R., Richards, G.J., Campfield, L.A., Clark, F.T., and Deeds, J. (1995). Identification and expression cloning of a leptin receptor, OB-R. *Cell*, 83:1263-1271.

39. Chua, S.C., White, D.W., Wupeng, X.S., Liu, S.M., Okada, N., Kershaw, E.E., Chung, W.K., Powerkehoe, L., Chua, M., Tartaglia, L.A., and Leibel, R.L. (1996). Phenotype of fatty due to gln269pro mutation in the leptin receptor (lepr). *Diabetes*, 45:1141-1143.

40. Zelissen, P.M.J., Stenlof, K., Lean, M.E.J., Fogteloo, J., Keulen, E.T.P., Wilding, J., Finer, N., Rossner, S., Lawrence, E., Fletcher, C., McCamish, M., De Bruin, T.W.A., Koppeschaar, H.P.F., Sjostrom, L., and Williams, G. (2005). Effect of three treatment schedules of recombinant methionyl human leptin on body weight in obese adults: A randomized, placebo-controlled trial. *Diabetes Obes Metabol*, 7:755-761.

41. Montague, C.T., Farooqi, I.S., Whitehead, J.P., Soos, M.A., Rau, H., Wareham, N.J., Sewter, C.P., Digby, J.E., Mohammed, S.N., Hurst, J.A., Cheetham, C.H., Earley, A.R., Barnett, A.H., Prins, J.B., and O'Rahilly, S. (1997). Congenital leptin deficiency is associated with severe early-onset obesity in humans. *Nature*, 387:903-908.

42. Farooqi, I.S., Jebb, S.A., Langmack, G., Lawrence, E., Cheetham, C.H., Prentice, A.M., Hughes, I.A., McCamish, M.A., and O'Rahilly, S. (1999). Effects of recombinant leptin therapy in a child with congenital leptin deficiency. *New Eng J Med*, 341:879-884.

43. Farooqi, I.S. and O'Rahilly, S. (2000). Recent advances in the genetics of severe childhood obesity. *Arch Dis Child*, 83:31-34.

44. Schwartz, M.W., Figlewicz, D.P., Baskin, D.G., Woods, S.C., and Porte, D.J. (1992). Insulin in the brain: A hormonal regulator of energy balance. *Endocr Rev*, 13:387-414.

45. Port, D. and Woods, S.C. (1981). Regulation of food intake and body weight by insulin. *Diabetologia*, 20:274-280.

46. Wilding, J.P.H. (2002). Neuropeptides and appetite control. *Diabet Med*, 19:619-627.

47. Jackson, R.S., Creemers, J.W.M., Ohagi, S., RaffinSanson, M.L., Sanders, L., Montague, C.T., Hutton, J.C., and O'Rahilly, S. (1997). Obesity and impaired prohormone processing associated with mutations in the human prohormone convertase 1 gene. *Nat Genet*, 16:303-306.

48. Kim, K.S., Larsen, N., Short, T., Plastow, G., and Rothschild, M.F. (2000). A missense variant of the porcine melanocortin-4 receptor (MC4R) gene is associated with fatness, growth, and feed intake traits. *Mamm Genome*, 11:131-135.

49. Nakazato, M., Murakami, N., Date, Y., Kojima, M., Matsuo, H., Kangawa, K., and Matsukura, S. (2001). A role for ghrelin in the central regulation of feeding. *Nature*, 409:194-198.

50. Halford, J.C.G., Harrold, J.A., Lawton, C.L., and Blundell, J.E. (2005). Serotonin (5-HT) drugs: Effects on appetite expression and use for the treatment of obesity. *Curr Drug Target*, 6:201-213.

51. Dourish, C.T. and Hutson, P.H. (1989). The role of 5-HT1B receptors in the paraventricular nucleus of the hypothalamus in the control of feeding [letter]. *Neurobiol Aging*, 10:209.

52. Grignaschi, G., Sironi, F., and Samanin, R. (1996). Stimulation of 5-HT2a receptors in the para-ventricular hypothalamus attenuates neuropeptide Y-induced hyperphagia through activation of corticotropin-releasing factor. *Brain Res*, 708:173-176.

53. Nonogaki, K., Strack, A.M., Dallman, M.F., and Tecott, L.H. (1998). Serotonin 5-HT2c receptor mutant mice develop obesity a prediabetic state and moderate leptin insensitivity. *Diabetes*, 47:1237.

54. Vickers, S.P., Easton, N., Webster, L.J., Wyatt, A., Bickerdike, M.J., Dourish, C.T., and Kennett, G.A. (2003). Oral administration of the 5-HT2C receptor agonist, mCPP, reduces body weight gain in rats over 28 days as a result of maintained hypophagia. *Psychopharmacology*, 167:274-280.

55. Shire, D., Calandra, B., RinaldiCarmona, M., Oustric, D., Pessegue, B., BonninCabanne, O., LeFur, G., Caput, D., and Ferrara, P. (1996). Molecular cloning, expression and function of

the murine CB2 peripheral cannabinoid receptor. *Biochimica et Biophysica Acta-Gene Structure and Expression*, 1307:132–136.

56. Williams, C.M. and Kirkham, T.C. (1999). Anandamide induces overeating: mediation by central cannabinoid (CB1) receptors. *Psychopharmacology*, 143:315–317.

57. Trillou, C.R., Delgorge, C., Menet, C., Arnone, M., and Soubrie, P. (2004). CB1 cannabinoid receptor knockout in mice leads to leanness, resistance to diet-induced obesity and enhanced leptin sensitivity. *Int J Obes*, 28:640–648.

58. Han, T.S., van Leer, E.M., Seidell, J.C., and Lean, M.E.J. (1995). Waist circumference action levels in the identification of cardiovascular risk factors: Prevalence study in a random sample. *BMJ*, 311:1401–1405.

59. Jebb, S.A. (1998). Measuring body composition: From the laboratory to the clinic. In Kopelman, P. and Stock, M. (eds.), *Clinical Obesity*. Oxford, Blackwell Science, pp. 18–49.

60. Noreen, E.E. and Lemon, P.W.R. (2006). Reliability of air displacement plethysmography in a large, heterogeneous sample. *Med Sci Sports Exerc*, 38:1505–1509.

61. Ward, L.C., Doman, D., and Jebb, S.A. (2000). Evaluation of a new bioelectrical impedance instrument for the prediction of body cell mass independently of height or weight. *Nutrition*, 16:745–750.

62. Kamel, E.G., McNeill, G., and Van Wijk, M.C.W. (2000). Change in intra-abdominal adipose tissue volume during weight loss in obese men and women: Correlation between magnetic resonance imaging and anthropometric measurements. *Int J Obes*, 24:607–613.

63. Krssak, M., Falk, P.K., Dresner, A., DiPietro, L., Vogel, S.M., Rothman, D.L., Roden, M., and Shulman, G.I. (1999). Intramyocellular lipid concentrations are correlated with insulin sensitivity in humans: A 1H NMR spectroscopy study. *Diabetologia*, 42:113–116.

64. Yang, R.Z., Lee, M.J., Hu, H., Pray, J., Wu, H.B., Hansen, B.C., Shuldiner, A.R., Fried, S.K., McLenithan, J.C., and Gong, D.W. (2006). Identification of omentin as a novel depot-specific adipokine in human adipose tissue: Possible role in modulating insulin action. *Am J Physiol Endocrinol Metabol*, 290:E1253–E1261.

65. Batista, C.M.D.S., Yang, R.Z., Lee, M.J., Glynn, N.M., Yu, D.Z., Pray, J., Ndubuizu, K., Patil, S., Schwartz, A., Kligman, M., Fried, S.K., Gong, D.W., Shuldiner, A.R., Pollin, T.I., and McLenithan, J.C. (2007). Omentin plasma levels and gene expression are decreased in obesity. *Diabetes*, 56:1655–1661.

66. Marti, A., de Miguel, C., Jebb, S.A., Lafontan, M., Laville, M., Palou, A., Remesar, X., Trayhurn, P., and Martinez, J.A. (2000). Methodological approaches to assess body-weight regulation and aetiology of obesity. *Proc Nutr Soc*, 59:405–411.

67. Gibney, E.R., Murgatroyd, P., Wright, A., Jebb, S., and Elia, M. (2003). Measurement of total energy expenditure in grossly obese women: Comparison of the bicarbonate–urea method with whole-body calorimetry and free-living doubly labelled water. *Int J Obes*, 27:641–647.

68. DeFronzo, R.A., Tobin, J.D., and Andres, R. (1979). The glucose clamp technique: A method of quantifying insulin secretion and resistance. *Am J Physiol*, 237:E214–E223.

69. Matthews, D.R., Hosker, J.P., Rudenski, A.S., Naylor, B.A., Treacher, D.F., and Turner, R.L. (1985). Homeostasis model assessment: Insulin resistance and beta-cell function from fasting plasma glucose and insulin concentrations in man. *Diabetologia*, 28:412–419.

70. Halford, J.C.G. and Blundell, J.E. (2001). Separate systems for serotonin and leptin in appetite control. *Ann Med*, 32:222–232.

71. Blundell, J.E., Levin, F., King, N.A., Barkeling, B., ustafson, T., Hellstrom, P.M. Holst, J., Nåsland, E. (2008) Over-consumption and obesity: Peptides and susceptiblity to weight gain. Peptides (in press).

72. Farooqi, S.F., Matarese, G., Lord, G.M., Keogh, J.M., Lawrence, E., Agwu, C. *et al.* (2002). Beneficial effects of leptin on obesity, T cell hyporesponsiveness, and neuroendocrine/metabolic dysfunction of human congenital leptin deficiency. *J Clin Investigat*, 110:1093–1103.

73. Allison, D.B., Kapiro, J., Korkeila, M., Koskenvuo, M., Neale, M.C., and Hayakawa, K. (1996). The heritability of body mass index among an international sample of monozygotic twins reared apart. *Int J Obes*, 20:501-506.

74. Sorensen, T.I.A., Holst, C., and Stunkard, A.J. (1998). Adoption study of environmental modifications on the generic influences of obesity. *Int J Obes*, 22:73-81.

75. Segal, N.L. and Allison, D.B. (2002). Twins and virtual twins: Bases of relative body weight revisited. *Int J Obes*, 26:437-441.

76. Mutch, D.M. and Clément, K. (2006). Genetics of human obesity. *Best Pract Res Clin Endocrinol Metabol*, 20:647-664.

77. Rankinen, T., Zuberi, A., Chagnon, Y.C., Weisnagal, S.J., Argyropoulos, G., Walts, B., and Prérusse C. Bouchard, L. (2006). The human obesity gene map: The 2005 update. *Obesity*, 14:529-644.

78. O'Rahilly, S., Farooqi, S.I. (2006). Genetics of obesity, philosophical transactions of the royal B society. doi:10.1098/rsrb.2006.1850.

79. Stunkard, A. and Kaplan, D. (1977). Eating in public places: A review of reports of the direct observation of eating behavior. *Int J Obes*, 1:89-101.

80. Blundell, J.E., Stubbs, R.J., Golding, C., Croden, F., Alam, R., Whybrow, S., Le Noury, J., and Lawton, C.L. (2005). Resistance and susceptibility to weight gain: Individual variability in response to a high fat diet. *Physiol Behav*, 86:614-622.

81. Meyer, J.E. and Pudel, V. (1972). Experimental studies on food-intake in obese and normal weight subjects. *J Psychosom Res*, 16:305-308.

82. Halford, J.C.G., Harrold, J.A., Boyland, E.J., Lawton, C.L., and Blundell, J.E. (2007a). Serotonergic drugs – Effects on appetite expression and use for the treatment of obesity. *Drugs*, 67:27-55.

83. Halford, J.C.G., Boyland, E., Dovey, T.M., Huda, M., Dourish, C.T., Dawson, G.R., and Wilding, J.P.H. (2007b). A double-blind, placebo-controlled crossover study to quantify the effects of sibutramine on energy intake and energy expenditure in obese subjects during a test meal using a Universal Eating Monitor (UEM) method. *Int J Obes*, 31(supplement 1). T3PO.188s151

84. Pearcey, S.M. and de Castro, J.M. (2002). Food intake and meal patterns of weight-stable and weight-gaining persons. *Am J Clin Nutr*, 76:107-112.

85. Rissanen, A., Hakala, P., Lissner, L., Mattlar, C.E., Koskenvuo, M., and Ronnemaa, T. (2002). Acquired preference especially for dietary fat and obesity: A study of weight discordant twin pairs. *Int J Obes*, 26:973-977.

86. Bray, G.A. and Popkin, B.M. (1998). Dietary fat intake does affect obesity! *Am J Clin Nutr*, 68:1157-1173.

87. Bray, G.A., Paeratakul, S., and Popkin, B.M. (2004). Dietary fat and obesity: A review of animal, clinical and epidemiological studies. *Physiol Behav*, 83:549-555.

88. Bell, E.A. and Rolls, B.J. (2001). Energy density of foods affects energy intake across multiple levels of fat content in lean and obese women. *Am J Clin Nutr*, 73:1010-1018.

89. Kral, T.V.E., Roe, L.S., and Rolls, B.J. (2004). Combined effects of energy density and portion size on energy intake in women. *Am J Clin Nutr*, 79:962-968.

90. Geliebter, A. and Hashim, S.A. (2001). Gastric capacity in normal obese, and bulimic women. *Physiol Behav*, 74:743-746.

91. Geliebter, A., Yahav, E.K., Gluck, M.E., and Hashim, S.A. (2004). Gastric capacity, test meal intake, and appetitive hormones in binge eating disorder. *Physiol Behav*, 81:735-740.

92. De Graaf, C., Blom, W.A.M., Smeets, P.A.M., Staflue, A., and Hendricks, H.F.J. (2004). Biomarkers of satiation and satiety. *Am J Clin Nutr*, 17:949-961.

93. Huda, M.S.B., Wilding, J.P.H., and Pinkney, J.H. (2006). Gut peptides and the regulation of appetite. *Obes Rev*, 7:163-182.

94. Lowe, M.R. and Levine, A.S. (2005). Eating motives and the controversy over dieting: Eating less than needed versus less than wanted. *Obes Res*, 31:797-806.

95. Hill, A.J., Magson, L.D., Lambers, A.C., and Blundell, J.E. (1984). Hunger and palatability: Tracking ratings of subjective experience before, during and after the consumption of preferred and unpreferred foods. *Appetite*, 5:361–371.

96. Yeomans, M.R., Gray, R.W., Mitchel, C.J., and True, A. (1997). Independent effects of palatability and within-meal pauses on intake and appetite ratings in human volunteers. *Appetite*, 29:61–76.

97. De Graaf, C., de Jon, L.S., and Lambers, A.C. (1999). Palatability affects satiation but not satiety. *Physiol Behav*, 66:681–688.

98. Yeomans, M.R., Blundell, J.E., and Lesham, M. (2004). Palatability: Response to nutritional need or need-free stimulation of appetite. *Br J Nutr*, 92:s3–s14.

99. Drewnowski, A., Brunzell, J.D., Sande, K., Iverise, K., and Greenwood, M.R.C. (1985). Sweet tooth reconsidered: Taste responsiveness in human obesity. *Physiol Behav*, 35:617–622.

100. Rissanen, A., Hakala, P., Lissner, L., Mattlar, C.E., Koskenvuo, M., and Ronnemaa, T. (2002). Acquired preference especially for dietary fat and obesity: A study of weight discordant twin pairs. *Int J Obes*, 26:973–977.

101. Mela, D.J. and Sacchetti, D.A. (1991). Sensory preference for fats: Relationship with diet and body composition. *Am J Clin Nutr*, 53:908–915.

102. Blundell, J.E. and Finlayson, G. (2004). Is susceptibility to weight gain characterised by homeostatic or hedonic risk factors for over-consumption? *Physiol Behav*, 82:21–25.

103. Karlsson, J., Persson, L-O., Sjostrom, L., and Sullivan, M. (2000). Psychometric properties and factor structure of the three-factor eating questionnaire (TFEQ) in obese men and women, results from the Swedish obese subjects (SOS) study. *Int J Obes*, 24:1715–1725.

104. Barkeling, B., King, N.E., Näslund, E., and Blundell, J.E. (2007). Characterization of obese individuals who claim to detect no relationship between their eating pattern and sensations of hunger or fullness. *Int J Obes*, 31:45–439.

105. Provencher, V., Drapeau, V., Tremblay, A., Després, J.-P., and Lemieux, S. (2003). Eating behaviors and indexes of body composition in mean and women form the Quebéc family study. *Obes Res*, 11:783–792.

106. Dykes, J., Brunner, E.J., Martikeainene, P.T., and Wardle, J. (2000). Socioeconomic gradient in body size and oboist among women: The role of dietary restraint, disinhibition and hunger in the Whitehall II study. *Int J Obes*, 28:262–268.

107. Lawson, O.J., Williamson, D.A., Champagne, C.M., DeLany, J.P., Brooks, E.R., Howat, P.M., Wozniak, P.J., Bray, G.A., and Ryan, D.H. (1995). The association of body weight, dietary intake, and energy expenditure with dietary restraint and disinhibition. *Obes Res*, 3:153–161.

108. Williamson, D.A., Lawson, O.J., Brooks, E.R., Wozniak, P.J., Ryan, D.H., Bray, G.A., and Duchmann, E.G. (1995). Association of body mass with dietary restraint and disinhibition. *Appetite*, 25:31–41.

109. Hays, N.P., Bathaloon, G.P., McCrory, M.A., Roubeboff, R., Lipman, R., and Roberts, S.R. (2002). Eating behaviour correlates of adult weight gain and obesity in healthy women aged 55–65 y. *Am J Clin Nutr*, 75:476–483.

110. Bellisle, F., Clément, K., Le Barzic, M., Le Call, A., Guy-Grand, B., and Basdevant, A. (2004). Eating inventory and body adiposity from leanness to massive obesity: A study of 2509 adults. *Obes Res*, 12:2023–2030.

111. Borg, P., Fogelholm, M., and Kukkonen-Harjula, K. (2004). Food selection and eating behaviour during weight maintenance intervention and 2-y follow-up in obese men. *Int J Obes*, 28:148–1554.

112. Provencher, V., Pérusse, L., Buchard, L., Drapeau, V., Bouchards, C., Rice, T., Rao, D.C., Trembley, A., Després, J.-P., and Lemieux, S. (2005). Familial resemblance in eating behaviors in men and women form the Quebéc family study. *Obes Res*, 13:1624–1629.

113. Bouchard, L., Drapeau, V., Provencher, V., Lemieux, Y., Rice, T., Rao, D.C., Vohl, M-C., Tremblay, A., Bouchards, C., and Pérusse, L. (2004). Neuromedin ß: A strong candidate gene linking eating behaviours and susceptibility to obesity. *Am J Clin Nutr*, 80:1478–1486.

114. Stunkard, A.J. (1959). Eating patterns in obesity. *Psychiatry Quart*, 33:284–292.

115. Dingemans, A.E., Bruna, M.J., and van Furth, E.F. (2002). Binge eating disorder: A review. *Int J Obes*, 26:299–307.

116. Hsu, L.K.G., Mulliken, B., McDonagh, B., Das, S.K., Rand, W., Fairburn, C.G., Rolls, B., McCrory, M.A., Saltzman, E., Shikora, S., Dwyer, J., and Roberts, S. (2002). Binge eating disorder in extreme obesity. *Int J Obes*, 26:1398–1403.

117. Devlin, M.J., Goldstein, J.A., and Dobrow, I. (2003). What is this thing called BED? Current status of binge eating disorder nosology. *Int J Eat Disord*, 34:s2–s18.

118. De Zwaan, M., Mitchell, J.E., Howell, M., Monson, N., Swan-Kremeier, L., Crosby, R.D., and Seim, H.C. (2003). Characteristics of morbidly obese patients before gastric bypass surgery. *Comp Psychiatry*, 44:428–434.

119. Grucza, R.A., Przybeck, T.R., and Cloninger, C.R. (2007). Prevalence and correlates of binge eating disorder in a community sample. *Comp Psychiatry*, 48:124–131.

120. Walsh, B.T. and Boudreau, G. (2003). Laboratory studies of binge eating disorder. *Int J Eat Disord*, 34:S30–S38.

121. Crow, S., Kendell, D., Praus, B., and Thuras, P. (2001). Binge eating and other psychopathology in patients with type II diabetes mellitus. *Int J Eat Disord*, 30:222–226.

122. Devlin, M.J., Walsh, B.T., Guss, J.L., Kissileff, H.R., Liddle, R.A., and Petkova, E. (1997). Postprandial cholecystokinin release and gastric emptying in patients with bulimia nervosa. *Am J Clin Nutr*, 65:114–120.

123. Appolinario, J.C., Godoy-Matos, A., Fontenelle, L.F., Carraro, L., Cabral, M., Vieira, A., and Coutinho, W. (2002). An open-label trial of sibutramine in obese patients with binge-eating disorder. *J Clin Psychiatry*, 63:28–30.

124. Mitchell, J.E., Gosnell, B.A., Roerig, J.L., de Zwaan, M., Wonderlich, S.A., Crosby, R.D., Burgard, M.A., and Wambach, B.N. (2003). Effects of sibutramine on binge eating, hunger, and fullness in a laboratory human feeding paradigm. *Obes Res*, 11:599–602.

125. Appolinario, J.C. and McElroy, S.L. (2004). Pharmacological approaches in the treatment of binge eating disorder. *Curr Drug Target*, 5:301–307.

126. Milano, W., Petrella, C., Casella, A., Capasso, A., Carrino, S., and Milano, L. (2005). Use of sibutramine, an inhibitor of the reuptake of serotonin and noradrenaline, in the treatment of binge eating disorder: A placebo-controlled study. *Adv Ther*, 22:25–31.

127. Dobrain, A.D., Davies, M.J., Prewitt, R.L., and Lauterio, T.L. (2000). Development of hypertension in a rat model of diet-induced obesity. *Hypertension*, 35:1009–1015.

128. Keenan, K.P., Hoe, C.M., Mixson, L., McCoy, C.L., Coleman, J.B., Mattison, B.A., Ballam, G.A., Gumprecht, L.A., and Soper, K.A. (2005). Diabesity: A polygenic model of dietary induced obesity form ad libitum overfeeding of Sprague-Dawley rats and its modulation by moderate and marked dietary restriction. *Toxicol Pathol*, 33:650–674.

129. Woods, S.C., Seeley, R.J., Riching, P.A., D'Aessio, D., and Tso, P. (2003). A controlled high-fat diet induces an obese syndrome in rats. *J Nutr*, 133:1081–1087.

130. Levin, B.E., Dunn, A., Meynell, A., Balkan, B., and Keesey, R.E. (1997). Selective breeding for diet-induced obesity and resistance in Sprague-Dawley rats. *Am J Physiol (R. I. C. P.)*, 273:R725–R730.

131. Harrold, J.A., Widdowson, P.S., Clapham, J.C., and Williams, G. (2000). Individual severity of dietary obesity in unselected Wistar rats. *Am J Physiol (E.M.)*, 279:E340–E347.

132. Vasselli, J.R., Wendruch, R., Heymsfield, S., Pi-Sunyer, X., Boozer, C.N., Yi, N., Wang, C., Pietrobelli, A., and Allison, D.B. (2005). Intentional weight loss reduced mortality rate in a rodent model of obesity. *Obes Res*, 13:693–702.

133. Blundell, J.E. and Latham, C.J. (1979). Serotonergic influences on food intake: Effect of 5-HT on parameters of feeding behaviour in deprived and free-feeding rats. *Pharmacol Biochem Behav*, 11:431–437.

134. Blundell, J.E. (1986). Serotonin manipulations and the structure of feeding-behaviour. *Appetite*, 7:39–56.

135. Blundell, J.E., Rogers, P.J., and Hill, A.J. (1985). Behavioural structure and mechanisms of anorexia: Calibration of normal and abnormal inhibition of eating. *Brain Res Bull*, 15:319–326.

136. Antin, J., Gibbs, J., Holt, J., Young, R.C., and Smith, G.P. (1975). Cholecystokinin elicits the complete behavioural sequence of satiety in rats. *J Comp Physiol Psychol*, 89:760–784.

137. Gibbs, J. and Smith, G.P. (1982). Gut peptides and food in the gut produce similar satiety effects. *Peptides*, 3:553–557.

138. Kitchener, S.J. and Dourish, C.T. (1994). An examination of the behavioural specificity of hypophagia induced by 5-HT1B, 5-HT1C and 5-HT2 receptor agonists using the post-prandial satiety sequence in rats. *Psychopharmacology*, 113:369–377.

139. Phillips, G.D., Howes, S.R., Whitelaw, R.B., Robbins, T.W., and Everitt, B.J. (1995). Analysis of the effects of intraaccumbens SKF-38393 and LY-171555 upon the behavioral satiety sequence. *Psychopharmacology*, 117:82–90.

140. Ishii, Y., Blundell, J.E., Halford, J.C.G., Upton, N., Porter, R., John, A., Jeffrey, P., Summerfield, S., and Rodgers, R.J. (2005). Anorexia and weight loss in male rats 24h following single dose treatment with orexin-1 receptor antagonist SB-334867. *Behav Brain Res*, 157:331–341.

141. Scott, V., Kimura, N., Stark, J.A., and Luckman, S.M. (2005). Intravenous peptide YY3-36 and Y-2 receptor antagonism in the rat: Effects on feeding behaviour. *J Neuroendocrinol*, 17:452–457.

142. Tallett, A.J., Blundell, J.E., and Rodgers, R.J. (2007). Grooming, scratching and feeding: Role of response competition in acute anorectic response to rimonabant in male rats. *Psychopharmacology*, 195:27–39.

143. Hill, A.J., Rogers, P.J., and Blundell, J.E. (1995). Techniques for the experimental measurement of human feeding behaviour and food intake: A practice guide. *Int J Obes*, 19:361–375.

144. Rogers, P.J. and Blundell, J.E. (1979). Effect of anorexic drugs on food-intake and the micro-structure of eating in human-subjects. *Psychopharmacology*, 66:159–165.

145. Blundell, J.E. and Rogers, P.J. (1980). Effects of anorexic drugs on food intake, food selections and preferences and hunger motivation and subjective experiences. *Appetite*, 1:151–165.

146. Drapeau, V., King, N., Hetherington, M., Doucet, E., Blundell, J., and Tremblay, A. (2007). Appetite sensations and satiety quotient: Predictors of energy intake and weight loss. *Appetite*, 48:159–166.

147. Blundell, J.E. and Hill, A.J. (1988). On the mechanism of action of dexfenfluramine: Effect on alliesthesia and appetite motivation in lean and obese subjects. *Clin Neuropharmacol*, 11(Suppl 1):S121–S134.

148. Goodall, E. and Silverstone, T. (1988). Differential effect of d-fenfluramine and metergoline on food intake in human subjects. *Appetite*, 11:215–288.

149. Rolls, B.J., Shide, D.J., Thorwart, M.L., and Ulbrecht, J.S. (1998). Sibutramine reduces food intake in non-dieting women with obesity. *Obes Res*, 6:1–11.

150. McGuirk, J. and Silverstone, T. (1990). The effect of 5-HT re-uptake inhibitor fluoxetine on food intake and body weight in healthy male subjects. *Int J Obes*, 14:61–372.

151. Lawton, C.L., Wales, J.K., Hill, A.J., and Blundell, J.E. (1995). Serotoninergic manipulation meal-induced satiety and eating pattern: Effects of fluoxetine in obese female subjects. *Obes Res*, 3:345–356.

152. Erondu, N., Grantz, I., Suryawanshi, S., Malick, M., Addy, C., Cote, J., Bray, G., Fujoka, K., Bays, H., Hollander, P., Sanabria-Bohórquez, S.M., Eng, A.S., Längström, B., Hargeaves, R.J., Burns, S.,

Kanatani, A., Fukami, T., MacNeil, D.J., Gottesdiener, K.M., Kaufman, K.D., and Heymsfield, S.B. (2006). Neuropeptide Y5 receptor antagonist does not induce clinically meaningful weight loss in overweight and obese adults. *Cell Metabol*, 4:275-282.

153. Kissileff, H.R., Klinsberg, G., and Van Itallie, T.B. (1980). Universal eating monitor for the continuous recording of solid or liquid consumption in man. *Am J Physiol (R.I.C.P.)*, 238:R14-R22.

154. Guss, J.L. and Kissileff, H.R. (2000). Microstructural analysis of human ingestive patterns: From description to mechanistic hypotheses. *Neurosci Biobehav Rev*, 24:261-268.

155. Westerterp-Plantenga, M.S. (2000). Eating behaviour in humans, characterized by cumulative food intake curves – a review. *Neurosci Biobehav Rev*, 24:239-248.

156. Hubel, R., Laessle, R.G., Lehrke, S., and Jass, J. (2006). Laboratory measurement of cumulative food intake in humans: Results on the reliability. *Appetite*, 46:57-62.

157. Stunkard, A., Coll, M., Lundquist, S., and Meyers, A. (1980). Obesity and eating style. *Arch Gen Psychiatry*, 37:1127-1129.

158. Laessle, R.G., Uhl, H., and Lindel, B. (2001a). Parental influences on eating behaviour in obese and nonobese preadolescents. *Int J Eat Disord*, 30:447-453.

159. Laessle, R.G., Uhl, H., Lindel, B., and Müller, A. (2001b). Parental influences on eating behaviour in obese and nonobese children. *Int J Obes*, 25(Suppl 1):S60-S62.

160. Lindgren, A.C., Barkeling, B., Hägg, A., Ritzén, E.M., Marcus, C., and Rössner, S. (2000). Eating behaviour in Prader-Willi syndrome, normal weight and obese control groups. *J Pediatr*, 137:50-55.

161. Näslund, E., Gutniak, M., Skogar, S., Rössner, S., and Hellstrom, P.M. (1998). Glucagon-like peptide 1 increases the period of postprandial satiety and slows gastric emptying in obese men. *Am J Clin Nutr*, 68(3):525-530.

162. Barkeling, B., Elfhag, K., Rooth, P., and Rössner, S. (2003). Short-term effects of sibutramine (Reductil TM) on appetite and eating behaviour and the long-term therapeutic outcomes. *Int J Obes*, 27:693-700.

163. Yeomans, M.R. and Gray, R.W. (1997). Effects of naltrexone on food intake and changes in subjective appetite during eating: Evidence for opioid involvement in the appetizer effect. *Physiol Behav*, 62:15-21.

164. Kopelman, P., Bryson, A., Hickling, R., Rissanen, A., Rossner, S., Toubro, S., and Valensi, P. (2007). Cetilistat (ATL-962), a novel lipase inhibitor: A 12-week randomized placebo-controlled study of weight reduction in obese patients. *International Journal of Obesity*, 31:494-499.

165. Alizyme media release (2006). Alizyme announces successful end of II Phase meeting FDA with for cetilistat. http://www.alizyme.com, 2 August 2006.

166. Hauser, R. A., Salin, L., Juhel, N., Konyago, V. L. and the NS 2330 Monotherapy PD Study Group (2007). Randomized trial of the triple monoamine reuptake inhibitor NS 2330 (Tesofensine) in early Parkinson's disease. Movement Disorders, 22 359-365.

167. NeuroSearch media release (2007). NeuroSearch announces breakthrough tesofensine results from clinical Phase IIb study in obesity ("TIPO-1"). http://www.neurosearch.com, 17 September 2007.

168. Ramage, A.G. (2001). Central cardiovascular regulation and 5-hydroxytryptamine receptors. *Brain Res Bull*, 56:425-439.

169. Murphy, D.L., Lesch, K.P., Aulakh, C.S., and Pigott, T.A. (1991). III. Serotonin-selective aryl-piperazines with neuroendocrine, behavioral, temperature, and cardiovascular effects in humans. *Pharmacol Rev*, 43:527-552.

170. Seletti, B., Benkelfat, C., Blier, P., Annable, L., Gilbert, F., and de Montigny, C. (1995). Serotonin 1A receptor activation by flesinoxan in humans. Body temperature and neuroendocrine responses. *Neuropsychopharmacology*, 13:93-104.

171. Pitchot, W., Wauthy, J., Legros, J.-J., and Ansseau, M. (2004). Hormonal and temperature responses to flesinoxan in normal volunteers: An antagonist study. *Eur Neuropsychopharmacology*, 14:151–155.

172. Babbs, A.J., Cheetham, S.C., Guillaume, P., Thomas, G.H., and Widdowson, P.S. (2006). PSN S1 and PSN S2 are novel 5-HT(1A) agonists and norepinephrine reuptake inhibitors which show anti-feeding efficacy and superior cardiovascular safety to sibutramine in rats. 36th Annual Meeting of the Society for Neuroscience: abstr. 62.3/X4.

173. Thomas, G.H., Babbs, A.J., Jackson, H.C., Sørensen, R.V., Widdowson, P.S. (2006). PSN S1 and PSN S2 are novel 5-HT(1A) agonists and norepinephrine reuptake inhibitors that reduce food intake and body weight in animal models of obesity. 36th Annual Meeting of the Society for Neuroscience: abstract 62.4/X5.

174. Hoyer, D., Hannon, J.P., and Martin, G.R. (2002). Molecular pharmacological and functional diversity of 5-HT receptors. *Pharmacol Biochem Behav*, 71:533–554.

175. Dourish, C.T. (1995). Multiple serotonin receptors: Opportunities for new treatments for obesity? *Obes Res*, 3(Suppl 4):449S–462S.

176. Bickerdike, M.J., Vickers, S.P., and Dourish, C.T. (1999). 5-HT2C receptor modulation and the treatment of obesity. *Diabetes Obes Metabol*, 1:207–214.

177. Vickers, S.P. and Dourish, C.T. (2004). Serotonin receptor ligands and the treatment of obesity. *Curr Opin Investig Drugs*, 5:377–388.

178. Hirst, W.D., Abrahamsen, B., Blaney, F.E., Calver, A.R., Aloj, L., Price, G.W., and Medhurst, A.D. (2003). Differences in the central nervous system distribution and pharmacology of the mouse 5-hydroxytryptamine-6 receptor compared with rat and human receptors investigated by radioligand binding, site-directed mutagenesis, and molecular modeling. *Mol Pharmacol*, 64:1295–1308.

179. Fisas, A., Codony, X., Romero, G., Dordal, A., Giraldo, J., Mercé, R., Holenz, J., Vrang, N., Sørensen, R.V., Heal, D., Buschmann, H., and Pauwels, P.J. (2006). Chronic 5-HT6 receptor modulation by E-6837 induces hypophagia and sustained weight loss in diet-induced obese rats. *Br J Pharmacol*, 148:973–983.

180. Vickers, S.P., Dourish, C.T., and Kennett, G.A. (2001). Evidence that hypophagia induced by D-fenfluramine and D-norfenfluramine in the rat is mediated by 5-HT2C receptors. *Neuropharmacology*, 41:200–209.

181. Vickers, S.P., Clifton, P.G., Dourish, C.T., and Tecott, L.H. (1999). Reduced satiating effect of D-fenfluramine in serotonin 5-HT2C receptor mutant mice. *Psychopharmacology*, 43:309–314.

182. Hewitt, K.N., Lee, M.D., Dourish, C.T., and Clifton, P.G. (2002). Serotonin 2C receptor agonists and the behavioural satiety sequence in mice. *Pharmacol Biochem Behav*, 71:691–700.

183. Goodall, E.M., Cowen, P.J., Franklin, M., and Silverstone, T. (1993). Ritanserin attenuates anorectic, endocrine and thermic responses to d-fenfluramine in volunteers. *Psychopharmacology*, 112:461–466.

184. Sargent, P.A., Sharpley, A.L., Williams, C., Goodall, E.M., and Cowen, P.J. (1997). 5-HT2C receptor activation decreases appetite and body weight in obese subjects. *Psychopharmacology*, 133:309–312.

185. Vickers, S.P., Benwell, K.R., Porter, R.H., Bickerdike, M.J., Kennett, G.A., and Dourish, C.T. (2000). Comparative effects of continuous infusion of mCPP, Ro 60-0175 and D fenfluramine on food intake, water intake, body weight and locomotor activity in rats. *Br J Pharmacol*, 130:1305–1314.

186. Bickerdike, M.J., Adams, D.R., Bentley, J., Benwell, K.R., Cliffe, I.A., Kennett, G.A., Knight, A.R., Malcolm, C.S., Misra, A., Quirk, K., Roffey J.R.A., and Dourish, C.T. (2002) Radioligand binding profile and in vitro functional efficacy of VER-3323, a novel 5-HT2C/5-HT2B receptor agonist. 5th IUPHAR Satellite Meeting on Serotonin, Acapulco.

187. Vickers, S.P., Bass, C., Bickerdike, M.J., Kennett, G.A., and Dourish, C.T. (2002). The effects of the orally active 5-HT$_{2C}$ receptor agonist, VER-3323, on food intake. 5th IUPHAR Satellite Meeting on Serotonin, Acapulco.

188. Bentley, J.M., Adams, D.R., Bebbington, D., Benwell, K.R., Bickerdike, M.J., Davidson, J.E.P., Dawson, C.E., Dourish, C.T., Duncton, M.A.J., Gaur, S., George, A.R., Giles, P.R., Hamlyn, R.J., Kennett, G.A., Knight, A.R., Malcolm, C.S., Mansell, H.L., Misra, A., Monck, N.J.T., Pratt, R.M., Quirk, K., Roffey, J.R.A., Vickers, S.P., and Cliffe, I.A. (2004). Indoline derivatives as 5-HT2C receptor agonists. *Bioorg Med Chem Lett*, 14:2367–2370.

189. Vernalis media release (2002). Roche and Vernalis to commence Phase I clinical trials of selective 5-HT2C receptor agonist in obesity. http://www.vernalis.com

190. Biovitrum media release (2003). Biovitrum and GlaxoSmithKline to focus efforts on developing highly selective 5-HT2C receptor agonists. http://www.biovitrum.com

191. Biovitrum media release (2007). Biovitrum terminates preclinical program in obesity. All rights are returned by GlaxoSmithKline. http://www.biovitrum.com

192. Arena Pharmaceuticals media release (2006). Arena Pharmaceuticals initiates lorcaserin phase 3 obesity clinical trial. http://www.arenapharm.com

193. Arena Pharmaceuticals website. Lorcaserin hydrochloride for obesity. Product candidates. http://www.arenapharm.com/wt/page/lho

194. Russell, M.G.N. and Dias, R. (2002). Memories are made of this (perhaps): A review of serotonin 5-HT6 receptor ligands and their biological functions. *Curr Top Med Chem*, 2(6):643–654.

195. Sleight, A.J., Boess, F.G., Bos, M., Levet-Trafit, B., Riemer, C., and Bourson, A. (1998). Characterization of Ro 04-6790 and Ro 63-0563: Potent and selective antagonists at human and rat 5-HT6 receptors. *Br J Pharmacol*, 124(3):556–562.

196. Bentley, J.C., Marsden, C.A., Sleight, A.J., and Fone, K.C.F. (1999). Effect of the 5-HT antagonist Ro 04-6790 on food consumption in rats trained to a fixed feeding regime. *Br J Pharmacol*, 126:66p.

197. Bentley, J.C., Sleight, A.J., Marsden, C.A., and Fone, K.C.F. (1997). 5-HT6 antisense oligonucleotide i.c.v. affects rat performance in the water maze and feeding. *J Psychopharmacol*, 11(Suppl):A64.

198. Woolley, M.L., Bentley, J.C., Sleight, A.J., Marsden, C.A., and Fone, K.C. (2001). A role for 5-HT6 receptors in retention of spatial learning in the Morris water maze. *Neuropharmacology*, 41(2):210–219.

199. Woolley, M.L., Marsden, C.A., and Fone, K.C. (2004). 5-HT6 receptors. *Curr Drug target, CNS Neurol Disord*, 3:59–79.

200. Perez-Garcia, G. and Meneses, A. (2005). Oral administration of the 5-HT6 receptor antagonists SB-357134 and SB-399885 improves memory formation in an autoshaping learning task. *Pharmacol Biochem Behav*, 81:673–682.

201. Holenz, J., Merce, R., Diaz, J.L., Guitart, X., Codony, X., Dordal, A., Romero, G., Torrens, A., Mas, J., Andaluz, B., Hernandez, S., Monroy, X., Sanchez, E., Hernandez, E., Perez, R., Cubi, R., Sanfeliu, M.O., and Buschmann, H. (2005). Medicinal chemistry driven approaches toward novel and selective serotonin 5-HT6 receptor ligands. *J Med Chem*, 48:1781–1795.

202. Caldirola, P. (2003). 5-HT6 receptor antagonism, a novel mechanism for the management of obesity. SMi Conf Obesity and Related Disorders, London, UK.

203. Biovitrum media release (2006). Biovitrum's project portfolio within obesity advances: Biovitrum enters clinical trials with one additional candidate drug. http://www.biovitrum.com

204. EPIX Pharmaceuticals media release (2007). EPIX Pharmaceuticals reports preliminary findings from Phase 1b clinical trial of novel 5-HT6 drug candidate. http://www.epixpharma.com

205. Smith, G.P. and Gibbs, J. (1992). The development and proof of the CCK hypothesis of satiety. In Dourish, C.T., Cooper, S.J., Iversen, S.D., and Iversen, L.L. (eds.), *Multiple Cholecystokinin Receptors in the CNS.* Oxford University Press.

206. Jones, M.C. (2007). Therapies for diabetes: Pramlintide and exenatide. *Am Fam Physician*, 75(12):1831–1835.

207. Parker, E., Van Heek, M., and Stamford, A. (2002). Neuropeptide Y receptors as targets for anti-obesity drug development: Perspective and current status. *Eur J Pharmacol*, 440:173–187.

208. 7TM Pharma media release (2007a). Results from a Phase I/II clinical study with the drug candidate Obinepitide for the treatment of obesity, http://www.7tm.com

209. 7TM Pharma media release (2007b). 7TM Pharma advances the next first-in-class obesity compound into clinical development, http://www.7tm.com

210. Luthin, D.R. (2007). Anti-obesity effects of small molecule melanin-concentrating hormone receptor 1 (MCHR1) antagonists. *Life Sci*, 81:423–440.

211. Takekawa, S., Asami, A., Ishihara, Y., Terauchi, J., Kato, K., Shimomura, Y., Mori, M., Murakoshi, H., Suzuki, N., Nishimura, O., and Fujino, M. (2002). T226296: A novel orally active and selective melanin-concentrating hormone receptor antagonist. *Eur J Pharmacol*, 438:129–135.

212. Borowsky, B., Durkin, M.M., Ogozalek, K., Marzabadi, M.R., DeLeon, J., Lagu, B., Heurich, R., Lichtblau, H., Shaposhnik, Z., Daniewska, I., Blackburn, T.P., Branchek, T.A., Gerald, C., Vaysse, P.J., and Forray, C. (2002). Antidepressant, anxiolytic and anorectic effects of a mela-nin concentrating hormone-1 receptor antagonist. *Nature Med*, 8:825–830.

213. Haynes, A.C., Stevens, A.J., Muir, A.I., Avenell, K.Y., Clapham, J.C., Tadayyon, M., Darker, J.G., Wroblowski, B., O'Toole, C., Witty, D.R., and Arch, J.R.S. (2001). A melanin-concentrating hormone receptor (SLC-1) antagonist reduces food intake and body weight gain in rats and mice. *Obes Res*, 9:90S.

214. Lewis, D.R., Joppa, M.A., Markison, S., Gogas, K.R., Schwarz, D.A., Zhu, F., Maki, R.A., and Foster, A.C. (2002). Melanin-concentrating hormone receptor antagonists decrease food intake in rodents. *Appetite*, 39:89.

215. Batterham, R.L., Cohen, M.A., Ellis, S.M., Le Roux, C.W., Withers, D.J., Frost, G.S., Ghatel, M.A., and Bloom, S.R. (2003). Inhibition of food intake in obese subjects by peptide YY 3-36. *New Eng J Med*, 349(10):941–948.

216. Lush, C., Chen, K., Hompesch, M., Tropin, B., Lacerte, C., Burns, C., Ellero, C., Kornstein, J., Vayser, I., Wintle, M., Blundell, J. *et al.* (2005). A phase 1 study to evaluate the safety, toler-ability, and pharmacokinetics of rising doses of AC162352 (synthetic PYY human 3-36) in lean and obese subjects. *Obesity Reviews*, 6(Suppl 1). AbsO051

217. Nastech media release (2006). Nastech initiates dose ranging study of PYY(3–36) nasal spray for obesity. http://www.nastech.com

218. Kola, I. and Landis, J. (2004). Can the pharmaceutical industry reduce attrition rates?. *Nat Rev Drug Discov*, 3:711–715.

219. Wentzlaff, T.H., Guss, J.L., and Kissileff, H.R. (1995). Subjective ratings as a function of amount consumed: A preliminary report. *Physiol Behav*, 57:1209–1214.

220. Yeomans, M.R. (1996). Palatability and the micro-structure of feeding in humans: The appe-tizer effect. *Appetite*, 27:119–133.

221. Zimmermann, U., Kraus, T., Himmerich, H., Schuld, A., and Pollmächer, T. (2003). Epidemiology implications and mechanisms underlying drug-induced weight gain in psy-chiatric patients. *J Psychiatry Res*, 37:193–220.

222. Tuthill, A., Slawik, H., O'Rahilly, S., and Finer, N. (2006). Psychiatric co-morbidities in patients attending specialist obesity services in the UK. *Quart J Med*, 99:317–325.

223. Hill, M.N. and Gorzalka, B.B. (2005). Is there a role for the endocannabinoid system in the etiology and treatment of melancholic depression? *Behav Pharmacol*, 16:333–352.

224. Gadde, K.M. (2006). Effect of rimonabant on weight and cardiometabolic risk factors. *JAMA*, 296:649-650.

225. Harmer, C.J., Hill, S.A., Taylor, M.J., Cowen, P.J., and Goodwin, G.M. (2003a). Toward a neuropsychological theory of antidepressant drug action: Increase in positive emotional bias after potentiation of norepinephrine activity. *Am J Psychiatry*, 160:990-992.

226. Harmer, C.J., Bhagwagar, Z., Perrett, D.I., Vollm, B.A., Cowen, P.J., and Goodwin, G.M. (2003b). Acute SSRI administration affects the processing of social cues in healthy volunteers. *Neuropsychopharmacology*, 28:148-152.

227. Harmer, C.J., Shelley, N.C., Cowen, P.J., and Goodwin, G.M. (2004). Increased positive versus negative affective perception and memory in healthy volunteers following selective serotonin and norepinephrine reuptake inhibition. *J Am Psychiatry*, 161:1256-1263.

228. Attenburrow, M.J., Williams, C., Odontiadis, J., Reed, A., Powell, J., Cowen, P.J., and Harmer, C.J. (2003). Acute administration of nutritionally sourced tryptophan increases fear recognition. *Psychopharmacology*, 169(1):104-107.

229. Harmer, C.J., Rogers, R.D., Tunbridge, E., Cowen, P.J., and Goodwin, G.M. (2003c). Tryptophan depletion decreases the recognition of fear in female volunteers. *Psychopharmacology*, 167(4):411-417.

230. Wong, E.H., Sonders, M.S., Amara, S.G., Tinholt, P.M., Piercey, M.F., Hoffmann, W.P., Hyslop, D.K., Franklin, S., Porsolt, R.D., Bonsignori, A., Carfagna, N., and McArthur, R.A. (2000). Reboxetine: A pharmacologically potent, selective, and specific norepinephrine reuptake inhibitor. *Biol Psychiatry*, 47(9):818-829.

231. Amylin Pharmaceuticals website. Amylin PYY pipeline 3-36. http://www.amylin.com/pipeline/pyy336.cfm

232. Browning, M., Reid, C., Cowen, P.J., Goodwin, G.M., and Harmer, C.J. (2007). A single dose of citalopram increases fear recognition in healthy subjects. *J Psychopharmacology*, 21:684-690.

233. Beuttner, R., Schölmerich, J., and Bollheimer, C. (2007). High-fat diets: Modeling the metabolic disorders of human obesity in rodents. *Obesity*, 15:798-808.

234. Dourish, C.T., Boyland, E., Halford, J.C.G., Dawson, G.R., and Wilding, J.P.H. (2006). Assessment of the potential efficacy and CNS side-effects of novel serotonergic anti-obesity agents in Phase 1 clinical studies. *J Pharmacol Sci*, 101(Suppl 1):17.

235. GlaxoSmithKline plc media release (2005). Preliminary announcement of results for the year ended 31st December 2004 GSK delivers 2004 EPS in line with guidance and confirms return to growth in 2005. http://www.gsk.com

236. Hayward, G., Goodwin, G.M., Cowen, P.J., and Harmer, C.J. (2005). Low-dose tryptophan depletion in recovered depressed patients induces changes in cognitive processing without depressive symptoms. *Biol Psychiatry*, 57:517-524.

237. Neurogen media release (2007). Neurogen announces results of first-in-human trial for new approach to treating obesity, http://www.neurogen.com

238. Shionogi & Co media release (2007). R&D presentation, http://www.shionogi.co.jp

239. Vernalis media release (2002). Vernalis Group plc, interim results for the six months ended 30 June 2002. http://www.vernalis.com

Current Concepts in the Classification, Treatment, and Modeling of Pathological Gambling and Other Impulse Control Disorders

Wendol A. Williams[1], Jon E. Grant[2], Catharine A. Winstanley[3] and Marc N. Potenza[4]

[1]Departments of Psychiatry and Diagnostic Radiology, Yale PET Center, Yale University School of Medicine, Yale New Haven Psychiatric Hospital, New Haven, CT, USA
[2]Department of Psychiatry, University of Minnesota Medical School, Minneapolis, MN, USA
[3]Department of Psychology, University of British Columbia, Vancouver, BC, Canada
[4]Departments of Psychiatry and Child Study Center, Yale University School of Medicine, Connecticut Mental Health Center, New Haven, CT, USA

Animal and Translational Models for CNS Drug Discovery,
Vol. 3 of 3: Reward Deficit Disorders
Robert McArthur and Franco Borsini (eds), Academic Press, 2008

INTRODUCTION

In the *Diagnostic and Statistical Manual of Mental Disorders* 4th edition (DSM-IV),[1] impulse control disorders (ICDs) are grouped as a heterogeneous cluster of disorders linked by a "failure to resist" impulses to engage in harmful, disturbing or distressing behaviors. The formal group of ICDs not elsewhere classified includes pathological gambling (PG), intermittent explosive disorder (IED), kleptomania, pyromania, trichotillomania, and ICDs not otherwise specified. Criteria for other ICDs have been proposed including compulsive sexual behavior, compulsive shopping, compulsive skin picking, and compulsive computer use.[2] Other disorders characterized by impaired impulse control (e.g., substance use disorders, Tourette syndrome, attention deficit hyperactivity disorder (ADHD)) are categorized in other sections of DSM-IV-TR. ICDs have been conceptualized as "obsessive–compulsive spectrum disorders" and "behavioral addictions." Consistent with the latter formulation, ICDs have core clinical features including: (1) compulsive and repetitive performance of the problematic behavior despite adverse consequences; (2) loss of control over the behavior; (3) an appetitive urge or "craving" state before starting the behavior; and (4) a pleasurable quality associated with its performance.[3]

Despite relatively high prevalence rates in the general population[4] and in psychiatric cohorts,[5] ICDs have been relatively understudied. Controlled treatment trials have not been performed for many ICDs.[6] This chapter reviews data on the neurobiology and treatment of ICDs including PG, trichotillomania, compulsive buying, IED, and kleptomania. As most neurobiological studies of ICDs to date have investigated PG, the following sections will focus on this disorder.

PG is characterized by persistent and recurrent maladaptive patterns of gambling behavior and is associated with impaired functioning, reduced quality of life, and high rates of bankruptcy, divorce, and incarceration.[6] PG usually begins in adolescence or early adulthood, with males tending to start at an earlier age.[7,8] Although prospective studies are largely lacking, PG appears to follow a similar trajectory as substance dependence, with high rates in adolescence and young adult groups, lower rates in older adults, and periods of abstinence and relapse.[5]

NEUROBIOLOGY OF PATHOLOGICAL GAMBLING

Biochemistry of Neurotransmitters

Disordered monoamine neurotransmission has been implicated in the pathophysiology of PG and other ICDs. Three neurotransmitter systems will be discussed: (1) serotonin (5-HT) function in the initiation and cessation of the problematic behavior;[2,9,10] (2) dopamine (DA) function contributing to rewarding and reinforcing behaviors; and, (3) norepinephrine (NE) function associated with arousal and excitement. Of the formal ICDs, PG is arguably the best studied to date from a neurobiological perspective. Therefore, this review will focus on PG. Data from other ICDs, other non-ICDs characterized by impaired impulse control, and studies of impulsivity will be integrated as deemed appropriate.[i]

Hypotheses that drive the use of medications in the treatment of PG are based on the neurobiology of PG and other ICDs,[6] and effective pharmacological treatments provide insight into the pathophysiology of PG.[11-14] Abnormalities in DA, serotonin, and noradrenergic neurotransmitter activity have been reported in PG.[15,16]

Serotonin

5-HT is involved in the regulation of mood states, sleep, and appetitive behaviors. Reduced 5-HT neurotransmission has been associated with increased impulsivity in humans and animal models.[17-20] A role for 5-HT in PG and ICDs is supported by data from pharmacological challenge studies that suggest decreased 5-HT synaptic activity in PG,[21] postsynaptic 5-HT receptor hypersensitivity and reduced 5-HT availability,[9] and decreased platelet monoamine oxidase B (MAO-B) activity in PG.[22,23,15] In a study of PG, intravenous clomipramine (CMI) (12.5 mg) was used to target the 5-HT transporter, yielding a blunted prolactin response, which suggested diminished 5-HT transporter binding.[21]

---

[i]Please refer to Heidbreder, Impulse and reward deficit disorders: Drug discovery and Development, in this volume and Joel *et al.*, Animal models of obsessive–compulsive disorder: From bench to bedside via endophenotypes and biomarkers, in Volume 1, Psychiatric Disorders, for further discussion of the neurobiology of impulse control disorders.

Meta-chlorophenylpiperazine (m-CPP), a metabolite of trazodone (Desyrel®) acting as a partial 5-HT receptor agonist (predominantly at the 2C subtype), produces euphoria in individuals with PG.[24] This response is similar to that reported in individuals with alcohol use disorders exposed to the agonist.[25,24] Subjects with PG who were administered 0.5 mg/kg of m-CPP had a significantly increased prolactin response, a reported "high" sensation, and a neuroendocrine response that correlated with gambling severity.[26] Similar behavioral responses to m-CPP have been reported in individuals with other disorders in which impulsive or compulsive behaviors are prominent; for example, antisocial personality disorder, borderline personality disorder, trichotillomania, and alcohol abuse/dependence.[27-29,25]

Decreased levels of the 5-HT metabolite and 5-hydroxyindoleacetic acid (5-HIAA) have been found in cerebrospinal fluid (CSF) of individuals with impulsive characteristics, people attempting suicide, impulsive alcoholic criminals, fire setters, and alcohol use disorders.[24,30-35] When correcting for CSF flow rates, decreased CSF 5-HIAA concentration was observed in men with PG.[35] Clinical trials of selective serotonin reuptake inhibitors (SSRIs) have produced equivocal results. SSRIs studied have included citalopram (Celexa®), escitalopram (Lexapro®), fluvoxamine (Luvox®), paroxetine (Seroxat®/Paxil®), and sertraline (Zoloft®/Lustral®), and are described in greater detail later in the chapter.

Dopamine

The DA system influences reward and reinforcement behaviors and has long been implicated in substance dependence,[36] with more recent suggestions of abnormalities in PG.[37] Dopaminergic pathways may contribute to the seeking of rewards related to gambling and drugs that trigger release of DA and produce pleasurable feelings.[38] A subset of dopaminergic neurons projects from the ventral tegmental area (VTA) to the nucleus accumbens (NAc). These neurons are important for reward reinforcement because interruptions in DA impulse trafficking along axonal routes or at the receptor level decrease the rewarding influences of VTA DA stimulation.[39,40]

A proposed mechanism of addiction, termed the "reward deficiency syndrome," suggests that diminished DA function places vulnerable individuals at high risk for addictive, impulsive, and compulsive behaviors.[38] However, no ligand-based imaging studies of individuals with PG or other ICDs have been published to support this hypothesis, and thus, this hypothesis remains largely speculative as it relates to ICDs.

It has been hypothesized that discrete areas of frontal cortex involved in impulse control are divided into functionally "dissociable areas".[41-43,20] For example, the ventral and dorsal pre-frontal regions are thought to represent distinct neuroanatomical substrates with respect to impulsive behavior, and these regions show a differential relationship between impulsivity and monoamine neurotransmission. In support of this hypothesis, DA, 5-HT, and metabolite concentrations were measured in rat medial prefrontal cortex (mPFC) and orbitofrontal cortex (OFC) using *in vivo* microdialysis during a delay-discounting paradigm, which tested impulsive choice.[20] During task performance, the investigators observed significant increases in mPFC – but not OFC-related 5-HT efflux. 3,4-Di-hydroxyphenylacetic acid (DOPAC, a DA metabolite) levels increased in OFC during the performance task but not under control conditions, whereas mPFC-DOPAC levels increased in all animals. These data suggest a "double dissociation" in frontocortical 5-HT and DA neuromodulation during impulsive decision-making.[20]

Cortico-limbic brain activation is seen in subjects with cocaine dependence after a cocaine-induced rush[44] or following viewing of cocaine-related videotapes,[45] and occupancy of the DA transporter is correlated with cocaine's euphorigenic effect.[46] However, a different pattern of cortico-limbic activation has been observed in individuals with PG.[47] These findings raise the possibility that chronic or acute effects of cocaine exposure may contribute to cortico-limbic activations seen in cocaine dependent subjects. Haloperidol (Haldol®), a DA D_2-like receptor antagonist, significantly increased self-reported rewarding effects of gambling, post-game priming of desire to gamble, and gambling-induced elevation in blood pressure.[48]

Peripheral measures of DA have been observed to be elevated in problem gamblers during casino gambling[49] and in people playing Pachinko, a form of gambling combining elements of pinball and slot machines.[50] Data on central DA function in PG indicate abnormalities including altered CSF levels of DA (decreased) and its metabolite 3,4-dihydroxyphenylacetic acid (DHPA) (increased), together suggesting an increase in DA neurotransmission.[37] However, when corrected for flow rate of CSF, levels of homovanillic acid (HVA) were not decreased.[35] Drugs with D_2-like receptor antagonism [e.g., olanzapine (Zyprexa®)] have not shown efficacy in treating symptoms of PG in the limited controlled trials performed to date.[6] Taken together, these data raise questions regarding the precise role for DA in PG.

DA involvement in PG and other ICDs is also suggested in some studies of individuals with Parkinson's disease (PD).[51] Psychiatric behaviors and disorders reported in PD may be related to PD pathology, the treatment of PD, a combination of both, or some other factors.[52,53] Repetitive and reward-seeking behaviors have been reported in patients with PD, and these behaviors include compulsive gambling,[54,55] pathological hypersexuality,[56] binge eating,[57] and compulsive shopping.[58,59] Although initial publications involved case reports, case series, and retrospective chart reviews, two recent studies screened larger samples of individuals with PD for ICDs. A study of 297 patients with PD found estimates of pathological hypersexuality of 2.4% and compulsive shopping of 0.7%.[60] The estimate of ICDs in the sample (excessive or interfering patterns of shopping, sex, and gambling) was 6.1% among all subjects and 13.7% in patients taking DA agonists.[60] Concurrent levodopa therapy was also associated with the presence of an ICD.[60] An independent study of 272 patients with PD identified an association between DA agonist treatment and presence of an ICD (compulsive gambling, compulsive shopping, or compulsive sexual behaviors).[61] In this study, 6.6% of subjects experienced an ICD at some point during treatment for PD and levodopa equivalency dose was higher in individuals with an ICD as compared to those without. In both studies, an association between DA agonist treatment and ICD occurrence was reported, and, in contrast to several case series, no difference was observed between specific DA agonists and their associations with ICDs.

From a biological perspective, these findings suggest the involvement of specific DA receptor subtypes (e.g., D_2-like receptors) in the pathophysiology of ICDs in PD. The involvement of specific receptors (e.g., the D_3 receptor that is primarily localized to brain limbic regions[62,63,56]) should be entertained cautiously in the absence of data directly investigating the underlying brain biology of PD subjects with and without ICDs. Additionally, younger age of onset of PD, personal or family history positive for alcohol use disorders, and higher novelty seeking traits have also been associated with ICDs in PD.[64] As such, a precise role for DA in the pathophysiology of ICDs in PD requires further investigation and substantiation. The extent to which findings in a PD population extend to the general population also warrants direct investigation.

Norepinephrine

NE is involved in cognitive processes, particularly with regard to attention and arousal, and has been implicated in PG.[9,65-68] Central noradrenergic activity is increased in PG.[49,65,66] Other findings suggest that the noradrenergic system mediates selective attention in PG and is related to heightened arousal, readiness for gambling, or risk-taking.[65,66,9,69,50,22,49] Higher measures of NE and its metabolites have been found in urine and CSF samples from men with PG as compared to control comparison subjects without PG. Measures of extraversion in PG subjects correlated positively with CSF, plasma, and urinary levels of NE and NE metabolites.[70] In an independent study examining physiological changes in Pachinko, NE levels were found to increase as winning commenced.[50] Although NE has been implicated in multiple studies of PG, the subjects in these studies have been predominantly or exclusively male. The extent to which these findings are applicable to women requires further investigation.

MAO Activity

The MAOs, subtypes MAO-A and MAO-B, are enzymes that metabolize NE, 5-HT, and DA.[71] Peripheral MAO derived from platelets is of the MAO-B subtype and has been suggested to be an indicator of 5-HT function,[72,73] although MAO-B also binds with high affinity to and catabolizes DA.[71] Decreased platelet MAO activity has been reported in association with impulsive behaviors,[74,75] high levels of sensation-seeking,[23,76,77] and other disorders characterized by impaired impulse control, including PG and eating disorders.[22,23,78]

In one study, men with PG as compared to those without were found to have MAO activities 26% lower.[23] A separate study with a similar male PG cohort found MAO activity levels 41% lower than matched control subjects.[22] Each group investigated personality and sensation-seeking characteristics of the PG and control groups, and found statistically significant between-group differences. However, no clear picture emerged regarding correlation of the characteristics with MAO levels: no associations persisted after Bonferroni correction in one study[22] and a positive correlation between MAO and several measures of sensation-seeking was observed in the other study.[23] An independent investigation demonstrated a significant interaction between childhood maltreatment and reduced MAO-A enzymatic activity in modulating the risk for antisocial behavior, aggressiveness, and violence during adolescence.[79] Future studies are needed to investigate the relationship between MAO-A enzymatic activity, childhood maltreatment, and ICDs, particularly as childhood trauma has been found in association with ICDs such as PG.[80]

Stress Response Systems

Cortisol, adrenergic, and heart rate measures represent key components of stress responses. Stress responsiveness has been implicated in ICDs and other disorders characterized by impaired impulse control.[24,68,81] Amongst Australian Aboriginals, urinary epinephrine and cortisol hormone output were approximately twofold higher in individuals on days during which gambling was concentrated.[69] Non-pathological or recreational gamblers demonstrate increases in salivary cortisol during casino gambling.[82] In a study analyzing the influence of casino gambling on cardiovascular and neuroendocrine activity in problem gambling, elevated epinephrine and NE levels were found in problem gamblers at baseline and throughout a session of gambling.[49] In problem gamblers, adrenocorticotropic hormone (ACTH) and cortisol levels were transiently increased during gambling. Positive correlations were found between gambling outcomes and NE plasma concentrations. The desire for starting and continuing gambling was positively correlated with plasma NE at baseline and following gambling sessions.[49] In a group of recreational, problem, and pathological gamblers, researchers found that heart rate and cortisol levels significantly increased with the onset of gambling and remained elevated throughout the gambling sessions.[83] Subjects with high levels of impulsivity showed significantly higher heart rate compared to those with low levels of impulsivity. Correlation analyses in this study revealed a positive relationship between impulsivity scores and severity of PG.[83] Together, these findings implicate stress response pathways in PG and subsyndromal levels of gambling.

Opioidergic Pathways

Gambling or drug use may trigger brain synaptic DA release and produce feelings of pleasure.[84,85] Opioid receptor antagonists indirectly inhibit DA release in the NAc and ventral pallidum through the disinhibition of gamma-aminobutyric acid (GABA) input to the DA neurons in the VTA.[84,85] Opioid antagonists are hypothesized to work by decreasing DA neurotransmission in the NAc component of motivational neurocircuitry, thus attenuating euphoria and cue intensity related to PG.[84,85]

Data suggest an important role for endogenous opioids in mediating hedonia and for μ-opioid receptors (μORs) in mediating reward and reinforcement.[86,87][ii] In an investigation of β-endorphin function in gambling behaviors, blood levels of β-endorphins were elevated during Pachinko play, which peaked at the start of high-pitched play.[2,50] However, β-endorphin levels did not change during casino gambling sessions in problem and non-problem gamblers, and problem gamblers tended to have lower levels than did non-problem gamblers.[49] Opioidergic involvement in PG is supported by clinical studies demonstrating the efficacy of the opioid antagonists naltrexone (Revia®) and nalmefene in the treatment of ICDs.[24,88,89] As allelic variants of the μOR gene have been related to naltrexone treatment outcome in alcoholism,[90] further work is needed to investigate the impact of specific genetic factors on treatment outcome in PG and other ICDs. Taken together, although multiple findings implicate opioidergic factors in PG, the precise nature of their involvement remains incompletely understood.

[ii] For further discussion of hedonia as a putative behavioral phenotype for ICD and mood disorders, please refer to Koob, The role of animal models in reward deficit disorders: Views of academia, in this volume and to Cryan *et al.*, Developing more efficacious antidepressant medications: Improving and aligning preclinical and clinical assessment tools, in Volume 1, Psychiatric Disorders.

Neuroimaging

There exist relatively few imaging studies of ICDs and most have focused on PG. Studies suggest similarities and differences between PG and other psychiatric disorders. In one investigation, gambling cues were presented to men with PG ($n = 10$) and control subjects ($n = 11$).[91] Two gambling scenarios and two emotional scenarios were presented. For each scenario, subjects were instructed to push a button at the onset of an emotional (e.g., happiness, sadness, or anger) or motivational (e.g., desire to eat, drink, or gamble) response. After each scenario, participants were asked to describe the quality and rate of the peak and average intensities of their emotions and motivations (including gambling urges) using visual analog scales from 0 to 10. During three time periods, comparisons in brain activation were made between PG and control subjects: (1) initial portion of videotape viewing as compared with pre-tape baseline; (2) immediately after a subjective response as compared with immediately before a subjective response; and (3) the final period of videotape viewing when the most provocative stimuli were presented, as compared with the post-tape baseline.

As compared to control subjects, PG subjects reported stronger gambling urges after viewing gambling scenarios. The most pronounced between-group differences in brain activations were observed during the initial period of viewing of the gambling scenarios: PG subjects displayed relatively decreased activity in frontal and orbitofrontal cortices, basal ganglia, and thalamus. The relatively decreased activation in PG as compared with control subjects of the cortico-basal-ganglionic-thalamo-cortical circuitry is different from the relatively increased activation of this network observed in cue provocation studies in obsessive–compulsive disorder (OCD).[92] During the period of videotape viewing corresponding to the most intense gambling cues, individuals with PG showed relatively diminished activation in the ventromedial prefrontal cortex (vmPFC), a brain region previously implicated in disadvantageous decision-making and disorders characterized by impaired impulse control (e.g., impulsive aggression).[93-96] Together, the data suggest that a complex network of brain regions distinguishes PG and control subjects during gambling-related motivational states, and that these neural processes are dynamic over time.

The neural correlates of cognitive control were examined with functional magnetic resonance imaging (fMRI) using an event-related Stroop paradigm in men with and without PG.[91] Following the presentation of infrequent incongruent stimuli (mismatched color-word pairs), both subject groups demonstrated similar activity changes in multiple brain regions, including activation of the dorsal anterior cingulate and dorsolateral frontal cortex. As compared with control subjects, those with PG demonstrated greater deactivation of the vmPFC, resulting in a between-group difference in left vmPFC. A similar region of left vmPFC was found to distinguish subjects with bipolar disorder from those without during the performance of the same fMRI task.[97] These findings suggest a common neurocircuitry underlying impaired impulse control across diagnostic boundaries. An independent group studied 12 subjects with PG and matched controls without PG using an fMRI task that simulates gambling and involves the processing of monetary rewards and losses.[98] Significantly lower activation of the right ventral striatum was observed in subjects with PG as compared to those without in winning versus losing contrasts. The PG cohort also showed relatively diminished activation in the vmPFC, consistent with prior studies of PG subjects.[91,47] Severity of gambling in PG was inversely correlated with ventral striatal and vmPFC activation. Potential confounding factors (e.g., depression or smoking) did not account for the findings. In healthy subjects, the ventral striatum and vmPFC activate during the selection of immediate rather than delayed rewards, and ventral striatum activation is associated with anticipation of working for monetary reward and activation of vmPFC with receipt of monetary reward. Relatively diminished activation of the ventral striatum during anticipation of working for monetary reward has been observed in individuals at risk for alcoholism or those with alcoholism,[99-101] suggesting that similar neurocircuits operate in PG and substance use disorders, and that diminished striatal activation during reward processing might represent a meaningful endophenotype across addictive disorders.[102,95]

Finally, an fMRI study examined whether subjects with PG exhibited differential brain activity when exposed to gambling cues.[103] The investigators found that PG subjects exhibited greater activity in the right dorsolateral prefrontal cortex (DLPFC), right parahippocampal gyrus, and left occipital cortex. They also reported that after the study, PG subjects experienced a significant increase in craving for gambling.[103] Hollander and *et al.* (2005) performed two [^{18}F]FDG PET scans 7 days apart on subjects with PG who were playing computerized blackjack under two different reward conditions: monetary reward and computer game points only. The investigators observed significantly higher relative metabolic rate in primary visual cortex, cingulate gyrus, putamen, and prefrontal areas during the monetary reward condition versus the point reward condition. The authors interpreted this pattern of activation as indicative of heightened limbic and sensory activity with regard to risk versus reward incentives. These data highlight the salience of monetary reward in PG.[104]

Genetic Considerations

Data from studies of twin samples suggest that a substantial degree of the risk for PG is heritable.[105] Eisen *et al.*[105] determined that the prevalence of PG in the Vietnam Era Twin (VET) Registry was 1.4%. Of twins reporting gambling at least 25 times in a year in their lifetime, 29% (7.6% of the total cohort) also reported at least one inclusionary criterion of PG.[105,106] Familial factors contribute substantially to syndromal and sub-syndromal PG. More recent investigations of the same sample indicate that genetic and environmental factors contribute to PG, that there exists overlap in the genetic and environmental contributions between PG and alcohol dependence and PG and adult antisocial behaviors, and that the majority of the co-occurrence between PG and major depression appears to be determined by common genetic factors.[106-109] As this sample is comprised of a unique group of male twins, the extent to which these findings extend to other groups, particularly women, warrants additional investigation.

Differential frequencies of allelic variants involving mainly serotonergic and dopaminergic genes have been implicated in preliminary studies of PG. Specific genes implicated include those coding for MAO-A,[110] the serotonin transporter,[111] and the DA D_1, D_2, and D_4 receptors.[111-113] The D_2A1 allele of the DA D_2 receptor gene has been implicated in compulsive and addictive behaviors, including drug abuse, compulsive eating, and smoking. In 171 non-Latino whites with PG, 51% carried the D_2A1 allele as compared with 26% of controls.[15,112] Frequency of homozygosity of the DA Dde I allele of the D_1 receptor has also been found to be elevated in PG, tobacco smokers, and Tourette syndrome probands.[113] Allelic variants of the DRD4 gene have also been associated with PG.[114,115] Some differences in allelic variation related to PG appear influenced by sex status, raising the possibility that the genetic contributions to PG differ for men and women.[15]

One should view the findings from these association studies cautiously, particularly as methodological limitations often exist (e.g., in the areas of diagnostic evaluation and lack of stratification by racial/ethnic identity).[15] The extent to which these preliminary findings generalize to other ICDs, and the precise manner in which allelic distributions of these and other genes may contribute to the development of PG require further investigation. Ongoing studies using more comprehensive diagnostic evaluations, larger samples, and genome-wide investigative approaches should provide important information with respect to the genetic contributions to PG and other ICDs.

Conclusions

Multiple neurotransmitter systems may contribute to behavioral initiation, arousal, reward and reinforcement, and behavioral disinhibition in PG.[11,24,16,95,116] PG shares core features with substance use disorders including tolerance, withdrawal, repeated attempts to cut back or stop, and impairment in major areas of life functioning.[117] Phenomenological similarities (e.g., age-related prevalence estimates and "telescoping" patterns in women)[118,95] also exist between PG and substance addictions.[16,119,120] Further research is needed to understand the molecular and biochemical factors underlying specific behavioral features seen in PG and other ICDs. An improved understanding of the neurobiologies of ICDs will facilitate clinical advances in the identification, prevention, and treatment of PG and other ICDs.

TREATMENT

Pathological Gambling

Several medications have been investigated as treatments for PG. These have included antidepressants, mood stabilizers, and opioid antagonists. A meta-analysis of post-treatment pharmacological interventions in PG showed that the pharmacological interventions were more effective than no treatment/placebo, yielding an overall effect size of 0.78 (95% confidence interval, 0.64–0.92). No difference was observed in outcome measures between three main classes of pharmacological interventions: antidepressants, opiate antagonists, and mood stabilizers.[121]

Seven open-labeled studies using various medications (citalopram, escitalopram, carbamazepine (Tegretol®), nefazodone (Serzone®), bupropion (Wellbutrin®), valproate (Depacon®), and naltrexone) of periods of 8–14 weeks have demonstrated efficacy in 73% of PG subjects.[6] Although the meta-analysis found largely comparable efficacies across studies,[121] the nine double-blinded, placebo-controlled studies completed in PG have shown some variation in efficacy and tolerability, as described below.

Psychopharmacology

Serotonin reuptake inhibitors

Based on results that SSRIs are effective in OCD,[122,123][iii] and the observed association between low serotonin and disinhibited behaviors,[124-128] it was hypothesized that SSRIs might be an effective treatment in ICDs. Thus, these drugs were examined in the treatment of PG.[6]

Six placebo-controlled studies using SSRIs have yielded equivocal results. The SSRIs studies included citalopram, fluvoxamine, paroxetine, and nefazodone, a $5\text{-}HT_1/5\text{-}HT_2$ receptor antagonist with mixed noradrenergic/serotonergic reuptake inhibition.

Clomipramine   CMI (Anafranil®) is an SSRI that also inhibits NE uptake. It was described as being of potential clinical utility for PG in a case report.[129] In this one-subject, double-blind study, placebo drug was administered for 10 weeks without response. The initiation of 125 mg/day resulted in significant improvement (90% improvement) in gambling symptoms, which was sustained over a 28-week period on a dose of 175 mg/day.[129] In a study of PG, intravenous CMI (12.5 mg) was used to target the 5-HT transporter, yielding a blunted prolactin response which suggested diminished 5-HT transporter binding.[21] CMI can produce dizziness, drowsiness, dry mouth, constipation, nausea, and vomiting. Among the more serious side effects are cardiac conduction problems and drug–drug interactions.[6] No additional controlled studies using CMI have been conducted to confirm or extend these results.[6]

Sertraline   Sixty subjects with PG were treated for 6 months in a double-blind, placebo-controlled study using sertraline (mean dose = 95 mg/day);[130,6] 74% ($n = 23$) of sertraline-treated subjects and 73% ($n = 21$) of placebo-treated subjects were rated as responders on an outcome measure assessing urges to gamble and gambling behavior. Thus, the placebo treatment was rated as effective as sertraline. In the study, sertraline was well tolerated but the side effect profile does include nausea, dizziness, drowsiness, dry mouth, loss of appetite, diarrhea, upset stomach, or insomnia.[6]

Fluvoxamine   Fluvoxamine has been examined in multiple open-label and randomized, placebo-controlled trials of PG. First, in an 8-week open-label trial of fluvoxamine, 70% of patients who completed the study were determined to be treatment responders as judged by decreases in gambling behavior (Yale-Brown Obsessive Compulsive Scale modified for PG, PG-YBOCS) and improvement in clinical status (Clinical Global Impression score).[131] Second, in a double-blind study of 15 patients with PG, clinical response to drug was more frequent among those treated with fluvoxamine (41%) than among those treated with placebo (41% versus 17%; $F(1,8) = 14.31, P < 0.05$).[132] Third, 32 PG patients were treated with fluvoxamine (200 mg/day) for 6 months in a double-blind, placebo-controlled study. On outcome measures of reduction in money and time

---

[iii] Joel *et al.*, Animal models of obsessive–compulsive disorder: From bench to bedside via endophenotypes and biomarkers, in Volume 1, Psychiatric Disorders, for further discussion of drug treatment of OCD.

spent gambling per week, fluvoxamine was not statistically significantly different from placebo, except in males and younger patients where it proved to be statistically significantly superior to placebo.[133,6] The power of the study was limited by the high (59%) placebo-response rate and the limited number ($n = 3$) of subjects on medication completing the study.[133]

Paroxetine   Studies examining the efficacy of paroxetine have produced inconsistent results.[6] A double-blind placebo-controlled study using outcome measures including the Gambling Symptom Assessment Scale (G-SAS) and the Clinical Global Impressions scale (CGI) to gauge responses found statistically significantly greater reductions in the G-SAS score compared with placebo ($P = 0.003$) and a greater proportion of patient responders at weeks 7 and 8 ($P = 0.011$ and 0.010, respectively) as determined by CGI. A 16-week, double-blind, placebo-controlled trial was conducted at five outpatient research centers in the United States and Spain. Outpatients with PG were randomized to acute paroxetine treatment (dosages = 10–60 mg/day; $n = 36$) or placebo ($n = 40$). The paroxetine- and placebo-treated groups both demonstrated improvement at 16 weeks (paroxetine 59% response rate, placebo 49%; $P = 0.39$). On the GGI, the PG-YBOCS, or the G-SAS scores, paroxetine did not result in statistical superiority compared to placebo. However, on the CGI, paroxetine consistently demonstrated a non-significant statistical advantage over placebo.[134]

Escitalopram   Thirteen subjects with DSM-IV PG and co-morbid anxiety were treated in a 12-week open-label trial of escitalopram followed by double-blind discontinuation. Assessment scales used were the PG-YBOCS, which was the primary outcome measure, the Hamilton Anxiety Rating Scale (HAM-A), the CGI, and measures of psychosocial functioning and quality of life. PG-YBOCS scores were significantly decreased ($P = 0.002$), with 62% classified as responders. Scores on the HAM-A also decreased by 83% over the 12-week period ($P < 0.001$). The mean end-of-study dose of escitalopram was 25.4 ± 6.6 mg/day.[135] Open-label escitalopram treatment was associated with improvements in gambling and anxiety symptoms and measures of psychosocial functioning and quality of life.[135] In the double-blind discontinuation phase, active drug administration was associated with maintenance of symptom reduction in the gambling and anxiety domains whereas placebo administration was associated with symptom worsening.

Mood stabilizers

Hollander *et al.*[131] observed that two of three non-responders in a fluvoxamine study had cyclothymia. This result suggested that a different class of drugs, such as mood stabilizers, might work better for certain individuals with PG. Although a case report[136] and open-label study[137] suggest that carbamazepine and valproate, respectively, may be efficacious in PG treatment, there has been only one randomized trial of a mood stabilizer tested in PG. In that study, 40 patients with PG and bipolar spectrum disorders participated in a 10-week randomized, double-blind, placebo-controlled treatment study of lithium sustained release. Using outcome measures including gambling and mania scales, 83% of patients on lithium carbonate sustained release (mean lithium level 0.87 mEq/L) (compared to 23% on placebo) displayed significant decreases in gambling urges, thoughts, and behaviors as measured by the PG-YBOCS.[138] Concurrent decreases in manic symptoms were observed using the Clinician Administered Rating Scale for Mania.[138] An open-label study examining treatment response to lithium (up to 1800 mg/day) in three males with bipolar spectrum mood disorders and PG was found to be at least partially effective in controlling risk-taking behavior, gambling, and mood symptoms.[139] A case report of carbamazepine in the treatment of a male with PG showed a decrease in gambling behavior 2 weeks into active carbamazepine treatment, sustained over a 30-month period.[136]

DA antagonists

Few studies have examined DA antagonists or antipsychotic drugs in individuals with PG. In an open-label case study, it was reported that olanzapine was effective in decreasing gambling and psychotic symptoms in a patient with schizophrenia and co-occurring PG.[140] No statistically significant difference was found in treatment outcome in a double-blind, placebo-controlled study of olanzapine in hospitalized PG patients without psychosis who were treated with active drug or placebo.[141,6]

Opioid antagonists

Opioid antagonists are effective in treating substance use disorders (alcohol dependence, opioid dependence) and their mechanism of action is hypothesized to involve reduced mesolimbic dopaminergic neurotransmission. Factors influencing the regulation of gambling motivation and gambling-specific behaviors by DA, endorphins, and GABA have been examined, but no unifying mechanism has been advanced explaining μOR antagonism in PG.[142,143,6]

Naltrexone, a US Food and Drug Administration (FDA)-approved treatment for alcohol and opioid dependence, was investigated in an open-label study and shown to have efficacy in reducing gambling and thoughts and urges to gamble when given in doses ranging from 50 to 250 mg/day.[144,85] A 12-week double-blind placebo-controlled trial of naltrexone (mean end-of-study dose: 188 mg/day) demonstrated superiority to placebo in 45 subjects with PG, particularly in subjects with strong urges to gamble.[144,85] Naltrexone's adverse effect profile, including nausea and elevated liver enzymes, especially in combination with nonsteroidal anti-inflammatory drugs, may limit its clinical usage, particularly at high doses.[145,85] In a 16-week double-blind multi-center trial studying 207 individuals with PG, the opioid antagonist, nalmefene (not currently available in oral form in the United States), demonstrated statistically significant improvement in gambling symptoms compared with a placebo.[146] As compared with placebo, nalmefene effectively reduced gambling-related ideations, urges, and behaviors. Like naltrexone, nalmefene may cause nausea.[85,146]

Psychological Treatments

Psychological treatment strategies have used cognitive, behavioral, and motivational approaches in the treatment of individuals with PG. Behavioral strategies include isolating gambling triggers and developing non-gambling sources to compete with reinforcers associated with gambling.[6] Cognitive strategies include increased awareness techniques and cognitive restructuring.[6] One study[147] suggested that cognitive treatment (corrective cognition) can significantly decrease PG, with 88% of treated participants having greater perception of control over their gambling problem. This study emphasized that gambling outcomes are based on random events that cannot be controlled, and that correcting these erroneous perceptions by gamblers can constitute an important component of treatment. Earlier findings[148] suggested that when a gambler's erroneous perceptions and understanding of randomness were corrected, the motivation to gamble significantly decreased. In another study, 29 men with PG were randomly assigned to treatment or to a wait-list control group where treatment consisted of cognitive correction of erroneous perceptions about gambling, problem solving and social skills training, and relapse prevention. There were significant gains realized on all outcome measures from 6- to 12-month analyses.[149] In the latter two studies, treatment discontinuation was high, only data on those participants completing the study was reported, and despite manualized treatment sessions, no rigorous analysis of therapist competence was undertaken.[6]

In evaluating the efficacy of psychotherapy, researchers randomly assigned 231 individuals with PG to a workbook-based cognitive–behavioral (CB) treatment, 8-invidualized CB sessions, or a referral to Gamblers Anonymous (GA) alone (all three conditions included a referral to GA). During the treatment period, CB treatment reduced gambling relative to GA referral alone, with clinically significant improvement, and partial maintenance. Other researchers have assigned problem gamblers to cognitive–behavioral therapy (CBT) plus motivational enhancement (telephone interview), CBT workbook alone, or wait-list conditions. At 6 months, rates of abstinence did not differ between groups, although the frequency of gambling, money lost gambling, and South Oaks Gambling Screen scores were lower in the motivational interview group.[150,6] A significant reduction in gambling was observed in 84% of study participants ($n = 102$) over a 12-month follow-up period. At 2-year follow-up, an advantage was observed for participants ($n = 67$) who received a motivational telephone intervention plus a self-help workbook compared with participants who received only the workbook.[151]

Imaginal desensitization (ID) has been compared to treatment with other behavioral procedures. In this technique, participants are taught relaxation methods and instructed to imagine experiencing and resisting gambling triggers. In an initial study, 20 compulsive gamblers who were followed for 1 year were randomly allocated, half to receive aversion-relief therapy and half to receive ID. Compared with those who received aversion-relief, gamblers receiving ID reported a significantly greater reduction in gambling urges and behaviors, and showed significant reductions in trait and state anxiety.[152,6] In a subsequent larger study of PG ($n = 120$, 60 patients per group, procedures administered over 1 week),[153] 26/33 who received ID reported control or cessation of gambling compared with 16/30 who received other behavioral procedures, a difference that reached statistical significance.

Conclusion

Initial studies indicate that multiple pharmacological and behavioral therapies appear promising in the treatment of PG. Nonetheless, most studies are relatively small and often contain multiple design limitations.[6] While both drug and behavioral treatments have been shown to be effective in PG, few studies have performed systematic comparisons of interventions examining which treatments are more beneficial to which patients. Outcome measures have not been consistently used or adopted. Predictive markers for those individuals with PG who might respond to therapy have yet to be identified. There are limited data regarding effective medication usage in patients with co-morbid psychiatric disorders.

Trichotillomania

Trichotillomania is characterized by repetitive hair pulling that results in noticeable hair loss and clinically significant distress or functional impairment.[1] The mean age at onset is approximately 13 years.[154,6] Hair pulling, like compulsions in OCD, may reduce feelings of tension, boredom, and anxiety.[155] Trichotillomania is relatively common with an estimated prevalence between 1% and 3%.[154,6] Trichotillomania is more common in women[156,6] and is associated with generalized anxiety disorder and depression (30% to over 60%).[157,156,6]

Pharmacological Treatments

Antidepressants, particularly SSRIs, represent the class of agents most studied in tricho-tillomania. There have been double-blinded pharmacological studies that have examined the efficacy of the antidepressants, including CMI, desipramine, and fluoxetine.[6] CMI, an SSRI, was compared with desipramine (conventional tricyclic) in a 10-week double-blind, crossover design. With CMI, 92% of PG trichotillomania cohort ($n = 13$) achieved significant improvement in symptoms ($P = 0.006$), less clinical impairment ($P = 0.03$), and reduced severity of symptoms.[158] In a study comparing CMI with fluoxetine[159] and using a randomized 20-week crossover design (10 weeks per agent), both agents achieved a similar positive treatment effect.[160]

In addition to the study mentioned above, fluoxetine has been studied in two other randomized trials.[6] Using a 5-week washout period between agents, fluoxetine was compared with placebo in a 6-week double-blind, crossover study. Chronic hair pull-ers ($n = 21$) were recruited and took doses up to 80 mg/day. Short-term efficacy could not be demonstrated in 15 subjects (14 female and 1 male) completing the study, since no significant impact on urge to pull hair, number of episodes, or the amount of hair pulled per week was found.[154] In a double-blind, placebo-controlled crossover design, the 31-week trial consisted of a 2-week washout phase, an initial treatment phase of 12 weeks, intervening 5-week washout phase, and a 12-week crossover second treat-ment phase. The researchers found no statistically significant difference between fluoxetine and placebo.[161]

A randomized double-blind, placebo-controlled of the opioid antagonist naltrex-one was conducted over a 6-week period. Seven of 17 subjects completing the study received 50 mg/day of naltrexone and achieved significant symptomatic improvement on one measure of trichotillomania and a non-significant trend toward symptomatic improvement on two other clinical measures.[160]

Psychological Treatments

Multiple techniques, including relaxation training, aversion, stimulus control, awareness training, self-monitoring, negative practice, habit reversal training (HRT), overcorrection, and covert sensitization (CS), have been examined in the treatment of trichotilloma-nia.[162,163] HRT,[164] one of the central behavioral approaches to treating trichotilloma-nia, consists of awareness training (AT), competing response (CR) training, and social support.[165]

While some researchers have modified HRT by integrating cognitive elements that replicate cues for hair pulling episodes,[166-168] only two randomized controlled studies have examined the efficacy of CBT in treatment of trichotillomania.[163] Azrin et al.[169] first evaluated HRT for trichotillomania (n = 34). They showed that HRT was superior (as assessed by hairs pulled by self-report) to massed negative practice for reduction of hair pulling in adults. However, this study was limited by a number of methodologi-cal problems, including assessments not having been conducted by blinded evaluators, diagnosis of trichotillomania not being clearly established, substantial attrition dur-ing the follow-up phase, and no formalized treatment manual to facilitate study rep-lication.[170] Woods et al.[170] conducted a randomized trial that compared a combined acceptance and commitment therapy/habit reversal training (ACT/HRT) to wait-list control in the treatment of adults with trichotillomania. They found that ACT/HRT was superior to wait-list, with significant reduction in hair pulling severity, impairment rat-ings, and hairs pulled, and significant decreases in experiential avoidance, depression, and anxiety. These investigators found that treatment compliance was highly corre-lated with symptom reduction and that the gains realized with ACT/HRT were at least partially maintained at 3-month follow-up.[170]

Comparison Studies

In the first controlled trial directly comparing pharmacological interventions with psychotherapy, Ninan et al.[171] compared CMI, CBT, and placebo. CBT, which included HRT, stimulus control, relaxation training, thought stopping, cognitive restructuring, and guided self-dialog, produced greater changes in severity of hair pulling and in asso-ciated impairment and a higher rate of response than did either double-blinded CMI or placebo. The small sample sizes in each cell represent important limitations of the study, and differences between CMI and placebo approached but did not achieve sta-tistical significance. In a randomized-control evaluation of CBT for older adolescents (>16 years old) and adults with trichotillomania, van Minnen et al.[172] compared behav-ior therapy (BT) versus fluoxetine and wait-list control in randomly assigned subjects (n = 43) using psychometrically sound assessments (Massachusetts General Hospital Hairpulling Scale[173]) and blinded clinical evaluators. BT consisted of six, 45-min manu-alized sessions involving self-monitoring, stimulus control, learned disruption habitual motor patterns, and relapse prevention. Results showed that the BT group showed a greater decrease in hair pulling when compared to the other two control groups.

Conclusions

There are few pharmacological and psychological randomized, controlled trials exam-ining the efficacy of treatments in trichotillomania. The findings from the few available studies are equivocal, with several of the studies containing methodological issues that hamper interpretation and replication of results. Adequate testing of pharmacothera-peutic interventions utilizing SSRIs and opioid antagonists awaits future investigation since currently available data are inconclusive.

Compulsive Buying/Shopping

Compulsive buying or shopping is similar to other ICDs in that it is characterized by the failure to resist an impulse, drive, or temptation to perform an act (shopping) that is harmful to the person or to others.[174] This buying behavior leads to significant distress or impairment.[175] For example, compulsive shoppers experience repetitive, intrusive urges to shop,[6] and shopping thoughts and behaviors are time-consuming and significantly interfere with occupational functioning. Triggers for shopping impulses include environmental stimuli such as being in stores, stress, emotional difficulties, or boredom. These behaviors contribute to poor quality of life with marital discord, severe financial dislocations, and legal entanglements.[6] Guilt, shame, and embarrassment may come in the aftermath of such buying sprees, with the objects of the compulsive buyer's interest at times being unopened,[6] returned or given away, and less often, sold.[176,177] Lifetime prevalence in the United States is approximately 6%,[175] and the disorder appears to be more common in females.[178,176] Compulsive shopping appears familial, with high prevalence in families with mood and substance use disorders.[175]

Pharmacological Treatments

There have been three double-blind, randomized, placebo-controlled trials. The efficacy of fluoxetine (Prozac®) has been investigated in two studies: in the first, 37 patients (responders: 9/20 patients on mean dose 215 mg/day; 8/17 patients, placebo) were treated for 13 weeks and a high placebo-response rate was observed.[179] In an independent double-blind study, Black *et al.*[180] employed a 1-week placebo lead-in phase after which 23 patients were treated for 9 weeks. There was no difference in response rates between active medication (mean dose = 200 mg/day) and placebo.

A third controlled study consisted of a 7-week, open-label trial followed by a 9-week, double-blind, placebo-controlled discontinuation trial. Fifteen subjects meeting criteria for response entered the double-blind treatment discontinuation phase. Five of eight subjects (63%) randomized to placebo relapsed compared with none of seven randomized to citalopram ($P = 0.019$), with the latter group showing decreased shopping-related thoughts and behaviors.[181]

Psychological Treatments

Examinations of psychological treatments for compulsive shopping have been limited to case studies.[182,183] Investigators have sought to isolate patterns of cognition which trigger compulsive buying such as negative and euphoric emotional states.[184] One theoretical framework has suggested conceptualizing compulsive buying on the basis of: (a) depressed mood; (b) compromised self-perceptions and perfectionist expectations; (c) erroneous beliefs about the nature of objects, potential purchases, and purchasing opportunities; (d) erroneous beliefs about the psychological benefits of buying; and (e) decision-making difficulties.[185] Such efforts suggest that CBT centered on alternative coping strategies would be effective.[186] A pilot trial comparing the efficacy of CBT in compulsive buyers ($n = 28$) to a wait-list control group ($n = 11$) revealed significant advantages to CBT resulting in decreases in buying episodes and time spent buying, with gains maintained at 6 months.[186] However, the absence of randomized, controlled studies examining the efficacies of psychotherapies for compulsive buying highlights the need for more research in this area.

Conclusions

Given the scarcity of empirical evidence regarding effective treatments for compulsive buying, there are limited recommendations for pharmacotherapeutic interventions. Citalopram may offer some clinical efficacy but the small sample and use of non-validated outcome measures hinder an equivocal recommendation.[6] Current data do not support CBT or other psychological treatments for compulsive buying.[6]

Intermittent Explosive Disorder

IED is defined as a verbal or physical act directed against a person or object that can potentially cause physical or emotional harm, occurs in a deliberate and non-premeditated fashion,[187-189] and is characterized by repeated episodes of aggressive, violent behavior that is grossly out of proportion to the situation. It may affect as many as 6.3%[190] to 7.3%[191] of adults in the United States. Reactive aggression triggered by anger is impulsive, and can result in sudden, heightened, and enduring or inappropriate aggressive responses.[191] These aggressive outbursts are relatively short-lived (<30 min) and occur frequently, on the order of multiple times per month.[192,6] Symptoms typically start in adolescence and tend to evolve into a chronic illness.[190,192,6]

Pharmacological Treatments

Despite methodological limitations,[189] pharmacotherapeutic interventions with lithium, anticonvulsants, antipsychotics, and SSRIs have shown efficacy in decreasing aggression in multiple patient populations, although ones not always characterized as having IED. In an 8-week (placebo lead-in, 2 weeks) double-blind, placebo-controlled, randomized clinical trial, lithium was administered to children diagnosed with conduct disorder who were hospitalized for treatment-refractory aggressiveness and explosiveness. Lithium (mean serum level = 1.12 mEq/L) was superior to placebo in reducing aggression in children and adolescents diagnosed with conduct disorder.[193,194,189] Similarly, lithium has been shown to reduce aggressive behavior in adults[191] with borderline personality disorder[195,189] and inmates without apparent psychiatric diagnoses.[196,197,189]

There have been several double-blind, placebo-controlled studies demonstrating the efficacy of anticonvulsants (Depacon®, divalproex (Depakote®[198-200]), phenytoin (Dilantin®[201]), and carbamazepine[202]) in the treatment of impulsive aggression and hostility.[189] For example, divalproex, used in a series of studies spanning a range of psychiatric symptoms and disorders (including open-label trials, case reports), has been shown to be effective in treating irritability and aggression in patients with personality disorders,[198,199,203-205,189] traumatic brain injury,[206] temper outbursts,[207-209,189] post-traumatic stress disorder,[210-212,189] and agitation in the elderly.[200,213,189] In a double-blind, placebo-controlled, crossover design study testing the hypothesis that phenytoin would decrease impulsive aggressive acts, incarcerated inmates ($n = 60$) were divided into two groups characterized by: (a) impulsive aggressive acts or (b) premeditated aggressive acts. Phenytoin was found to significantly reduce impulsive aggressive acts but not premeditated aggressive acts.[201]

Conventional antipsychotics have been used to decrease aggressive behaviors in patients with alcoholism and organic brain syndromes,[214] psychosis and borderline personality disorder,[215] and in patients who are elderly patients with cognitive impairment.[216] Most recently, drug treatments of aggression have used the "atypical" antipsychotics, which produce less sedation and have a different side effect profile.[191] There is preliminary evidence suggesting that atypical antipsychotic agents may be useful in treating impulsive aggression,[217] and at least one agent, olanzapine, has been evaluated in a double-blind, placebo-controlled study of 28 female subjects meeting DSM-IV criteria for borderline personality disorder. Subjects were randomly assigned to olanzapine or placebo for a 6-month duration. Olanzapine was associated with a significantly greater rate of improvement than placebo in four core areas of borderline psychopathology: affect, cognition, impulsivity, and interpersonal relationships.[218] Risperidone (Risperdal®), which antagonizes 5-HT$_2$ and D$_2$-like receptors, has proven effective in children with autism spectrum disorder who display maladaptive aggression.[191] The SSRI fluoxetine has undergone three double-blind, placebo-controlled studies that suggest that it may decrease aggressive symptoms in patients with personality disorders.[219-221]

In contrast to the considerable clinical effort put forth in drug treatment of impulsive aggression, only one controlled pharmacotherapeutic clinical trial specific to IED has been completed. In a double-blind, placebo-controlled study of individuals with cluster B disorders ($n = 96$), IED ($n = 116$), and posttraumatic stress disorder ($n = 34$), subjects were randomized to divalproex or placebo for duration of 12 weeks. There was no treatment effect on aggression observed in the IED cohort.[189]

Psychological Treatments

Although case reports suggest that insight-oriented psychotherapy and behavioral therapy may be beneficial, there are no controlled psychological treatment studies in IED.[192,6]

Conclusions

The data currently available for treatment studies on IED do not support clear treatment recommendations. One controlled pharmacological study did not produce a treatment benefit and there are no rigorous psychological treatment studies (only case reports) available. Given the significance of IED as a public health concern, treatment studies are very much needed.[6]

Kleptomania

Kleptomania, or compulsive shoplifting, is characterized by repetitive, uncontrollable stealing of items not needed for personal use.[6] Kleptomania may result in personal and occupational disruptions, and potential legal consequences.[222] The course of illness may be chronic, with many individuals trying unsuccessfully to stop stealing and experiencing feelings of shame, guilt, remorse, or depression.[223] Women appear twice as likely to suffer from kleptomania as do men.[223-226] Epidemiological data regarding the prevalence of kleptomania in the community are not available.[222]

Pharmacological Treatments

There have been no randomized, placebo-controlled studies of medication in the treatment of kleptomania. Small case series and case reports constitute the majority of published treatment data, and there are no medications approved by the FDA in the United States for this disorder. Case reports examining drug efficacy in kleptomania suggest multiple candidates including SSRIs (paroxetine,[227] fluvoxamine,[228] escitalopram[229,222]) and combination therapies (sertraline/methylphenidate,[230] imipramine/fluoxetine,[231] and valproic acid[232,222]). Data on the efficacies of pharmacotherapeutic interventions in the treatment of kleptomania have been equivocal.[224]

In a case series of five patients with kleptomania, each subject was successfully treated with fluoxetine or paroxetine.[233] Three patients achieved remission of symptoms when treated with topiramate in doses ranging from 100 to 150 mg/day.[234] Of these three patients, two were being augmented with SSRIs and two were diagnosed with co-morbid ADHD and panic disorder. In a case series of two patients with kleptomania evaluated for treatment efficacy with naltrexone (50 and 100 mg/day), both urges to steal and stealing behavior achieved remission.[235]

There have been only two small, open-label trials of medication for kleptomania. In a 12-week open-label trial examining the efficacy and tolerability of naltrexone (mean dose 145 mg/day) in the treatment of kleptomania ($n = 10$), eight patients reported significant reduction in urges to steal along with 20% who had complete symptom remission.[88] Naltrexone was well tolerated among kleptomania subjects and was associated with minimal nausea and no elevated liver enzymes.[6] A 3-year retrospective, longitudinal study of naltrexone monotherapy in 17 individuals with kleptomania found that 77% reported decreased urges to steal, 41% reported cessation of stealing, and 53% reported very mild or no significant symptoms.[236]

In another open-label study, escitalopram was evaluated in 20 individuals with kleptomania. Of those receiving medication, a positive response was seen in 79% who reported improvement in stealing behavior. Subsequently, positive responders were randomized to receive medication or placebo. During this double-blind phase, 43% taking medication and 50% taking placebo failed to maintain their treatment response, suggesting that there was not a substantial drug effect.[237]

Psychological Treatments

While no controlled studies of psychological treatments have been validated for kleptomania, behavioral, psychoanalytic, psychodynamic, and CBT treatments may be of benefit. Therapies involving systematic desensitization, aversion therapy, and CS have some preliminary support in the treatment of kleptomania.[3] In case studies, CS has shown promise.[6] Combined with exposure and response prevention, CS was effectively applied in a male kleptomaniac who was able to decrease frequency of stealing, although his urges to steal remained static.[238] In a second case study, CS was used in 5 weekly sessions with a female patient who was instructed to practice CS whenever the urge to steal arose. Over a 14-month period, she stopped stealing (with one lapse), and reported no urges to steal.[239] One female patient was instructed to focus on images of vomiting associated with stealing in order to experience nausea when experiencing stealing urges.[240,6] With four sessions over 8 weeks, she had only one lapse in behavior over a 19-month period. A strategy involving diary-keeping of urges to steal and aversive breath holding was associated with reduced frequency of stealing.[241] ID has also shown promise in two subjects for a 2-year period.[242]

There have also been successful case reports of combination treatments using CBT with medication in kleptomania. A 43-year-old male with frontotemporal traumatic brain injury and kleptomania symptoms was treated with citalopram and CBT and experienced remission of symptoms.[243] A 77-year-old woman with late-onset klepto-mania (at age 73 years) reported remission of stealing behaviors with CBT and sertra-line (50 mg/day).[244,222]

Conclusions

Empirically validated treatments for kleptomania are currently lacking. Current research is based primarily on case reports, with some promise for specific pharmaco-therapies and behavioral treatments.[6]

INSIGHT FROM ANIMAL MODELS

In terms of modeling ICDs, various animal models of illnesses such as ADHD and OCD are being tested and developed as discussed elsewhere in this work.[iv] Such research demonstrates that it is possible to measure aspects of impulsive and compulsive behaviors in standard laboratory animals. However, considerably less is known about the ICDs discussed in the current chapter, which has hampered the development of adequate animal models. It is also fundamentally difficult to elicit some of these symp-toms in animals. This section will largely focus on modeling PG, as there are promising indications that gambling behavior may be reproduced in animals. Although we will focus on rodent research, some of the concepts and experimental paradigms are also applicable in non-human primates.

Regarding the other ICDs discussed in this chapter, developing an animal model with strong face or construct validity for disorders such as compulsive buying/shop-ping or kleptomania is inherently problematic. It may be possible to tap into processes associated with IED through the study of abnormal aggression in rats.[245] However, such an approach is largely in its infancy, and future research will reveal whether such a model will exhibit predictive validity.

Animal Models of Gambling Behavior

At the outset of this discussion, it is worth noting that there are currently no estab-lished animal models of gambling. This stands in contrast to the many reports in the literature concerning the measurement of impulsivity. The questions faced by research-ers attempting to model gambling behavior are manifold, but can be broadly divided into two camps. Firstly, is there any reason to expect that animals will gamble, and, if

---

[iv] Please refer to Joel *et al.*, Animal models of obsessive–compulsive disorder: From bench to bedside via endophenotypes and biomarkers, and Tannock *et al.*, Towards a biological understanding of ADHD and the discovery of novel therapeutic approaches, in Volume 1, Psychiatric Disorders.

so, how can we measure it? Secondly, can we ever hope to model the pathological drive to engage in gambling behavior, and what criteria would have to be met in order to validate such a model? These questions are somewhat distinct, and will be dealt with sequentially.

The Feasibility of Modeling Gambling in Animals

Much of the success attained in modeling aspects of impulsivity in laboratory animals has stemmed from the design of behavioral paradigms analogous to the neuropsychological tests used to assess impulsive responding in both healthy and clinical populations. Impulsivity itself covers a range of behaviors that can be loosely described as action without foresight, and which often result in undesirable consequences.[246] Through fractionating impulsivity into its component parts, it has proved possible to model the different psychological processes involved in aspects of impulse control within a laboratory setting. For example, motoric impulsivity or impulsive action can be conceptualized as the inability to withhold from making a prepotent motor response. This form of impulsivity is captured by the stop-signal reaction time task commonly used to assess ADHD patients,[247] and the continuous performance test (CPT) widely used to probe attentional function and impulsive action in clinical populations.[248] Rodent analogs of both of these tests have been designed[249-251] and are being successfully used to investigate the neurobiological basis of these behaviors.[252,253]

Measuring impulsive decision-making has proved more difficult. Tasks such as the matching familiar figures test (MFFT,[254]) assess the tendency of human subjects to respond to incoming information without allotting sufficient time to analyze its content or meaning, a process known as reflection impulsivity. Modeling such behavior in animals is fundamentally difficult, and although it has been attempted, the results were difficult to interpret.[255] The most successful model of impulsive decision-making developed to date is the delay-discounting paradigm, where subjects choose between a small reward available immediately and a larger reward delivered after a delay.[256] Impulsive choice is defined as selection of the small immediate reward. In parallel to human choices when faced with such decisions, rats fail to maximize their expected returns, and gradually shift their preference from the large to the small reward as the delay to the large reward increases. A number of different tasks have been developed for rats based on this concept (e.g.,[257-259]), and have been used to further understand the mechanism underlying the therapeutic action of amphetamine in ADHD.[260-263] Such experiments have identified areas of the brain involved in making delay-discounting judgments and have been used to make predictions about the way in which brain circuits may interact during this behavior.[253,264-266] Some of these predictions have been subsequently validated in human fMRI studies (e.g.,[267]).

Researchers interested in modeling gambling in rats can draw on this body of data to inform and guide their experiments. However, one thing that is becoming increasingly apparent is that distinct aspects of impulsivity are governed by similar yet dissociable neural circuitries, and are affected differently by various neurochemical and pharmacological manipulations (see [268,253] for a more in-depth discussion of this issue). Generalizing from one type of impulse control to another is therefore ill-advised without empirical observation. Although intolerance to delayed reinforcement and behavioral disinhibition may relate to gambling behavior, these models tap into distinct psychological processes and are not equivalent to gambling *per se*. In humans, laboratory-based measures of risk-reward decision-making have been developed, the original and most widely used of which is the Iowa Gambling Task (IGT,[269]). On each trial, subjects choose cards from four decks to accumulate points. The optimal strategy is to choose cards from the two decks associated with small immediate gains but also low and infrequent losses, an approach which healthy volunteers learn during the course of the session. Persistent selection from the two disadvantageous decks leads to large immediate gain but heavy losses in the long term. This pattern of preferential risky decision-making is observed in pathological gamblers,[270] substance abusers,[271] and patients with damage to the OFC or basolateral amygdala.[272] However, despite the considerable literature that has accumulated using this, and similar, tasks to assess impulsive choice, there is currently no rodent analog of gambling paradigms, although progress is being made toward this goal.

Focusing on the process of gambling, the behavior can be broken down into a series of decisions with outcomes of variable degrees of certainty. There is an opportunity to make significant gains by choosing the uncertain or risky option, but this is offset by the risk of losing the gains already made. In terms of optimal foraging theory, rats should be able to solve such computational dilemmas as comparable problems could arise in a naturalistic setting (e.g., visit location Y for a virtually guaranteed small sum of food, or risk visiting location X where a larger amount of food is sometimes available). The value of the large reward is progressively discounted as the probability of its delivery decreases, such that a small certain reward becomes preferable to a larger uncertain reward. Aspects of gambling can therefore be reduced to a problem of probability discounting. Researchers have built on the delay-discounting task structure, substituting delay to reward delivery with probability of reward delivery to model risky behavior (e.g.,[273,274]). Rats therefore choose between small certain rewards and larger, increasingly uncertain rewards. In contrast to the delay-discounting task, risky or impulsive choice in probability-discounting paradigms is indicated by choice of the larger reward.

Although this line of research shows considerable promise, there is more to gambling behavior than simple choices between certain and uncertain rewards. One of the concerns is that in probability-discounting tasks, rats cannot lose the reward they have earned in the same way that gamblers can lose the money they are playing with. Is "not winning" psychologically equivalent to "losing what you have won"? The latter process may be particularly difficult to build into animal behavioral models. One approach currently under development is to add time-to-earn-reward into the "currency" with which rats are gambling. In such models, the animal has a set period of time in which to earn as much reward as it can. A new trial can be initiated on average every 5–6 s. On each trial, the animal chooses between four different response options, analogous to the four decks of cards in the IGT, which can lead to the delivery of small or large rewards under different probability schedules. For example, responding in one location could result in delivery of one pellet 90% of the time, whereas responding in another location could lead to delivery of three pellets 50% of the time. In this task, failure to win pellets results in both the absence of reward delivery plus a time-out period during which the animal cannot initiate any more trials to earn more reward. Such time-outs are used as punishments in some cognitive tasks, and there is evidence to suggest that rats find signaled periods of non-reward frustrating and aversive. Hence, repeated choice of the risky options not only results in the lack of reward delivery on that trial ("not winning"), but also in a loss of time in which more rewards could be earned (perhaps equivalent to "losing ground"). Preliminary data indicates that rats are capable of learning these discriminations and optimize their performance on this task very effectively.[v]

Another approach to represent this concept of loss has been to add the bitter substance quinine to the sugar pellets that rats typically earn during behavioral paradigms.[275] In this model, rats chose between frequent delivery of one un-doctored sugar pellet within a set number of quinine-treated pellets or the rare delivery of three un-doctored sugar pellets within a set number of quinine-treated pellets. As in the IGT, rats learned to avoid the options associated with the larger but less certain palatable rewards. Although innovative, there are some confounds within the experimental design, including the degree to which individual rats find the bitter taste aversive. All animals continued to eat the pellets regardless of whether they had been treated with quinine or not; therefore, it is difficult to determine whether the quinine treatment adequately captures the concept of "loss." The task is also run in a custom-built maze-like box rather than in standard operant chambers which makes the task difficult to apply in most laboratories, and both the preparation of the quinine pellets and the running of the experimental trials are labor-intensive. Nevertheless, further work with such a paradigm might provide useful insight into gambling behavior.

Another key symptom in PG is the compulsion to continue gambling in order to recover losses. This behavior is known as "loss-chasing" and appears important in the switch from recreational to PG, as reflected in the diagnostic criteria for the disorder.[276] However, little is known about the neurobiological basis underlying this behavior. Recent fMRI data indicate that engaging in loss-chasing behavior activates areas such as the vmPFC, a brain region that is activated by the expectancy of positive outcomes[277] and is relatively underactive in PG patients performing the Stroop task.[278] In contrast, activity in the anterior cingulate cortex is implicated in the decision to quit chasing after experiencing further losses.[279] Whether such a process could be modeled in animals remains an open question; that is, if a rat "loses" in one of the gambling-like paradigms outlined above, might it try and recover that loss given the option "double-or-quits"? Although such a concept sounds complicated for a rat to grasp, initial attempts to model this process have been encouraging.[vi]

Considerations for a Model of Pathological Gambling

As illustrated in the discussion above, core features of ICDs such as PG incorporate ideas of compulsive persistence in certain behaviors despite aversive consequences, loss of control, craving, and the experience of pleasure when engaged in the problem behavior. The first of these may be reflected in persistent choice of risky options

---

[v] Unpublished observations by F.D. Zeeb, D.E.H. Theobald, R.N. Cardinal, T.W. Robbins and C.A. Winstanley.
[vi] Unpublished observations by C.A. Winstanley and R.D. Rogers.

despite a reduction in the amount of reward earned. Indeed, it is possible that increased risky decision-making of this kind may be the only output variable that can be adequately obtained from animal models. Modeling the pleasurable aspect of gambling behavior will have inherent difficulties. All of the behavioral tasks so far discussed culminate in the earning of a food reward; therefore, food-restricted rats should always be motivated to engage in "the game," although individual differences related to hunger and food preference might confound findings. Rather than preference for "the thrill of the chase," choice of risky options may be attributed to an association between those actions and larger rewards. One potential way to overcome this confound would be to make the risky and non-risky options equivalent in terms of the reward earned and observe whether a proportion of animals simply prefer the more unpredictable option. Whether such a model would be useful is unclear. It is well established that rats will respond for more novel or complex stimuli and that individual differences can be observed in the degree of neophilia,[280] but whether animals will likewise differ in their preference for unpredictable or complex operant schedules remains strictly hypothetical.

Despite these concerns, animal models of gambling behavior may still be useful in terms of identifying which brain regions are critically involved in risky decision-making and how these structures are modulated by neurotransmitter systems, and thus they may stimulate the development of novel pharmacological therapies for PG. This may be achieved if the models proposed can be validated in so far as risky behavior in rats can be increased by the same pharmacological and environmental manipulations that increase risky decision-making in humans. As outlined in earlier sections of this chapter, despite some progress concerning the role of the monoaminergic neurotransmitters in PG, considerably more work is needed to determine if and how abnormalities in the DA, 5-HT, and NE systems contribute significantly to the disorder. Nevertheless, from the data available to date, one might predict that manipulations which decrease 5-HT and drugs which increase NE levels might exacerbate risky decision-making across a range of putative gambling paradigms. Stress might also be anticipated to exacerbate multiple risky behaviors. Bearing in mind the *caveats* mentioned previously when considering treatment options, opioid antagonists and lithium treatment may also be expected to reduce maladaptive gambling behavior.

In terms of the brain regions controlling gambling behavior, the neuroimaging data from PG patients discussed earlier indicate that damage to the basal ganglia and orbital/ventromedial prefrontal cortex should affect and potentially enhance risky choice or willingness to gamble. Likewise, given the suggestion that PG may represent a behavioral addiction, the neural structures involved could potentially overlap with those implicated in drug abuse. Damage to striatal regions such as the NAc should theoretically increase risky decisions, and, consistently, diminished activity in the striatum has been observed in subjects with PG (see previous discussion).

Given that many of the behavioral approaches discussed above have only recently been developed or are still in the pilot stages, many of the key experiments have yet to be performed. However, some data are available concerning the neural and neurochemical basis of probability discounting. Lesions to the NAc core decrease risky decision-making on a probability-discounting task; that is, damage to this area increases choice of the small, certain reward.[273] Given that lesions to this structure increase impulsive choice in delay-discounting paradigms, this result was somewhat unexpected. It would seem that in both cases, lesions to the NAc are promoting choice of smaller, more immediately accessible rewards. Damage to the OFC also increases choice of the smaller and more certain reward,[274] which does not match human data demonstrating that patients with damage localized to this region are more risk-seeking on the IGT. In conjunction with future data from some of the newer gambling paradigms currently under development, such findings may provide insight into which aspects of gambling or risky behavior are mediated by specific cortical and subcortical areas.

From a pharmacological perspective, administration of amphetamine and the D_2 receptor agonist bromocriptine (Parlodel®) increases risky choices on this simple probability-discounting model.[281] However, the D_2 receptor antagonist haloperidol has been associated with increased gambling-related desires in individuals with PG,[48] raising questions about a precise role for DA in PG. In animal models, globally decreasing levels of 5-HT through the intracerebroventricular administration of the serotonergic toxin, 5,7-dihydroxytryptamine did not affect choice of the larger uncertain rewards on another probability-discounting task,[282] which also may contradict some clinical findings (see earlier discussion). In terms of future experiments, the influences of noradrenergic manipulations warrant investigation, and the potentially negative effects of stress or the positive effects of clinically used treatments for PG have yet to be examined systematically in animal models or gambling or risky behaviors.

Summary

Interest is growing in developing animal models of gambling behavior, although some aspects of PG may be inherently difficult to capture in laboratory animals. As research into the manifestation of PG progresses, it will hopefully be possible to isolate specific psychological processes relevant to the clinical condition and build heuristically valid behavioral models in animals to investigate these aspects of risky decision-making. Through building on research into related forms of impulsivity, behavioral models of gambling are being developed in rodents, and it seems that these animals are capable of "playing the odds." From the data currently reported in the literature, probability-discounting models may prove particularly useful for investigating the dopaminergic component of risky decision-making that is seen in gambling and the role of the ventral striatum in these processes.

CONCLUSIONS AND RECOMMENDATIONS

Given the current lack of systematic evaluation of treatment efficacy in ICD research, it is not possible to make definitive treatment recommendations. Although several drug therapies hold promise, there are no drugs approved by the FDA for the treatment of any of the formal ICDs. Behavioral therapies have similarly received little empirical testing. Many studies are not controlled, and the randomized clinical trials typically involve small samples, short treatment durations, and limitations in diagnostic characterization of subjects. Definition of response measures and rater standards are non-uniform and variable. Given frequent study discontinuation and placebo responses, identification of the relevant factors influencing outcome would be of significant clinical relevance. Given the prevalence of ICDs in the community, the identification of empirically validated treatments would have significant impact on the health of the public. Through the continuing translation of information between clinicians and basic researchers, heuristically useful animal models of gambling behavior are currently in development. Progress on this front will hopefully lead to advances in our understanding of the neurobiological basis of maladaptive gambling and risk-taking behaviors, and ultimately stimulate research into novel pharmacotherapies for PG and other ICDs.

REFERENCES

1. American Psychiatry Association (1994). *Diagnostic and Statistical Manual of Mental Disorders*, 4th edition, Text revision edition, American Psychiatry Association.
2. Potenza, M. and Hollander, E. (2002). Pathologic gambling and impulse control disorders. In Davis, K., Charney, D., Coyle, J., and Nemeroff, C. (eds.), *Neuropsychopharmacology: The Fifth Generation of Progress*. American College of Neuropsychopharmacology, Nashville, pp. 1726-1741.
3. Grant, J. and Potenza, M. (2004). Impulse control disorders: Clinical characteristics and pharmacological management. *Ann Clin Psychiatry*, 16:27-34.
4. Kessler, R., Berglund, P., Demler, O., Jin, R., Merikangas, K., and Walters, E. (2005). Lifetime prevalence and age-of-onset distributions of DSM-IV disorders in the National Comorbidity Survey Replication. *Arch Gen Psychiatry*, 62:593-602.
5. Grant, J., Levine, L., Kim, D., and Potenza, M. (2005). Impulse control disorders in adult psychiatric inpatients. *Am J Psychiatry*, 162:2184-2188.
6. Grant, J. and Potenza, M. (2007). Treatments for pathological gambling and other impulse control disorders. In Nathan, P. and Gorman, J. (eds.), *A Guide to Treatments That Work, 3rd edition*. Oxford University Press, Oxford.
7. Ibáñez, A., Blanco, C., Moreryra, P., and Sáiz-Ruiz, J. (2003). Gender differences in pathological gambling. *J Clin Psychiatry*, 64:295-301.
8. Shaffer, H., Hall, M.N., and Vander Bilt, J. (1999). Estimating the prevalence of disordered gambling behavior in the United States and Canada: A research synthesis. *Am J Publ Health*, 89:1369-1376.
9. DeCaria, C., Begaz, T., and Hollander, E. (1998). Serotonergic and noradrenergic function in pathological gambling. *CNS Spectrums*, 3:38-47.
10. Hollander, E., Buchalter, A., and DeCaria, C. (2000). Pathological gambling. *Psychiatry Clin North Am*, 23:629-642.
11. DeCaria, C., Hollander, E., Grossman, R., Wong, C., Mosovich, S., and Cherkasky, S. (1996). Diagnosis, neurobiology, and treatment of pathological gambling. *J Clin Psychiatry*, 57 (Suppl 8):80-83. discussion 3-4
12. Raylu, N. and Oei, T. (2002). Pathological gambling. A comprehensive review. *Clin Psychol Rev*, 22:1009-1061.
13. Kim, S. and Grant, J. (2001). The psychopharmacology of pathological gambling. *Semin Clin Neuropsychiatry*, 6:184-194.
14. Sood, E., Pallanti, S., and Hollander, E. (2003). Diagnosis and treatment of pathologic gambling. *Curr Psychiatry Rep*, 5:9-15.

15. Ibáñez, A., Blanco, C., Pérez de Castro, I., Fernandez-Piqueras, J., and Sáiz-Ruiz, J. (2003). Genetics of pathological gambling. *J Gambl Stud*, 19:11-22.

16. Potenza, M. (2001). The neurobiology of pathological gambling. *Semin Clin Neuropsychiatry*, 6(3):217-226.

17. Fletcher, P. (1995). Effects of combined or separate 5,7-dihydroxytryptamine lesions of the dorsal and median raphe nuclei on responding maintained by a DRL 20s schedule of food reinforcement. *Brain Res*, 675(1-2):45-54.

18. Harrison, A., Everitt, B., and Robbins, T. (1997). Central 5-HT depletion enhances impulsive responding without affecting the accuracy of attentional performance: Interactions with dopaminergic mechanisms. *Psychopharmacology*, 133(4):329-342.

19. Crean, J., Richards, J., and deWit, H. (2002). Effect of tryptophan depletion on impulsive behavior in men with or without a family history of alcoholism. *Behav Brain Res*, 136:349-357.

20. Winstanley, C., Theobald, D., Dalley, J., Cardinal, R., and Robbins, T. (2006). Double dissociation between serotonergic and dopaminergic modulation of medial prefrontal and orbitofrontal cortex during a test of impulsive choice. *Cereb Cortex*, 16(1):106-114.

21. Moreno, I., Sáiz-Ruiz, J., and Lopez-Ibor, J. (1991). Serotonin and gambling dependence. *Hum Psychopharmacol*, 6(Suppl):9-12.

22. Blanco, C., Orensanz-Munoz, L., Blanco-Jerez, C., and Sáiz-Ruiz, J. (1996). Pathological gambling and platelet MAO activity: A psychobiological study. *Am J Psychiatry*, 153:119-121.

23. Carrasco, J., Sáiz-Ruiz, J., Hollander, E., Cesar, Jr., J., López-Ibor Jr., J.J. (1994). Low platelet monoamine oxidase activity in pathological gambling. *Acta Psychiatr Scand*. 90:427-431.

24. Grant, J., Brewer, J., and Potenza, M. (2006). The neurobiology of substance and behavioral addictions. *CNS Spectrums*, 11:924-930.

25. Benkelfat, C., Murphy, D., Hill, J., George, D., Nutt, D., and Linnoila, M. (1991). Ethanollike properties of the serotonergic partial agonist m-chlorophenylpiperazine in chronic alcoholic patients. *Arch Gen Psychiatry*, 48:383.

26. Pallanti, S., Bernardi, S., Quercioli, L., Decaria, C., and Hollander, E. (2006). Serotonin dysfunction in pathological gamblers: Increased prolactin response to oral m-CPP versus placebo. *CNS Spectrums*, 11:956-964.

27. Moss, H., Yao, J., and Panzak, G. (1990). Serotonergic responsivity and behavioral dimensions in antisocial personality associated with substance abuse. *Biol Psychiatry*, 28:325-338.

28. Hollander, E., Stein, D., DeCaria, C., Cohen, L., Saoud, J., Skodol, A. *et al.* (1994). Serotonergic sensitivity in borderline personality disorder: Preliminary findings. *Am J Psychiatry*, 151:277-280.

29. Stein, D., Hollander, E., DeCaria, C. *et al.* (1997). Behavioral responses to m-chlorophenylpiperazine and clonidine in trichotillomania. *J Serotonin Res*, 4:11-15.

30. Asberg, M., Traskman, L., and Thoren, P. (1976). 5-HIAA in the cerebrospinal fluid: A biochemical predictor. *Arch Gen Psychiatry*, 33:1193-1197.

31. Linnoila, M., Virkunnen, M., Scheinen, M., Nuutila, A., Rimon, R., and Goodwin, F. (1983). Low cerebrospinal fluid 5 hydroxyindoleacetic acid concentrations differentiates impulsive from non-impulsive violent behavior. *Life Sci*, 33:2609-2614.

32. Virkunnen, M., Nuutila, A., Goodwin, F. *et al.* (1987). Cerebrospinal fluid monoamine metabolites in male arsonists. *Arch Gen Psychiatry*, 44:241-247.

33. Coccaro, E., Siever, L., Klar, H. *et al.* (1990). Serotonergic studies in patients with affective and personality disorders: Correlates with suicidal and impulsive aggressive behavior. *Arch Gen Psychiatry*, 46:587-599.

34. Virkunnen, M., Rawlings, R., Tokola, R. *et al.* (1994). CSF biochemistries, glucose metabolism, and diurnal activity rhythms in alcoholic violent offenders, fire setters, and healthy volunteers. *Arch Gen Psychiatry*, 51:20-27.

35. Nordin, C. and Eklundh, T. (1999). Altered CSF 5-HIAA disposition in pathologic male gamblers. *CNS Spectrums*, 4:25–33.
36. Kalivas, P.W. and Volkow, N.D. (2005). The neural basis of addiction: A pathology of motivation and choice. *Am Psychiatry*, 162:1403–1413.
37. Bergh, C., Eklund, T., Sodersten, P., and Nordin, C. (1997). Altered dopamine function in pathological gambling. *Psychol Med*, 27(2):473–475.
38. Blum, K., Cull, J., Braverman, E., and Comings, D. (1996). Reward deficiency syndrome. *Am Sci*, 84:132–145.
39. Galanter, J. and Wartenberg, A. *et al.* (eds.), (2005). Pharmacology of chemical dependence and addiction. In Golan, D., Tashjian, A., Armstrong, E., Galanter, J., Armstrong, A., Arnaout, R. *Principles of Pharmacology: The Physiologic Basis of Drug Therapy*. Lippincott, Williams, & Wilkins, Philadelphia, PA, pp. 247–263.
40. Nestler, E. and Aghajanian, G. (1997). Molecular and cellular basis of addiction. *Science*, 278:58–62.
41. Dalley, J., Cardinal, R., and Robbins, T. (2004). Prefrontal executive and cognitive functions in rodents: Neural and neurochemical substrates. *Neurosci Biobehav Rev*, 28:771–784.
42. Goldman-Rakic, P. (1998). The prefrontal landscape: implications of frontal architecture for understanding human mentiation and the central executive. In Roberts, A., Robbins, T., and Weiskrantz, L. (eds.), *The Prefrontal Cortex: Executive and Cognitive Functions*. Oxford University Press, Oxford.
43. Zilles, K. and Wree, A. (1995). Cortex: Areal and laminar structure. In Paxinos, G. (ed.), *The Rat Nervous System, 2nd edition*. Academic Press, London.
44. Breiter, H., Gollub, R., Weisskopf, R., Kennedy, D., Makris, N., Burke, J. *et al.* (1997). Acute effects of cocaine on human brain activity and emotion. *Neuron*, 19:591–611.
45. Maas, L., Lukas, S., Kaufman, M., Weiss, R., Daniels, S., Rogers, V. *et al.* (1998). Functional magnetic resonance imaging of human brain activation during cue-induced cocaine craving. *Am J Psychiatry*, 155:124–126.
46. Volkow, N., Wang, G., Fischman, M., Foltin, R., Fowler, J., Abumrad, N. *et al.* (1997). Relationship between subjective effects of cocaine and dopamine transporter occupancy. *Nature*, 386:827–830.
47. Potenza, M., Steinberg, M., Skudlarski, P., Fulbright, R., Lacadie, C., Wilber, M. *et al.* (2003). Gambling urges in pathological gambling: A functional magnetic resonance imaging study. *Arch Gen Psychiatry*, 60:828–836.
48. Zack, M. and Poulos, C. (2007). A D2 antagonist enhances the rewarding and priming effects of a gambling episode in pathological gamblers. *Neuropsychopharmacology*, 32(8):1678–1686.
49. Meyer, G., Schwertfeger, J., Exton, M.S., Janssen, O.E., Knapp, W., Stadler, M.A. *et al.* (2004). Neuroendocrine response to casino gambling in problem gamblers. *Psychoneuroendocrinology*, 29(10):1272–1280.
50. Shinohara, K., Yanagisawa, A., Kagota, Y., Gomi, A., Nemoto, K., Moriya, E. *et al.* (1999). Physiological changes in Pachinko players; beta-endorphin, catecholamines, immune system substances and heart rate. *Appl Human Sci*, 18(2):37–42.
51. Weintraub, D. and Potenza, M. (2006). Impulse control disorders in Parkinson's disease. *Curr Neurol Neurosci Rep*, 6:302–306.
52. Bejjani, B., Damier, P., Arnulf, I., Thivard, L., Bonnet, A., Dormont, D. *et al.* (1999). Transient acute depression induced by high-frequency deep-brain-stimulation. *New Engl J Med*, 340:1476–1480.
53. Stocchi, F. (2005). Pathological gambling in Parkinson's disease. *Lancet Neurol*, 4:590–592.
54. Molina, J., Sainz-Artiga, M., Fraile, A., Jimenez-Jimenez, F., Villanueva, C., Orti-Pareja, M. *et al.* (2000). Pathologic gambling in Parkinson's disease: A behavioral manifestation of pharmacologic treatment? *Mov Disord*, 15:869–872.

55. Voon, V., Hassan, K., Zurowski, M., de Souza, M., Thomsen, T., Fox, S. *et al.* (2006). Prevalence of repetitive and reward-seeking behaviors in Parkinson disease. *Neurology*, 67:1254-1257.

56. Klos, K., Bower, J., Josephs, K.A., Matsumoto, J., and Ahlskog, J. (2005). Pathological hypersexuality predominantly linked to adjuvant dopamine agonist therapy in Parkinson's disease and multiple system atrophy. *Parkinsonism Relat Disord*, 11:381-386.

57. Nirenberg, M. and Waters, C. (2006). Compulsive eating and weight gain related to dopamine agonist use. *Mov Disord*, 21:524-529.

58. Pezzella, F., Colosimo, C., Vanacore, N., Rezze, S.D., Chianese, M., Fabbrini, G. *et al.* (2005). Prevalence and clinical features of hedonistic homeostatic dysregulation in Parkinson's disease. *Mov Disord*, 20:77-81.

59. Giovannoni, G., O'Sullivan, J.D., Turner, K., Manson, A., and Lees, A. (2000). Hedonistic homeostatic dysregulation in patients with Parkinson's disease on dopamine replacement therapies. *J Neurol Neurosurg Psychiatry*, 68:423-428.

60. Voon, V., Hassan, K., Zurowski, M., Duff-Canning, S., de Souza, M., Fox, S. *et al.* (2006). Prospective prevalence of pathologic gambling and medication association in Parkinson disease. *Neurology*, 66:1750-1752.

61. Weintraub, D., Siderowf, A., Potenza, M., Goveas, J., Morales, K., Duda, J. *et al.* (2006). Association of dopamine agonist use with impulse control disorders in Parkinson disease. *Arch Neurol*, 63:969-973.

62. Sokoloff, P., Giros, B., Martres, M., Bouthenet, M., and Schwartz, J.C. (1990). Molecular cloning and characterization of a novel dopamine receptor (D3) as a target for neuroleptics. *Nature*, 347:146-151.

63. Murray, A., Ryoo, H., Gurevich, E., and Joyce, J. (1994). Localization of dopamine D3 receptors to mesolimbic and D2 receptors to mesostriatal regions of human forebrain. *PNAS*, 91:11271-11275.

64. Voon, V., Thomsen, T., Miyasaki, J., Souza, M.d., Shafro, A., Fox, S. *et al.* (2007). Factors associated with dopaminergic drug-related pathological gambling in Parkinson disease. *Arch Neurol*, 64:212-216.

65. Roy, A., Adinoff, B., Roehrich, L., Lamparski, D., Custer, R., Lorenz, V. *et al.* (1988). Pathologic gambling: A psychobiological study. *Arch Gen Psychiatry*, 45:369-373.

66. Roy, A., Custer, R., Lorenz, V., and Linnoila, M. (1989). Personality factors and pathological gambling. *Acta Psychiatry Scand*, 80:37-39.

67. Zuckerman, M. (1979). *Sensation seeking: Beyond the optimal level of arousal.* Lawrence Erlbaum Associates, Hillsdale, NJ.

68. Potenza, M. (2002). Pathological gambling: Clinical aspects and neurobiology. In Soares, J. and Gershon, S. (eds.), *Handbook of Medical Psychiatry*. Marcel Dekker, Inc., New York, pp. 683-699.

69. Schmitt, L., Harrison, G., and Spargo, R. (1998). Variation in epinephrine and cortisol excretion rates associated with behavior in an Australian Aboriginal community. *Am J Phys Anthropol*, 106:249-253.

70. Roy, A., Jong, J.D., and Linnoila, M. (1989). Extraversion in pathological gambles: Correlates with indexes of noradrenergic function. *Arch Gen Psychiatry*, 46:679-681.

71. Deutch, A.Y., Roth, R.H. (1999). Neurochemical systems in the central nervous system In: D.S. Charney, E.J. Nesler, B.S. Bunney (Eds.), Neurobiology of Mental Illness, Oxford University Press, New York, pp. 10-25.

72. Levitt, P., Pintar, J., and Brackefield, X. (1982). Immunochemical demonstration of monoamine oxidase B in brain astrocytes and serotonin neurons. *PNAS*, 79:6385-6389.

73. Oreland, L., Wiberg, A., Asberg, M., Traskman, L., Sjostrand, L., Thoren, P. *et al.* (1981). Platelet MAO activity and monoamine metabolites in cerebrospinal fluid in depressed and suicidal patients and in healthy controls. *Psychiatry Res*, 4:21-29.

74. Buchsbaum, M., Haier, R., and Murphy, D. (1977). Suicide attempts, platelet monoamine oxidase and the average evoked response. *Acta Psychiatry Scand*, 56:69-79.

75. von Knorring, L., Oreland, L., and von Knorring, A.L. (1987). Personality traits and platelet MAO activity in alcohol and drug abusing teenage boys. *Acta Psychiatry Scand*, 75:307-314.

76. Fowler, C., von Knorring, L., and Oreland, L. (1980). Platelet monoamine activity in sensation seekers. *Psychiatry Res*, 3:273-279.

77. Ward, P., Catts, S., Norman, T., Burrows, G., and McConaghy, N. (1987). Low platelet monoamine oxidase and sensation-seeking in males: An established relationship? *Acta Psychiatry Scand*, 75:86-90.

78. Hallman, J., Sakurai, E., and Oreland, L. (1990). Blood platelet monoamine oxidase activity, serotonin uptake, and release rates in anorexia and bulimia patients and in healthy controls. *Acta Psychiatry Scand*, 81:73-77.

79. Caspi, A., McClay, J., Moffitt, T., Mill, J., Martin, J., Craig, I. *et al.* (2002). Role of genotype in the cycle of violence in maltreated children. *Science*, 297:851-854.

80. Scherrer, J., Xian, H., Kapp, J., Waterman, B., Shah, K., Volberg, R. *et al.* (2007). Association between exposure to childhood and lifetime traumatic events and lifetime pathological gambling in a twin cohort. *J Nerv Mental Dis*, 195:72-78.

81. Sinha, R. (2005). Stress and drug abuse. In Steckler, T., Kalin, N., and Reul, J. (eds.), *Handbook of Stress and the Brain. Part 2: Stress: Integrative and Clinical Aspects*. Elsevier, Amsterdam, pp. 333-356.

82. Meyer, G., Hauffa, B., Schedlowski, M., Pawlak, C., Stadler, M., and Exton, M. (2000). Casino gambling increases heart rate and salivary cortisol in regular gamblers. *Biol Psychiatry*, 48:948-953.

83. Krueger, T.H.C., Schedlowski, M., and Meyer, G. (2005). Cortisol and heart rate measures during casino gambling in relation to impulsivity. *Neuropsychobiology*, 52:206-211.

84. Kim, S. (1998). Opioid antagonists in the treatment of impulse-control disorders. *J Clin Psychiatry*, 59:159-162.

85. Grant, J. and Kim, S. (2006). Medication management of pathological gambling. *Minn Med*, 89:44-48.

86. Phillips, A. and LePiane, F. (1980). Reinforcing effects of morphine microinjection on to the ventral tegmental area. *Pharmacol Biochem Behav*, 12:965-968.

87. von Wolfswinkel, L. and van Ree, J.M. (1985). Effects of morphine and naloxone on thresholds of ventral tegmental electrical self-stimulation. *Naunyn Schmiedebergs Arch Pharmacol*, 330(2):84-92.

88. Grant, J. and Kim, S. (2002). An open label study of naltrexone in the treatment of kleptomania. *J Clin Psychiatry*, 63:349-356.

89. Kim, S., Grant, J., Adson, D., and Shin, Y. (2001). Double-blind naltrexone and placebo comparison study in the treatment of pathological gambling. *Biol Psychiatry*, 49:914-921.

90. Oslin, D., Berrettini, W., and O'Brien, C. (2006). Targeting treatments for alcohol dependence: The pharmacogenetics of naltrexone. *Addict Biol*, 11(3-4):397-403.

91. Potenza, M., Leung, H., Blumberg, H., Peterson, B., Fulbright, R., Lacadie, C. *et al.* (2003). An fMRI Stroop study of ventromedial prefrontal cortical function in pathological gamblers. *Am J Psychiatry*, 160:1990-1994.

92. Saxena, S. and Rauch, S.L. (2000). Functional neuroimaging and the neuroanatomy of obsessive-compulsive disorder. *Psychiatry Clin North Am*, 23:1-19.

93. Dougherty, D., Rauch, S., Deckersbach, T., Marci, C., Loh, R., Shin, L. *et al.* (2004). Ventromedial prefrontal cortex and amygdala dysfunction during an anger induction positron emission tomography study in patients with major depressive disorder with anger attacks. *Arch Gen Psychiatry*, 61:795-804.

94. New, A., Hazlett, E., Buchsbaum, M., Goodman, M., Reynolds, D., Mitropoulou, V. *et al.* (2002). Blunted prefrontal cortical 18-fluorodeoxyglucose positron emission tomography response to meta-chlorophenylpiperazine in impulsive aggression. *Arch Gen Psychiatry*, 59:621–629.

95. Potenza, M. (2006). Should addictive disorders include non-substance-related conditions? *Addiction*, 101(Suppl 1):142–151.

96. Siever, L., Buchsbaum, M., New, A., Spiegel-Cohen, J., Wei, T., Hazlett, E. *et al.* (1999). D,L-fenfluramine response in impulsive personality disorder assessed with [18F]fluorodeoxyglucose positron emission tomography. *Neuropsychopharmacology*, 20:413–423.

97. Blumberg, H., Leung, H., Skudlarski, P., Lacadie, C., Fredericks, C., Harris, B. *et al.* (2003). A functional magnetic resonance imaging study of bipolar disorder: State- and trait-related dysfunction in ventral prefrontal cortices. *Arch Gen Psychiatry*, 60:601–609.

98. Reuter, J., Raedler, T., Rose, M., Hand, I., Glascher, J., and Buchel, C. (2005). Pathological gambling is linked to reduced activation of the mesolimbic reward system. *Nat Neurosci*, 8:147–148.

99. Bjork, J., Knutson, B., Fong, G., Caggiano, D., Bennett, S., and Hommer, D. (2004). Incentive-elicited brain activation in adolescents: Similarities and differences from young adults. *J Neurosci*, 24:1793–1802.

100. Hommer, D. (Ed.) (2004). Motivation in alcoholism. *International Conference on Applications of Neuroimaging to Alcoholism* (accessed 12 January 2005), New Haven, CT.

101. Hommer, D., Bjork, J., Knutson, B., Caggiano, D., Fong, G., and Danube, C. (2004). Motivation in children of alcoholics. *Alcohol Clin Exp Res*, 28(5):22A.

102. Gottesman, I. and Gould, T. (2003). The endophenotype concept in psychiatry: Etymology and strategic intentions. *Am J Psychiatry*, 160:636–645.

103. Crockford, D., Goodyear, B., Edwards, J., Quickfall, J., and el-Guebaly, N. (2005). Cue-induced brain activity in pathological gamblers. *Biol Psychiatry*, 58:787–795.

104. Hollander, E., Pallanti, S., Baldini, R., Sood, E., Baker, B., and Buchsbaum, M. (2005). Imaging monetary reward in pathological gamblers. *World J Biol Psychiatry*, 6:113–120.

105. Eisen, S., Lin, N., Lyons, M., Scherrer, J., Griffith, K., True, W. *et al.* (1998). Familial influences on gambling behavior: An analysis of 3359 twin pairs. *Addiction*, 93(9):1375–1384.

106. Shah, K., Eisen, S., Xian, H., and Potenza, M. (2005). Genetic studies of pathological gambling: A review of methodology and analyses of data from the Vietnam Era Twin Registry. *J Gambl Stud*, 21:179–203.

107. Slutske, W., Eisen, S., True, W., Lyons, M., Goldberg, J., and Tsuang, M. (2000). Common genetic vulnerability for pathological gambling and alcohol dependence in men. *Arch Gen Psychiatry*, 57:666–673.

108. Slutske, W., Eisen, S., Xian, H., True, W., Lyons, M., Goldberg, J. *et al.* (2005). A twin study of the association between pathological gambling and antisocial personality disorder. *J Abnorm Psychol*, 110:297–308.

109. Potenza, M., Xian, H., Shah, K., Scherrer, J., and Eisen, S. (2005). Shared genetic contributions to pathological gambling and major depression in men. *Arch Gen Psychiatry*, 62:1015–1021.

110. Ibanez, A., de Castro, I.P., Fernandez-Piqueras, J., Blanco, C., and Sáiz-Ruiz, J. (2000). Pathological gambling and DNA polymorphic markers at MAO-A and MAO-B genes. *Mol Psychiatry*, 20:105–109.

111. Pérez de Castro, I., Ibanez, A., Sáiz-Ruiz, J., and Fernandez-Piqueras, J. (1999). Genetic contribution to pathological gambling: Possible association between a functional DNA polymorphism at the serotonin transporter gene (5-HTT) and affected men. *Pharmacogenetics*, 9(3):397–400.

112. Comings, D., Rosenthal, R., Lesieur, H., Rugle, L., Muhleman, D., Chiu, C. *et al.* (1996). A study of the dopamine D2 receptor gene in pathological gambling. *Pharmacogenetics*, 6(3):223–234.

113. Comings, D., Gade, R., Wu, S., Chiu, C., Dietz, G., Muhleman, D. *et al.* (1997). Studies of the potential role of the dopamine D1 receptor gene in addictive behaviors. *Mol Psychiatry*, 2(1):44–56.

114. Comings, D., Gonzalez, N., Wu, S., Gade, R., Muhleman, D., Saucier, G. *et al.* (1999). Studies of the 48 bp repeat polymorphism of the DRD4 gene in impulsive, compulsive, addictive behaviors: Tourette syndrome, ADHD, pathological gambling, and substance abuse. *Am J Med Gen*, 88:358–368.

115. Pérez de Castro, I., Ibanez, A., Torres, P., Sáiz-Ruiz, J., and Fernandez-Piqueras, J. (1997). Genetic association study between pathological gambling and a functional DNA polymorphism at the D4 receptor. *Pharmacogenetics*, 7:345–348.

116. WHO International Statistical Classification of Diseases and Related Health Problems: World Health Organization (2003). Report No.: 10th Revision.

117. Blanco, C., Moreyra, P., Nunes, E., Sáiz-Ruiz, J., and Ibanez, A. (2001). Pathological gambling: Addiction or compulsion?. *Semin Clin Neuropsychiatry*, 6:167–176.

118. Chambers, R., Taylor, J., and Potenza, M. (2003). Developmental neurocircuitry of motivation in adolescence: A critical period of addiction vulnerability. *Am J Psychiatry*, 160:1041–1052.

119. Tavares, H., Martins, S., Lobo, D., Silveira, C., Gentil, V., and Hodgins, D. (2003). Factors at play in faster progression for female pathological gamblers: An exploratory analysis. *J Clin Psychiatry*, 64:433–438.

120. Tavares, H., Zilberman, M., Beites, F., and Gentil, V. (2001). Gender differences in gambling progression. *J Gambl Stud*, 17:151–160.

121. Pallesen, S., Molde, H., Arnestad, H., Laberg, J., Skutle, A., Iversen, E. *et al.* (2007). Outcome of pharmacological treatments of pathological gambling: A review and meta-analysis. *J Clin Psychopharmacol*, 27:357–364.

122. Goodman, W., Price, L., Delgado, P., Palumbo, J., Krystal, J., Nagy, L. *et al.* (1990). Specificity of serotonin reuptake inhibitors in the treatment of obsessive–compulsive disorder, comparison of fluvoxamine and desipramine. *Arch Gen Psychiatry*, 47(6):577–585.

123. Leonard, H., Swedo, S., Rapoport, J., Koby, E., Lenane, M., Cheslow, D. *et al.* (1989). Treatment of obsessive–compulsive disorder with clomipramine and desipramine in children and adolescents. A double-blind crossover comparison. *Arch Gen Psychiatry*, 46:1088–1092.

124. Doudet, D., Hommer, D., Higley, J., Andreason, P., Moneman, R., Suomi, S. *et al.* (1995). Cerebral glucose metabolism, CSF 5-HIAA levels, and aggressive behavior in rhesus monkeys. *Am J Psychiatry*, 152:1782–1787.

125. Virkkunen, M., Goldman, D., Nielsen, D., and Linnoila, M. (1995). Low brain serotonin turnover rate (low CSF 5-HIAA) and impulsive violence. *J Psychiatry Neurosci*, 20:271–275.

126. Mehlman, P., Higley, J., Faucher, I., Lilly, A., Taub, D., Vickers, J. *et al.* (1995). Low CSF 5-HIAA concentrations and severe aggression and impaired impulse control in nonhuman primates. *Am J Psychiatry*, 151:1485–1491.

127. Coccaro, E. (1996). Neurotransmitter correlates of impulsive aggression in humans. *Ann NY Acad Sci*, 794:82–89.

128. Davidson, R., Putnam, K., and Larson, C. (2000). Dysfunction in the neural circuitry of emotion regulation – a possible prelude to violence. *Science*, 289:591–594.

129. Hollander, E., Frenkel, M., Decaria, C., Trungold, S., and Stein, D. (1992). Treatment of pathological gambling with clomipramine. *Am J Psychiatry*, 149:710–711.

130. Sáiz-Ruiz, J., Blanco, C., Ibáñez, A., Masramon, X., Gómez, M., Madrigal, M. *et al.* (2005). Sertraline treatment of pathological gambling: A pilot study. *J Clin Psychiatry*, 66(1):28–33.

131. Hollander, E., DeCaria, C., Mari, E., Wong, C., Mosovich, S., Grossman, R. *et al.* (1998). Short-term single-blind fluvoxamine treatment of pathological gambling. *Am J Psychiatry*, 155:1781–1783.

132. Hollander, E., DeCaria, C., Finkell, J., Begaz, T., Wong, C., and Cartwright, C. (2000). A randomized double-blind fluvoxamine/placebo crossover trial in pathologic gambling. *Biol Psychiatry*, 47:813–817.

133. Blanco, C., Petkova, E., Ibáñez, A., and Sáiz-Ruiz, J. (2002). A pilot placebo-controlled study of fluvoxamine for pathological gambling. *Ann Clin Psychiatry*, 14:9–15.

134. Grant, J., Kim, S., and Potenza, M. (2003). Advances in the pharmacological treatment of pathological gambling. *J Gambl Stud*, 19:85–109.

135. Grant, J. and Potenza, M. (2006). Escitalopram treatment of pathological gambling with co-occurring anxiety: An open-label pilot study with double-blind discontinuation. *Int Clin Psychopharmacol*, 21:203–209.

136. Haller, R. and Hinterhuber, H. (1994). Treatment of pathological gambling with carbamazepine. *Pharmacopsychiatry*, 27(3):129.

137. Pallanti, S., Quercioli, L., Sood, E., and Hollander, E. (2002). Lithium and valproate treatment of pathological gambling: A randomized single-blind study. *J Clin Psychiatry*, 63:559–564.

138. Hollander, E., Pallanti, S., Allen, A., Sood, E., and Rossi, N. (2005). Does sustained-release lithium reduce impulsive gambling and affective instability versus placebo in pathological gamblers with bipolar spectrum disorders? *Am J Psychiatry*, 162:137–145.

139. Pallanti, S. Lithium and valproate treatment of pathological gambling: A randomized, single-blind study. *15th National Council on Problem Gambling Conference*, 2005, Seattle, Washington.

140. Chambers, R. and Potenza, M. (2001). Schizophrenia and pathological gambling. *Am J Psychiatry*, 158(3):497–498.

141. Rugle, L. A double-blind, placebo-controlled trial of olanzapine in the treatment of pathological gambling. *National Center for Responsible Gambling Conference on Comorbidity*, 2000, Las Vegas, NV.

142. Kalivas, P. and Barnes, C. (1993). *Limbic Motor Circuits and Neuropsychiatry*. CRC Press, Boca Raton, FL.

143. Koob, G. (1992). Drugs of abuse: Anatomy, pharmacology and function of reward pathways. *Trends Pharmacol Sci*, 13(5):177–184.

144. Kim, S. and Grant, J. (2001). An open naltrexone treatment study in pathological gambling disorder. *Int Clin Psychopharmacol*, 16:285–289.

145. Kim, S., Grant, J., Adson, D., and Remmel, R. (2001). A preliminary report on possible naltrexone and nonsteroidal analgesic interactions. *J Clin Psychopharmacol*, 21:632–634.

146. Grant, J., Potenza, M., Hollander, E., Cunningham-Williams, R., Nurminen, T., Smits, G. *et al.* (2006). A multicenter investigation of the opioid antagonist nalmefene in the treatment of pathologic gambling. *Am J Psychiatry*, 163:303–312.

147. Ladouceur, R., Sylvain, C., Boutin, C., Lachance, S., Doucet, C., and Leblond, J. (2003). Group therapy for pathological gamblers: A cognitive approach. *Behav Res Ther*, 41:587–596.

148. Ladouceur, R., Sylvain, C., Boutin, C., Lachance, S., Doucet, C., Leblond, J. *et al.* (2001). Cognitive treatment of pathological gambling. *J Nerv Ment Dis*, 189:774–780.

149. Sylvain, C., Ladouceur, R., and Boisvert, J. (1997). Cognitive and behavioral treatment of pathological gambling: A controlled study. *J Consult Clin Psychol*, 65:727–732.

150. Hodgins, D., Currie, S., and el-Guebaly, N. (2001). Motivational enhancement and self-help treatments for problem gambling. *J Consult Clin Psychol*, 69:50–57.

151. Hodgins, D., Currie, S., el-Guebaly, N., and Peden, N. (2004). Brief motivational treatment for problem gambling: A 24-month follow-up. *Psychol Addict Behav*, 18:293–296.

152. McConaghy, N., Armstrong, M., Blaszczynski, A., and Allcock, C. (1983). Controlled comparison of aversive therapy and imaginal desensitization in compulsive gambling. *Br J Psychiatry*, 142:366–372.

153. McConaghy, N., Blaszczynski, A., and Frankova, A. (1991). Comparison of imaginal desensitisation with other behavioural treatments of pathological gambling. A two- to nine-year follow-up. *Br J Psychiatry*, 159:390-393.

154. Christenson, G., Mackenzie, T., Mitchell, J., and Callies, A. (1991). A placebo-controlled double-blind crossover study of fluoxetine in trichotillomania. *Am J Psychiatry*, 148:1566-1571.

155. Grant, J., Odlaug, B., and Potenza, M. (2007). Addicted to hair pulling? How an alternate model of trichotillomania may improve treatment outcome. *Harv Rev Psychiatry*, 15:80-85.

156. Swedo, S. and Leonard, H. (1992). Trichotillomania. An obsessive compulsive spectrum disorder? *Psychiatry Clin North Am*, 15:777-790.

157. Christenson, G. and Mansueto, C. (1999). Trichotillomania: Descriptive characteristics and phenomenology. In Stein, D., Christenson, G., and Hollander, E. (eds.), *Trichotillomania*. American Psychiatric Publishing, Washington, DC, pp. 1-42.

158. Swedo, S., Leonard, H., Rapoport, J., Lenane, M., Goldberger, E., and Cheslow, D. (1989). A double-blind comparison of clomipramine and desipramine in the treatment of trichotillomania (hair pulling). *New Engl J Med*, 321:497-501.

159. Pigott, T., L'Heueux, F., Grady, T. Controlled comparison of clomipramine and fluoxetine in trichotillomania, in *Abstracts of Panels and Posters of the 31st Annual Meeting of the American College of Neuropsychopharmacology*, San Juan, Puerto Rico, December 1992.

160. O'Sullivan, R., Christenson, G., and Stein, D. (1999). Pharmacotherapy of trichotillomania. In Stein, D., Christenson, G., and Hollander, E. (eds.), *Trichotillomania*. American Psychiatric Press, Washington, DC, pp. 93-123.

161. Streichenwein, S. and Thornby, J. (1995). A long-term, double-blind, placebo-controlled crossover trial of the efficacy of fluoxetine for trichotillomania. *Am J Psychiatry*, 152(8): 1192-1196.

162. Elliot, A. and Fuqua, R. (2000). Trichotillomania: Conceptualization, measurement, and treatment. *Behav Ther*, 31:529-545.

163. Woods, D., Flessner, C., Franklin, M., Wetterneck, C., Walther, M., Anderson, E. *et al.* (2006). Understanding and treating trichotillomania: What we know and what we don't know. *Psychiatry Clin North Am*, 29:487-501.

164. Azrin, N. and Nunn, R. (1973). Habit reversal: A method of eliminating nervous habits and tics. *Behav Res Ther*, 11:619-628.

165. Rapp, J., Miltenberger, R., Long, E., Elliott, A., and Lumley, V. (1998). Simplified habit reversal treatment for chronic hair pulling in three adolescents: A clinical replication with direct observation. *J Appl Behav Anal*, 31:299-302.

166. Lerner, J., Franklin, M., Meadows, E., Hembree, E., and Foa, E. (1998). Effectiveness of a cognitive behavioral treatment program for trichotillomania: An uncontrolled evaluation. *Beh Ther*, 29:157-171.

167. Pelissier, M. and O'Connor, K. (2004). Cognitive-behavioral treatment of trichotillomania, targeting perfectionism. *Clin Case Stud*, 3:57-69.

168. Rangaswami, K. (1997). Management of a case of trichotillomania by cognitive behaviour therapy. *Indian J Clin Psychol*, 24:89-92.

169. Azrin, N., Nunn, R., and Frantz, S. (1980). Treatment of hairpulling (trichotillomania): A comparative study of habit reversal and negative practice training. *J Behav Ther Exp Psychiatry*, 11:13-20.

170. Woods, D., Wetterneck, C., and Flessner, C. (2006). A controlled evaluation of acceptance and commitment therapy plus habit reversal for trichotillomania. *Behav Res Ther*, 44(5):639-656.

171. Ninan, P., Rothbaum, B., Martsteller, F., Knight, B., and Eccard, M. (2000). A placebo-controlled trial of cognitive-behavioral therapy and clomipramine in trichotillomania. *J Clin Psychiatry*, 61:47-50.

172. van Minnen, A., Hoogduin, K., Keijsers, G., Hellenbrand, I., and Hendriks, G. (2003). Treatment of trichotillomania with behavioral therapy or fluoxetine. *Arch Gen Psychiatry*, 60:517-522.

173. Keuthen, N., O'Sullivan, R., Ricciardi, J., Shera, D., Savage, C., and Borgmann, A. (1995). The Massachusetts General Hospital (MGH) Hairpulling Scale. 1: Development and factor analysis. *Psychother Psychosom*, 64:141-145.

174. Koran, L., Bullock, K., Hartston, H., Elliott, M., and D'Andrea, V. (2002). Citalopram treatment of compulsive shopping: An open-label study. *J Clin Psychiatry*, 63:704-708.

175. Black, D. (2007). A review of compulsive buying disorder. *World Psychiatry*, 6(1):14-18.

176. Christenson, G., Faber, R., Zwaan, M.D., Raymond, N., Specker, S., Ekern, M. *et al.* (1994). Compulsive buying: Descriptive characteristics and psychiatric comorbidity. *J Clin Psychiatry*, 55:5-11.

177. Schlosser, S., Black, D., Repertinger, S., and Freet, D. (1994). Compulsive buying: Demography, phenomenology, and comorbidity in 46 subjects. *Gen Hosp Psychiatry*, 16:205-212.

178. Black, D. (1996). Compulsive buying: A review. *J Clin Psychiatry*, 57(Suppl 8):50-54.

179. Ninan, P., McElroy, S., Kane, C., Knight, B., Casuto, L., Rose, S. *et al.* (2000). Placebo-controlled study of fluvoxamine in the treatment of patients with compulsive buying. *J Clin Psychopharmacol*, 20:362-366.

180. Black, D., Gabel, J., Hansen, J., and Schlosser, S. (2000). A double-blind comparison of fluvoxamine versus placebo in the treatment of compulsive buying disorder. *Ann Clin Psychiatry*, 12:205-211.

181. Koran, L., Chuong, H., Bullock, K., and Smith, S.C. (2003). Citalopram for compulsive shopping disorder: An open-label study followed by double-blind discontinuation. *J Clin Psychiatry*, 64:793-798.

182. Bernik, M., Akerman, D., Amaral, J., and Brayn, R. (1996). Cue exposure in compulsive buying. *J Clin Psychiatry*, 57:90.

183. Lawrence, L. (1990). The psychodynamics of the compulsive female shopper. *Am J Psychoanal*, 50:67-70.

184. Miltenberger, R., Redlin, J., Crosby, R., Stickney, G., Mitchell, J., Wonderlich, S. *et al.* (2003). Direct and retrospective assessment of factors contributing to compulsive buying. *J Behav Ther Exp Psychiatry*, 34:1-9.

185. Kyrios, M., Frost, R., and Steketee, G. (2004). Cognitions in compulsive buying and acquisition. *Cognit Ther Res*, 28:241-258.

186. Mitchell, J., Burgard, M., Faber, R., Crosby, R., and Zwaan, M.D. (2006). Cognitive behavioral therapy for compulsive buying disorder. *Behav Res Ther*, 44:1859-1865.

187. Coccaro, E. (1998). Impulsive aggression: A behavior in search of clinical definition. *Harv Rev Psychiatry*, 5:336-339.

188. Moeller, F., Barratt, E., Dougherty, D., Schmitz, J., and Swann, A. (2001). Psychiatric aspects of impulsivity. *Am J Psychiatry*, 158:1783-1793.

189. Hollander, E., Tracy, K., Swann, A.C., Coccaro, E., McElroy, S., Wozniak, P. *et al.* (2003). Divalproex in the treatment of impulsive aggression: Efficacy in cluster B personality disorders. *Neuropsychopharmacology*, 28:1186-1197.

190. Coccaro, E., Schmidt, C., Samuels, J., and Nestadt, G. (2004). Lifetime and 1-month prevalence rates of intermittent explosive disorder in a community sample. *J Clin Psychiatry*, 65:820-824.

191. Nelson, R. and Trainor, B. (2007). Neural mechanisms of aggression. *Nat Rev Neurosci*, 8:536-546.

192. McElroy, S., Soutullo, C., Beckman, D., Taylor Jr., P., Keck Jr., P.E. (1998). DSM-IV intermittent explosive disorder: A report of 27 cases. *J Clin Psychiatry*, 59(4):203-210.

193. Campbell, M., Adams, P., Small, A., Kafantaris, V., Silva, R., Shell, J. *et al.* (1995). Lithium in hospitalized aggressive children with conduct disorder: A double-blind and placebo-controlled study. *J Am Acad Child Adolesc Psychiatry*, 34:445-453.

194. Malone, R., Delaney, M., Luebbert, J., Cater, J., and Campbell, M. (2000). A double-blind placebo-controlled study of lithium in hospitalized aggressive children and adolescents with conduct disorder. *Arch Gen Psychiatry*, 57:649–654.

195. Hori, A. (1998). Pharmacotherapy for personality disorders. *Psychiatry Clin Neurosci*, 52:13–19.

196. Tupin, J., Smith, D., Clanon, T., Kim, L., Nugent, A., and Groupe, A. (1973). The long-term use of lithium in aggressive prisoners. *Compr Psychiatry*, 14:311–317.

197. Sheard, M., Marini, J., Bridges, C., and Wagner, E. (1976). The effect of lithium on impulsive aggressive behavior in man. *Am J Psychiatry*, 133:1409–1413.

198. Frankenburg, F. and Zanarini, M. (2002). Divalproex sodium treatment of women with borderline personality disorder and bipolar II disorder: A double-blind, placebo controlled pilot study. *J Clin Psychiatry*, 63:442–446.

199. Hollander, E., Allen, A., Lopez, R., Bienstock, C., Grossman, R., Siever, L. *et al.* (2001). A preliminary double-blind, placebo-controlled trial of divalproex sodium in borderline personality disorder. *J Clin Psychiatry*, 62:199–203.

200. Tariot, P., Schneider, L., Mintzer, J., Cutler, A., Cunningham, M., Thomas, J. *et al.* (2001). Safety and tolerability of divalproex sodium in the treatment of signs and symptoms of mania in elderly patients with dementia: Results of a double-blind, placebo-controlled trial. *Curr Ther Res*, 62:51–67.

201. Barratt, E., Stanford, M., Felthous, A., and Kent, T. (1997). The effects of phenytoin on impulsive and premeditated aggression: A controlled study. *J Clin Psychopharmacol*, 5:341–349.

202. Cowdry, R. and Gardner, D. (1988). Pharmacotherapy of border-line personality disorder: Alprazolam, carbamazepine, trifluoperazine, and tranylcypromine. *Arch Gen Psychiatry*, 45:111–119.

203. Kavoussi, R. and Coccaro, E. (1998). Divalproex sodium for impulsive–aggressive behavior in patients with personality disorder. *J Clin Psychiatry*, 59:676–680.

204. Wilcox, J. (1995). Divalproex sodium as a treatment for borderline personality disorder. *Ann Clin Psychiatry*, 7:3–7.

205. Stein, D., Simeon, D., Frenkel, M., Islam, M., and Hollander, E. (1995). An open-trial of valproate in borderline personality disorder. *J Clin Psychiatry*, 56:506–510.

206. Horne, M. and Lindley, S. (1995). Divalproex sodium in the treatment of aggressive behavior and dysphoria in patients with organic brain syndrome. *J Clin Psychiatry*, 56:430–431.

207. Donovan, S., Susser, E., Nunes, E., Stewart, J., Quitkin, F., and Klein, D. (1997). Divalproex treatment of disruptive adolescents: A report of 10 cases. *J Clin Psychiatry*, 58:12–15.

208. Donovan, S., Stewart, J., Nunes, E., Quitkin, F., Parides, M., Daniel, W. *et al.* (2000). Divalproex treatment for youth with explosive temper and mood lability: A double-blind, placebo-controlled crossover design. *Am J Psychiatry*, 157:818–820.

209. Giakas, W., Seibyl, J., and Mazure, C. (1990). Valproate in the treatment of temper outbursts. *J Clin Psychiatry*, 51:525.

210. Petty, F., Davis, L., Nugent, A., Kramer, G., Teten, A., Schmitt, A. *et al.* (2002). Valproate therapy for chronic combat-induced posttraumatic stress disorder. *J Clin Psychopharmacol*, 22:100–101.

211. Fesler, F. (1991). Valproate in combat related posttraumatic stress disorder. *J Clin Psychiatry*, 52:361–363.

212. Szymanski, H. and Olympia, J. (1991). Divalproex in posttraumatic stress disorder. *Am J Psychiatry*, 148:1086–1087.

213. Haas, S., Vincent, K., Holt, J., and Lippman, S. (1997). Divalproex: A possible treatment alternative for demented elderly aggressive patients. *Ann Clin Psychiatry*, 9:145–147.

214. Itil, T. and Wadud, A. (1975). Treatment of human aggression with major tranquilizers, antidepressants, and newer psychotropic agents. *J Nerv Ment Dis*, 160:83–99.

215. Chengappa, K., Ebeling, T., Kang, J., Levine, J., and Parepally, H. (1999). Clozapine reduces severe self-mutilation and aggression in psychotic patients with borderline personality disorder. *J Clin Psychiatry*, 60:477-484.
216. Allain, H., Dautzenberg, P., Maurer, K., Schuck, S., Bonhomme, D., and Gerard, D. (2000). Double blind study of tiapride versus haloperidol and placebo in agitation and aggressiveness in elderly patients with cognitive impairment. *Psychopharmacology (Berl)*, 148:361-366.
217. Khouzam, H.R. and Donnelly, N.J. (1997). Remission of self-mutilation in a patient with borderline personality disorder during risperidone therapy. *J Nerv Mental Dis*, 185:348-349.
218. Zanarini, M. and Frankenburg, F. (2001). Olanzapine treatment of female borderline personality disorder patients: A double-blind, placebo-controlled pilot study. *J Clin Psychiatry*, 62:849-854.
219. Fava, M., Rosenbaum, J.F., Pava, J., McCarthy, M., Steingard, R., and Bouffides, E. (1993). Anger attacks in unipolar depression. Part 1: Clinical correlates and response to fluoxetine treatment. *Am J Psychiatry*, 150:1158-1163.
220. Salzman, C., Wolfson, A., Schatzberg, A., Looper, J., Henke, R., Albanese, M. *et al.* (1995). Effect of fluoxetine on anger in symptomatic volunteers with borderline personality disorder. *J Clin Psychopharmacol*, 15:23-29.
221. Coccaro, E. and Kavoussi, R. (1997). Fluoxetine and impulsive aggressive behavior in personality-disordered subjects. *Arch Gen Psychiatry*, 54:1081-1088.
222. Grant, J., Odlaug, B. (2008). Kleptomania: clinical characteristics and treatment. *Rev Bras Psiquiatr,* 30(1):S11-S15.
223. Grant, J. and Kim, S. (2002). Clinical characteristics and associated psychopathology of 22 patients with kleptomania. *Compr Psychiatry*, 43:378-384.
224. McElroy, S., Pope, H.G., Jr., Hudson, J.I., Keck P.E., Jr., White, K.L. (1991). Kleptomania: a report of 20 cases. *Am J Psychiatry*. 148:652-657.
225. Presta, S., Marazziti, D., Dell'Osso, L., Pfanner, C., Pallanti, S., and Cassano, G.B. (2002). Kleptomania: Clinical features and comorbidity in an Italian sample. *Compr Psychiatry*, 43:7-12.
226. Sarasalo, E., Bergman, B., and Toth, J. (1996). Personality traits and psychiatric and somatic morbidity among kleptomaniacs. *Acta Psychiatry Scand*, 94:358-364.
227. Kraus, J. (1999). Treatment of kleptomania with paroxetine. *J Clin Psychiatry*, 60(11):793.
228. Chong, S. and Low, B. (1996). Treatment of kleptomania with fluvoxamine. *Acta Psychiatry Scand*, 93:314-315.
229. Aboujaoude, E., Gamel, N., and Koran, L. (2004). A case of kleptomania correlating with premenstrual dysphoria. *J Clin Psychiatry*, 65:725-726.
230. Feeney, D. and Klykylo, W. (1997). Treatment for kleptomania. *J Am Acad Child Adolesc Psychiatry*, 36:723-724.
231. Durst, R., Katz, G., Teitelbaum, A., Zislin, J., and Dannon, P. (2001). Kleptomania: Diagnosis and treatment options. *CNS Drugs*, 15:185-195.
232. Kmetz, G., McElroy, S., and Collins, D. (1997). Response of kleptomania and mixed mania to valproate. *Am J Psychiatry*, 154:580-581.
233. Lepkifker, E., Dannon, P., Ziv, R., Iancu, I., Horesh, N., and Kotler, M. (1999). The treatment of kleptomania and serotonin reuptake inhibitors. *Clin Neuropharmacol*, 22:40-43.
234. Dannon, P. (2003). Topiramate for the treatment of kleptomania: A case series and review of the literature. *Clin Neuropharmacol*, 26:1-4.
235. Dannon, P., Iancu, I., and Grunhaus, L. (1999). Naltrexone treatment in kleptomanic patients. *Hum Psychopharmacol*, 14:583-585.
236. Grant, J. (2005). Outcome study of kleptomania patients treated with naltrexone: A chart review. *Clin Neuropharmacol*, 28:11-14.
237. Aboujaoude, E., Koran, L., Gamel, N. (eds.). Escitalopram in the treatment of kleptomania. *25th Meeting of the New Clinical Drug Evaluation Unit*, 6-9 June, Boca Raton, FL.

238. Guidry, L. (1968). Use of a covert punishing contingency in compulsive stealing. *J Behav Ther Exp Psychiatry*, 6:169.

239. Gauthier, J. and Pellerin, D. (1982). Management of compulsive shoplifting through covert sensitization. *J Behav Ther Exp Psychiatry*, 13:73-75.

240. Glover, J. (1985). A case of kleptomania treated by covert sensitization. *Br J Clin Psychiatry*, 24:213-214.

241. Keutzer, C. (1972). Kleptomania: A direct approach to treatment. *Br J Med Psychol*, 45:159-163.

242. McConaghy, N. and Blaszczynski, A. (1988). Imaginal desensitization: A cost-effective treatment in two shoplifters and a binge-eater resistant to previous therapy. *Aust NZ J Psychiatry*, 22:78-82.

243. Aizer, A., Lowengrub, K., and Dannon, P. (2004). Kleptomania after head trauma: Two case reports and combination treatment strategies. *Clin Neuropharmacol*, 27:211-215.

244. McNeilly, D. and Burke, W. (1998). Stealing lately: A case of late-onset kleptomania. *Int J Geriatr Psychiatry*, 13:116-121.

245. Haller, J. and Kruk, M.R. (2006). Normal and abnormal aggression: Human disorders and novel laboratory models. *Neurosci Biobehav Rev*, 30:292-303.

246. Evenden, J.L. (1999). Varieties of impulsivity. *Psychopharmacology*, 146:348-361.

247. Logan, G.D. (1994). On the ability to inhibit thought and action. A users' guide to the stop signal paradigm. In Dagenbach, D. and Carr, T.H. (eds.), *Inhibitory Processes in Attention, Memory and Language*. Academic Press, San Diego, CA, pp. 189-236.

248. Rosvold, H.E., Mirsky, A.F., Sarason, I., Bransome, E.D., and Beck, L.H. (1956). A continuous performance test of brain damage. *J Consult Psychol*, 20:343-350.

249. Carli, M., Robbins, T.W., Evenden, J.L., and Everitt, B.J. (1983). Effects of lesions to ascending noradrenergic neurons on performance of a 5-choice serial reaction time task in rats – implications for theories of dorsal noradrenergic bundle function based on selective attention and arousal. *Behav Brain Res*, 9:361-380.

250. Eagle, D.M. and Robbins, T.W. (2003). Lesions of the medial prefrontal cortex or nucleus accumbens core do not impair inhibitory control in rats performing a stop-signal reaction time task. *Behav Brain Res*, 146:131-144.

251. Feola, T.W., de Wit, H., and Richards, J.B. (2000). Effects of d-amphetamine and alcohol on a measure of behavioral inhibition in rats. *Behav Neurosci*, 114:838-848.

252. Robbins, T.W. (2002). The 5-choice serial reaction time task: Behavioural pharmacology and functional neurochemistry. *Psychopharmacology (Berl)*, 163:362-380.

253. Winstanley, C.A., Eagle, D.M., and Robbins, T.W. (2006). Behavioral models of impulsivity in relation to ADHD: Translation between clinical and preclinical studies. *Clin Psychol Rev*, 26(4):379-395.

254. Kagan, J. (1966). Reflection - impulsivity: The generality and dynamics of conceptual tempo. *J Abnorm Psychol*, 71:17-24.

255. Evenden, J.L. (1999). The pharmacology of impulsive behaviour in rats VII: The effects of serotonergic agonists and antagonists on responding under a discrimination task using unreliable visual stimuli. *Psychopharmacology*, 146:422-431.

256. Ainslie, G. (1975). Specious reward: A behavioral theory of impulsiveness and impulse control. *Psychol Bull*, 82:463-498.

257. Evenden, J.L. and Ryan, C.N. (1996). The pharmacology of impulsive behaviour in rats: The effects of drugs on response choice with varying delays of reinforcement. *Psychopharmacology*, 128:161-170.

258. Mazur, J. (1987). An adjusting procedure for studying delayed reinforcement. In Commons, M.L., Mazur, J.E., Nevin, J.A., Rachlin, H.C. (ed.), *Quantitative Analyses of Behaviour: The Effect of Delay and Intervening Events on Reinforcement Value*. Erlbaum, Hillsdale, NJ, pp. 55-73.

259. Wogar, M.A., Bradshaw, C.M., and Szabadi, E. (1991). Evidence for an involvement of 5-hydroxytryptaminergic neurons in the maintenance of operant-behavior by positive reinforcement. *Psychopharmacology*, 105:119–124.

260. Cardinal, R.N., Robbins, T.W., and Everitt, B.J. (2000). The effects of d-amphetamine, chlordiazepoxide, alpha-flupenthixol and behavioural manipulations on choice of signalled and unsignalled delayed reinforcement in rats. *Psychopharmacology (Berl)*, 152:362–375.

261. van Gaalen, M.M., van Koten, R., Schoffelmeer, A.N., and Vanderschuren, L.J. (2006). Critical involvement of dopaminergic neurotransmission in impulsive decision making. *Biol Psychiatry*, 60:66–73.

262. Winstanley, C.A., Theobald, D.E., Dalley, J.W., and Robbins, T.W. (2003). Global 5-HT depletion attenuates the ability of amphetamine to decrease impulsive choice in rats. *Psychopharmacology*, 170:320–331.

263. Winstanley, C.A., Theobald, D.E., Dalley, J.W., and Robbins, T.W. (2005). Interactions between serotonin and dopamine in the control of impulsive choice in rats: Therapeutic implications for impulse control disorders. *Neuropsychopharmacology*, 30(4):669–682.

264. Cardinal, R.N., Pennicott, D.R., Sugathapala, C.L., Robbins, T.W., and Everitt, B.J. (2001). Impulsive choice induced in rats by lesions of the nucleus accumbens core. *Science*, 292:2499–2501.

265. Winstanley, C.A., Baunez, C., Theobald, D.E., and Robbins, T.W. (2005). Lesions to the subthalamic nucleus decrease impulsive choice but impair autoshaping in rats: The importance of the basal ganglia in Pavlovian conditioning and impulse control. *Eur J Neurosci*, 21(11):3107–3116.

266. Winstanley, C.A., Theobald, D.E., Cardinal, R.N., and Robbins, T.W. (2004). Contrasting roles for basolateral amygdala and orbitofrontal cortex in impulsive choice. *J Neurosci*, 24:4718–4722.

267. McClure, S.M., Laibson, D.I., Loewenstein, G., and Cohen, J.D. (2004). Separate neural systems value immediate and delayed monetary rewards. *Science*, 306:503–506.

268. Winstanley, C.A. (2007). The orbitofrontal cortex, impulsivity and addiction: Probing orbitofrontal dysfunction at the neural, neurochemical and molecular level. *Ann NY Acad Sci*, 1121:639–655.

269. Bechara, A., Damasio, A.R., Damasio, H., and Anderson, S.W. (1994). Insensitivity to future consequences following damage to human prefrontal cortex. *Cognition*, 50:7–15.

270. Cavedini, P., Riboldi, G., Keller, R., D'Annucci, A., and Bellodi, L. (2002). Frontal lobe dysfunction in pathological gambling patients. *Biol Psychiatry*, 51(4):334–341.

271. Bechara, A., Dolan, S., Denburg, N., Hindes, A., Anderson, S.W., and Nathan, P.E. (2001). Decision-making deficits, linked to a dysfunctional ventromedial prefrontal cortex, revealed in alcohol and stimulant abusers. *Neuropsychologia*, 39:376–389.

272. Bechara, A., Damasio, H., Damasio, A.R., and Lee, G.P. (1999). Different contributions of the human amygdala and ventromedial prefrontal cortex to decision-making. *J Neurosci*, 19:5473–5481.

273. Cardinal, R.N. and Howes, N.J. (2005). Effects of lesions of the nucleus accumbens core on choice between small certain rewards and large uncertain rewards in rats. *BMC Neurosci*, 6(1):37.

274. Mobini, S., Body, S., Ho, M.Y., Bradshaw, C.M., Szabadi, E., Deakin, J.F.W. *et al.* (2002). Effects of lesions of the orbitofrontal cortex on sensitivity to delayed and probabilistic reinforcement. *Psychopharmacology (Berl)*, 160:290–298.

275. van den Bos, R., Lasthuis, W., den Heijer, E., van der Harst, J., and Spruijt, B. (2006). Toward a rodent model of the Iowa gambling task. *Behav Res Meth*, 38(3):470–478.

276. Lesieur, H.R. (1979). The compulsive gambler's spiral of options and involvement. *Psychiatry*, 42(1):79–87.

277. Knutson, B., Fong, G.W., Bennett, S.M., Adams, C.M., and Hommer, D. (2003). A region of mesial prefrontal cortex tracks monetarily rewarding outcomes: Characterization with rapid event-related fMRI. *Neuroimage*, 18(2):263–272.

278. Potenza, M.N., Leung, H.C., Blumberg, H.P., Peterson, B.S., Fulbright, R.K., Lacadie, C.M. *et al.* (2003). An FMRI Stroop task study of ventromedial prefrontal cortical function in pathological gamblers. *Am J Psychiatry*, 160(11):1990–1994.

279. Campbell-Meiklejohn, D.K., Woolrich, M.W., Passingham, R.E., Rogers, R.D. (2008). Knowing when to stop: The brain mechanisms of chasing losses. *Biol Psychiatry*, 63(3):293–300.

280. Hughes, R.N. (2007). Neotic preferences in laboratory rodents: Issues, assessment and substrates. *Neurosci Biobehav Rev*, 31:441–464.

281. St. Onge, J.R., Floresco, S.B. (2007). Dopaminergic regulation of risky decision-making. *Neuroscience Meeting Planner*. Society for Neuroscience, San Diego, CA, Program No. 741.4 [Online].

282. Mobini, S., Chiang, T.J., Ho, M.Y., Bradshaw, C.M., and Szabadi, E. (2000). Effects of central 5-hydroxytryptamine depletion on sensitivity to delayed and probabilistic reinforcement. *Psychopharmacology*, 152:390–397.

Translational Models for the 21st Century: Reminiscence, Reflections, and Some Recommendations

Paul Willner[1], Franco Borsini[2] and Robert A. McArthur[3]

[1]Department of Psychology, Swansea University, Swansea, Wales, UK
[2]sigma-tau S.p.A., Pomezia, Roma, Italy
[3]McArthur and Associates GmbH, Ramsteinerstrasse, Basel, Switzerland

INTRODUCTION

This series has provided a systematic overview of translational research in psychopharmacology that integrates the perspectives of academics, industrial pharmacologists, and clinicians, in therapeutic areas representing three broad therapeutic domains, neurological disorders, psychiatric disorders, and reward/impulse control disorders. An earlier attempt to systematize translational research in psychopharmacology summarized the scope of the endeavor as follows:

> *The problem of using animal behavior to model human mental disorders is, explicitly or implicitly, the central preoccupation of psychopharmacology. The idea that psychiatric disorders might be modeled in animals provides the basis for a substantial proportion of current research; and research that does not employ behavioral models directly is usually justified by reference to its eventual benefits, in terms of an understanding of the human brain and the development of more effective and safer therapies.[1]*

In effect, the principle that translation from animals to humans is both possible and necessary underpins and justifies the whole field of preclinical psychopharmacology.

The forerunner to the present series, *Behavioural Models in Psychopharmacology: Academic, Theoretical and Industrial Perspectives*[2] was intended to generalize an approach first outlined in a more focused earlier review of animal models of depression.[3] These publications introduced four significant ideas: that drug screening tests should not be thought of as models and differ in their practical and evidential requirements; that academics, industrial pharmacologists, and clinicians view preclinical behavioral models from different perspectives, based on their different professional aspirations; that it is important to establish, or at least, estimate, the validity of animal models of psychiatric disorders; and that this is best achieved by considering validity from different aspects that reflect different bodies of evidence.

In the 20 or so years since these ideas were first introduced, they have achieved a wide degree of acceptance, and the first two are very apparent in the structure and

Animal and Translational Models for CNS Drug Discovery,
Vol. 3 of 3: Reward Deficit Disorders
Robert McArthur and Franco Borsini (eds), Academic Press, 2008

content of the present series. The three approaches to the validation of animal models that were outlined in these publications, using terms borrowed from the realm of psychometric testing, were predictive validity ("performance in the test predicts performance in the condition being modeled"), face validity ("there are phenomenological similarities between the two"), and construct validity ("the model has a sound theoretical rationale").[2] This analysis has since become standard, as seen, for example, in recent reviews of animal models of anxiety[4] and schizophrenia.[5] Indeed, these principles are now so well established that many contributors to the present series begin their discussions by considering the face, predictive, and construct validity of the models used and being developed in their therapeutic areas.[i] The extent to which these concepts have become absorbed into the language of psychopharmacology is demonstrated by the fact that they are now almost always used without attribution to the original publications.

Much has changed in terms of model development in the last 20 years; particularly in neurology. As has been discussed in the neurological therapeutic areas surveyed in this series, much of the work has been developed more recently, and it is noticeable that these chapters tend to have a different flavor to those chapters surveying psychiatric and reward deficit disorders, being heavily dominated by genetic and genomic models. In psychiatry, on the other hand, while these areas have also been heavily influenced by the genomic revolution, in behavioral terms, models of depression, anxiety, schizophrenia, or substance abuse have seen relatively little development. That is not to say that there have not been developments in the procedures by which models of depressed-like, anxious-like, schizophrenic-like or substance-abuse-like behaviors can be assessed. Indeed, there is now a significantly greater choice of procedures. However, the newer developments are largely more sophisticated variants on procedures that were well established prior to 1990.

Schizophrenia represents perhaps the one major therapeutic area where a significant shift in emphasis has been evident over the past 20 years. From the discovery of neuroleptics in the early 1950s[6,7] and the establishment of the dopamine hypothesis of schizophrenia,[8] drug discovery and development of pharmaceuticals used to treat the positive, that is, psychotic, symptoms of schizophrenia have centered around impairment of excessive dopaminergic function. Effects of mesolimbic dopaminergic imbalance have been relatively straightforward to model, mainly using changes in locomotion.[ii]

---

[i] For further and detailed discussions regarding criteria of validity, the reader is invited to refer to discussions, for example, by Steckler et al., Developing novel anxiolytics: Improving preclinical detection and clinical assessment; Large et al., Developing therapeutics for bipolar disorder: From animal models to the clinic; Joel et al., Animal models of obsessive–compulsive disorder: From bench to bedside via endophenotypes and biomarkers, in Volume 1, *Psychiatric Disorders*; Lindner et al., Development, optimization and use of preclinical behavioral models to maximize the productivity of drug discovery for Alzheimer's Disease; Merchant et al., Animal models of Parkinson's Disease to aid drug discovery and development, in Volume 2, *Neurologic Disorders*; Koob, The role of animal models in reward deficit disorders: Views from academia; Markou et al., Contribution of animal models and preclinical human studies to medication development for nicotine dependence, in Volume 3, *Reward Deficit Disorders*.

[ii] Please refer to Jones et al., Developing new drugs for schizophrenia: From animals to the clinic, in Volume 1, Psychiatric Disorders. For further discussion regarding models and procedures in schizophrenia research.

Notwithstanding the importance of dopamine in schizophrenia, awareness of the influence of other neurotransmitters such as serotonin, glutamate and GABA has led to considerable research into the role of these and their interactions with dopamine.[9,10] Pharmaceutical interest in glutamate in schizophrenia has been particularly fruitful with the development of the metabotropic glutamate$_{2/3}$ (mGlu$_{2/3}$) receptor agonist LY2140023, which appears to be as effective as olanzapine (Zyprexa®) as an antipsychotic in a Phase II trial.[11,12] mGlu$_{2/3}$ receptor agonists have been discovered and characterized using classical locomotor activity procedures or pharmacological interaction studies,[13-15] but it is interesting to note that they also show positive effects on newer procedures that have been developed to assess drug effects on models of negative (i.e., social withdrawal, lack of affect and drive, poverty of speech) and cognitive symptoms of schizophrenia.[15,16] Such procedures include sensory gating and pre-pulse inhibition,[17] sustained or focused attention,[18,19] and impaired social interaction.[20] Furthermore, greater awareness of neurodevelopmental and environmental factors of schizophrenia has led to the establishment of models such as neo-natal hippocampal lesions,[21] and to attempts to model effects of peri-partum complications as risk factors of schizophrenia.[22]

Stress as a predisposing factor, the role of the hypothalamo-pituitary-adrenal (HPA) axis and reactions to stress influenced by the environment and development, are themes discussed not only in the chapters of the Psychiatry volume, such as depression,[23] anxiety,[24] bipolar disorders,[25] and attention deficit hyperactivity disorder (ADHD),[26] but also the chapters representing the reward and impulse deficit disorders; disorders that share a high psychiatric component.[27-29] Consequently, major efforts have been made to model the effects of chronic stress and influences of development on psychiatric, reward deficit, and impulse control disorders. Behavioral developments in these areas are concerned almost exclusively with the ways in which stressors are applied and their effects evaluated. The procedures used to evaluate these changes have not changed markedly in the past 20 years. However, the way in which these procedures are used has been greatly influenced by the "endophenotypic" approach to considering psychiatric disorders and how to model aspects of these disorders.[23,24,iii] Similarly, models of reward and impulse disorders continue to be dominated by self-administration and place conditioning procedures. There is certainly a greater use of more sophisticated procedures, such as second order or progressive ratio reinforcement schedules[30,iv] but the procedures themselves have been well established for decades.[31]

Nevertheless, despite the relatively minor changes in the behavioral methodologies, research in these "traditional" areas of translational modeling is markedly different

---

[iii] Please refer to McArthur and Borsini, preface to this volume, "What do *you* mean by translational research? An enquiry through animal and translational models for CNS drug discovery," and references within for further discussion on endophenotypes and deconstructing behavioral syndromes.

[iv] See also Koob, The role of animal models in reward deficit disorders: Views from academia; and Rochas *et al.*, Development of medications for heroin and cocaine addiction and regulatory aspects of abuse liability testing in Volume 3, *Reward Deficits Disorders*.

from 20 years, ago, since they too have been transformed by the genomic revolution, which now dominates and largely determines directions of change, dwarfing developments in behavioral technologies. Almost every chapter in this series bears witness to the fact that current research is highly productive in proposing new models of aspects of psychiatric and impulse control disorders that, like the new neurological models, are based almost exclusively on the use of genetically modified animals.[32,33] Molecular biology has also had a huge impact on the understanding of intracellular processes, particularly in relation to those intracellular sequelae of transmitter–receptor interactions that result in changes in gene expression. These developments were driven in part by the recognition that successful pharmacotherapy for psychiatric disorders – depression in particular – often requires chronic drug treatment. Consequently, there is an increasing use in preclinical studies of chronic dosing strategies modeled on the time course of clinical action, leading to a search for the biochemical mechanisms that underlie slowly developing neuroadaptive changes and the identification of a plethora of novel intracellular targets.[34-38]

A similar dynamic is apparent in relation to the understanding of brain structure and function, which serves as the context for translational research. Twenty years ago, psychopharmacology was preoccupied almost exclusively with neurotransmitters; specifically the monoamines dopamine, norepinephrine and serotonin, acetylcholine and GABA, and their receptors. There had been a shift away from the original paradigm that saw the brain as a "biochemical soup," with a recognition that the anatomical site of action was also a crucial consideration, but little attention was paid to the underlying neural circuitry. Over the past 20 years, progress in understanding neurotransmitters and their receptors has been slowly incremental, particularly in relation to glutamatergic and peptidergic systems, as well as the cannabinoid systems, which represent the only major truly novel development. On the other hand, the understanding of brain circuitry has increased exponentially. The importance of dopamine and the mesolimbic and nigrostriatal systems in terms of movement (Parkinson's) and mood (schizophrenia), helped focus research on a more "neural systems" approach. Subsequently, examination of the relationships and interactions between neurotransmitter and neuromodulators within discrete neural circuits has helped to clarify neural substrates of behavioral processes, with the result that the basic operating principles of the brain are being increasingly well understood.[39-41] The parallel development of cognitive neuroscience has been accompanied by technological innovations, most notably, brain scanning methodologies, which have promoted a substantial growth of studies in human participants that complement the traditional work in animal models.

Over the past two decades, then, it is the dramatic developments in molecular biology and in behavioral and cognitive neuroscience, rather than developments in psychopharmacology itself, that have had the major impact on both the theory and practice of translational research, as exemplified by the contributions that comprise the present series. And the fundamental dilemma of psychopharmacology was already apparent 20 years ago, and remains so today: that following an early "golden age" all of the investment in academic and industrial research has produced tiny returns by way of novel pharmacotherapies. For 30, perhaps 40, years now, psychopharmacological research has not substantially changed the pharmacological armamentarium with which to treat devastating and life-threatening behavioral disorders. It has often been pointed out, but bears repeating, that it was 1958, now half a century ago, when the first publication appeared on the clinical efficacy of tricyclic[42] and monoamine oxidase inhibiting[43] antidepressants which, together with neuroleptics a few years earlier[6,7] and benzodiazepines a few years later,[44,45] completed the triad of innovations that have dominated clinical psychopharmacology ever since. Shortly afterwards, basic pharmacotherapy for neurologic disorders was revolutionized by the elucidation of the role of dopamine in Parkinson's disease[46,47] followed by the use of *L*-dopa in its treatment.[48] The concept of neurologic disorders related to neurotransmitter deficiency and its treatment by replacement of the neurotransmitter was one of the driving forces behind the research into acetylcholine potentiation by cholinesterase inhibitors in Alzheimer's disease,[49] which culminated with the registration of tacrine (Cognex®), a pro-cholinergic drug for dementia, in 1993.

Subsequent clinical developments following all of these innovations have been disappointingly unimpressive. For example, second or third generation antidepressants cannot be claimed as pharmacological innovations, since they target known properties of tricyclics, that is, serotonin and/or noradrenergic inhibition; furthermore, the efficacy of these drugs as antidepressants is increasingly questioned.[50] Atypical neuroleptics retain as an essential feature the major property of typical neuroleptics, dopamine receptor antagonism, tempered by additional properties, typically, serotonin receptor antagonism; they offer no new mechanism and very limited clinical advantage, and the prototype of this class of drug, clozapine (Clozaril®), was introduced as long ago as 1989. As indicated above, the early signs of efficacy of the mGlu$_{2/3}$ receptor agonist LY404039,[11] offers some hope of moving beyond dopaminergic compounds in the treatment of schizophrenia,[12] but this development is still at a very early stage. Acetylcholinesterase inhibition has been the standard treatment for dementia since the introduction of tacrine in 1993. Tacrine is hardly ever prescribed now: donepezil (Aricept®) is one of the drugs of choice, but even this "gold standard" is of limited effectiveness and duration.[51,52] Glutamatergic drugs offer some hope of a genuinely novel approach but the clinical efficacy of, in particular, memantine (Ebixa®/Namenda®) is extremely modest.[53,54] In Parkinson's disease, there has been no improvement, as monotherapy at least, on *L*-dopa. In contrast, new drugs have been approved for the treatment of reward deficit disorders such as alcoholism [naltrexone (Vivitrol®/Revia®) and acamprosate (Campral®)], nicotine abuse [varenicline (Chantix®/Champix®), bupropion (Zyban®)], and heroin abuse [methadone (Dolophine®), buprenorphine (Subutex®), and buprenorphine/naloxone (Suboxone®)]. These drugs, however, treat the symptoms of these disorders, and in the case of heroin abuse, the treatment is limited to replacement of heroin. We are still waiting for more effective drugs to be developed for all therapeutic indications surveyed in this book series.

And the certainties of 20–30 years ago have been questioned. Two major issues have emerged in the development and use of psychotherapeutics for the treatment of behavioral disorders: the placebo response, or the ability of certain patients to recover without medication,[50,55,56] and treatment-resistance,[57-59] which has led to initiatives such as the STAR*D trial.[60] For psychiatric indications the traditional triad of anxiety, depression, and schizophrenia are no longer viewed as self-contained entities: comorbidity is recognized as a major factor in the phenomenology of these disorders.[61] Traditional medications for one indication have now become drugs of choice for another. For example, selective serotonin reuptake inhibitor (SSRI) anti-depressants such as fluoxetine (Prozac®), paroxetine (Paxil®), sertraline (Zoloft®), fluvoxamine (Luvox®), or escitalopram (Lexapro®) are all approved by the US Food and Drug Administration (FDA) for the treatment of various anxiety disorders. SSRIs as well as tricyclic antidepressants such as imipramine (Tofranil®) and desipramine (Norpramin®) indeed are now considered first-line treatment for general anxiety disorder.[62] Anti-epileptics such as lamotrigine (Lamictal®) are being used to treat depression, while antipsychotics such as olanzapine, as well as lamotrigine, are treatments of choice for bipolar disorder.[v] Research into previously under-treated disorders such as obsessive–compulsive disorder (OCD), ADHD, or autistic spectrum disorders has also contributed to a blurring of the boundaries of traditional psychiatry.[vi] The gradual realization that psychiatric disorders are etiologically complex,[63] and that it is unrealistic to take a simple view that one pill might "cure" depression, or anxiety, or schizophrenia, has important implications for the modeling of these disorders.

What has gone wrong? Perhaps the most frustrating thing of all is that there is no clear answer to this question. One potential explanation lies in the wide anatomical distribution of most drug receptors, and the likelihood that systemically administered drugs have multiple effects that may in some cases be mutually antagonistic. Studies in genomic models employing regionally specific conditional knockouts and other genetic techniques[64-71] may go some way to addressing this issue. However, another, more disturbing, potential explanation arises from a consideration of the validity of psychiatric diagnoses. The predominant contemporary approach to diagnosis, based on the use of diagnostic criteria, defines disorders in terms of lists of symptoms, with patients typically required to display one or more items from each of a list of core and subsidiary symptoms.[72,73] It has frequently been pointed out that this can lead to a situation in which two patients who share the same diagnosis may not share any symptoms in common.[74] There have been attempts to introduce a different approach based around a systematic focus on specific symptoms,[75-77] but the failure of this eminently sensible alternative testifies to the hegemony of diagnostic criteria in determining what constitutes permissible clinical research design.[vii]

There have, of course, been numerous instances where compounds are identified in animal models as having clinical potential. However, it is now almost routine for novel pharmacotherapies, identified in models that are assessed as having a high

---

[v] Please refer to Large *et al.*, Developing therapeutics for bipolar disorder: From animal models to the clinic, in Volume 1, *Psychiatric Disorders*, for further discussion of psychotherapeutics for bipolar disorder.

[vi] Please refer to Joel *et al.*, Animal models of obsessive–compulsive disorder: From bench to bedside via endophenotypes and biomarkers; and Tannock *et al.*, An integrative assessment of attention deficit hyperactivity disorder: From biological comprehension to the discovery of novel therapeutic agents for a neurodevelopmental disorder, in Volume 1, *Psychiatric Disorders*.

[vii] Please refer to McEvoy and Freudenreich, Issues in the design and conductance of clinical trials, in Volume 1, *Psychiatric Disorders*; and Schneider, Issues in design and conduct of clinical trials for cognitive-enhancing drugs, in Volume 2, *Neurologic Disorders*, for further discussion regarding changes to standard clinical trial design.

degree of validity, to fail in the clinic. Just as examples, ondansetron, a 5-HT$_3$ receptor antagonist, was reported to antagonize scopolamine-induced effects in marmosets[78] but not in humans;[79] talsaclidine, a muscarinic receptor agonist, did not exert the same effects on muscarinic M$_2$ and M$_3$ receptors in animals and humans,[80] while repinotan, a 5-HT$_{1A}$ receptor agonist, reduced the neurological deficit in animal models of stroke but not in humans.[81] Other 5-HT$_{1A}$ receptor agonists, flibanserin, and gepirone[23,82,83] were found to be active in the majority of animal models sensitive to antidepressants but failed in clinical trials. TCH346 and CEP1347 seemed very promising as antiparkinsonian agents in preclinical experiments but failed in clinical trials.[84] Tramiprosate (Alzhemed) was found to reduce β-amyloid brain and plasma levels in animals[85] and cerebrospinal levels in humans,[86] but failed to meet its primary endpoint in a Phase III clinical study,[87,viii] although there are claims that significant reductions in hippocampal atrophy were observed with Alzhemed and that the drug itself will be distributed as a food supplement.[88]

How are we to understand these predictive failures? They might represent a failure of the model to predict clinical efficacy, but there are several other potential explanations. One, suggested above, is that a clinical trial could fail to recognize a successful compound because the population of participants identified by the use of diagnostic criteria is clinically heterogeneous. Another explanation, given that drugs are usually administered to animals at much higher doses than they are to people, even after taking into account inter-species scaling factors, is that a clinical failure could reflect the use of an inefficient clinical dose.[23,25,28,84,89–93] Or there could be pharmacokinetic/metabolism issues; in particular, problems with bioavailability, which account for a high proportion of failures at an early stage of clinical testing.[89,91,93] Also, other considerations that go beyond scientific matters, involving corporate policies and politics, may contribute to these predictive failures.[94,95] On the other side of the coin, it is also important to recognize that successful clinical trials may not always be all they seem. In particular, recent meta-analyses of antidepressant efficacy have drawn attention to the fact that, for a variety of reasons, positive outcomes are much more likely to be published than negative outcomes, and that when all of the data are taken into account, antidepressants appear significantly less efficacious than has been generally assumed.[55,96,97] There is no reason to suppose that this is a problem peculiar to antidepressants.

In the search for alternative ways forward, two new concepts have been embraced within translational research: endophenotypes and biomarkers.[ix] Biomarkers are essential to follow the progress of a disorder and the effects of therapeutic intervention. In terms of translational research, the same biomarker of the disorder should be mirrored in the animal model. However, so far very few biomarkers have been validated and this

---

viii See Schneider, Issues in design and conduct of clinical trials for cognitive-enhancing drugs, in Volume 2, *Neurologic Disorders*, for further discussion on primary and secondary endpoints in Alzheimer clinical trials.

ix Please refer to McArthur and Borsini, preface to this volume, "What do *you* mean by translational research? An enquiry through animal and translational models for CNS drug discovery," and references within for further discussion on endophenotypes and biomarkers.

remains a focus of intensive research and of great concern (cf.,[50,84,91,93,98-100]). As discussed in the preface to this volume, there is a notable proposal that the concept of validity for biomarkers should be deconstructed into face, predictive and construct dimensions, as used routinely to evaluate the validity of translational models.[101]

As discussed in many chapters of this book series, endophenotypes refer to the heritable, state-independent and family-associated characteristics of a disorder that can be linked to a candidate gene or gene region, which can help in the genetic analysis of the disorder.[102-104] An endophenotypic approach to modeling requires as a first step the identification of "… critical components of behavior (or other neurobiological traits) that are more representative of more complex phenomena" ([102] p. 641). To date, there are very few putative behavioral endophenotypes that actually meet the five criteria proposed by Gottesman and Gould[102] of: (1) a specific association with the illness of interest; (2) heritability; (3) state-independence; (4) greater prevalence among ill relatives than in the general population (familial association) and (5) than in well relatives (co-segregation). For example, Hasler and colleagues[105] reviewed eight putative behavioral endophenotypes of depression, and concluded that only one of them, anhedonia, met all five of these criteria, as well as a sixth criterion, plausibility. (A second candidate, increased stress sensitivity, also scored highly in Hasler's analysis, but does not show a specific association with depression.) Two putative biological endophenotypes (REM sleep abnormalities and a depressive response to tryptophan depletion) also appeared to meet all six criteria.

For the purposes of the present book series, it is important to distinguish between clinical endophenotypes and translational models of clinical endophenotypes. Taking anhedonia as a putative endophenotype for depression, two distinct considerations clearly apply to a putative model of this putative endophenotype. Consider first Hasler's plausibility criterion. The question of plausibility is reflected in Miczek's 1st principle for translational models, *The translation of preclinical data to clinical concerns is more successful when the development of experimental models is restricted in their scope to a cardinal or core symptom of a psychiatric disorder.*[120] Plausibility is intimately related to construct validity, so it is also essential to verify the construct. The anhedonia endophenotype (an internal construct) is inferred in a translational model from the presence of a variety of different exophenotypes (observable behaviors) such as decreased sucrose intake, increased brain stimulation reward threshold, or impaired place conditioning;[106,107] therefore it is important to evaluate a range of appropriate behavioral endpoints rather than relying on a single procedure.[x] The second, and perhaps more obvious, but also more demanding, consideration is that the Gottesman and Gould criteria imply a change in the focus of model building, moving away from the traditional preoccupation with environmental precipitants of behavioral change to encompass a search for susceptible individuals. Strategies to achieve this objective include inbreeding[108,109] or genetic modification[110,111] to produce putatively anhedonic strains, and the identification of susceptible and resilient individuals

---

[x] See also Miczek's 3rd principle, *Preclinical data are more readily translated to the clinical situation when they are based on converging evidence from at least two, and preferably more, experimental procedures, each capturing cardinal features of the modeled disorder.*

within strains.[112,113] What needs to be recognized is that having identified susceptible individuals, a rigorous program of further research is needed to demonstrate face validity, which in this context means that the model conforms to the range of criteria that apply to a clinical endophenotype. For example, a genetically produced trait is by definition heritable, but may not show individual differences in response to stress; conversely, traits identified as individual differences may not be heritable; and both of these approaches raise questions of specificity and state-independence.

While recognizing that the nature of translational modeling is changing, it remains important to retain the strengths of the traditional approach, such as the emphasis on validation of models. Indeed, as discussed in the preface to this volume, concepts of validity are also intimately involved in the establishment of human models of behavioral disorders in experimental medicine, as well as the establishment of biomarkers and endophenotypes.[101,114] There are several reasons to maintain this focus. Firstly, the use of a taxonomy of predictive, face and construct validity,[3,115] or similar alternative taxonomies,[116,117] ensures that when the merits of different models are compared, like is compared with like. Secondly, the practice of assessing validity along several dimensions challenges the predominant view within the pharmaceutical industry that pharmacological isomorphism is the paramount consideration, which incorporates the potentially unhelpful implication that novel agents must resemble known drugs in their spectrum of action. Thirdly, attention to validity forces the proponents of a particular procedure (perhaps one developed in their own laboratory) to engage a self-critical awareness of the model's weaknesses in addition to its strengths, and discourages the publication of over-zealous or biased claims that could be misleading to others. And fourthly, an overt discussion of validity will serve as a context for the selection and adoption of models by the increasing population of workers entering the field of translational modeling from molecular biology rather than from behavioral sciences.

In addition to the translational initiatives of training non-behavioral scientists in behavioral methods and analyses,[xi] a further issue that merits some reflection is the – largely unrecognized – increase in the volume of translational research emanating from the world's developing economies. This can be illustrated by the pattern of submissions to one of the major journals in the field, *Behavioral Pharmacology*, with similar trends in submissions to at least one other major journal (Personal Communication from the editor). In 2007, 20% of submissions to *Behavioral Pharmacology* were from four non-traditional sources: China, India, Brazil, and Iran, the majority of these papers reporting translational research. To take India as an example, submissions during the 1990s were only 0.2% of the total, rising to 2% in the 5 years from 2000 to 2004, and almost 4% in 2005–2007. The most dramatic change has been in submissions from China, mirroring the emergence of the Chinese economy, with submissions languishing at well below 0.5% for the 15 years to 2004 (and these largely from one laboratory), rising suddenly to well over 5% in 2005–2007. However, the growth

---

[xi] Please refer to http://www.ncrr.nih.gov/clinical_research_resources/clinical_and_translational_science_awards/index.asp, http://www.mrc.ac.uk/ApplyingforaGrant/InternationalOpportunites/fp7/index.htm for government-sponsored translational research training initiatives.

of research in developing economies is not well integrated into the wider scientific community. With rare exceptions[118] scientists in developing countries have little direct contact with the mainstream scientific community, and therefore miss out on both the apprenticeship model of scientific training and the background of informal discourse, particularly the conversations that take place outside the formal sessions at scientific meetings. One important consequence of this dislocation is that claims made in the literature may be accepted more readily by scientists outside the mainstream, where investigators with access to the oral tradition and the informal chatter might be more skeptical. Behavioral pharmacology in developing countries also is typically home grown, lacking the scientific cultural roots that come from a tradition of research and development in the area. There is a clear need for positive steps to enable engagement and integration between scientists in developing and developed economies. The European Behavioral Pharmacology Society has for many years achieved some success in meeting these objectives by offering financial support for attendance at scientific meetings to scientists in the developing economies of Eastern Europe. Models of outreach are now needed that can be applied globally, to support the development of behavioral pharmacology in general and translational research in particular, in emerging scientific communities. Academic institutions, scientific journals, learned societies, and the pharmaceutical industry will all have a role to play in shaping the new landscape.

Some considerations on the future DSM-V are also appropriate. Limitations to the current diagnostic paradigms described in the DSM-IV indicate that description of syndromes may never successfully uncover their underlying etiologies. Thus, in the agenda for preparing DSM-V (http://www.dsm5.org), a series of events is planned to try to overcome this limitation. First, a series of "white papers" is contemplated, with the aim of encouraging a research agenda that goes beyond our current thinking to integrate pieces of information from a wide variety of sources and technology. Neuroscience is the subject of one of these "white papers," aimed at developing a basic, clinical neuroscience, and genetics research agenda to guide the development of a future pathophysiological-based diagnostic classification. The working group focused on four main domains: (1) better animal models for the major psychiatric disorders; (2) genes that help determine abnormal behavior in animal models; (3) imaging studies in animals to understand better the nature of imaged signals in humans; and (4) functional genomics and proteomics involved in psychiatric disorders. Second, research conferences are planned to discuss several topics, among which is "stress-induced and fear circuitry disorders." A special session will also be dedicated to gender effects. Thus, we hope that the present series of books may also serve as another basis for discussion of how neuroscience can improve the diagnostic criteria of mental and nervous illnesses.

We end with some reflections on the scope of this project. Its predecessor[2] aimed to integrate academic, industrial, and clinical perspectives through a sequential construction process. Each section of the book had three chapters. The introductory chapter, which set out the concept of the project[1] was written first and forwarded to the academic authors in each therapeutic area, whose chapters were provided to the industrial authors, and finally, the academic and industrial chapters were forwarded to the clinical authors. The present project has taken a more adventurous approach, in which groups of academic, industrial, and clinical authors were established by the editors and were asked to write collaboratively. The introductory chapters to each volume, "What do *you* mean by translational research?" explain that this was done in order "to simulate the conditions of the creation of an industrial Project team" in which the combined talents and expertise of the project members are called together for a clearly defined goal. We can apply the, by now conventional, methodology to assess the validity of this attempt at simulation. There is certainly some face validity: in both the literary and industrial settings, the members of project teams "are all committed to achieving the goals set out by consensus" but "need not know each other...". Also, in some therapeutic areas "the authors were not able to establish an effective team", which reflects the reality that not all project teams are successful. In common with the majority of newly introduced models, predictive validity is less compelling, and only time will tell whether "what appears to be a novel and unusual way of working will become the norm." And as is so often the case, construct validity is elusive. The observation that "for many, this has been a challenging and exhilarating experience, forcing a paradigm shift from how they have normally worked" implies a theory of the role of paradigm shift in drug discovery. New paradigms are the bedrock of scientific revolutions[119] and the current explosion of knowledge that results from the transformations of psychopharmacology by systems neuroscience and behavioral genomics, as discussed earlier in this chapter, can certainly be understood in this light. We nurture the hope that adoption of the project-team approach might inject a comparable creative impetus into drug discovery!

REFERENCES

1. Willner, P. (1991). Behavioural models in psychopharmacology. In Willner, P. (ed.), *Behavioural Models in Psychopharmacology: Theoretical, Industrial and Clinical Perspectives*. Cambridge University Press, Cambridge, pp. 3–18.
2. Willner, P. (ed.) (1991). *Behavioural Models in Psychopharmacology: Theoretical, Industrial and Clinical Perspectives*. Cambridge University Press, Cambridge.
3. Willner, P. (1984). The validity of animal models of depression. *Psychopharmacology (Berl)*, 83(1):1–16.
4. Fendt, M. (2005). Animal models of fear and anxiety. In Koch, M. (ed.), *Animal Models of Neuropsychiatric Diseases* Imperial College Press, London, pp. 293–336
5. Koch, M. (2005). Animal models of schizophrenia. In Koch, M. (ed.), *Animal Models of Neuropsychiatric Diseases*. Imperial College Press, London, pp. 337–402.
6. Delay, J. and Deniker, P. (1955). Neuroleptic effects of chlorpromazine in therapeutics of neuropsychiatry. *J Clin Exp Psychopathol*, 16(2):104–112.
7. Delay, J., Deniker, P., and Harl, J.M. (1952). Therapeutic use in psychiatry of phenothiazine of central elective action (4560 RP). *Ann Med Psychol (Paris)*, 110(2:1):112–117.
8. Carlsson, A. (1977). Does dopamine play a role in schizophrenia? *Psychol Med*, 7(4):583–597.
9. Carlsson, M. and Carlsson, A. (1990). Schizophrenia: A subcortical neurotransmitter imbalance syndrome? *Schizophr Bull*, 16(3):425–432.
10. Carlsson, A. (1995). Neurocircuitries and neurotransmitter interactions in schizophrenia. *Int Clin Psychopharmacol*, 10(Suppl 3):21–28.
11. Patil, S.T., Zhang, L., Martenyi, F., Lowe, S.L., Jackson, K.A., Andreev, B.V. *et al.* (2007). Activation of mGlu2/3 receptors as a new approach to treat schizophrenia: A randomized Phase 2 clinical trial. *Nat Med*, 13(9):1102–1107.

12. Conn, P.J., Tamminga, C., Schoepp, D.D., and Lindsley, C. (2008). Schizophrenia: Moving beyond monoamine antagonists. *Mol Interv*, 8(2):99–107.

13. Rorick-Kehn, L.M., Perkins, E.J., Knitowski, K.M., Hart, J.C., Johnson, B.G., Schoepp, D.D. *et al.* (2006). Improved bioavailability of the mGlu2/3 receptor agonist LY354740 using a prodrug strategy: *In vivo* pharmacology of LY544344. *J Pharmacol Exp Ther*, 316(2):905–913.

14. Woolley, M.L., Pemberton, D.J., Bate, S., Corti, C., and Jones, D.N. (2008). The mGlu2 but not the mGlu3 receptor mediates the actions of the mGluR2/3 agonist, LY379268, in mouse models predictive of antipsychotic activity. *Psychopharmacology (Berl)*, 196(3):431–440.

15. Harich, S., Gross, G., and Bespalov, A. (2007). Stimulation of the metabotropic glutamate 2/3 receptor attenuates social novelty discrimination deficits induced by neonatal phencyclidine treatment. *Psychopharmacology (Berl)*, 192(4):511–519.

16. Greco, B., Invernizzi, R.W., and Carli, M. (2005). Phencyclidine-induced impairment in attention and response control depends on the background genotype of mice: Reversal by the mGLU(2/3) receptor agonist LY379268. *Psychopharmacology (Berl)*, 179(1):68–76.

17. Geyer, M.A. and Braff, D.L. (1987). Startle habituation and sensorimotor gating in schizophrenia and related animal models. *Schizophr Bull*, 13(4):643–668.

18. Robbins, T.W. (2002). The 5-choice serial reaction time task: Behavioural pharmacology and functional neurochemistry. *Psychopharmacology (Berl)*, 163(3–4):362–380.

19. Robbins, T.W. (2000). Animal models of set-formation and set-shifting deficits in schizophrenia. In Myslobodsky, M. and Weiner, I. (eds.), *Contemporary Issues in Modeling Psychopathology*. Kluwer Academic, Boston, MA, pp. 247–258.

20. Sams-Dodd, F. (1996). Phencyclidine-induced stereotyped behaviour and social isolation in rats: A possible animal model of schizophrenia. *Behav Pharmacol*, 7(1):3–23.

21. Lipska, B.K., Swerdlow, N.R., Geyer, M.A., Jaskiw, G.E., Braff, D.L., and Weinberger, D.R. (1995). Neonatal excitotoxic hippocampal damage in rats causes post-pubertal changes in prepulse inhibition of startle and its disruption by apomorphine. *Psychopharmacology (Berl)*, 122(1):35–43.

22. Jones, D.N.C., Gartlon, J.E., Minassian, A., Perry, W., and Geyer, M.A. (2008). Developing new drugs for schizophrenia: From animals to the clinic. In McArthur, R.A. and Borsini, F. (eds.), *Animal and Translational Models for CNS Drug Discovery: Psychiatric Disorders*. Academic Press, Elsevier, New York.

23. Cryan, J.F., Sánchez, C., Dinan, T.G., and Borsini, F. (2008). Developing more efficacious antidepressant medications: Improving and aligning preclinical and clinical assessment tools. In McArthur, R.A. and Borsini, F. (eds.), *Animal and Translational Models for CNS Drug Discovery: Psychiatric Disorders*. Academic Press, Elsevier, New York.

24. Steckler, T., Stein, M.B., and Holmes, A. (2008). Developing novel anxiolytics: Improving preclinical detection and clinical assessment. In McArthur, R.A. and Borsini, F. (eds.), *Animal and Translational Models for CNS Drug Discovery: Psychiatric Disorders*. Academic Press, Elsevier, New York.

25. Large, C.H., Einat, H., and Mahableshhwarkar, A.R. (2008). Developing new drugs for bipolar disorder (BPD): From animal models to the clinic. In McArthur, R.A. and Borsini, F. (eds.), *Animal and Translational Models for CNS Drug Discovery: Psychiatric Disorders*. Academic Press, Elsevier, New York.

26. Tannock, R., Campbell, B., Seymour, P., Ouellet, D., Soares, H., Wang, P. *et al.* (2008). Towards a biological understanding of ADHD and the discovery of novel therapeutic approaches. In McArthur, R.A. and Borsini, F. (eds.), *Animal and Translational Models for CNS Drug Discovery: Psychiatric Disorders*. Academic Press, Elsevier, New York.

27. Little, H.J., McKinzie, D.L., Setnik, B., Shram, M.J., and Sellers, E.M. (2008). Pharmacotherapy of alcohol dependence: Improving translation from the bench to the clinic. In McArthur, R.A. and Borsini, F. (eds.), *Animal and Translational Models for CNS Drug Discovery: Reward Deficit Disorders*. Academic Press, Elsevier, New York.

28. Markou, A., Chiamulera, C., and West, R.J. (2008). Contribution of animal models and preclinical human studies to medication development for nicotine dependence. In McArthur, R.A. and Borsini, F. (eds.), *Animal and Translational Models for CNS Drug Discovery: Reward Deficit Disorders*. Academic Press, Elsevier, New York.

29. Williams, W.A., Grant, J.E., Winstanley, C.A., and Potenza, M.N. (2008). Currect concepts in the classification, treatment and modelling of pathological gambling and other impulse control disorders. In McArthur, R.A. and Borsini, F. (eds.), *Animal and Translational Models for CNS Drug Discovery: Reward Deficit Disorders*. Academic Press, Elsevier, New York.

30. Czachowski, C.L. and Samson, H.H. (1999). Breakpoint determination and ethanol self-administration using an across-session progressive ratio procedure in the rat. *Alcohol Clin Exp Res*, 23(10):1580–1586.

31. Goudie, A.J. (1991). Animal models of drug abuse and dependence. In Willner, P. (ed.), *Behavioural Models in Psychopharmacology: Theoretical, Industrial and Clinical Perspectives*. Cambridge University Press, Cambridge, pp. 453–484.

32. Cryan, J.F. and Holmes, A. (2005). Model organisms: The ascent of mouse: advances in modelling human depression and anxiety. *Nat Rev Drug Discov*, 4(9):775–790.

33. Holmes, A., le Guisquet, A.M., Vogel, E., Millstein, R.A., Leman, S., and Belzung, C. (2005). Early life genetic, epigenetic and environmental factors shaping emotionality in rodents. *Neurosci Biobehav Rev*, 29(8):1335–1346.

34. Nibuya, M., Nestler, E.J., and Duman, R.S. (1996). Chronic antidepressant administration increases the expression of cAMP response element binding protein (CREB) in rat hippocampus. *J Neurosci*, 16(7):2365–2372.

35. Blom, J.M., Tascedda, F., Carra, S., Ferraguti, C., Barden, N., and Brunello, N. (2002). Altered regulation of CREB by chronic antidepressant administration in the brain of transgenic mice with impaired glucocorticoid receptor function. *Neuropsychopharmacology*, 26(5):605–614.

36. Fujimaki, K., Morinobu, S., and Duman, R.S. (2000). Administration of a cAMP phosphodiesterase 4 inhibitor enhances antidepressant-induction of BDNF mRNA in rat hippocampus. *Neuropsychopharmacology*, 22(1):42–51.

37. Warner-Schmidt, J.L. and Duman, R.S. (2007). VEGF is an essential mediator of the neurogenic and behavioral actions of antidepressants. *Proc Natl Acad Sci USA*, 104(11):4647–4652.

38. Nestler, E.J., Gould, E., Manji, H., Buncan, M., Duman, R.S., Greshenfeld, H.K. *et al.* (2002). Preclinical models: Status of basic research in depression. *Biol Psychiatr*, 52(6):503–528.

39. Goldman-Rakic, P.S. (1987). Development of cortical circuitry and cognitive function. *Child Dev*, 58(3):601–622.

40. LeDoux, J. (2003). The emotional brain, fear, and the amygdala. *Cell Mol Neurobiol*, 23(4–5):727–738.

41. Clark, A.S., Schwartz, M.L., and Goldman-Rakic, P.S. (1989). GABA-immunoreactive neurons in the mediodorsal nucleus of the monkey thalamus. *J Chem Neuroanat*, 2(5):259–267.

42. Kuhn, R. (1958). The treatment of depressive states with G 22355 (imipramine hydrochloride). *Am J Psychiatr*, 115(5):459–464.

43. Kline, N.S. (1958). Clinical experience with iproniazid (marsilid). *J Clin Exp Psychopathol*, 19(2, Suppl 1):72–78. discussion 8–9.

44. Tobin, J.M., Bird, I.F., and Boyle, D.E. (1960). Preliminary evaluation of librium (Ro 5-0690) in the treatment of anxiety reactions. *Dis Nerv Syst*, 21(Suppl 3):11–19.

45. Kerry, R.J. and Jenner, F.A. (1962). A double blind crossover comparison of diazepam (Valium, Ro5-2807) with chlordiazepoxide (Librium) in the treatment of neurotic anxiety. *Psychopharmacologia*, 3:302–306.

46. Birkmayer, W. and Hornykiewicz, O. (1962). The *L*-dihydroxyphenylalanine (*L*-DOPA) effect in Parkinson's syndrome in man: On the pathogenesis and treatment of Parkinson akinesis. *Arch Psychiatr Nervenkr Z Gesamte Neurol Psychiatr*, 203:560–574.

47. Hornykiewicz, O. (1966). Dopamine (3-hydroxytyramine) and brain function. *Pharmacol Rev*, 18(2):925–964.

48. Hornykiewicz, O.D. (1970). Physiologic, biochemical, and pathological backgrounds of levodopa and possibilities for the future. *Neurology*, 20(12):1–5.

49. Bartus, R.T., Dean, R.L., Beer, B., and Lippa, A.S. (1982). The cholinergic hypothesis of geriatric memory dysfunction. *Science*, 217(4558):408–414.

50. Kirsch, I., Deacon, B.J., Huedo-Medina, T.B., Scoboria, A., Moore, T.J., and Johnson, B.T. (2008). Initial severity and antidepressant benefits: A meta-analysis of data submitted to the Food and Drug Administration. *PLoS Med*, 5(2(e45)):260–268.

51. Lindner, M.D., McArthur, R.A., Deadwyler, S.A., Hampson, R.E., and Tariot, P.N. (2008). Development, optimization and use of preclinical behavioral models to maximise the productivity of drug discovery for Alzheimer's disease. In McArthur, R.A. and Borsini, F. (eds.), *Animal and Translational Models for CNS Drug Discovery: Neurologic Disorders*. Academic Press, Elsevier, New York.

52. Schneider, L.S. (2008). Issues in design and conduct of clinical trials for cognitive-enhancing drugs. In McArthur, R.A. and Borsini, F. (eds.), *Animal and Translational Models for CNS Drug Discovery: Neurologic Disorders*. Academic Press, Elsevier, New York.

53. McShane, R., Areosa Sastre, A., and Minakaran, N. (2006). Memantine for dementia. *Cochrane Database Syst Rev*, 2. CD003154.

54. National Institute for Health and Clinical Excellence. (2006). NICE technology appraisal guidance 111. *Donepezil, Galantamine, Rivastigmine (Review) and Memantine for the Treatment of Alzheimer's Disease*. National Institute for Health and Clinical Excellence, London.

55. Kirsch, I. and Sapirstein, G. (1998). Listening to Prozac but hearing placebo: A meta-analysis of antidepressant medication. *Prev Treat*, 1(Article 0002a):1–13.

56. Khan, A., Kolts, R.L., Rapaport, M.H., Krishnan, K.R., Brodhead, A.E., and Browns, W.A. (2005). Magnitude of placebo response and drug-placebo differences across psychiatric disorders. *Psychol Med*, 35(5):743–749.

57. Peuskens, J. (1999). The evolving definition of treatment resistance. *J Clin Psychiatr*, 60(Suppl 12):4–8.

58. Souery, D., Oswald, P., Massat, I., Bailer, U., Bollen, J., Demyttenaere, K. *et al.* (2007). Clinical factors associated with treatment resistance in major depressive disorder: Results from a European Multicenter Study. *J Clin Psychiatr*, 68(7):1062–1070.

59. Fava, G.A., Ruini, C., and Rafanelli, C. (2005). Sequential treatment of mood and anxiety disorders. *J Clin Psychiatr*, 66(11):1392–1400.

60. Rush, A.J., Trivedi, M.H., Wisniewski, S.R., Nierenberg, A.A., Stewart, J.W., Warden, D. *et al.* (2006). Acute and longer-term outcomes in depressed outpatients requiring one or several treatment steps: A STAR*D report. *Am J Psychiatr*, 163(11):1905–1917.

61. Kessler, R.C., McGonagle, K.A., Zhao, S., Nelson, C.B., Hughes, M., Eshleman, S. *et al.* (1994). Lifetime and 12-month prevalence of DSM-III-R psychiatric disorders in the United States. Results from the National Comorbidity Survey. *Arch Gen Psychiatr*, 51(1):8–19.

62. Davidson, J.R. (2001). Pharmacotherapy of generalized anxiety disorder. *J Clin Psychiatr*, 62(Suppl 11):46–50. discussion 1–2.

63. Tsuang, M.T., Glatt, S.J., and Faraone, S.V. (2006). The complex genetics of psychiatric disorders. In: Runge, M.S. and Patterson, C. (eds.), *Principles of Molecular Medicine*, Humana Press, Totowa, NJ, pp.1184–1190.

64. Zeller, A., Crestani, F., Camenisch, I., Iwasato, T., Itohara, S., Fritschy, J.M. *et al.* (2008). Cortical glutamatergic neurons mediate the motor sedative action of diazepam. *Mol Pharmacol*, 73(2):282–291.

65. Gaveriaux-Ruff, C. and Kieffer, B.L. (2007). Conditional gene targeting in the mouse nervous system: Insights into brain function and diseases. *Pharmacol Ther*, 113(3):619–634.

66. Monteggia, L.M., Luikart, B., Barrot, M., Theobold, D., Malkovska, I., Nef, S. *et al.* (2007). Brain-derived neurotrophic factor conditional knockouts show gender differences in depression-related behaviors. *Biol Psychiatr*, 61(2):187–197.
67. Chen, A.P., Ohno, M., Giese, K.P., Kuhn, R., Chen, R.L., and Silva, A.J. (2006). Forebrain-specific knockout of B-raf kinase leads to deficits in hippocampal long-term potentiation, learning, and memory. *J Neurosci Res*, 83(1):28–38.
68. Xiao, D., Bastia, E., Xu, Y.H., Benn, C.L., Cha, J.H., Peterson, T.S. *et al.* (2006). Forebrain adenosine A2A receptors contribute to *L*-3,4-dihydroxyphenylalanine-induced dyskinesia in hemiparkinsonian mice. *J Neurosci*, 26(52):13548–13555.
69. Nguyen, N.K., Keck, M.E., Hetzenauer, A., Thoeringer, C.K., Wurst, W., Deussing, J.M. *et al.* (2006). Conditional CRF receptor 1 knockout mice show altered neuronal activation pattern to mild anxiogenic challenge. *Psychopharmacology (Berl)*, 188(3):374–385.
70. Valverde, O., Mantamadiotis, T., Torrecilla, M., Ugedo, L., Pineda, J., Bleckmann, S. *et al.* (2004). Modulation of anxiety-like behavior and morphine dependence in CREB-deficient mice. *Neuropsychopharmacology*, 29(6):1122–1133.
71. Bingham, N.C., Anderson, K.K., Reuter, A.L., Stallings, N.R., and Parker, K.L. (2008). Selective loss of leptin receptors in the ventromedial hypothalamic nucleus results in increased adiposity and a metabolic syndrome. *Endocrinology*, 149(5):2138–2148.
72. American Psychiatric Association (ed.). (1994). *Diagnostic and Statistical Manual of Mental Disorders*. 4th edition. American Psychiatric Association, Washington, DC.
73. World Health Organization (2007). *International Statistical Classification of Diseases*, 10th revision, 2nd edition. World Health Organization, Geneva.
74. Fibiger, H.C. (1991). The dopamine hypotheses of schizophrenia and mood disorders: Contradictions and speculations. In Willner, P. and Scheel-Kruger, J. (eds.), *The Mesolimbic Dopamine System: From Motivation to Action*. John Wiley, Chichester, UK, pp. 615–638.
75. Costello, C.G. (ed.) (1993). *Symptoms of Depression*. John Wiley, New York.
76. Parker, G., Roy, K., Wilhelm, K., Mitchell, P., and Hadzi-Pavlovic, D. (2000). The nature of bipolar depression: Implications for the definition of melancholia. *J Affect Disord*, 59(3):217–224.
77. Parker, G., Hadzi-Pavlovic, D., Wilhelm, K., Hickie, I., Brodaty, H., Boyce, P. *et al.* (1994). Defining melancholia: Properties of a refined sign-based measure. *Br J Psychiatr*, 164(3):316–326.
78. Carey, G.J., Costall, B., Domeney, A.M., Gerrard, P.A., Jones, D.N., Naylor, R.J. *et al.* (1992). Ondansetron and arecoline prevent scopolamine-induced cognitive deficits in the marmoset. *Pharmacol Biochem Behav*, 42(1):75–83.
79. Broocks, A., Little, J.T., Martin, A., Minichiello, M.D., Dubbert, B., Mack, C. *et al.* (1998). The influence of ondansetron and m-chlorophenylpiperazine on scopolamine-induced cognitive, behavioral, and physiological responses in young healthy controls. *Biol Psychiatr*, 43(6):408–416.
80. Wienrich, M., Meier, D., Ensinger, H.A., Gaida, W., Raschig, A., Walland, A. *et al.* (2001). Pharmacodynamic profile of the M1 agonist talsaclidine in animals and man. *Life Sci*, 68(22–23):2593–2600.
81. Lutscp, II.L. (2002). Repinotan bayer. *Curr Opin Investig Drugs*, 3(6):924–927.
82. Blier, P. and Ward, N.M. (2003). Is there a role for 5-HT1A agonists in the treatment of depression? *Biol Psychiatr*, 53(3):193–203.
83. Scrip. (2007). *FDA Rejects Fabre Kramer Antidepressant Gepirone ER*. Scrip. 3310:24.
84. Merchant, K.M., Chesselet, M.-F., Hu, S.-C., and Fahn, S. (2008). Animal models of Parkinson's disease to aid drug discovery and development. In McArthur, R.A. and Borsini, F. (eds.), *Animal and Translational Models for CNS Drug Discovery: Neurologic Disorders*. Academic Press, Elsevier, New York.
85. Wright, T.M. (2006). Tramiprosate. *Drugs Today (Barc)*, 42(5):291–298.

86. Aisen, P.S., Gauthier, S., Vellas, B., Briand, R., Saumier, D., Laurin, J. *et al.* (2007). Alzhemed: A potential treatment for Alzheimer's disease. *Curr Alzheimer Res*, 4(4):473–478.

87. Scrip. (2007). *Neurochem Shares Fall after Alzheimer Disappointment.* Scrip. 3289/90:24.

88. Neurochem Inc. (2007). We are Neurochem Quarterly Report. Third Quarter ended September 30, 2007. Laval, Quebec, Canada.

89. Winsky, L. and Brady, L. (2005). Perspective on the status of preclinical models for psychiatric disorders. *Drug Discovery Today: Disease Models*, 2(4):279–283.

90. Dourish, C.T., Wilding, J.P.H., and Halford, J.C.G. (2008). Anti-obesity drugs: From animal models to clinical efficacy. In McArthur, R.A. and Borsini, F. (eds.), *Animal and Translational Models for CNS Drug Discovery: Reward Deficit Disorders.* Academic Press, Elsevier, New York.

91. Hunter, A.J. (2008). Animal and translational models of neurological disorders: An industrial perspective. In McArthur, R.A. and Borsini, F. (eds.), *Animal and Translational Models for CNS Drug Discovery: Neurologic Disorders.* Academic Press, Elsevier, New York.

92. McEvoy, J.P. and Freudenreich, O. (2008). Issues in the design and conductance of clinical trials. In McArthur, R.A. and Borsini, F. (eds.), *Animal and Translational Models for CNS Drug Discovery: Psychiatric Disorders.* Academic Press, Elsevier, New York.

93. Millan, M.J. (2008). The discovery and development of pharmacotherapy for psychiatric disorders: A critical survey of animal and translational models, and perspectives for their improvement. In McArthur, R.A. and Borsini, F. (eds.), *Animal and Translational Models for CNS Drug Discovery: Psychiatric Disorders.* Academic Press, Elsevier, New York.

94. Cuatrecasas, P. (2006). Drug discovery in jeopardy. *J Clin Invest*, 116(11):2837–2842.

95. McArthur, R. and Borsini, F. (2006). Animal models of depression in drug discovery: A historical perspective. *Pharmacol Biochem Behav*, 84(3):436–452.

96. Kirsch, I., Moore, T.J., Scoboria, A., and Nicholls, S.S. (2002). The emperor's new drugs: An analysis of antidepressant medication data submitted to the US Food and Drug Administration. *Prev Treat*, 5(1):1–23.

97. Turner, E.H., Matthews, A.M., Linardatos, E., Tell, R.A., and Rosenthal, R. (2008). Selective publication of antidepressant trials and its influence on apparent efficacy. *N Engl J Med*, 358(3):252–260.

98. Bartz, J., Young, L.J., Hollander, E., Buxbaum, J.D., and Ring, R.H. (2008). Preclinical animal models of autistic spectrum disorders (ASD). In McArthur, R.A. and Borsini, F. (eds.), *Animal and Translational Models for CNS Drug Discovery: Psychiatric Disorders.* Academic Press, Elsevier, New York.

99. Montes, J., Bendotti, C., Tortarolo, M., Cheroni, C., Hallak, H., Speiser, Z. *et al.* (2008). Translational research in ALS. In McArthur, R.A. and Borsini, F. (eds.), *Animal and Translational Models for CNS Drug Discovery: Neurologic Disorders.* Academic Press, Elsevier, New York.

100. Wagner, L.A., Menalled, L., Goumeniouk, A.D., Brunner, D.P., and Leavitt, B.R. (2008). Huntington disease. In McArthur, R.A. and Borsini, F. (eds.), *Animal and Translational Models for CNS Drug Discovery: Neurologic Disorders.* Academic Press, Elsevier, New York.

101. Lesko, L.J. and Atkinson, A.J.J. (2001). Use of biomarkers and surrogate endpoints in drug development and regulatory decision making: Criteria, validation, strategies. *Annu Rev Pharmacol Toxicol*, 41:347–366.

102. Gottesman, I.I. and Gould, T.D. (2003/4/1). The endophenotype concept in psychiatry: Etymology and strategic intentions. *Am J Psychiatr*, 160(4):636–645.

103. Bearden, C.E. and Freimer, N.B. (2006). Endophenotypes for psychiatric disorders: Ready for primetime?. *Trends Genet*, 22(6):306–313.

104. Cannon, T.D. and Keller, M.C. (2006). Endophenotypes in the genetic analyses of mental disorders. *Annu Rev Clin Psychol*, 2(1):267–290.

105. Hasler, G., Drevets, W.C., Manji, H.K., and Charney, D.S. (2004). Discovering endophenotypes for major depression. *Neuropsychopharmacology*, 29(10):1765–1781.

106. Willner, P. (1997). Validity, reliability and utility of the chronic mild stress model of depression: A 10-year review and evaluation. *Psychopharmacology (Berl)*, 134(4):319–329.

107. Willner, P., Muscat, R., and Papp, M. (1992). Chronic mild stress-induced anhedonia: A realistic animal model of depression. *Neurosci Biobehav Rev*, 16(4):525–534.

108. Bekris, S., Antoniou, K., Daskas, S., and Papadopoulou-Daifoti, Z. (2005). Behavioural and neurochemical effects induced by chronic mild stress applied to two different rat strains. *Behav Brain Res*, 161(1):45–59.

109. Pucilowski, O., Overstreet, D.H., Rezvani, A.H., and Janowsky, D.S. (1993). Chronic mild stress-induced anhedonia: Greater effect in a genetic rat model of depression. *Physiol Behav*, 54(6):1215–1220.

110. Martin, M., Ledent, C., Parmentier, M., Maldonado, R., and Valverde, O. (2002). Involvement of CB1 cannabinoid receptors in emotional behaviour. *Psychopharmacology (Berl)*, 159(4):379–387.

111. Mormede, C., Castanon, N., Medina, C., Moze, E., Lestage, J., Neveu, P.J. *et al.* (2002). Chronic mild stress in mice decreases peripheral cytokine and increases central cytokine expression independently of IL-10 regulation of the cytokine network. *Neuroimmunomodulation*, 10(6):359–366.

112. Bergstrom, A., Jayatissa, M.N., Mork, A., and Wiborg, O. (2008). Stress sensitivity and resilience in the chronic mild stress rat model of depression; an *in situ* hybridization study. *Brain Res*, 1196:41–52.

113. Strekalova, T., Gorenkova, N., Schunk, E., Dolgov, O., and Bartsch, D. (2006). Selective effects of citalopram in a mouse model of stress-induced anhedonia with a control for chronic stress. *Behav Pharmacol*, 17(3):271–287.

114. Littman, B.H. and Williams, S.A. (2005). The ultimate model organism: Progress in experimental medicine. *Nat Rev Drug Discov*, 4(8):631–638.

115. Willner, P. (1991). Methods for assessing the validity of animal models of human psychopathology. In Boulton, A., Baker, G., and Martin-Iverson, M. (eds.), *Neuromethods: Animal Models in Psychiatry I*, Vol. 18. Humana Press, Inc, pp. 1–23.

116. Geyer, M.A. and Markou, A. (1995). Animal models of psychiatric disorders. In Bloom, F.E. and Kupfer, D.J. (eds.), *Psychopharmacology, The Fourth Generation of Progress*. Raven Press, New York, pp. 787–798.

117. Geyer, M.A. and Markou, A. (2002). *The role of preclinical models in the development of psychotropic drugs*. American College of Neuropsychopharmacology, New York. pp. 445–455.

118. Li, Q., Zhao, D., and Bezard, E. (2006). Traditional Chinese medicine for Parkinson's disease: A review of Chinese literature. *Behav Pharmacol*, 17(5–6):403–410.

119. Kuhn, T.S. (1996). *The Structure of Scientific Revolutions*, 3rd edition. The University of Chicago Press, Chicago, IL.

120. Miczek, K.A. (2008). Challenges for translational psychopharmacology research – the need for conceptual principles. In McArthur, R.A. and Borsini, F. (eds.), *Animal and Translational Models for CNS Drug Discovery: Psychiatric Disorders*. Academic Press: Elsevier, New York.

Index

Printed and bound by CPI Group (UK) Ltd, Croydon, CR0 4YY

03/10/2024

01040314-0018